Smart Nanomaterials Technology

Series Editors

Azamal Husen, Department of Biotechnology, Smt. S. S. Patel Nootan Science & Commerce College, Sankalchand Patel University, Visnagar, Gujarat, India
Department of Biotechnology, Graphic Era (Deemed to be University), Dehradun, Uttarakhand, India
Laboratory of Bio Resource Management, Institute of Tropical Forestry and Forest Product, University Putra Malaysia, UPM Serdang, Selangor, Malaysia

Mohammad Jawaid, Laboratory of Biocomposite Technology, Universiti Putra Malaysia, INTROP, Serdang, Selangor, Malaysia

Nanotechnology is a rapidly growing scientific field and has attracted a great interest over the last few years because of its abundant applications in different fields like biology, physics and chemistry. This science deals with the production of minute particles called nanomaterials having dimensions between 1 and 100 nm which may serve as building blocks for various physical and biological systems. On the other hand, there is the class of smart materials where the material that can stimuli by external factors and results a new kind of functional properties. The combination of these two classes forms a new class of smart nanomaterials, which produces unique functional material properties and a great opportunity to larger span of application. Smart nanomaterials have been employed by researchers to use it effectively in agricultural production, soil improvement, disease management, energy and environment, medical science, pharmaceuticals, engineering, food, animal husbandry and forestry sectors.

This book series in Smart Nanomaterials Technology aims to comprehensively cover topics in the fabrication, synthesis and application of these materials for applications in the following fields:

- Energy Systems—Renewable energy, energy storage (supercapacitors and electrochemical cells), hydrogen storage, photocatalytic water splitting for hydrogen production
- Biomedical—controlled release of drugs, treatment of various diseases, biosensors,
- Agricultural—agricultural production, soil improvement, disease management, animal feed, egg, milk and meat production/processing,
- Forestry—wood preservation, protection, disease management
- Environment—wastewater treatment, separation of hazardous contaminants from wastewater, indoor air filters

Alle Madhusudhan · Shiv Dutt Purohit ·
Rajendra Prasad · Azamal Husen
Editors

Functional Smart Nanomaterials and Their Theranostics Approaches

Springer

Editors
Alle Madhusudhan
Department of Chemistry
The University of Memphis
Memphis, TN, USA

Shiv Dutt Purohit
School of Chemical Engineering
Yeungnam University
Gyeongsan, Korea (Republic of)

Rajendra Prasad
School of Biochemical Engineering
Indian Institute of Technology (BHU)
Varanasi, Uttar Pradesh, India

Azamal Husen
Department of Biotechnology
Smt. S. S. Patel Nootan Science &
Commerce College, Sankalchand Patel
University
Visnagar, Gujarat, India

Department of Biotechnology
Graphic Era (Deemed to be University)
Dehradun, Uttarakhand, India

Laboratory of Bio Resource Management
Institute of Tropical Forestry and Forest
Product, University Putra Malaysia
UPM Serdang, Selangor, Malaysia

ISSN 3004-8273 ISSN 3004-8281 (electronic)
Smart Nanomaterials Technology
ISBN 978-981-99-6596-0 ISBN 978-981-99-6597-7 (eBook)
https://doi.org/10.1007/978-981-99-6597-7

This Springer imprint is published by the registered company Springer Nature Singapore Pte Ltd.
The registered company address is: 152 Beach Road, #21-01/04 Gateway East, Singapore 189721, Singapore

Preface

In recent years, nanotechnology has become an epoch-making area with enormous potential and can revolutionize every aspect of our lives. One of the most interesting applications of nanotechnology is the development of functional intelligent nanomaterials, which interact with nano-class biological systems with remarkable capabilities. These data will open a new era of medical care by increasing the space for the development of diagnosis, treatment, and customized medical care.

The book *Functional Smart Nanomaterials and Their Theranostics Approaches* explores their potential in the field of nanotechnology, which combines therapy and diagnosis, to explore the fascinating world of nanomaterials and realize accurate and personalized treatment strategies. This book outlines the latest development and future prospects of functional intelligent nanomaterials in the medical field with the contribution of leading researchers and experts in this field.

This book is divided into 14 chapters and focuses on the specific aspects and applications of functional intelligent nanomaterials. Chapters "Basic Principles of Functional Materials for Biomedical Applications" and "Functional Biomaterials for Targeted Drug Delivery Applications" introduce the basic principles of functional materials, outline the unique characteristics of nanomaterials, and make them ideal candidates for biomedical applications and drug delivery applications. These chapters highlight the importance of exploring various materials and their manufacturing techniques. Chapters "Metallic Nanoparticles for Imaging and Therapy" and "Theranostic Applications of Functional Nanomaterials Using Microscopic and Spectroscopic Techniques" comprise the information related to nanoparticles and their applications in the imaging and spectroscopy-based theranostic applications. After that, Chaps. "Next-Generation Therapies for Breast Cancer"–"Functional Biomaterials for Image-Guided Therapeutics of Solid Tumor" provide insights into the application of functional nanomaterials in diagnosis and treatment of the cancer.

Chapters "Nanostructured Electrodes as Electrochemical Biosensors for Biomedical Applications" to "Screen-Printed Electrode (SPE)-Based Biosensor for Point-Of-Care (POC) Diagnostic in Medical Applications, Their Scope, and Challenges" emphasize the role and significance of the use of nanostructures for biosensor-based theranostics applications. The development in biosensor-based diagnostics for bacterial diseases has been also explained (Chap. "Development in Biosensor-Based Diagnostics for Bacterial Diseases: Opportunities and Challenges"). The sensing applications range from electrochemical biosensors, screen printed electrode-based point-of-care devices to the organ on chip systems. Finally, the book provides insights into the next generations of therapy for various applications including type 2 diabetes mellitus using natural products, quantum dots, and nanomaterials (Chaps. "Next-Generation Therapies for Type 2 Diabetes Mellitus" and "Quantum Dots for Theranostic Applications").

The ultimate goal of this book is to provide comprehensive resources to researchers, students, and experts in nanotechnology and biomedical fields to emphasize the potential and importance of functional intelligent nanomaterials in nanotechnology. The text aims to deepen understanding of the principles and applications of these materials and to encourage further research and innovation in this interesting and rapidly evolving field.

We sincerely thank all authors for their contribution to the expertise and knowledge of this book. Their valuable insight and groundbreaking research enabled publications. Furthermore, we would like to thank the reviewers for their thoughtful opinions and suggestions.

We hope this book will be a valuable resource for readers who want to explore the state-of-the-art development of functional smart nanomaterials and their rational methods. We hope to contribute to improving the level of medical service and quality of life by providing new ways of thinking, cooperation, and innovation to the global public.

Memphis, USA — Alle Madhusudhan
Gyeongsan, Korea (Republic of) — Shiv Dutt Purohit
Varanasi, India — Rajendra Prasad
Visnagar, India — Azamal Husen

Contents

About the Editors

Alle Madhusudhan is currently working as a Postdoc-fellow at The University of Memphis, TN, U.S.A. Earlier he worked as a Research Professor at Department of Biomedical Science & Institute of Bioscience and Biotechnology, Kangwon National University, South Korea and also worked as an Associate Professor at Department of Chemistry, Gondar University, Gondar, Ethiopia (2014–2018). During this period, he served as a Coordinator of MSc (Chemistry) Programs and as the Head of the Chemistry Department. Four students submitted their master's thesis under his supervision. He received his doctorate from Department of Chemistry, Osmania University, India (2013), where he worked on the research projects sponsored by University Grants Commission, India. He has more than 18 years of experience in teaching and research in both academia and industry. He has published more than 70 scientific papers in SCI/SCIE-grade journals (>1900 citations; h-index = 22) as a first and corresponding author and published 1 book and 11 book chapters (Springer and Elsevier). He is also listed as a potential reviewer for many reputed international journals. He is also a member of Ethiopian chemical society and Indian chemical society. He has presented his work in several national and international conferences in India, Ethiopia, South Korea, and U.S.A. He has conducted two research projects sponsored by University Grants Commission (UGC), New Delhi, India. His current research interests include liquid biopsy, cancer biology, polymeric nano-drug delivery systems, biomaterials for sustainability and catalysis. Extraction and chemical modification of nanocellulose from lignocellulosic biomass, green synthesis of metal and metal oxide nanoparticles on nanocellulose support and their novel applications in biomedicine, biosensors, and catalysis.

Shiv Dutt Purohit is currently working as International Research Professor (Assistant Professor) at the School of Chemical Engineering, Yeungnam University, Gyeongsan, South Korea. He obtained his Ph.D. from the Department of Polymer and Process Engineering, Indian Institute of Technology Roorkee, in 2021. He is Materials Scientist, and his research interests lie in biomedical and related areas such as tissue engineering and wound healing. Besides, he is also intrigued to work in other areas such as food packaging materials. He has published more than 20 research articles and chapters and has more than 500 citations on his name with an h-index of 10. He is also Reputed Reviewer for several peer-reviewed international journals and working as Academic Editor for Scientifica-Hindawi. He is also involved in the editing of a few special issues in SCIE journals.

Rajendra Prasad is currently working as Assistant Professor at the School of Biochemical Engineering, Indian Institute of Technology (BHU), Varanasi, India. Prior to this position, Rajendra Prasad had postdoctoral trainings from Tufts University, Medford, Boston, MA, USA, NOVA Medical School, Faculdade de Ciências Médicas Campo Mártires da Pátria, Lisboa, Portugal and Technion-Israel Institute of Technology, Haifa, Israel, and Indian Institute of Technology, Bombay, India, in the area of Cancer NanoMedicine and targetable cancer theranostics. Moreover, Rajendra Prasad had a chance to help few master students during my postdoctoral trainings, which gave me a confidence to handle scientific challenges and lead independent research projects. Rajendra Prasad has also cracked prestigious Postdoctoral Fellowships such as IPDF, PBC, and JSPS and hold my membership for Royal Society of Chemistry.

Azamal Husen is presently working as a Professor at Sankalchand Patel University, Visnagar, India; and Adjunct Professor at Graphic Era (Deemed to be University), Dehradun, Uttarakhand, India. He is also working as a Visiting Professor at University Putra Malaysia, Selangor, Malaysia. Previously, he served as Professor and Head of the Department of Biology, University of Gondar, Ethiopia; and worked as a Foreign Delegate at Wolaita Sodo University, Wolaita, Ethiopia. He also worked as a Visiting Faculty of the Forest Research Institute and the Doon College of Agriculture and Forest at Dehra Dun, India. His research and teaching experience of 25 years encompasses biogenic nanomaterial fabrication and application; plant responses to nanomaterials; plant adaptation to harsh environments at the physiological, biochemical, and molecular levels; herbal medicine; and clonal propagation for improvement of tree species. He has conducted research sponsored by the World Bank, the National Agricultural Technology Project, the Indian Council of Agriculture Research, the Indian Council of Forest Research Education, and the Japan Bank for International Cooperation. Husen has published extensively (over 250) and served on the Editorial Board and as reviewer of reputed journals published by Elsevier, Frontiers Media, Taylor & Francis, Springer Nature, RSC, Oxford University Press, Sciendo, the Royal Society, CSIRO, PLOS, MDPI, John Wiley & Sons, and UPM Journals. He is on the advisory board of Cambridge Scholars Publishing, UK. He is a fellow of the Plantae group of the American Society of Plant Biologists, and a member of the International Society of Root Research, Asian Council of Science Editors, and International Natural Product Sciences. He is Editor-in-Chief of the *American Journal of Plant Physiology*, and a Series Editor of *Exploring Medicinal Plants* (Taylor & Francis Group, USA); *Plant Biology, Sustainability, and Climate Change* (Elsevier, USA); and *Smart Nanomaterials Technology* (Springer Nature, Singapore).

Basic Principles of Functional Materials for Biomedical Applications

Vaskuri G S Sainaga Jyothi, Valamla Bhavana, and Nagavendra Kommineni

Abstract Functional materials have distinctly engrossed in the field of biomedical applications playing critical role in drug delivery design. Currently, numerous materials of synthetic (metallic and nonmetallic systems) and natural origin are extensively studied to surpass the limitations of conventional approaches. In a broader context, functional nanomaterials are used to accomplish targeted delivery and reduce toxic effects. They exhibit an extensive variability of physical and chemical properties, allowing for the fine-tuning of biocompatibility, biodegradability, stimuli-responsiveness, and bioactivities. These functional materials, with different patterns and versatile functionality, are exhibiting considerable applications in diagnosis, controlled delivery of therapeutic agents, efficient adjuvants of immunotherapy, regenerative medicine, and medical devices. This book chapter provides an insight on the basic principles and applications of the revolutionized functional materials.

Keywords Functional materials · Targeted delivery · Biomedical applications · Stimuli-responsiveness

1 Introduction

Material science is indispensable in the field of biomedical applications including targeted delivery, bioimaging, biosensing, and tissue engineering. Advancement of nanotechnology in the field of material science has revolutionized the field of biomedical sciences which enabled the ease of targeted drug delivery and diagnosis [1–3]. Notwithstanding, customization of nanomaterials is to be accomplished to liberate

V. G. S. S. Jyothi · V. Bhavana
Department of Pharmaceutics, National Institute of Pharmaceutical Education and Research, Hyderabad, Telangana, India

N. Kommineni (✉)
Center for Biomedical Research, Population Council, New York NY-10065, USA
e-mail: nkommineni@popcouncil.org

A. Madhusudhan et al. (eds.), *Functional Smart Nanomaterials and Their Theranostics Approaches*, Smart Nanomaterials Technology,
https://doi.org/10.1007/978-981-99-6597-7_1

the desired properties. Functionalization is the strategic method to engineer the nanomaterials that eventually leads to improved efficacy eventually diminishing its side effects, due to targeted delivery and active uptake into the cells. Functionalization leads to the variability of the physical and chemical properties of the nanomaterials, rendering the incorporation of desired function to the material [4].

Functionalization is the deliberate bonding of atoms or molecules to material customizing the physicochemical properties of the carrier system [5]. The purpose of functionalization of the carrier material is to incorporate the required properties enabling it to be multifunctional [6]. Some of the primary objectives of functionalization are to reduce cytotoxicity, enhance stability, and to achieve targeting and diagnosis (Fig. 1). However, progress of chemical methods in the arena of nanomaterials is the utmost requirement for its advancement. Numerous methods have been explored in the chemical conjugation and functionalization of nanomaterials specifically to achieve targeted delivery, in tissue engineering and for diagnostic purposes [7]. The continuous efforts in the research of functional materials help in the technological advancement in diagnosis and pharmacotherapy. Functional materials with different patterns and versatile functionality are manifested with a variety of nanomaterials.

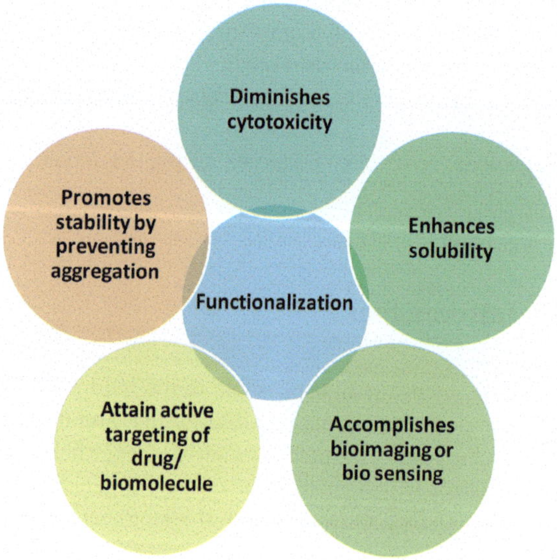

Fig. 1 Primary objectives of functionalized materials

2 Different Categories of Functional Materials Explored

Functional materials are easily processed into various device modalities, such as liposomes, polymeric nanoparticles, inorganic nanoparticles for instance quantum dots (QDs), magnetic nanoparticles, and carbon nanotubes (CNTs) (Fig. 2). So, the most widely explored functional materials in the field of biomedical applications including QDs, CNTs, magnetic nanoparticles, silica nanoparticles, lipid-based nanocarriers, and polymeric nanoparticles are being discussed here (Table 1).

2.1 Carbon Nanotubes

CNT is one of the allotropes of carbon, for example graphite, diamond, amorphous carbon, and fullerene [30]. They are considered as one of the novel and promising nanomaterials in biomedical relevance with prominent applications [31, 32]. They are widely utilized in the delivery of drugs, biomolecules, and for achieving active targeting to the site of action. Additionally, unparalleled electrical and optical properties of CNTs enable them as essential candidates in biosensing and diagnostic applications [33]. Nevertheless, CNTs were found to be cytotoxic which limit their utilization in biomedical field [34]. To surmount the cytotoxicity and facilitate targeting, functionalization of CNTs is being explored in the field of biomedical applications [35]. Functionalized CNTs were reported for their use in bone regeneration [18] cancer therapy, and diagnosis [19, 20] and also in the biosensing [23], angiographic imaging [24], and molecular targeting [25]. Functionalization also enabled CNT biocompatible and enhanced the permeability across the biological membranes. Functionalized CNTs have gained immense popularity in the field of theranostics due to their unique

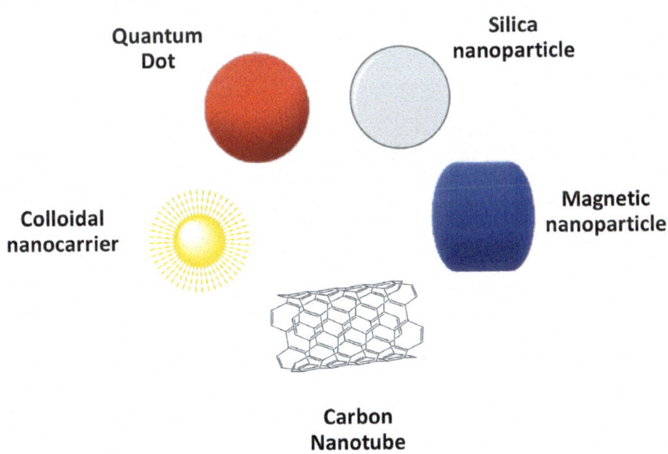

Fig. 2 Most relevant functional materials used in biomedical field

Table 1 Biomedical applications of functionalized materials

Application	Functional material	Active ingredient	Description	References
Wound healing	Carbon quantum dots	Gentamicin sulfate and diammonium citrate	Peilili et al. synthesized antibacterial carbon quantum dots with low-drug resistance self-healing hydrogel with potent wound healing for treating bacterial infections	Li et al. [8]
Biosensor for cholesterol sensing	Nanocomposites of quantum dots	–	Adel et al. synthesized nitrogen graphene quantum dots and copper indium sulfide/zinc sulfide QDs for cholesterol sensing. The fluorescent nanocomposite showed a highly sensitive, selective nonenzymatic cholesterol optical biosensor in 0.312–5 mM cholesterol	Adel et al. [9]
Imaging of microbial cells	Nitrogen-doped carbon quantum dots	–	Jain et al. synthesized nitrogen-doped grapheme quantum dots with photoluminescence ability for bioimaging of microbial cells including E. coli and Saccharomyces cerevisiae yeast	Jain et al. [10]
Targeted drug delivery and bioimaging tool	Folic acid-based carbon dot	Doxorubicin	Sarkar et al. developed folic acid-based carbon dot functionalized stearic acid-g-polyethyleneimine amphiphilic nanomicelle for targeted doxorubicin delivery and imaging for triple-negative breast cancer	Sarkar et al. [11]

(continued)

Table 1 (continued)

Application	Functional material	Active ingredient	Description	References
Targeted delivery of doxorubicin	Maltose-functionalized dendrimer/graphene quantum dots	Doxorubicin	Karimi et al. prepared maltose-functionalized dendrimer/graphene quantum dots as a pH-sensitive biocompatible carrier for targeted delivery of doxorubicin	Karimi and Namazi [12]
Targeted delivery of telmisartan	Chitosan Fe_3O_4 magnetic nanoparticles	Telmisartan	Dhavale et al. grafted chitosan on Fe_3O_4 magnetic nanoparticles and used as a carrier of poorly water-soluble anticancer drug telmisartan	Dhavale et al. [13]
Imaging and targeted delivery in breast cancer cell	FITC functionalized magnetic nanoparticles	Cinnamaldehyde	Folate mediated targeted delivery of cinnamaldehyde loaded and FITC functionalized magnetic nanoparticles in breast cancer: in vitro, in vivo and pharmacokinetic studies	Shetty et al. [14]
Quercetin drug delivery	Functionalized magnetite nanoparticle by chitosan	Quercetin	Quercetin-loaded magnetite nanoparticles exhibited greater toxicity against MCF-7 cells compared to quercetin-free magnetic nanoparticles	Askar et al. [15]
Functionalized boron-dipyrromethene magnetic nanoparticles	Magnetic nanoparticles	–	Scanone et al. developed functionalized magnetic nanoparticles with boron-dipyrromethene for bioimaging and antimicrobial activity. They used the modified nanoparticles to produce fluorescent images of bacterial cells and photoinactivate pathogens	Scanone et al. [16]

(continued)

Table 1 (continued)

Application	Functional material	Active ingredient	Description	References
Bioimaging of liver	Polymeric micelles-coated magnetic nanoparticles	–	Popescu Din et al. developed polymeric micelles-coated magnetic nanoparticles for in vivo bioimaging of liver	Popescu Din et al. [17]
Bone regeneration	Functionalized multiwall carbon nanotubes	–	Shrestha et al. engineered a novel fibrous scaffold with the aid of zein and chitosan into polyurethane with functionalized multiwall carbon nanotubes as a bone cell repair material for bone regeneration	Shrestha et al. [18]
SWCNT bioconjugate	Annexin A5 functionalized single-walled CNTs	Anti-CTLA-4	Annexin A5 functionalized single-walled CNTs bioconjugate synergistically enhances an anti-cytotoxic T-lymphocyte-associated protein 4 dependent abscopal response	McKernan et al. [19]
Anticancer drug carriers	Functionalized single-walled carbon nanotubes	5-Fluorouracil	Ershadi et al. investigated the electronic properties of the combination metallic ((4,0)) or semiconductive ((8,0)) single-wall carbon nanotubes plus 5-Fluorouracil	Ershadi et al. [20]
Bioimaging and biosensing of acute myeloid leukemia	Sgc8-aptamer with luminescent terbium (III) complexes with thiacalix[4]arenesulfonate-doped silica nanoparticles	–	The Sgc8 aptamer conjugated silica nanoparticles showed affinity to CCRF-CEM and Jurkat cells with flow cytometry selectively. The resulted nanoparticles were promising agent for diagnosis and therapy of acute myeloid leukemia creating an opportunity for bioimaging and biosensing	Grechkin, [21], Gubala [22]

(continued)

Table 1 (continued)

Application	Functional material	Active ingredient	Description	References
Biosensors	Functionalized multiwalled carbon nanotubes	–	Wang et al. developed functionalized multiwalled carbon nanotubes biosensors that mimic the hierarchically helical assembly of tissues for chronic chemical monitoring in vivo	Wang et al. [23]
Angiographic imaging	Oxygen-doped carbon nanotubes	–	Takeuchi et al. developed oxygen-doped carbon nanotubes with phospholipid-polyethylene glycol coating to provide bioaffinity and showed prominent fluorescence and Raman signals in spleen and liver; the signals remained for 1 month	Takeuchi et al. [24]
Nanosensors for molecular targeting	Biotin functionalized single-walled carbon nanotubes	–	Chio et al. developed dual functionalized single-walled carbon nanotube nanosensors for molecular targeting and evaluated its functionality by attaching biotin as an affinity pair with avidin protein	Chio et al. [25]
Gene delivery	Protein liposomes	–	Wang et al. prepared transferrin-modified liposomes for the targeted delivery of acetylcholinesterase therapeutic gene to liver cancer	Wang et al. [26]

(continued)

Table 1 (continued)

Application	Functional material	Active ingredient	Description	References
Targeted delivery for spinal cord injury	Functionalized liposomes	–	Wang et al. prepared orally administered targeted delivery system in which the neuropeptide apamin, stabilized by sulfur replacement with selenium, was adopted as a targeting moiety, and the liposome surface was protected with a noncovalent cross-linked chitosan oligosaccharide lactate layer	Wang et al. [26]
Targeted delivery of toll-like receptor (TLR) agonists	C3 targeted liposomes	–	Francian et al. encapsulated toll-like receptor agonists, within the C3-liposomes, including monophosphoryl lipid A, R848, and CpG 1826, specific for TLR4, TLR9 and TLR7/8	Francian et al. [27]

(continued)

Table 1 (continued)

Application	Functional material	Active ingredient	Description	References
Drug delivery to colon cancer cells	Polymeric nanoparticles	Alpha mangostin	Andrade et al. formulated genipin and Eudragit® L100 modified α-Mangostin-loaded mucoadhesive thiolated chitosan nanoparticles for colon-targeted drug delivery against colorectal cancer cells using pH-dependent composite mucoadhesive nanoparticles	Andrade et al. [28]
Targeted nanodelivery	Polymeric nanoparticle	–	Treekoon et al. synthesized iodo-substituted aza-BODIPY encapsulated nanoparticles via the nanoprecipitation method using the amphiphilic poly(ethylene glycol)-*block*-poly(ε-caprolactone) polymer (PEG-*b*-PCL) for a targeted nanodelivery system	Treekoon et al. [29]

properties. One of the significant benefits of functionalized carbon nanotubes is their ability to act as multifunctional agents for both imaging and therapeutic applications. They can be used as diagnostic agents to detect diseases or pathological conditions and simultaneously act as therapeutic agents to treat the disease [36]. Additionally, the functionalized CNTs have high surface area, excellent biocompatibility, and tunable optical and magnetic properties, which make them suitable for targeted drug delivery and imaging [37]. Overall, functionalized carbon nanotubes hold great potential for theranostic applications and can play a crucial role in the development of personalized medicine in the future.

2.2 *Silica Nanoparticles*

Silica nanoparticles are extensively investigated in biomedical applications owing to their low toxicity, exceptional biocompatibility, and high surface area [38–40]. It is feasible to control the size, crystallinity, shape, and porosity of the silica nanoparticles. Besides, surface chemistry of the silica nanoparticles can be fabricated to amend loading of drug, circulation half-life, and enable targeted delivery [41]. Silica nanoparticles doped with dye can be used as a potential fluorescent nanoprobe with fluorescent capabilities similar to heavy metal. The ability to amalgamate these properties renders silica nanoparticles an enviable arena for biosensing, drug delivery, monitoring, and ablative therapies [42]. Sgc8 aptamer conjugated silica nanoparticles were used in bioimaging and biosensing of acute myeloid leukemia [22]. Folic acid-conjugated polyglycerol-grafted $Fe_3O_4@SiO_2$ nanoparticles were prepared for targeting ovarian cancer. The results showed the preferential uptake of nanoparticles by the carcinoma cells (SKOV-3) derived from human ovary. A study on the functionalization of silica nanoparticles with a glypican-3 (GPC3) targeting peptide for ultrasound molecular imaging of human hepatocarcinoma cells was performed. A simple co-precipitation method was used to synthesize the nanoparticles and then conjugate them with the targeting peptide. The results of the study showed that the GPC3-targeting nanoparticles had a high binding affinity to hepatocarcinoma cells and could effectively enhance the ultrasound imaging signal. The study also found that the nanoparticles had a low cytotoxicity and were biocompatible [43].

2.3 *Quantum Dots*

QDs are nanoscale particles which are semiconductor crystals with distinctive electronic and optical properties such as bright and intensive fluorescence [44]. QDs discriminate the conventional organic label dyes by high absorption of light yielding high quantum and large extinction coefficient. They offer extensive observation period with good sensitivity during observation under fluorescence microscope. They are most widely implemented in theranostic purposes enabling diagnosis and therapy simultaneously in the multidisciplinary fields like physics, chemistry, nanotechnology, material sciences, drug delivery, and pharmacology [45]. However, the uncoated QDs tend to undergo photochemical degradation, surface oxidation, aggregation, and even leaching of metal ions on long-time exposure in the biological fluids [46]. This makes them unsuitable for biomedical applications. Functionalization of quantum dots is the emerging field in the biological applications which makes them conquer all the limitations associated with it. They are used in wound healing [8], imaging of microbial cells, [10] targeted delivery of doxorubicin in cancer therapy [11, 12], and also used as nanozymes [47].

2.4 Magnetic Nanoparticles

Magnetic nanoparticles are the promising materials in the biomedical application with the areas focusing on diagnosis, targeted therapy, hyperthermia, and tissue engineering [48]. Nonetheless, functionalization of magnetic nanoparticles is essential to execute the multifunctions in the biomedical field [49]. It empowers the magnetic nanoparticles biocompatible, increases water solubility, and is bio-specific, making them potential carriers in the pharmacotherapy and diagnosis. Functionalized magnetic nanoparticles were used in targeted delivery of telmisartan, cinnamaldehyde [13, 14]. Functionalized magnetic nanoparticles have been used for various applications including targeted drug delivery, magnetic hyperthermia, and magnetic resonance imaging (MRI). They have been shown to improve the efficacy of chemotherapy drugs and reduce the toxicity associated with these treatments. They can also be used to destroy cancer cells through magnetic hyperthermia, where the nanoparticles generate heat when exposed to a magnetic field [50]. In addition, it can be used as contrast agents in MRI, providing high-resolution images of tissues and organs.

2.5 Lipid-Based Carriers and Colloidal Systems

Liposomes, lipid, and polymeric nanoparticles have evolved as prominent carriers in the pharmaceutical formulation development. They have been vastly explored in the biomedical applications. They have demonstrated as highly effective drug carriers as compared to the conventional drug delivery system. However, fabrication of liposomes, lipids and polymeric nanoparticles with longer blood circulation and site-specific drug delivery is still an ongoing research objective which needs to be achieved [51]. One exciting means to accomplish targeting delivery is ligand tagging, such as peptides or monoclonal antibodies, to the carrier [52]. So, liposomes, lipid, and polymeric nanoparticles are extensively studied for the functionalization to achieve the targeted delivery and to escape from the reticular endothelial system.

Functionalized liposomes were reported to target the spinal cord injury [53]. Polymeric nanoparticles were prepared for colon targeting where the polymer exhibited pH-dependent release of Alpha-mangostin [28] and celecoxib [54].

2.6 Basic Principles of Functionalization

Functionalization is the key to incorporate the desired properties to the materials which can be characterized as an intentional addition of functional groups on the surface of nanoparticles to accomplish surface moderation that renders them viable by improving the desired physical, chemical, and mechanical properties. Functionalization can be achieved by binding the functionally active moiety on the surface of nanoparticles and conjugating or interacting the nanoparticle with the carrier or ligand molecule [55].

One of the beneficial outcomes of functionalizing is that the required characteristics in the nanomaterial can be superintended in a predictable approach to fit the specific applications. The procedure utilized to produce, customize, and organize functionalized materials presents exaggeratedly new opportunities for the advance of new multifunctional tools for biological and pharmaceutical relevance.

Functionalization of materials involves the molecular level of interaction with the functional moiety. Subsequently, numerous categories of materials are used for the biomedical applications. Thus, different materials involve different approaches to functionalize. Covalent and noncovalent interactions like electrostatic interactions, hydrogen bonding, and Van der Waals forces are widely involved in these reactions.

2.7 Carbon Nanotubes

CNTs, when generated originally, are not soluble in water and biological fluids and, therefore, elevate their need to enhance solubility and dispersion in the biorelevant fluids. Apart from solubility and dispersibility, reducing its cytotoxicity and improving biocompatibility and enabling them site specific and bioimaging is also essential. Hence, surface modification of CNTs is gaining importance in the biomedical field. CNT surface chemistry, as well as the methods used to purify and functionalize them, has a significant impact on both their biological activities and cytotoxic effects. It also aids in targeted delivery, biosensing, and bioimaging by tagging the CNTs with appropriate ligands.

Among all the methods followed for functionalization (Fig. 3), noncovalent and covalent surface modifications are two frequently used methods. Recently, a study explored the use of CNT for bone regeneration where the CNTs were functionalized with chitosan and zein by covalent bonding [18].

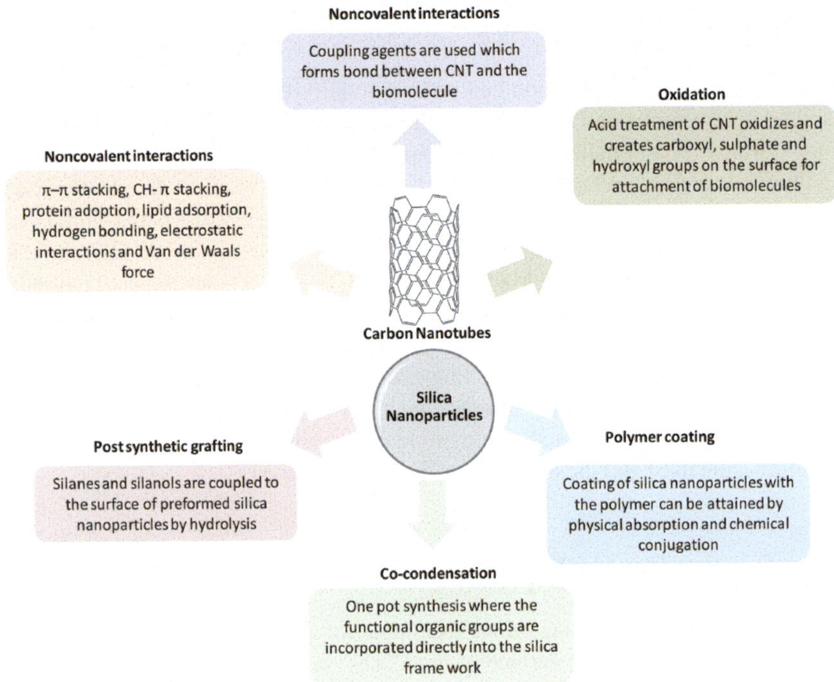

Fig. 3 Carbon nanotubes (CNTs) and silica nanoparticles functionalization

2.7.1 Noncovalent Functionalization

In order to improve solubility of CNTs making them less toxic, biocompatibility, protein assembly, and drug delivery, noncovalent functionalization is regarded as one of the vital means as no considerable alteration in structure and physical properties occurs [56]. CNTs interact noncovalently with various molecules via weak interactions such as surface adsorption onto the side walls of CNTs, π–π stacking, CH-π stacking, protein adoption, lipid adsorption, electrostatic interactions, hydrogen bonding, and van der Waals force [57, 58]. Many polymers, biomolecules, and surfactants have been explored for the noncovalent functionalization of CNTs to make them biocompatible. Sodium dodecyl sulfate (SDS) was the most widely used surfactant to yield individual single-walled CNTs (SWCNTs) under ultra-sonication and ultra-centrifugation conditions [59]. Instead of surfactants, polymers have been tried to enhance dispersion. However, the observations revealed that polymers are inefficient as compared to surfactants. Correspondingly, surfactants with different charges and nonionic charges are also strived for their effectiveness in enhancing dispersion of SWCNTs, however, the quantity of surfactants used is lower than 5% [60].

2.7.2 Covalent Functionalization

Direct covalent functionalization can notably modify the solubility, provide better stability, accessibility, and reduce leaching and render compatibility of CNTs [61]. Thus, covalent functionalization is one of the widely utilized methods for surface modification of CNTs. Diimide-activated amidation is the most frequently used means to couple the CNTs with proteins. Coupling agents are being employed in coupling of proteins and peptides to the CNTs where *N*-ethyl-*N*-(3-dimethyl-aminopropyl) carbodiimide hydrochloride (EDAC) or N, N'-dicyclohexyl carbodi-imide (DCC) is being widely used [62]. Biomolecules can be grafted covalently on the surface of CNTs. For example, multiwalled CNTs were covalently bonded with the polyetherketones in phosphoric acid media [63]. Also, the covalent modification of the nanotube sidewalls totally changes the electronic properties of the CNTs. The reliability of covalent functionalization implementation in CNTs is based on the type of biomolecule used and the desired function to be attained.

2.7.3 Functionalizing by Oxidation

Oxidation of CNTs is the stand-alone technique for the functionalization. It marks in addition of oxygen to the surface of the CNTs' side walls making them soluble in water and biological fluids. Oxidation leads to the introduction of carboxyl or hydroxyl groups on CNTs [64]. Sulfuric acid or nitric acid was used to oxidize CNTs. Acid treatment opens the tubes and creates holes on the surface to which carboxyl, sulfate, and hydroxyl groups are attached [65]. Functionalization makes the CNTs hydrophilic and soluble in aqueous solvents and diminishing its cytotoxicity. Various functional groups can be introduced onto the surface of CNTs. Oxidized CNTs can be functionalized further by esterification, amidation, etc. [66]. Various proteins were found to be absorbed spontaneously and nonspecifically onto the sidewalls of CNTs. They act as carriers and deliver the proteins and other biomolecules directly into the mammalian cells.

2.8 *Silica Nanoparticles*

Silica nanoparticles are usually functionalized to enhance the colloidal stability and solubility in the biological fluids, improve biocompatibility, enhance high cellular uptake, enable bioimaging, and promote site-specific delivery [67]. Thereby, surface modification of silica nanoparticles plays an important role in influencing the interaction of functionalized silica nanoparticles with the physiological environment, biodistribution, cellular internalization, and disease targeting [68]. Currently, there are three main chemical methods to functionalize nanosilica which are post-synthetic grafting, co-condensation, and polymer coating (Fig. 3).

2.8.1 Post-synthetic Grafting

Silica nanoparticles are usually functionalized by post-synthesis grafting for surface modification. Different organic and inorganic silane precursors, for instance, alkoxysilanes/halosilanes, are incorporated into the silica nanoparticles and are being used for cellular imaging, drug targeting, etc. [69]. Silanes and silanols are coupled to the surface of silica by hydrolysis. Nevertheless, the post-synthesis grafting often leads to heterogeneous chemistry, where pore opening shows higher density of grafting and interior framework shows lower efficiency.

Alkoxysilanes and halosilanes undergo condensation reaction with the silanol groups on the surface of silica forming 1-3 Si-O-Si links where the halosilanes classically hydrolyze substituting the halide for alcohol group forming 1-3 Si-O-Si links [70]. However, halosilanes react directly with silanol groups on the surface in the anhydrous condition. Most often, 3-mercaptopropyl trimethoxysilane (MPTS) and 3-aminopropyl triethoxysilane (ATPS) are being used for functionalization of silica nanoparticles [71]. Subsequently, PEGylation is the most frequently used functionalization method to improve the pharmacokinetics of drug molecules by alteration of circulation half-life and enhances the solubility, stability, and biocompatibility of nanoparticles. Bovine serum albumin (BSA) is also used more usually in low concentration to improve the stability of nanoparticles [72].

Silica is also conjugated with the antibodies via a linker group through the functional moiety on the surface of silica and amino acid residues. Linker group prevents the interaction between the antibody and silica protecting it from denaturation by creating a space between the silica and antibody. The linker groups are available as monovalent or multivalent based on the functional groups available on the surface of the linker. For instance, monovalent linker with glutaraldehyde functional group makes imine covalent bond between the amino residue on the biomolecule and the amine group on the surface of nanosilica [69]. However, these linkers are coupled with few limitations like aggregation and thereby stability in the biological fluid.

2.8.2 Co-condensation

In co-condensation method, functionalization of silica nanoparticles is carried out during the synthesis process. It is a one-pot synthesis process where the functional organic groups are incorporated directly into the silica framework via sol–gel method [73]. Organosilanes are the functional organic groups widely used for the functionalization of nanoparticles where they are added to the reaction mixture containing the tetrafunctional silane. This technique aids in the functionalization of internal surface of mesoporous silica nanoparticles [74].

However, the degree of functionalization depends on the nature of co-condensing agents and its concentration and size. Alkoxysilanes or halosilanes are used as condensing agents same as with the post-synthetic grafting. These alkoxy or haloxysilanes form 1-3 Si-O-Si links with the surface silanol groups [75]. However, halosilanes substitute the halide with alcohol by hydrolysis forming 1-3 Si-O-Si link.

APTS and MPTS are the most frequently used functional moieties for functionalization of silica. Co-condensing agents also impact the shape of the particles where hydrophobic agents yield rod-shaped particles while hydrophilic agents yield round particles.

2.9 Polymer Coating

Coating of silica nanoparticles with the polymer can be attained by physical absorption and chemical conjugation. Silica has negative charge on its surface which has the affinity toward the positively charged polymers [76]. The modified silica surface can accommodate various biomolecules which include polypeptides, carbohydrates, nucleic acids, and antibodies [77]. The versatile chemical nature of silica surface makes the conjugation of functional moieties reliably easy. Thus, the desired characteristics can be incorporated in the silica nanoparticles by functionalization approach.

2.10 Quantum Dots

QDs are the potential nanoprobes being explored in the biomedical applications such as drug delivery, biomolecules delivery, site-specific delivery, and theranostic approaches [78]. However, functionalization is the crucial step in utilizing the QDs for multifunctional purposes as QDs are suffering from stability issues, insolubility in aqueous solutions [79]. Even for achieving the targeted delivery, conjugation of ligands onto the surface of QDs is essential which is in turn achieved by functionalization [80]. Originally, synthesized QDs are hydrophobic in nature where hydrophilization of QDs is the first step in utilizing the QDs for biomedical applications. Consequently, bioconjugation of QDs after their hydrophilization results in a multifunctional nanoparticle that conglomerates the electrochemical/optical properties of QDs with the biological utility of the biomolecule.

2.10.1 Hydrophylization

Mainly three approaches are used to hydrophilize the QDs which include ligand exchange, silanization, and encapsulation (Fig. 4).

(i) *Ligand exchange*

In this method, the ligands with bifunctional molecules are exchanged with the original hydrophobic coating. These bifunctional ligands are water soluble with one end is being attached to the surface of QD and the other end is free to conjugate with

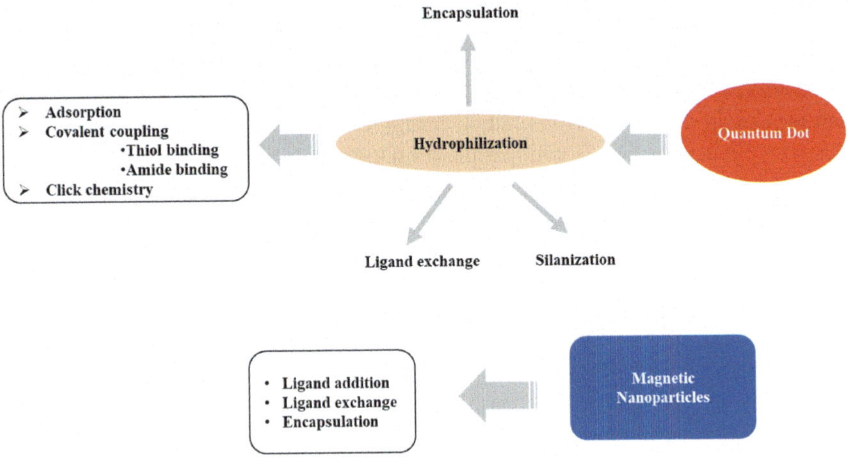

Fig. 4 Quantum dot and magnetic nanoparticles functionalization

a biomolecule [81, 82]. The most commonly used ligands used in this technique are trioctylphosphine oxide (TOPO), trioctylphosphine (TOP), tetradecyl phosphonic acid (TDPA), oleic acid, and hexadecyl amine (HDA) [83, 84]. These ligands prevent aggregation of nanoparticles and their growth and passivate its surface defects to preserve quantum yield [85]. Organic ligands offer exceptional stability and solubility to the QDs where the exchange process occurs in the organic noncoordinating solvents via simple mass action [86].

Ligands pre-existing on the surface of QD are in dynamic equilibrium with the surrounding solvent. Hence, the ligands are in continuous motion where the ligand on the surface leaves the QDs, whereas free ligands in the solution occupy and attach the available sites on the surface of QD. Thereby, the new hydrophilic ligands intended to attach to the QD are added to the solvent where the pre-existing ligands which are in dynamic equilibrium are exchanged with the new added ligands [87]. However, to achieve this ligand exchange process successfully, the concentration of the newly added ligand should be equal to the concentration of the pre-existing ligand and should have high affinity toward the QDs. Conversely, the newly added/ replacing ligand with low affinity to the surface of QD can also be used in this technique by increasing the concentration of ligand thus enhancing the probability of attachment of replacing ligand [88]. The most widely used ligands for the ligand exchange technique are polymers with functional groups such as carboxyl and amine and thiol, which can replace the existing ligands on the surface of the QDs.

(ii) *Silanization*

Silanization is the process of coating the QDs with a layer of amorphous silica [89]. This technique is similar to the ligand exchange process. However, it is regarded as a separate technique due to its prevalent use in the hydrophilization of QDs. In the first step, the surface of the QD is activated for attaching the silane molecules to the

QD. This step requires the exchange of TOPO and HDA on the surface of QD with the polar ligands that can be easily dispersed in water or ethanol [90]. This method is considered as highly advantageous as compared with the nanoparticles coated with organic compounds. This method results in a shell of silica on the surface of QDs where it achieves hydrophilization of QD. Apart from that, silica is considered as inert, nontoxic, and optically transparent. It also provides colloidal stability to the QDs and protects from leaching and protects physical and chemical processes. Additionally, silica is biocompatible, and its surface can be functionalized easily making it suitable for further conjugation with biomolecules.

(iii) *Encapsulation*

In this approach, the hydrophobic surface of QD is coated with a variety of carriers which includes liposomes, polymers with amphiphilic nature [91]. The QD with hydrophobic shell interacts directly with the coating material by physical interaction, which includes electronic or hydrophobic interactions. Large number of materials like amphiphilic copolymers and biopolymers were used for the encapsulation of QDs [92]. Surfactants can also be used to encapsulate the QDs, however, their weak interactions make them unsuitable for encapsulation. Thereby, amphiphilic polymers (PEG are widely used for encapsulation where they form very strong interaction with the surface of QD owing to its numerous hydrophilic and hydrophobic units [93].

Liposomes are also being explored in encapsulation of QDs. Liposomes are spherical vesicles with hollow structure enabling them for loading of hydrophobic QDs. The hollow structure of liposomes captivates high loading ability and loading of QDs into its hollow structure makes the QDs water soluble [91]. It also facilitates conjugation with a variety of biomolecules including proteins, enzymes, etc. and enhances its stability and amplifies its analytical signal. However, the use of liposomes as encapsulating agents is limited by their instability in physiological conditions and during storage. It also suffers from its sensitivity toward the slight variations in pH, temperature, and osmotic pressure [94]. Nevertheless, these drawbacks can be surmounted by templating or covering with different materials such as silica, polymers. But this method results in an increase in the hydrodynamic diameter of the particle which makes it unsuitable for biosensing and active targeting [95].

2.10.2 Functionalization with Biomolecules

Immobilization of biomolecules on to the surface of QD involves two approaches which include covalent linking and noncovalent binding where the biomolecules are coupled directly to the surface of QD (Fig. 4). Covalent linkage is attained by different bioconjugation chemistry using activated functional groups at the surface of QDs, whereas noncovalent binding is established by electrostatic, hydrophobic, or affinity interactions between biomolecules and the surface of the QD.

The common tactics for the attachment of biomolecules can be classified as covalent attachment and nonspecific adsorption.

(i) *Adsorption of functional molecules*

Adsorption is the simple attachment of chemical moieties on the surface of nanoparticles with any chemical reaction or bonds. This is the simple and highly utilized technique for the conjugation of ligands or molecules on the surface of nanoparticles [77, 96]. In adsorption, electrostatic interaction, hydrogen bonding, and hydrophobic interactions are mainly involved in this functionalization process. The large surface area on the nanoparticles can be used for the adsorption of large molecules for instance proteins, polymers, nucleic acids, and enzymes. There is no role of functional groups in this strategy and considered as a nonselective binding technique. The electrostatic interaction observed in this adsorption strategy is seen in molecules with charges opposite to the surface of QD. For example, proteins engineered with positively charged terminal are interacted electrostatically with the negatively charged QD resulted from the ligation of QD with carboxyl groups [97].

Hydrogen bonding is another weak chemical interaction in adsorption where the ligand on the surface of QD forms hydrogen bond with the biomolecules. Thereby, this adsorption technique depends on the type of molecule being involved in the conjugation process. For instance, oligonucleotides adsorb on the surface of QD depending on the pH and type of medium, ionic strength. It was reported that pH plays a key role in adsorption of oligonucleotides where it adsorbs strongly in the acidic pH when the carboxyl groups on the surface of QD are protonated. In the presence of alkaline or neutral pH conditions, the adsorption of oligonucleotide is considerably reduced. This suggested that the hydrogen bond is the primary mechanism involved in the adsorption of oligonucleotide on to the surface QD [98]. For further confirmation of the formation of hydrogen bond in the adsorption of oligonucleotide to the QD, formamide was added to the solution of QD. Formamide is a well-known molecule which disrupts the hydrogen bonding interaction of nucleic acid. It was noticed that with the increase in concentration of formamide, there is decrease in adsorption of oligonucleotide to the QD [99, 100]. Thus, it was evidenced that the hydrogen bond is the main driving force for the adsorption of oligonucleotides.

Despite its easiness in the adoption of adsorption strategy, it suffers from various drawbacks. These include lack of specificity in the adsorption of biomolecules leading to lack of control on the number of molecules conjugating, lack of control on the orientation of biomolecules attached to the QD and the presence of weaker interactions between the molecules and the QD leading to competitive binding from the host ligands with the surface of QD. Thus, assurance of the biomolecules attachment onto the surface of QD and their orientation is critical in this strategy and to be monitored carefully.

(ii) *Covalent coupling of biomolecules*

QDs can be coupled with the biomolecules by covalent bonding that involves different bonds.

Amide coupling

QDs are hydrophilized with attachment of various ligands which makes them solubilized in the aqueous solutions. The ligand attached in hydrophilization of QD is made in such a way that the terminal groups of ligands are –OH, –COOH, –SH or $-NH_2$. These functional groups aids in easy conjugating with the biomolecules such as peptides, proteins or antibodies [101]. The carboxyl-terminated functional group on the surface of QD forms the amide bond with the amine residues in the biomolecules. The amide bonds are formed easily in the alkaline buffered conditions which is a simple water-soluble process [77]. This simple reaction also retains the intact structure of the compounds and does not require any lengthy spacers. Thus, the QDs conjugated with the amide functionality preserve the structure and also maintain the hydrodynamic size of the particles as the size of the particle is directly proportional to the size of the functional moiety attached on the surface of QD [102].

Thiol binding

Functionalization of QDs involves two processes, where hydrophilization of QD is followed by conjugation with the ligand. However, this thiol binding technique does not require any solubilization method, and both functionalization and solubilization are achieved by this single step [45]. Thus, this is an alternative to the two-step process for the functionalization of QDs. Peptides can be customized in a way to act as both targeting and anchoring agent [103]. The same procedure can be used in QDs where the thiol-terminated functional group on the surface of QD can be attached to the sulfur-containing amino acid [104]. The thiol or amine group on the surface of QD forms thiol bond with the thiol or amine group on the proteins or DNA or peptide. Even the stability of thiol bonds is stronger when compared to disulfide bonds as the disulfide bonds can undergo exchange reaction in the cellular media with the competing thiol groups. Thus, the thiol-bonded QDs can be used for the targeted delivery of drugs, proteins, or antibodies where the release of conjugated molecule from the QD results from the cleavage of thiol bonds by selective enzymes present at a specific site in the body [105].

Click chemistry

Click chemistry is the most prevailing chemical synthesis technique used in the small molecule bioconjugation reaction. It is used to couple the ligands onto the surface of QDs. This reaction is also used to attach huge number of biomolecules and coupling agents to the QDs. The reaction was carried out at room temperature and formed a thermodynamically stable bond between the carbon and a heteroatom resulting in a higher yield. However, the click reaction with the biomolecules results in lower yield which is resulted by the synthesis of terminal azide or alkyne groups [106, 107]. The stability of these groups gives a stable final reactant which can be introduced to diverse linking groups and biomolecules. Moreover, multiple functionalities can be achieved with the same QDs possessing the azide or alkyne termination [108, 109].

2.11 *Magnetic Nanoparticles*

Chemical modification of magnetic nanoparticles is essential to realize its biomedical application which is achieved mainly by three different mechanisms which include ligand addition, ligand exchange, and encapsulation (Fig. 4).

2.11.1 Ligand Addition

This method is associated with the coupling of a ligand to the surface of magnetic nanoparticles. There is no involvement in the removal of pre-existing ligands on the surface of magnetic nanoparticles. Ligand addition is accomplished by four different ways. The first way of addition is to conjugate the ligand to the nanoparticles with no capping agent. The second way of addition is associated with indirect method where a layer of inorganic material is developed on the surface of magnetic nanoparticle and the ligand is adsorbed to the nanoparticle via ionic or nonspecific interactions with the added layer [110–112]. As the surface of magnetic nanoparticles is hydrophobic in nature, hydrophobic moieties are capped to the surface by the hydrophobic interaction which is the third way of ligand addition. Covalent bond formation between the newly added ligand and the pre-existing ligand also assists in the ligand addition.

2.11.2 Ligand Exchange

Ligand exchange is the exchange of pre-existing ligands with the newly added ligands. Here, the hydrophobic ligands on the surface of magnetic nanoparticles are replaced with the hydrophilic ligands. The detachment and attachment of ligands depend on their strength where the hydrophilic ligand needs to be stronger than the hydrophobic ligand in order to replace them [113–115]. This enables the magnetic nanoparticles soluble in the aqueous solution.

2.11.3 Encapsulation

Encapsulation is a process in which the magnetic nanoparticles capped with the hydrophobic ligands are coated with the amphiphilic materials. This coating of amphiphilic material on the surface of magnetic nanoparticles renders them soluble in aqueous solution and makes them biocompatible. The amphiphilic material interacts with the ligand on the surface of nanoparticles with the hydrophobic part and the hydrophilic part is exposed toward the solution thus forming a coat on the nanoparticle surface [116]. Furthermore, the functional groups on the encapsulated nanoparticles can be used for the bioconjugation of different biomolecules.

Table 2 Functional bonds involved in colloidal nanocarriers

Nanocarrier	Class of chemical interaction	Bonds or interactions involved
Liposomes	Covalent coupling	Thioether bond
		Disulfide bond
		Amide bond
		Hydrazide bond
		Amine–amine bond
	Noncovalent coupling	Weak interactions
Polymeric nanocarrier	Adsorption	Hydrogen bond
		Hydrophobic interactions
	Covalent binding	PEGylation
		Biotinylation

2.12 Lipid-Based Carriers and Colloidal Systems

Liposomes, lipid nanoparticles, and polymeric nanoparticles are widely explored in pharmaceutical technology and diagnosis. However, functionalization is the key to realize all the desired functions. Binding of ligand to the liposomes and the polymeric nanoparticles are discussed separately here (Table 2).

2.12.1 Lipid-Based Carriers

Ligands are attached to the surface of liposomes by two methods which enable them to execute multifunctional. The two coupling methods are covalent and noncovalent binding.

(i) *Covalent binding*

Ligands are attached to the liposomal surface via an anchor which aids in connecting the ligand and the liposome. The anchor with the functional groups forms covalent bonds with the ligands. The most widely used anchors are phospholipids such as phosphatidylinositol and phosphatidylethanolamine and fatty acids such as palmitic acid [117, 118]. Among these, phosphatidylethanolamine is the most widely used anchor. These anchors are easily incorporated into the liposomes. However, there are two approaches to conjugate the ligand to the liposomes via an anchor. In the first approach, the ligand is covalently bound to the anchor, and the resulted ligand is mixed with the constituents of the liposome during its synthesis [119]. In the second approach, the anchor is mixed with the constituents of liposomes during its fabrication, and thus, the anchor is incorporated in the liposome. The ligand is attached to the preformed liposome with the aid of anchor [120]. Thereby, functionalization of liposomes with the ligands is achieved by the anchor molecules. The most commonly employed bonds in tagging the ligands to the liposome are thioether bonds, disulfide bonds, amide bond, hydrazide bond, and amine–amine bond [121–123].

(ii) *Noncovalent binding*

Noncovalent binding is the alternative way to tag the ligand to the liposomes. It is the easiest way of coupling ligand to the liposomes where there is no need of any anchor molecule. The ligand is mixed with the constituents of the liposomes and the anchor is ligated to the liposome during its formation [124]. However, practical concerns associated with this technique are attachment of ligand to the liposome is low about 4–40%, and the control of ligand attachment is not possible where the orientation of biomolecules cannot be assured. It also results in physical instability of liposomes. However, detachment of ligand from the liposomes can be assured in the body as the bond involved is noncovalent in nature. Attempts have been made to attach antibodies to liposomes by noncovalent interactions via heptanes [125]. Nevertheless, the noncovalent technique for ligand attachment is still in infancy stage due to its weak interaction and random distribution of ligand on the surface of liposome and needs to be explored much.

2.12.2 Polymeric Nanoparticles

Polymeric nanoparticles are ligated with ligands to execute various functions, and it is achieved by two different methods (Table 2).

(i) *Adsorption*

Fabrication of liposomes with the ligands is achieved by the adsorption which is widely used approach to mount antibodies onto the surface of nanoparticles. The physical adsorption of ligand on the nanoparticle surface is attained by the hydrogen bond or by hydrophobic interaction or electrostatic interaction [126]. These interactions immobilize the ligands onto the surface of nanoparticles. To avoid the chemical reaction between the ligand and the nanoparticle, coating can be implemented which ensures the stability of the ligated polymeric nanoparticles.

(ii) *Covalent binding*

Covalent coupling of ligand to the polymeric nanoparticles is also one of the approaches to tag the ligand onto the polymeric nanoparticles. However, only very few studies have been reported with the covalent coupling. A study was reported on the poly(cyanoacrylate) nanoparticles where the nanoparticles are coated with PEG and the PEG is conjugated with the transferrin as a ligand [127]. This transferrin ligand was used to deliver the DNA to the targeted cells. In another study, poly(lactic acid) nanoparticles were conjugated with the biotin where the biotin and poly (methyl methacrylate-co-methacrylic acid) conjugate was precipitated with the poly(lactic acid) to synthesize stable polymeric nanoparticles for active targeting [128]. Thus, targeted delivery of nanoparticles can be attained by conjugating with the appropriate ligands.

3 Conclusion

Functional materials are widely explored in the arena of biomedical relevance. Various classes of materials are being used which include quantum dots, silica nanoparticles, carbon nanotubes, magnetic nanoparticles, etc. However, direct usage of these materials in biomedical applications is limited due to various reasons. Thereby, functionalization of these materials can be performed to surmount the issues associated with them. Consequently, functionalization is considered as a prerequisite for the versatile usage of nanomaterials in the diagnostic and therapeutic applications. Functionalization involves either the chemical reaction or a simple adsorption of nanomaterial and a functional moiety. It aids in the attachment of biomolecules or ligands or fluorescent dyes onto the surface of the material executing multifunctions in the biological applications.

References

1. Fenton OS, Olafson KN, Pillai PS, Mitchell MJ, Langer R (2018) Advances in biomaterials for drug delivery. Adv Mater 30(29):1705328. https://doi.org/10.1002/adma.201705328
2. Husen A, Siddiqi KS (2023) Advances in smart nanomaterials and their applications. Elsevier Inc., 50 Hampshire St., 5th Floor, Cambridge, MA 02139, USA https://doi.org/10.1016/C2021-0-02202-1
3. Jin-Chul K, Madhusudhan A, Husen A (2021) Smart Nanomaterials in Biomedical Applications. Springer Nature Switzerland AG, Gewerbestrasse 11, 6330 Cham, Switzerland. https://doi.org/10.1007/978-3-030-84262-8
4. Bao G, Mitragotri S, Tong S (2013) Multifunctional nanoparticles for drug delivery and molecular imaging. Annu Rev Biomed Eng 15:253–282. https://doi.org/10.1146/annurev-bioeng-071812-152409
5. Sapsford KE, Algar WR, Berti L, Gemmill KB, Casey BJ, Oh E, Stewart MH, Medintz IL (2013) Functionalizing nanoparticles with biological molecules: developing chemistries that facilitate nanotechnology. Chem Rev 113(3):1904–2074. https://doi.org/10.1021/cr300143v
6. Alle M, Sharma G, Lee S-H, Kim J-C (2022) Next-generation engineered nanogold for multimodal cancer therapy and imaging: a clinical perspectives. J Nanobiotechnol 20(1):1–34. https://doi.org/10.1186/s12951-022-01402-z
7. Navya P, Daima HK (2016) Rational engineering of physicochemical properties of nanomaterials for biomedical applications with nanotoxicological perspectives. Nano Convergence 3(1):1. https://doi.org/10.1186/s40580-016-0064-z
8. Li P, Liu S, Yang X, Du S, Tang W, Cao W, Zhou J, Gong X, Xing X (2021) Low-drug resistance carbon quantum dots decorated injectable self-healing hydrogel with potent antibiofilm property and cutaneous wound healing. Chem Eng J 403:126387. https://doi.org/10.1016/j.cej.2020.126387
9. Adel R, Ebrahim S, Shokry A, Soliman M, Khalil M (2021) Nanocomposite of CuInS/ZnS and nitrogen-doped graphene quantum dots for cholesterol sensing. ACS Omega 6(3):2167–2176. https://doi.org/10.1021/acsomega.0c05416
10. Jain R, Vithalani R, Patel D, Lad U, Modi CK, Suthar D, Solanki JD, Surati KR (2021) Highly fluorescent nitrogen-doped graphene quantum dots (N-GQDs) as an efficient nanoprobe for imaging of microbial cells. Fuller Nanotub Carbon Nanostructures 29(8):588–595. https://doi.org/10.1080/1536383X.2021.1872548

11. Sarkar P, Ghosh S, Sarkar K (2021) Folic acid based carbon dot functionalized stearic acid-g-polyethyleneimine amphiphilic nanomicelle: Targeted drug delivery and imaging for triple negative breast cancer. Colloids Surf B: Biointerfaces 197:111382. https://doi.org/10.1016/j.colsurfb.2020.111382
12. Karimi S, Namazi H (2020) Simple preparation of maltose-functionalized dendrimer/graphene quantum dots as a pH-sensitive biocompatible carrier for targeted delivery of doxorubicin. Int J Biol Macromol 156:648–659. https://doi.org/10.1016/j.ijbiomac.2020.04.037
13. Dhavale RP, Dhavale RP, Sahoo SC, Kollu P, Jadhav SU, Patil PS, Dongale TD, Chougale AD, Patil PB (2021) Chitosan coated magnetic nanoparticles as carriers of anticancer drug Telmisartan: pH-responsive controlled drug release and cytotoxicity studies. J Phys Chem Solids 148:109749. https://doi.org/10.1016/j.jpcs.2020.109749
14. Shetty V, Jakhade A, Shinde K, Chikate R, Kaul-Ghanekar R (2021) Folate mediated targeted delivery of cinnamaldehyde loaded and FITC functionalized magnetic nanoparticles in breast cancer: in vitro, in vivo and pharmacokinetic studies. New J Chem 45(3):1500–15. https://doi.org/10.1039/D0NJ04319B
15. Askar MA, El-Nashar HA, Al-Azzawi MA, Rahman SSA, Elshawi OE (2022) Synergistic effect of quercetin magnetite nanoparticles and targeted radiotherapy in treatment of breast cancer. Breast Cancer (Auckl) 16:11782234221086728. https://doi.org/10.1177/11782234221086728
16. Scanone AC, Santamarina SC, Heredia DA, Durantini EN, Durantini A, s. M. (2020) Functionalized magnetic nanoparticles with BODIPYs for bioimaging and antimicrobial therapy applications. ACS Appl Bio Mater 3(2):1061–1070. https://doi.org/10.1021/acsabm.9b01035
17. Popescu Din IM, Balas M, Hermenean A, Vander Elst L, Laurent S, Burtea C, Cinteza LO, Dinischiotu A (2020) Novel polymeric micelles-coated magnetic nanoparticles for in vivo bioimaging of liver: toxicological profile and contrast enhancement. Materials 13(12):2722. https://doi.org/10.3390/ma13122722
18. Shrestha S, Shrestha BK, Ko SW, Kandel R, Park CH, Kim CS (2021) Engineered cellular microenvironments from functionalized multiwalled carbon nanotubes integrating Zein/Chitosan@ Polyurethane for bone cell regeneration. Carbohydr Polym 251:117035. https://doi.org/10.1016/j.carbpol.2020.117035
19. McKernan P, Virani NA, Faria GN, Karch CG, Silvy RP, Resasco DE, Thompson LF, Harrison RG (2021) Targeted single-walled carbon nanotubes for photothermal therapy combined with immune checkpoint inhibition for the treatment of metastatic breast cancer. Nanoscale Res Lett 16(1):1–9. https://doi.org/10.1186/s11671-020-03459-x
20. Ershadi N, Safaiee R, Golshan M (2021) Functionalized (4, 0) or (8, 0) SWCNT as novel carriers of the anticancer drug 5-FU; A first-principle investigation. Appl Surf Sci 536:147718. https://doi.org/10.1016/j.apsusc.2020.147718
21. Grechkin YA, Grechkina SL, Zaripov EA, Fedorenko SV, Mustafina AR, Berezovski MV (2022) Aptamer-conjugated Tb (III)-doped silica nanoparticles for luminescent detection of leukemia cells. Biomedicines 8(1):14. https://doi.org/10.3390/biomedicines8010014
22. Gubala V, Giovannini G, Kunc F, Monopoli MP, Moore CJ (2020) Dye-doped silica nanoparticles: synthesis, surface chemistry and bioapplications. Cancer Nano 11(1):1–43. https://doi.org/10.1186/s12645-019-0056-x
23. Wang L, Xie S, Wang Z, Liu F, Yang Y, Tang C, Wu X, Liu P, Li Y, Saiyin H, Zheng S (2020) Functionalized helical fibre bundles of carbon nanotubes as electrochemical sensors for long-term in vivo monitoring of multiple disease biomarkers. Nat Biomed Eng 4(2):159–171. https://doi.org/10.1038/s41551-019-0462-8
24. Takeuchi T, Iizumi Y, Yudasaka M, Kizaka-Kondoh S, Okazaki T (2019) Characterization and biodistribution analysis of oxygen-doped single-walled carbon nanotubes used as in vivo fluorescence imaging probes. Bioconjugate Chem 30(5):1323–1330. https://doi.org/10.1021/acs.bioconjchem.9b00088
25. Chio L, Pinals RL, Murali A, Goh NS, Landry MP (2020) Covalent surface modification effects on single-walled carbon nanotubes for targeted sensing and optical imaging. Adv Funct Mater 30(17):1910556. https://doi.org/10.1002/adfm.201910556

26. Wang K, Shang F, Chen D, Cao T, Wang X, Jiao J, He S, Liang X (2021) Protein liposomes-mediated targeted acetylcholinesterase gene delivery for effective liver cancer therapy. J Nanobiotechnol 19(1):1–15. https://doi.org/10.1186/s12951-021-00777-9
27. Francian A, Widmer A, Olsson T, Ramirez M, Heald D, Rasic K, Adams L, Martinson H, Kullberg M (2021) Delivery of toll-like receptor agonists by complement C3-targeted liposomes activates immune cells and reduces tumor growth. J Drug Target 29(7):754–760. https://doi.org/10.1080/1061186X.2021.1878364
28. Andrade RG, Reis B, Costas B, Lima SAC, Reis S (2021) Modulation of macrophages M1/M2 polarization using carbohydrate-functionalized polymeric nanoparticles. Polymers 13(1):88. https://doi.org/10.3390/polym13010088
29. Treekoon J, Chansaenpak K, Tumcharern G, Zain ZSZ, Lee HB, Kue CS, Kamkaew A (2021) Aza-BODIPY encapsulated polymeric nanoparticles as effective nanodelivery system for photodynamic cancer treatment. Mater Chem Front 5(5):2283–2293. https://doi.org/10.1039/D0QM00891E
30. Okpalugo T, Papakonstantinou P, Murphy H, McLaughlin J, Brown N (2005) High resolution XPS characterization of chemical functionalised MWCNTs and SWCNTs. Carbon 43(1):153–161. https://doi.org/10.1016/j.carbon.2004.08.033
31. Bandi R, Tummala S, Dadigala R, Alle M, Lee S-H (2022) Role of metal-doped carbon dots in bioimaging and cancer therapy. Smart nanomaterials in biomedical applications. Springer, pp 101–123. https://doi.org/10.1007/978-3-030-84262-8_4
32. Dadigala R, Bandi R, Alle M, Gangapuram BR, Lee S-H (2022) Graphene-based smart nanomaterials for photothermal therapy. Smart nanomaterials in biomedical applications. Springer, pp 125–153
33. Liu Z, Tabakman S, Welsher K, Dai H (2009) Carbon nanotubes in biology and medicine: in vitro and in vivo detection, imaging and drug delivery. Nano Res 2(2):85–120. https://doi.org/10.1007/s12274-009-9009-8
34. Zhang Y, Ali SF, Dervishi E, Xu Y, Li Z, Casciano D, Biris AS (2010) Cytotoxicity effects of graphene and single-wall carbon nanotubes in neural phaeochromocytoma-derived PC12 cells. ACS Nano 4(6):3181–3186. https://doi.org/10.1021/nn1007176
35. Abidin MN, Goh PS, Ismail AF, Othman MH, Hasbullah H, Said N, Kadir SH, Kamal F, Abdullah MS, Ng BC (2017) Development of biocompatible and safe polyethersulfone hemodialysis membrane incorporated with functionalized multi-walled carbon nanotubes. Mater Sci Eng C 77:572–582. https://doi.org/10.1016/j.msec.2017.03.273
36. Mostafavi E, Iravani S, Varma RS, Khatami M, Rahbarizadeh F (2022) Eco-friendly synthesis of carbon nanotubes and their cancer theranostic applications. Mater Adv 3(12):4765–4782. https://doi.org/10.1039/D2MA00341D
37. Kohls A, Maurer Ditty M, Dehghandehnavi F, Zheng S-Y (2022) Vertically aligned carbon nanotubes as a unique material for biomedical applications. ACS Appl Mater Interfaces 14(5):6287–6306. https://doi.org/10.1021/acsami.1c20423
38. Fu C, Liu T, Li L, Liu H, Chen D, Tang F (2013) The absorption, distribution, excretion and toxicity of mesoporous silica nanoparticles in mice following different exposure routes. Biomaterials 34(10):2565–2575. https://doi.org/10.1016/j.biomaterials.2012.12.043
39. Liu T, Li L, Teng X, Huang X, Liu H, Chen D, Ren J, He J, Tang F (2011) Single and repeated dose toxicity of mesoporous hollow silica nanoparticles in intravenously exposed mice. Biomaterials 32(6):1657–1668. https://doi.org/10.1016/j.biomaterials.2010.10.035
40. Rahman I, Vejayakumaran P, Sipaut C, Ismail J, Chee C (2009) Size-dependent physicochemical and optical properties of silica nanoparticles. Mater Chem Phys 114(1):328–332. https://doi.org/10.1016/j.matchemphys.2008.09.068
41. Choi JY, Gupta B, Ramasamy T, Jeong JH, Jin SG, Choi HG, Yong CS, Kim JO (2018) PEGylated polyaminoacid-capped mesoporous silica nanoparticles for mitochondria-targeted delivery of celastrol in solid tumors. Colloids Surf B Biointerfaces 165:56–66. https://doi.org/10.1016/j.colsurfb.2018.02.015
42. Peng F, Su Y, Zhong Y, Fan C, Lee S-T, He Y (2014) Silicon nanomaterials platform for bioimaging, biosensing, and cancer therapy. Acc Chem Res 47(2):612–623. https://doi.org/10.1021/ar400221g

43. Di Paola M, Quarta A, Conversano F, Sbenaglia EA, Bettini S, Valli L, Casciaro S (2017) Human hepatocarcinoma cell targeting by glypican-3 ligand peptide functionalized silica nanoparticles: implications for ultrasound molecular imaging. Langmuir 33(18):4490–4499. https://doi.org/10.1021/acs.langmuir.7b00327
44. Ekimov A, Onuschenko A (1981) The quantum size effect in three-dimensional microscopic semiconductors. JETP Lett 34(6):363
45. Zrazhevskiy P, Sena M, Gao X (2010) Designing multifunctional quantum dots for bioimaging, detection, and drug delivery. Chem Soc Rev 39(11):4326–4354. https://doi.org/10.1039/B915139G
46. Schiffman JD, Balakrishna RG (2018) Quantum dots as fluorescent probes: Synthesis, surface chemistry, energy transfer mechanisms, and applications. Sens Actuators B Chem 258:1191–1214. https://doi.org/10.1016/j.snb.2017.11.189
47. Bandi R, Alle M, Dadigala R, Park CW, Han SY, Kwon GJ, Kim JC, Lee SH (2022) Integrating the high peroxidase activity of carbon dots with easy recyclability: immobilization on dialdehyde cellulose nanofibrils and cholesterol detection. Appl Mater Today 26:101286. https://doi.org/10.1016/j.apmt.2021.101286
48. Pankhurst Q, Thanh N, Jones S, Dobson J (2009) Progress in applications of magnetic nanoparticles in biomedicine. J Phys D: Appl Phys 42(22):224001. https://doi.org/10.1088/0022-3727/42/22/220301
49. Schladt TD, Schneider K, Schild H, Tremel W (2011) Synthesis and bio-functionalization of magnetic nanoparticles for medical diagnosis and treatment. Dalton Trans 40(24):6315–6343. https://doi.org/10.1039/C0DT00689K
50. Dash S, Das T, Patel P, Panda PK, Suar M, Verma SK (2022) Emerging trends in the nanomedicine applications of functionalized magnetic nanoparticles as novel therapies for acute and chronic diseases. Nanobiotechnol 20(1):1–24. https://doi.org/10.1186/s12951-022-01595-3
51. Riaz MK, Riaz MA, Zhang X, Lin C, Wong KH, Chen X, Zhang G, Lu A, Yang Z (2018) Surface functionalization and targeting strategies of liposomes in solid tumor therapy: a review. Int J Mol Sci 19(1):195. https://doi.org/10.3390/ijms19010195
52. Borrajo ML, Alonso MJ (2022) Using nanotechnology to deliver biomolecules from nose to brain—peptides, proteins, monoclonal antibodies and RNA. Drug Deliv Transl Res 1–19 https://doi.org/10.1007/s13346-021-01086-2
53. Wang X, Wu J, Liu X, Tang K, Cheng L, Li J, Tang Y, Song X, Wang X, Li C (2021) Engineered liposomes targeting the gut–CNS Axis for comprehensive therapy of spinal cord injury. J Control Release 331:390–403. https://doi.org/10.1016/j.jconrel.2021.01.032
54. Madhusudhan A (2018) The effect of colon targeted delivery of celecoxib loaded microspheres on experimental colitis induced by acetic acid in rats. Asian J Pharm (AJP) 12(01). https://doi.org/10.22377/ajp.v12i01.1919
55. Subbiah R, Veerapandian M, Subbiah R, Yun K (2010) Nanoparticles: functionalization and multifunctional applications in biomedical sciences. Curr Med Chem 17(36):4559–4577. https://doi.org/10.2174/092986710794183024
56. Antonucci A, Kupis-Rozmysłowicz J, Boghossian AA (2017) Noncovalent protein and peptide functionalization of single-walled carbon nanotubes for biodelivery and optical sensing applications. ACS Appl Mater Interfaces 9(13):11321–11331. https://doi.org/10.1021/acsami.7b00810
57. Bai L, Bai Y, Zheng J (2017) Improving the filler dispersion and performance of silicone rubber/multi-walled carbon nanotube composites by noncovalent functionalization of polymethylphenylsiloxane. J Mater Sci 52(12):7516–7529. https://doi.org/10.1007/s10853-017-0984-y
58. Mittal V (2011) Carbon nanotubes surface modifications: an overview Surface modification of nanotube fillers. Wiley Online Library
59. O'connell MJ, Bachilo SM, Huffman CB, Moore VC, Strano MS, Haroz EH, Rialon KL, Boul PJ, Noon WH, Kittrell C, Ma J (2002) Band gap fluorescence from individual single-walled carbon nanotubes. Science 297(5581):593–596. https://doi.org/10.1126/science.1072631

60. Moore VC, Strano MS, Haroz EH, Hauge RH, Smalley RE, Schmidt J, Talmon Y (2003) Individually suspended single-walled carbon nanotubes in various surfactants. Nano Lett 3(10):1379–1382. https://doi.org/10.1021/nl034524j
61. Daniel S, Rao TP, Rao KS, Rani SU, Naidu G, Lee H-Y, Kawai T (2007) A review of DNA functionalized/grafted carbon nanotubes and their characterization. Sens Actuators B Chem 122(2):672–682. https://doi.org/10.1016/j.snb.2006.06.014
62. Gao Y, Kyratzis I (2008) Covalent immobilization of proteins on carbon nanotubes using the cross-linker 1-ethyl-3-(3-dimethylaminopropyl) carbodiimide—a critical assessment. Bioconjugate Chem 19(10):1945–1950. https://doi.org/10.1021/bc800051c
63. Oh SJ, Lee HJ, Keum DK, Lee SW, Wang DH, Park SY, Tan LS, Baek JB (2006) Multiwalled carbon nanotubes and nanofibers grafted with polyetherketones in mild and viscous polymeric acid. Polymer 47(4):1132–1140. https://doi.org/10.1016/j.polymer.2005.12.064
64. Prato M, Kostarelos K, Bianco A (2008) Functionalized carbon nanotubes in drug design and discovery. Acc Chem Res 41(1):60–68. https://doi.org/10.1021/ar700089b
65. Carrero-Sanchez J, Elias A, Mancilla R, Arrellin G, Terrones H, Laclette J, Terrones M (2006) Biocompatibility and toxicological studies of carbon nanotubes doped with nitrogen. Nano Lett 6(8):1609–1616. https://doi.org/10.1021/nl060548p
66. Abuilaiwi FA, Laoui T, Al-Harthi M, Atieh MA (2010) Modification and functionalization of multiwalled carbon nanotube (MWCNT) via fischer esterification. Arab J Sci Eng 35(1):37–48
67. Cha BG, Kim J (2019) Functional mesoporous silica nanoparticles for bio-imaging applications. Wiley Interdiscip Rev Nanomed Nanobiotechnol 11(1):e1515. https://doi.org/10.1002/wnan.1515
68. Rosenholm JM, Sahlgren C, Lindén M (2010) Towards multifunctional, targeted drug delivery systems using mesoporous silica nanoparticles–opportunities and challenges. Nanoscale 2(10):1870–1883. https://doi.org/10.1039/C0NR00156B
69. Liberman A, Mendez N, Trogler WC, Kummel AC (2014) Synthesis and surface functionalization of silica nanoparticles for nanomedicine. Surf Sci Rep 69(2–3):132–158. https://doi.org/10.1016/j.surfrep.2014.07.001
70. Banga R, Yarwood J, Morgan AM, Evans B, Kells J (1995) FTIR and AFM studies of the kinetics and self-assembly of alkyltrichlorosilanes and (perfluoroalkyl) trichlorosilanes onto glass and silicon. Langmuir 11(11):4393–4399. https://doi.org/10.1021/la00011a036
71. Hermanson GT (2013) Bioconjugate techniques. Academic press.
72. Galdino FE, Picco AS, Sforca ML, Cardoso MB, Loh W (2020) Effect of particle functionalization and solution properties on the adsorption of bovine serum albumin and lysozyme onto silica nanoparticles. Colloids Surf B: Biointerfaces 186:110677. https://doi.org/10.1016/j.colsurfb.2019.110677
73. Huh S, Wiench JW, Yoo J-C, Pruski M, Lin VS-Y (2003) Organic functionalization and morphology control of mesoporous silicas via a co-condensation synthesis method. Chem Mater 15(22):4247–4256. https://doi.org/10.1021/cm0210041
74. Gottuso A, Armetta F, Cataldo A, Nardo VM, Parrino F, Saladino ML (2022) Functionalization of mesoporous silica nanoparticles through one-pot co-condensation in w/o emulsion. Microporous Mesoporous Mater 335:111833. https://doi.org/10.1016/j.micromeso.2022.111833
75. Bagwe RP, Hilliard LR, Tan W (2006) Surface modification of silica nanoparticles to reduce aggregation and nonspecific binding. Langmuir 22(9):4357–4362. https://doi.org/10.1021/la052797j
76. Buchman YK, Lellouche E, Zigdon S, Bechor M, Michaeli S, Lellouche J-P (2013) Silica nanoparticles and polyethyleneimine (PEI)-mediated functionalization: a new method of PEI covalent attachment for siRNA delivery applications. Bioconjug Chem 24(12):2076–2087. https://doi.org/10.1021/bc4004316
77. Sperling RA, Parak WJ (2010) Surface modification, functionalization and bioconjugation of colloidal inorganic nanoparticles. Philos Trans Math Phys Eng Sci 368(1915):1333–1383. https://doi.org/10.1098/rsta.2009.0273

78. Thanh NT, Green LA (2010) Functionalisation of nanoparticles for biomedical applications. Nano Today 5(3):213–230. https://doi.org/10.1016/j.nantod.2010.05.003
79. Snee PT (2018) The role of colloidal stability and charge in functionalization of aqueous quantum dots. Acc Chem Res 51(11):2949–2956. https://doi.org/10.1021/acs.accounts.8b00405
80. Graf C, Dembski S, Hofmann A, Rühl E (2006) A general method for the controlled embedding of nanoparticles in silica colloids. Langmuir 22(13):5604–5610. https://doi.org/10.1021/la060136w
81. Mattoussi H, Mauro JM, Goldman ER, Anderson GP, Sundar VC, Mikulec FV, Bawendi MG (2000) Self-assembly of CdSe−ZnS quantum dot bioconjugates using an engineered recombinant protein. J Am Chem Soc 122(49):12142–12150. https://doi.org/10.1021/ja002535y
82. Uyeda HT, Medintz IL, Jaiswal JK, Simon SM, Mattoussi H (2005) Synthesis of compact multidentate ligands to prepare stable hydrophilic quantum dot fluorophores. JACS 127(11):3870–3878. https://doi.org/10.1021/ja044031w
83. Bera D, Qian L, Tseng T-K, Holloway PH (2010) Quantum dots and their multimodal applications: a review. Materials 3(4):2260–2345. https://doi.org/10.3390/ma3042260
84. Ong W-L, Rupich SM, Talapin DV, McGaughey AJ, Malen JA (2013) Surface chemistry mediates thermal transport in three-dimensional nanocrystal arrays. Nature Mater 12(5):410–415. https://doi.org/10.1038/nmat3596
85. Hammer NI, Emrick T, Barnes MD (2007) Quantum dots coordinated with conjugated organic ligands: new nanomaterials with novel photophysics. Nanoscale Res Lett 2(6):282. https://doi.org/10.1007/s11671-007-9062-8
86. Mei BC, Susumu K, Medintz IL, Delehanty JB, Mountziaris T, Mattoussi H (2008) Modular poly (ethylene glycol) ligands for biocompatible semiconductor and gold nanocrystals with extended pH and ionic stability. J Mater Chem 18(41):4949–4958. https://doi.org/10.1039/B810488C
87. Olsson C (2009) Synthesis of bifunctional poly (ethylene oxides) and their use as ligands to nanoparticles (Doctoral dissertation, Staats-und Universitätsbibliothek Hamburg Carl von Ossietzky)
88. Zhang Y, Clapp A (2011) Overview of stabilizing ligands for biocompatible quantum dot nanocrystals. Sensors 11(12):11036–11055. https://doi.org/10.3390/s111211036
89. Correa-Duarte MA, Kobayashi Y, Caruso RA, Liz-Marzán LM (2001) Photodegradation of SiO2-coated CdS nanoparticles within silica gels. J Nanosci Nanotechnol 1(1):95–99. https://doi.org/10.1166/jnn.2001.010
90. Nann T, Riegler J, Nick P, Mulvaney P (2005) Quantum dots with silica shells. Paper presented at the nanobiophotonics and biomedical applications II, vol 5705. pp 77–84. https://doi.org/10.1117/12.582871
91. Beloglazova NV, Shmelin PS, Speranskaya ES, Lucas B, Helmbrecht C, Knopp D, Niessner R, De Saeger S, Goryacheva IY (2013) Quantum dot loaded liposomes as fluorescent labels for immunoassay. Anal Chem 85(15):7197–7204. https://doi.org/10.1021/ac401729y
92. Fernández-Argüelles M, Costa-Fernández JM, Pereiro R, Sanz-Medel A (2008) Simple bio-conjugation of polymer-coated quantum dots with antibodies for fluorescence-based immunoassays. Analyst 133(4):444–447. https://doi.org/10.1039/B802360N
93. Shen L (2011) Biocompatible polymer/quantum dots hybrid materials: current status and future developments. J Funct Biomater 2(4):355–372. https://doi.org/10.3390/jfb2040355
94. Ciobanu M, Heurtault B, Schultz P, Ruhlmann C, Muller C, Frisch B (2007) Layersome: development and optimization of stable liposomes as drug delivery system. Int J Pharm 344(1–2):154–157. https://doi.org/10.1016/j.ijpharm.2007.05.037
95. Banerjee A, Grazon C, Nadal B, Pons T, Krishnan Y, Dubertret B (2015) Fast, efficient, and stable conjugation of multiple DNA strands on colloidal quantum dots. Bioconjug Chem 26(8):1582–1589. https://doi.org/10.1021/acs.bioconjchem.5b00221
96. DeLong RK, Reynolds CM, Malcolm Y, Schaeffer A, Severs T, Wanekaya A (2010) Functionalized gold nanoparticles for the binding, stabilization, and delivery of therapeutic DNA,

RNA, and other biological macromolecules. Nanotechnol Sci Appl 3:53. https://doi.org/10.2147/NSA.S8984
97. Clapp AR, Medintz IL, Mattoussi H (2006) Förster resonance energy transfer investigations using quantum-dot fluorophores. ChemPhysChem 7(1):47–57. https://doi.org/10.1002/cphc.200500217
98. Algar WR, Krull UJ (2010) Developing mixed films of immobilized oligonucleotides and quantum dots for the multiplexed detection of nucleic acid hybridization using a combination of fluorescence resonance energy transfer and direct excitation of fluorescence. Langmuir 26(8):6041–6047. https://doi.org/10.1021/la903751m
99. Algar WR, Krull UJ (2011) Interfacial chemistry and the design of solid-phase nucleic acid hybridization assays using immobilized quantum dots as donors in fluorescence resonance energy transfer. Sensors 11(6):6214–6236. https://doi.org/10.3390/s110606214
100. Xiao Q, Huang S, Qi Z-D, Zhou B, He Z-K, Liu Y (2008) Conformation, thermodynamics and stoichiometry of HSA adsorbed to colloidal CdSe/ZnS quantum dots. BBA Proteins Proteomics 1784(7–8):1020–1027. https://doi.org/10.1016/j.bbapap.2008.03.018
101. Pereira M, Lai EP (2008) Capillary electrophoresis for the characterization of quantum dots after non-selective or selective bioconjugation with antibodies for immunoassay. J Nanobiotechnol 6(1):1–15. https://doi.org/10.1186/1477-3155-6-10
102. Kairdolf BA, Smith AM, Stokes TH, Wang MD, Young AN, Nie S (2013) Semiconductor quantum dots for bioimaging and biodiagnostic applications. Annu Rev Anal Chem 6:143–162. https://doi.org/10.1146/annurev-anchem-060908-155136
103. Koolpe M, Burgess R, Dail M, Pasquale EB (2005) EphB receptor-binding peptides identified by phage display enable design of an antagonist with ephrin-like affinity. J Biol Chem 280(17):17301–17311. https://doi.org/10.1074/jbc.M500363200
104. Wolcott A, Gerion D, Visconte M, Sun J, Schwartzberg A, Chen S, Zhang JZ (2006) Silica-coated CdTe quantum dots functionalized with thiols for bioconjugation to IgG proteins. J Phys Chem B 110(11):5779–5789. https://doi.org/10.1021/jp057435z
105. Derfus AM, Chen AA, Min D-H, Ruoslahti E, Bhatia SN (2007) Targeted quantum dot conjugates for siRNA delivery. Bioconjug Chem 18(5):1391–1396. https://doi.org/10.1021/bc060367e
106. Kolb HC, Sharpless KB (2003) The growing impact of click chemistry on drug discovery. Drug Discovery Today 8(24):1128–1137. https://doi.org/10.1016/S1359-6446(03)02933-7
107. Nwe K, Brechbiel MW (2009) Growing applications of "click chemistry" for bioconjugation in contemporary biomedical research. Cancer Biother Radiopharm 24(3):289–302. https://doi.org/10.1089/cbr.2008.0626
108. Huisgen R (1963) Kinetics and mechanism of 1, 3-dipolar cycloadditions. Angew Chem Int Ed 2(11):633–645. https://doi.org/10.1002/anie.196306331
109. Speers AE, Cravatt BF (2004) Profiling enzyme activities in vivo using click chemistry methods. Chem Biol 11(4):535–546. https://doi.org/10.1016/j.chembiol.2004.03.012
110. Al-Saadi A, Yu CH, Khutoryanskiy VV, Shih S-J, Crossley A, Tsang SC (2009) Layer-by-layer electrostatic entrapment of protein molecules on superparamagnetic nanoparticle: a new strategy to enhance adsorption capacity and maintain biological activity. J Phys Chem C 113(34):15260–15265. https://doi.org/10.1021/jp903100j
111. Sun B, Zhang Y, Gu K-J, Shen Q-D, Yang Y, Song H (2009) Layer-by-layer assembly of conjugated polyelectrolytes on magnetic nanoparticle surfaces. Langmuir 25(10):5969–5973. https://doi.org/10.1021/la804217w
112. Xie J, Peng S, Brower N, Pourmand N, Wang SX, Sun S (2006) One-pot synthesis of monodisperse iron oxide nanoparticles for potential biomedical applications. Pure Appl Chem 78(5):1003–1014. https://doi.org/10.1351/pac200678051003
113. Jun YW, Huh YM, Choi JS, Lee JH, Song HT, Kim S, Yoon S, Kim KS, Shin JS, Suh JS, Cheon J (2005) Nanoscale size effect of magnetic nanocrystals and their utilization for cancer diagnosis via magnetic resonance imaging. J Am Chem Soc 127(16):5732–5733. https://doi.org/10.1021/ja0422155

114. Patil R, Thorat N, Shete P, Otari S, Tiwale B, Pawar S (2016) In vitro hyperthermia with improved colloidal stability and enhanced SAR of magnetic core/shell nanostructures. Mater Sci Eng C 59:702–709. https://doi.org/10.1016/j.msec.2015.10.064
115. Xu C, Xu K, Gu H, Zheng R, Liu H, Zhang X, Guo Z, Xu B (2004) Dopamine as a robust anchor to immobilize functional molecules on the iron oxide shell of magnetic nanoparticles. J Am Chem Soc 126(32):9938–9939. https://doi.org/10.1021/ja0464802
116. Shin J, Anisur RM, Ko MK, Im GH, Lee JH, Lee IS (2009) Hollow manganese oxide nanoparticles as multifunctional agents for magnetic resonance imaging and drug delivery. Angew Chem Int Ed 48(2):321–324. https://doi.org/10.1002/anie.200802323
117. Huang A, Huang L, Kennel SJ (1980) Monoclonal antibody covalently coupled with fatty acid. A reagent for in vitro liposome targeting. J Biol Chem 255(17):8015–8018. https://doi.org/10.1016/S0021-9258(19)70595-X
118. Torchilin V, Klibanov A, Smirnov V (1982) Phosphatidylinositol may serve as the hydrophobic anchor for immobilization of proteins on liposome surface. FEBS Lett 138(1):117–120. https://doi.org/10.1016/0014-5793(82)80408-0
119. Harsch M, Walther P, Weder H, Hengartner H (1981) Targeting of monoclonal antibody-coated liposomes to sheep red blood cells. Biochem Biophys Res Commun 103(3):1069–1076. https://doi.org/10.1016/0006-291X(81)90917-7
120. Torchilin V, Omel'Yanenko V, Klibanov A, Mikhailov A, Gol'danskII V, Smirnov V (1980) Incorporation of hydrophilic protein modified with hydrophobic agent into liposome membrane. BBA-Biomembranes 602(3):511–521. https://doi.org/10.1016/0005-2736(80)90330-2
121. Chua M-M, Fan S-T, Karush F (1984) Attachment of immunoglobulin to liposomal membrane via protein carbohydrate. Biochim Biophys Acta Gen Subj 800(3):291–300. https://doi.org/10.1016/0304-4165(84)90408-2
122. Hansen CB, Kao GY, Moase EH, Zalipsky S, Allen TM (1995) Attachment of antibodies to sterically stabilized liposomes: evaluation, comparison and optimization of coupling procedures. Biochim Biophys Acta Biomembr 1239(2):133–144. https://doi.org/10.1016/0005-2736(95)00138-S
123. Nässander UK, Steerenberg PA, Poppe H, Storm G, Poels LG, De Jong WH, Crommelin DJ (1992) In vivo targeting of OV-TL 3 immunoliposomes to ascitic ovarian carcinoma cells (OVCAR-3) in athymic nude mice. Cancer Res 52(3):646–653
124. Huang L, Kennel SJ (1979) Binding of immunoglobulin G to phospholipid vesicles by sonication. Biochemistry 18(9):1702–1707. https://doi.org/10.1021/bi00576a011
125. Leserman LD, Weinstein JN, Moore JJ, Terry WD (1980) Specific interaction of myeloma tumor cells with hapten-bearing liposomes containing methotrexate and carboxyfluorescein. Cancer Res 40(12):4768–4774. http://www.ncbi.nlm.nih.gov/pubmed/7438109
126. Illum L, Jones P, Kreuter J, Baldwin R, Davis S (1983) Adsorption of monoclonal antibodies to polyhexylcyanoacrylate nanoparticles and subsequent immunospecific binding to tumour cells in vitro. Int J Pharm 17(1):65–76. https://doi.org/10.1016/0378-5173(83)90019-4
127. Li Y, Ogris M, Wagner E, Pelisek J, Rüffer M (2003) Nanoparticles bearing polyethyleneglycol-coupled transferrin as gene carriers: preparation and in vitro evaluation. Int J Pharm 259(1–2):93–101. https://doi.org/10.1016/S0378-5173(03)00211-4
128. Gautier S, Grudzielski N, Goffinet G, De Hassonville SH, Delattre L, Jérôme R (2001) Preparation of poly (D, L-lactide) nanoparticles assisted by amphiphilic poly (methyl methacrylate-co-methacrylic acid) copolymers. J Biomater Sci Polym Ed 12(4):429–450. https://doi.org/10.1163/156856201750195306

Functional Biomaterials for Targeted Drug Delivery Applications

Hemant Singh, Muzammil Kuddushi, Ramesh Singh, Sneha Sathapathi, Aniruddha Dan, Narayan Chandra Mishra, Dhiraj Bhatia, and Mukesh Dhanka

Abstract Targeted drug delivery approaches have been widely employed to regulate the absorption and biodistribution of small molecules and biologics at the targeted diseased sites to maximize therapeutic performances without affecting the healthy cells of the tissue or organ inside the body. For this purpose, it is essential to select effective biomaterial-based platforms that can release drugs in a sustained manner without altering their bioactivity and causing toxic effects in off-targeted tissues. Such biomaterials can be directly implanted/injected into the targeted diseased area of the body to enhance drug delivery efficiency. Various materials have been explored to fabricate targeted drug delivery systems. Functional biomaterials with desired physicochemical and biological properties are getting colossal attention because they can respond to their environmental cues, such as fluctuations in pH, temperature, or cell-associated enzymatic activity, and improve drug delivery integration and tissue regeneration. This chapter explores the progress of different approaches in functionalizing polymeric, metallic, and ceramic biomaterials for targeted drug delivery systems.

Keywords Targeted drug delivery · Biomaterials · Functionalized polymer · Functionalized metals and ceramics

---

H. Singh · M. Kuddushi · R. Singh · S. Sathapathi · A. Dan · D. Bhatia (✉) · M. Dhanka (✉)
Department of Biological Sciences and Engineering, Indian Institute of Technology Gandhinagar, Gandhinagar, Gujarat, India
e-mail: dhiraj.bhatia@iitgn.ac.in

M. Dhanka
e-mail: mukesh.d@iitgn.ac.in

M. Kuddushi
Department of Chemical and Materials Engineering, University of Alberta, Alberta, Canada

N. C. Mishra
Polymer and Process Engineering, Indian Institute of Technology Roorkee, Roorkee, Uttarakhand, India

A. Madhusudhan et al. (eds.), *Functional Smart Nanomaterials and Their Theranostics Approaches*, Smart Nanomaterials Technology,
https://doi.org/10.1007/978-981-99-6597-7_2

1 Introduction

Drug delivery is the process of delivering drugs to humans or animals to achieve a desirable therapeutic effect in treating illnesses [1]. Various drug delivery systems (e.g., tablets, capsules, syrups, ointments, and among others) have been developed and tested for therapeutics molecule delivery [1, 35]. However, these traditional drug delivery technologies are ineffective in providing a sustained drug release profile and suffer from low bioavailability and rapid fluctuation in plasma drug levels [1, 35]. Thus, the entire treatment procedure may be ineffective without an effective delivery method [35, 118]. Moreover, the drug must be administered at an actual controlled rate and at the target site to provide optimal efficacy and safety [35, 118]. Therefore, properly designed targeted drug delivery systems with controlled drug release are being developed to address the issues with traditional drug delivery technologies [5, 35, 65, 79, 118]. Presently, there is a high demand for developing effective targeted therapeutic delivery systems to reduce dosage frequency and boost drug efficiency at the required site to minimize its adverse side effects [118]. A collaborative effort by engineers, chemists, biologists, clinicians, and biomaterial scientists who have been using biomaterial for fabricating the targeted drug delivery system for specific cells has sparked a surge in interest in improving human health [1, 35, 118]. Various pharmacological substances, including antibodies, peptides, vaccines, drugs, and enzymes, have increased delivery and effectiveness thanks to the development of biomaterials-based targeted drug delivery systems [99]. Biomaterials are generally grouped into three classes: metals, ceramics, and polymers [1]. These biomaterials have been used as a carrier, and many of these materials with combinations have been created to release medicines over long periods of time [51, 58, 77, 78, 99]. They may be further customized to target certain cells/tissues of the body [99]. This phenomenon allows the reduction of dosage of pharmaceutical drugs while still huge gap in achieving the desired therapeutic effect and lowering patient toxicity [99]. An overview of available possibilities in the polymers, metals, and ceramics for fabricating targeted drug delivery systems has shown in Fig. 1, and this chapter explores the progress of different approaches in functionalizing polymeric, metallic, and ceramic biomaterials for the application in targeted drug delivery.

2 Functionalization of Polymeric Materials for Targeted Drug Delivery

The use of polymers and polymer-based nanomaterials in the biomedical field has risen dramatically in recent years [3, 35]. Identifying and understanding the intended applications that call for modifying these materials is crucial before designing or selecting polymers and their nanocomposites [35, 118]. Specific moieties and conjugated compounds that enhance material performance can be grafted onto polymers through chemical functionalization [1]. Since a wide range of inert polymers and

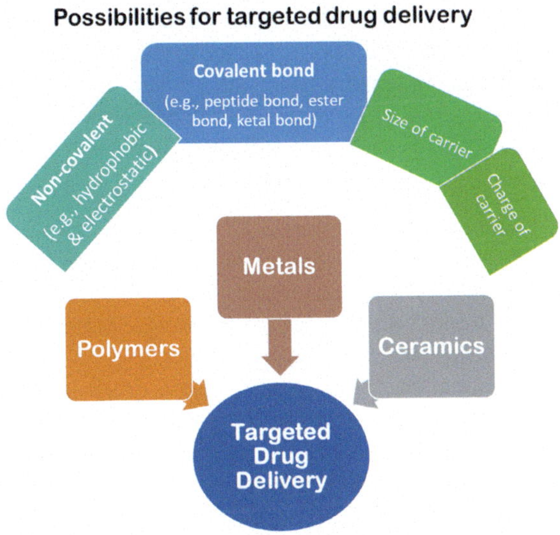

Fig. 1 Overview of available possibilities in the polymers, metals, and ceramics for fabricating targeted drug delivery systems

nanomaterials are commercially available, surface modification is a necessary step to increase their adherence and enable the development of these materials for specialized applications across a variety of sectors [35, 76, 99]. The properties of polymeric materials are typically determined by functional groups other than the backbone chains [35, 99]. Metal coordination is a type of noncovalent bond that has been employed in the functionalization of supramolecular polymers [35, 99]. Since numerous pyridine-based ligands are commercially accessible and can structurally be changed, pyridyl-based systems have a high probability of success as polymerizable metal complexes and polymers functionalized by metal coordination [118]. Bipyridines and terpyridines in particular are appealing because they may act as acceptors to maintain different metal oxidation states and are well known for coordinating a range of metals [118]. For example, hyaluronic acid derivatives were functionalized by the maleic anhydride ring opening process by reacting with awaiting hydroxyl groups [99, 118]. One such promising application of the functionalization of polymeric materials is the usage of the same for targeted drug delivery [1, 118]. For fabricating the drug delivery systems, the immobilization of bioactive substances on a polymeric surface through physical adsorption, i.e., noncovalent interactions, is desirable [1, 118]. These variations are crucial, especially for various drug delivery methods that are intended to deliver an adequate amount of medication to the targeted site while minimizing the adverse impacts of the drugs to the other body organs [1, 118]. In order to address the limitations of the traditional dosage forms, targeted and customized drug delivery systems were developed [1, 118]. Moreover, to enable an effective drug delivery, several materials have been created through employing nanotechnology techniques in polymeric materials for anticancer drug delivery [5, 6, 100, 107]. Various disciplines are actively developing products using carbon-based biomaterials and nanomaterials. In the biomedical industry, intensive research is being done to create new carbon nanotubes (CNT) for targeted disease diagnosis and their treatment [100, 106, 107]. Due to their smaller size (nanometers in diameter and micrometers in length), which allows them to interact with living biological tissues thereby improving the efficiency of drug delivery. Moreover CNTs also have some significant properties [100, 106, 107] like,

- Biocompatible with a variety of cell types, making them a safe material for use in drug delivery systems.
- CNTs can be functionalized with specific targeting molecules, such as antibodies or peptides, to enable targeted delivery of drugs to specific cells or tissues.
- CNTs can help to stabilize the drugs, protecting them from degradation and increasing their shelf life.
- CNTs can be engineered to release drugs in a controlled manner, which can improve the efficacy of the drug while reducing potential side effects.
- CNTs have been shown to be nontoxic in low concentrations, making them an attractive material for use in medical applications.

Therefore, drug delivery systems in the cancer treatment, CNTs can be employed. Targeted tissue distribution of CNTs via microencapsulation by employing biocompatible polymeric membranes like alginate-poly-l-lysine-alginate can be performed as a part of cancer therapy [100, 106, 107]. To address the issue of intrinsic toxicity of CNTs, it has been suggested that CNTs covalently functionalized with phenyl-SO_3H or phenyl $(COOH)_2$ groups provide less toxicity to cells than raw CNTs treated with surfactants [100, 106, 107]. Focusing on CNTs in particular, their low solubility in aqueous media severely hinders their use in almost all biological media [100]. Therefore, surface functionalization of the CNTs is equally important for their activity [100]. In addition to resolving the issue of CNTs' hydrophobic nature, which can lead to the formation of highly poisonous aggregates, surface functionalization also improves the material's biocompatibility [37, 100]. In the area of diagnostic and imaging, CNTs have another promising powerful use [100]. CNTs have been used to detect the prostate and colorectal cancer marker, and a hepatocarcinoma marker [106]. Drugs that could not ordinarily be given to cancer cells by microscale vehicles can now be supplied to them using a nanoscale vehicle, such as CNTs mixed with other composites [106]. CNTs are potential needle-like transporters of macromolecules like genes and proteins as well as tiny medicinal compounds [18, 38]. In the treatment of cancer, single-walled nanotubes (SWNTs) functionalized by Polyethylene glycol (PEG) and phospholipid, further coupled with the anticancer paclitaxel (PTX) successful used [73]. SWCNTs, graphene-based nanomaterials, and their prototypes have also been evaluated for their use in TB and cancer chemotherapy [105]. It is still essential to concentrate on the metabolism, biodistribution, clearance, and accumulation of nanomaterials, especially CNTs [100]. Additionally, CNTs have been discovered to be potential scaffold materials for the bone and nerve tissue regeneration [13]. By regulating its type, delivery site, and dosage, CNTs can be a secure, novel, and high-performance biomaterial [13]. Furthermore, cross-linked polymers known as hydrogels are essential in many biomedical fields including drug administration, sensor systems, and tissue engineering [85]. Recent progresses in the development of magnetic hydrogels as a drug carrier uses the natural polymers like polysaccharides, protein, and DNA have been documented [85]. According to studies, adding versatility to the hydrogel formulation by functionalizing PEG chains with hydrazone groups or chemically orthogonal ester and grafting Rhodamin B (RhB) allows for the release of the hydrophilic drug mimic load to last for longer [85]. In terms of drug delivery effectiveness, aggregating norfloxacin molecules in hydrogels and raising drug concentration both improved drug delivery performance [50]. In a number of soft tissue applications, thermosensitive injectable hydrogels have been employed to administer pharmacological and cellular therapy [39]. Magnetic hybrid hydrogels have also been constructed. They mix the magnetic characteristics of nanoparticles with the usual properties of the hydrogel to enhance the desired properties [19]. A substantial amount of medication can be put into the hybrid hydrogel before being administered [19]. Because the hydrogels contain magnetic nanoparticles, the hydrogel can be transferred and localized in the target spot using a magnet [19]. Therapeutic agent delivery to bladder lesion hope will soon be transformed by hydrogel-based drug delivery devices. The efficacy of therapy seems to be increased by combining a variety of treatment modalities based on hydrogels and combinatorial drug delivery systems with distinct mechanisms of action [102]. This section focuses on a couple of highly applicable polymeric biomaterials that can be applied to the delivery of certain drugs.

2.1 *Functionalized Collagen for Targeted Drug Delivery*

The discipline of tissue engineering is currently seeing a revival in the usage of collagen as a biomaterial [9]. In many different drug delivery systems and biomaterial applications, collagen plays a significant role [9]. Because it is widely available and has great biocompatibility and biodegradability qualities, collagen has been frequently used in drug delivery [9]. The most effective and exciting applications include mini-pellets containing protein medicines, injectable dispersions for treating local tumors, and sponges carrying antibiotics [9]. The subcutaneous injection of soluble collagen for the treatment of dermatological problems has been one of collagen's most successful commercial applications [9]. Some instances of frequently utilized collagen-based delivery vehicles are corneal shields, scaffolds for various chronic and burn wounds, nanoparticles for gene delivery, growth factor-loaded injectable hydrogels, and antibiotic dressings [9]. The physical or chemical cross-linking of collagen modulates biodegradability and drug release to control the impact of drug delivery [9]. It has been demonstrated that collagen nanoparticles are thermally stable particles that quickly reach sterilization and work well as carriers for medicinal substances and cytotoxicity agents [14]. Collagen-based polymeric materials also exhibit high efficacy in being used as a system for the controlled release of antibiotics/steroids [14]. In the upcoming years, smart collagen nanoparticles can direct the growth of in vivo tissues, engage stem cells, and direct them to particular body regions [14]. It has been reported that thermoresponsive drug delivery vehicles made of elastin-b-collagen-like peptide (ELP-CLP) vesicles can target collagen-containing matrices, as shown in Fig. 2 [74]. For the targeted therapy of various diseases like joint problems linked to high collagen remodeling activity, ELP-CLP vesicles may be viable contenders as drug carriers for such associated problems [74]. Using collagen in combination with biodegradable hydrophobic polyesters is a feasible approach for creating therapeutic biomaterials [74]. With the help of these polymeric materials, additional lipophilic medications supporting tissue repair can be delivered [74].

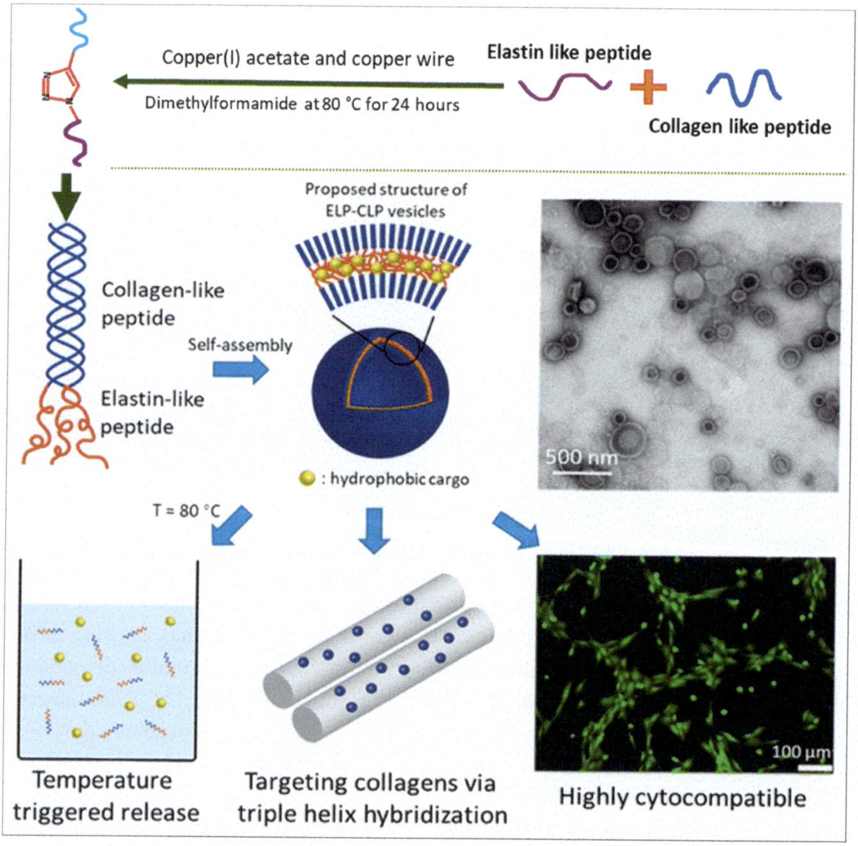

Fig. 2 Fabrication of elastin-b-collagen-like peptide (ELP-CLP) vesicles using collagen. (Adapted with permission from [74])

2.2 Functionalized Chitosan for Targeted Drug Delivery

Natural cationic biopolymer chitosan was developed by partial deacetylated of chitin [69]. Chitosan-modified metallic and polymer scaffolds have been the subject of extensive research during the past ten years [69]. Chitosan coatings have been created using a variety of techniques, including electrophoretic deposition, sol–gel processes, dip and spin coating, electrospinning, etc. Chitosan's surface is functionalized to improve the interaction with the scaffolds. CS-based coatings offer a promising alternative for promoting successful orthopedic applications [69]. The synthesis of chitosan nanoparticles using collagen peptide (CP) is particularly intriguing since the produced nanoparticles found stable under physiological settings, as shown in Fig. 3a [69]. Such nanoparticles offer a great deal of potential for application as intelligent drug delivery systems as cutting-edge cancer therapies [69]. In contrast to chemical processes that change the bulk properties of chitosan, employing collagen in chitosan films can preserve the native qualities of chitosan [10]. Although the creation of collagen-chitosan films and their application in the delivery of drugs have been previously reported, the fabrication of collagen peptide-chitosan nanoparticles and an examination of the fundamental bonding mechanisms on the size and morphology of nanoparticles have not yet been done [10, 69]. According to a study, the synthesized chitosan-based polymeric material had good cytocompatibility with 3T3 fibroblast cells [10]. The pH-controlled release of DOX (a medication) suggests that its potential for administering antitumor medications is based on the extracellular pH of the tumor environment. Hence, proper functionalization of chitosan-based polymeric material encapsulated with other suitable composites results in the better release of antitumor drugs at the targeted site [10].

2.3 Functionalized Gelatin for Targeted Drug Delivery

The irreversible denaturation of collagen protein yields gelatin also known as the molecular derivative of collagen [70]. Gelatin is widely being used in cell and tissue culture as an alternative of collagen for biomaterial purposes because it has a molecular structure and function that seem to be remarkably similar to those of collagen [70]. Gelatin is a frequently utilized natural macromolecule with outstanding biocompatibility and biodegradability in the biomedical field [70]. Nanoparticles-based drug delivery has drawn significant attention among numerous promising new drug delivery methods for the analysis and handling of a number of diseases, and at the moment, several formulations are being used in clinical settings [81]. Nanoparticles-based drug delivery methods have been created using a variety of biodegradable and biocompatible polymers from both natural and synthetic sources [81]. By employing exfoliated graphene oxide as a precursor, a straightforward and environmentally acceptable synthetic method for producing gelatin-functionalized

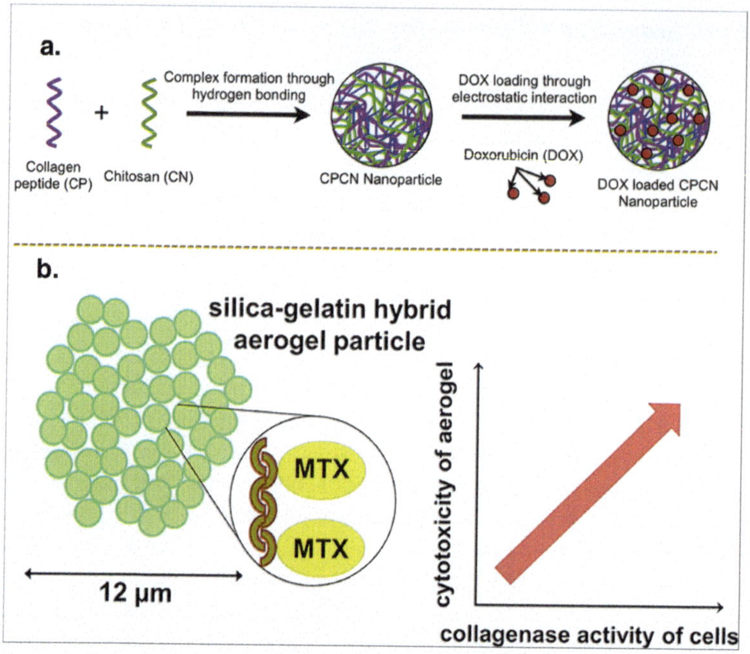

Fig. 3 **a** Systematic fabrication of an effective doxorubicin-loaded collagen peptide-functionalized chitosan nanoparticles for targeted cancer drug delivery (Adapted with permission from [10]) and **b** hybrid aerogel particles showed controlled release of methotrexate (MTX) against tumor cell growth (Adapted with permission from [92]

graphene nanosheets (gelatin-GNS) has been reported in a study [8]. Hybrid aerogels made of silica and gelatin are ideal choices for targeted drug delivery, as shown in Fig. 3b [92]. Aerogel particles made of a mixture of silica and gelatin like to interact closely with cells because of their porous surface and gelatin content [92]. As a result, functionalizing these particles with pharmacologically active components may enhance the extent to which drugs work to prevent tumor cell invasion [92]. The effectiveness of surface-functionalized gelatin nanoparticles (f-GNPs) in the treatment of visceral leishmaniasis by effectively delivering amphotericin B (AmB) to macrophages was also examined in a study (VL) [93]. Systems made on photosensitive gelatin have been studied as tissue adhesives and drug release mechanisms. It was suggested that photo-cross-linking Gel-NC into microgels is effective, particularly when functionalized for specific medical fields like targeted drug delivery systems [42]. A hydrophilic plug, such as cellulose or guar gum, can be used in hard gelatin capsules to achieve programmed drug delivery [17]. Hydrophilicity enables the protein nanostructures to defend the laden cargoes for its tailored delivery systems [17]. For the purpose of targeting the drug at the colon or any other anticipated site of gastrointestinal tract, gelatin capsules loaded with a pH-independent drug-loaded polymer matrix are chemically treated with an aqueous solution of formaldehyde that produces a delayed-release delivery system [17]. Most of the anticancer have shown rapid clearance from blood circulation that eventually leads low bioavailability in the targeted cancerous tumor in the body [52]. Anticancer medications would not be quickly eliminated by the body if they are encapsulated into suitable nano-based platforms, though, because NPs are too massive and continue to circulate in the blood [52]. Particularly when creating targeted drugs, functional groups of gelatin molecules are used for a range of chemical processes that could be carried out either directly or by using various linkers [52]. Another study that examined the distribution of GNPs upon intravenous injection indicated that they may circulate for a longer period of time in the blood. In order to facilitate and improve drug delivery and get past physiological constraints, the current method for employing GNPs in DDSs involves attaching a drug to a gelatin scaffold [52]. GNPs that are redox-responsive have the potential to be effective DDSs against a variety of cancer cells [52]. The final biological performance of gelatin is enhanced by the inclusion of amino acid sequences as Arg-Gly-Asp (RGD) over synthetic polymers devoid of these cell-recognition motifs [108]. As a drug transport carrier, gelatin has been tested to be flexible because of its intrinsic functions that allow the loading of charged biomolecules [108]. In addition, based on the electrostatic characteristics of the targeted drug molecule, the gelatin isoelectric point (IEP) can be tuned to optimize drug loading efficiency by employing either an alkaline or acidic treatment [108]. Gelatin has been extensively researched as a delivery system for several kinds of drugs due to its enormous potential for controlled release, among which are but not limited to antibacterial agents, anti-inflammatory pharmaceuticals, and many different antineoplastic substances [108].

2.4 Functionalized Poly (Ethylene Glycol) for Targeted Drug Delivery

Poly (ethylene glycol), or PEG, is a synthetic, nonbiodegradable polymer that is also known as a copolymer of poly (ethylene oxide) and poly (oxyethylene) [32]. PEG can be effectively combined with other polymers or nanoparticles to increase their biocompatibility [32]. The PEGylation procedure allows for the conjugation of biomolecules or nanoparticles with linear PEG via functional groups [32]. Surface modification with poly (ethylene glycol) (PEGylation) is a popular method for preventing nonspecific interactions since it can give drug carriers better biodistribution and pharmacokinetic properties [32]. The targeting of stealth PEG particles to tumor cells can be modified by PEG particles containing bispecific antibodies (BsAbs), according to a study [32]. In an investigation, an effective drug delivery system called GO-PEG-FA was created and loaded with anticancer medication [33]. Breast cancer cell lines (MCF-7) were used to test the system's cytotoxicity [33]. GO nanoparticles were initially created and coupled with PEG to improve its biocompatibility and enhance the drug solubility [33]. Hydrophobic anticancer medications would be delivered via a carrier that has been PEGylated onto graphene [125]. PEG helps in making these nanoparticles more biocompatible by limiting aggregation, oxidation of the particle surface, and surface degradation [120]. Due to resistance to plasma protein deposition, the main benefit of PEG coating is that it prolongs circulation time. PEG and FA were successful in functionalizing hydroxyapatite nanoparticles [120]. These FA-containing nanoparticles can be successfully employed to target tumor cells that express folate receptors. Paclitaxel, an anticancer medication, is joined to the HAp-PEG-FA molecule [120]. An early burst release of the drug is seen in the drug release profile, followed by a steady release [120]. A promising drug delivery method for the glioma tumor can be used with biotinylated PEG-PLA nanoparticles [30]. For the intracellular delivery of nucleic acids, a dually functionalized graphene oxide (GO)-based nanocarrier with the combination of aminated-polyethylene glycol (PEG-diamine) and octa-arginine (R8) was presented [53]. The improved nanocarrier formulation had exceptional cell viability and was stable in biological solutions [53]. Physical adsorption, such as electrostatic or hydrophobic contact, causes PEG to physically bind to NPs. By establishing a strong chemical bond, PEG is safely grafted onto the surface of NPs for targeted drug delivery [110]. PEG chains and a chromophore are coupled to stain the surface of NPs [110]. The PEG in NPs frequently has an impact on the effectiveness of loaded molecule encapsulation [110]. The amount of PEG-containing lipid and the molecular weight of PEG both affect how well water-soluble drugs are encapsulated [110]. The PEG on the surface of PEGylated NPs is typically linked to ligands to enhance endocytosis [110]. The internalization of PEGylated NPs into cells can be improved by the interaction of certain ligands with cell surface receptors, which can overcome the spatial restriction between PEG and cell membranes [110]. The primary malignant brain tumor that affects adults most frequently is the glioblastoma [104]. Temozolomide (TMZ) is a powerful alkylating drug used to treat brain tumors. This medication can significantly produce the adverse effects from large doses reducing its ability to reach the target spot effectively [104]. It has been discovered that functionalizing CNT with hydrophilic polymers like polyethylene glycol and chitosan enables the production of functional moieties that can be further bioconjugated with therapeutic drugs [104]. There are many benefits of functionalization with the PEG group, including faster blood circulation time [104]. They are also ideal for drug delivery due to their minimal cytotoxicity in vitro and in vivo [104]. Because of their longer half-lives, PEGylated molecules are slowly eliminated from the body [127]. By protecting them, PEGylation can lengthen the maintenance period of therapeutics such as proteins, liposomes, and nanoparticles. PEG immunogenicity, however, seems to be a developing problem that restricts the efficacy of both currently marketed drugs and upcoming therapeutics that are being developed [48].

2.5 *Functionalized N-(2-hydroxypropyl) Methacrylamide for Targeted Drug Delivery*

The monomer used to produce the polymer poly (N-(2-hydroxypropyl) methacrylamide) is N-(2-hydroxypropyl) methacrylamide, or HPMA [22]. N-(2-hydroxypropyl) methacrylamide (pHPMA) has drawn interest because of its hydrophilicity, lack of immunogenicity, biocompatibility, and potential for functionalization by different ligands [22]. Reversible addition–fragmentation chain–transfer polymerization (RAFT) controlled radical polymerization is commonly used to create this polymer [22]. The polymer-based drug conjugates based on poly (N-(2-hydroxypropyl) methacrylamide) (pHPMA), which offer both site-specific delivery and regulated drug release, are one of the extensively investigated systems [22]. It is possible to conjugate numerous targeted molecules to the same polymer chain because of the secondary alcohol functionality [22]. By altering HPMA-based copolymers with hydrophobic moieties, hydrophobic drugs can be solubilized and retained while remaining in the micellar core [22]. A promising method to control the drug biodistribution, drug release kinetics, and cell trafficking is the use of prodrug platform, which involves conjugating HPMA with drugs [22]. HPMA-based polymers have shown to be a flexible scaffold with a wide range of potential applications as targeted drug delivery systems in nanomedicine [22]. The anticancer medication Pirarubicin is attached to the HPMA copolymer conjugate via the pH-sensitive hydrazone bond, and its activity has been tested in human patients [67]. Two linear HPMA polymer chains joined together by a disulfide bond in an HPMA diblock conjugate with the anticancer drug pirarubicin recently demonstrated improved anti-tumor efficacy [67]. To meet the requirements for an efficient drug delivery system, HPMA polymer carriers with variety of structures, such as grafted, block, branching, and star polymer conjugates, have been developed [22]. A study has revealed that polymer coating has a significant impact on the in vivo fate and tumor targeting of the nanoparticles in tumor-bearing animals. The retention of NPs coated with poly [N-(2-hydroxypropyl methacrylamide] (pHPMA) was marginally greater [97]. With a strong therapeutic potential, galectin-3 (Gal-3) is a prospective target in cancer therapy [23]. Specific glyconanomaterials suitable for targeting biomedically important lectins like Gal-3 can be produced via HPMA conjugation with customized carbohydrate ligands [23]. According to research findings, lysosomotropic HPMA copolymers have the ability to deliver antileishmanial drugs to specific areas of the body to treat visceral leishmaniasis [94]. pHPMA-based copolymers containing cholesterol (pHPMA-Chol) and the anticancer medication doxorubicin (Dox) self-assemble into highly effective nanoparticles (NPs) in an aqueous solution [56]. The hydrophilic nature of pHPMA raises the energy barrier of protein or other biomacromolecules adsorption during the blood circulation, hence, this is an important reason why they are used in drug administration [31]. Additionally, the binding of active substances to the pHPMA can solubilize medications that are not water soluble, significantly enhance pharmacokinetics, and remove negative pharmacological side effects. The creation of pHPMA-based nanomedicines that take control of the enhanced and permeability retention (EPR) effect has received a great deal of attention [31]. The HPMA copolymer conjugates containing cancerostatic drugs have demonstrated reduced systemic toxicity, high antitumor activity, and even the capacity to cause total tumor regression with subsequent formation of antitumor immunity in experimental tumor models [31]. HPMA could be very promising for various kinds of targeted drug delivery systems (DDS) in the near future [31].

3 Functionalization of Metals for Targeted Drug Delivery

Metal and metal oxide nanoparticles have distinct physicochemical features and play an essential role in developing novel bio-nanotech routes, particularly in targeted drug delivery systems [29, 36]. Metal nanoparticles fulfill significant properties needed in an efficient drug delivery system, such as drug loading/absorption, good carrier/distribution, and release at a specific site [29, 36]. Metal nanoparticles have been functionalized to serve as carriers for essential therapeutic agents such as peptides, proteins, antibodies, nucleic acids, and chemotherapeutic drugs [7, 29]. Surface functionalization of metal nanoparticles inhibits agglomeration and increases biocompatibility, wettability, and adherence to a particular cell surface receptor [36, 47, 60, 61, 68, 98]. Multifunctional nanoparticles can boost medication water solubility and blood lifetime. Metal nanoparticles conjugated with active ligands easily access the specific site at the cell surface and transport the loaded component to the target region [7, 60, 98]. Metal-based inorganic nanoparticles, in general, comprise various metal and metal oxide nanomaterials [7, 60, 98]. Metallic nanomaterials including gold (AuNPs), silver (AgNPs), copper (CuNPs), zinc oxides (ZnO NPs), cerium oxide (CeO_2), silica (SiO_2), titanium dioxide (TiO_2), iron oxide (Fe_2O_3), and iron oxide (Fe_3O_4) received significant attention for their effective therapeutic effect in the biomedical field [36, 44, 96].

3.1 *Functionalized Gold Nanoparticles for Targeted Drug Delivery*

Gold nanoparticles are extensively used for drug delivery among all metal-based nanoparticles [7, 29]. There are several advantages of AuNPs (i) easy synthetic and functionalization process; (ii) large surface for payload (iii) easy conjugation of drug molecules; (iv) controlled release of payloads under physiological environment; (vi) nontoxic and high biocompatibility [7, 29]. Functionalized gold nanoparticles have outstanding optical behavior due to surface plasmon resonance (SPR) and other material properties such as stability, self-assembly, easy, and green synthesis [7, 61]. Thiol groups generally achieve surface functionalization/modification of AuNPs through Au-sulfur interaction; however, a wide range of functional molecules also used for desired functionality [80, 123]. Carboxylic and amine groups are present in proteins/peptides, and nucleic acid has been extensively used as a reducing as well as a capping agent [80, 123]. Gold nanoparticles are one of the potential materials for the efficient delivery of payloads, including drugs to biomolecules like DNA, RNA, and proteins. Effective and specific release of these payloads at the targeted site is critical for effective therapeutic applications [29, 61, 126]. Strategically designed functional molecules coated over the surface of AuNPs triggered the discharge of payloads from the AuNPs surface in the presence of internal and external stimuli such as pH, molecular receptors, endogenous gases, light [34, 54, 80, 89, 122].

There are various studies which have demonstrated the drug delivery action of functional AuNps [25]. A few of them are discussed here. AuNPs coated with polyethylene glycol (PEG)-loaded oxaliplatin exhibit enhanced therapeutic capability in lung and colon cancer cell lines [25]. PEG, poly (gamma-glutamic acid), and phospholipid showed resistance to pH variation and increased blood circulation. Human albumin coating of AuNPs helps to cross the blood–brain barrier [25, 89, 101]. Gold nanoparticles are synthesized and stabilized with short peptides, amino acids, and nucleic acid [7, 25, 63, 80, 109, 113, 123]. Amino acids such as lysine, poly-lysine, and glycine-coated gold nanoparticles easily bind DNA for gene delivery [7, 25, 63, 80, 109, 113, 123]. Functionalizing AuNPs with short peptides such as CALNN and its derivative uses target intracellular originals. Peptides are also used to control the shape and size distribution of gold nanoparticles and help to cross the cell membrane efficiently [7, 116]. Mass spectrometry of cell lysates also supports the cellular uptake of AuNPs functionalized with cationic and neutral peptides or other ligands [7, 40, 138]. Gold nanoparticles conjugated with hydrophobzide short peptide amphiphile itself exhibit antibacterial activity [26, 63]. However, antibiotic functionalization was also evaluated against various Gram-negative and Gram-positive strains [26, 115, 136]. Moreover, Radiac and his group [103] developed the biomimetic phosphatidylserine-caped gold nanoparticles (PS-AuNPs) to check the efficacy in breast and prostate cancer cells, as shown in Fig. 4. Both cell lines in their investigation showed morphological alterations consistent with apoptosis [103]. When comparing the effects of PS-AuNP therapy to the control treatment, it was shown that DNA fragmentation significantly increased [103]. Histone-associated DNA fragments are an indication of apoptosis. These results show the possibility of phosphatidylserine and gold nanoparticles as a prostate and breast cancer therapy [103].

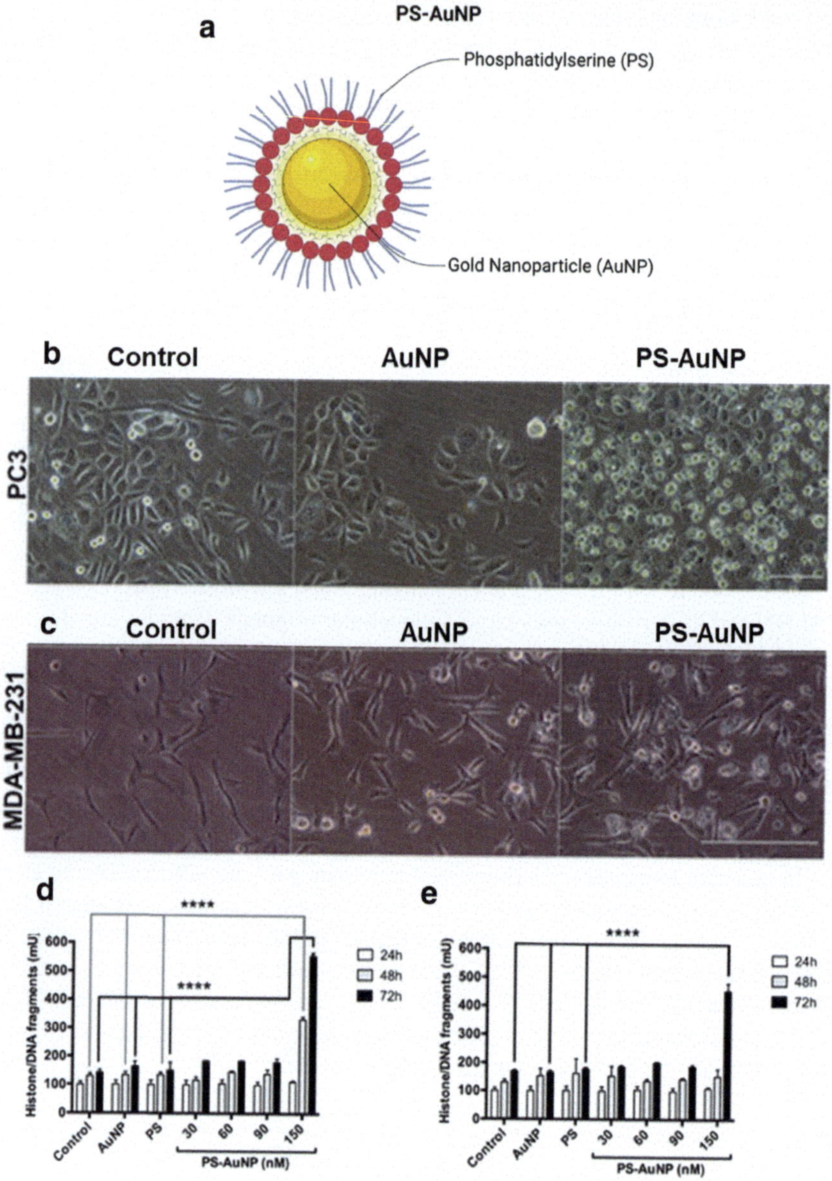

Fig. 4 **a** Phosphatidylserine functionalized gold nanoparticle (PS-AuNP) for targeted drug delivery, **b** PS-AuNPs significantly induced the morphology of prostate cancer (PC3) cells than PBS treated cells (control), **c** PS-AuNPs significantly induced the morphology of breast cancer (MDA-MB-231) cells than PBS treated cells (control), **d** PS-AuNPs induce the development of Histone/DNA fragments in PC3 and **e** MDA-MB-231 cell lines (Adapted with permission from [103])

3.2 Functionalized Silver Nanoparticles for Targeted Drug Delivery

Silver nanoparticles (AgNPs) are another extensively studied nanomaterial among the various metal nanoparticles and play a significant role in nanotechnology [28, 45, 91, 135]. The unique photophysical, chemical, and biological properties of AgNPs draw considerable attention in different research areas, including the biomedical field [28, 135]. AgNPs have impressive potential to develop advanced biomedical materials such as antimicrobial antiviral, anti-inflammatory, and antifungal agents, drug delivery formulations, cancer diagnosis, and enhanced therapeutic alternatives [28, 135]. Among the diverse potential biomedical applications, antibacterial properties have been extensively developed due to silver nanoparticles' large surface-to-volume ratios and crystalline surface structure [28, 135]. Due to the antibacterial properties of silver, AgNPs are used as an antibacterial agent for an extended period [119]. With the continuous increase in infectious diseases and the urgency of antibiotics against drug-resistant pathogens, AgNPs have gained close attention as potential antibiotics [119]. Plenty of studies evaluated and demonstrated that the AgNPs and their associated particles are accumulated in the cell wall of bacteria [119]. It was observed that AgNP accumulation causes oxidative stress, protein dysfunction, and membrane and DNA damage, followed by microbial cell damage [119]. Along with the excellent antimicrobial activity of silver nanoparticles, it affects human health [64, 131, 137]. A new approach to surface functionalization of silver nanoparticles with multifunction biomolecules such as peptides, amino acid, and other have been devolved to minimize this toxic effect [64, 131, 137]. Additionally, these biomolecules facilitate the delivery of nanoparticles through different pathways, for example, metal transportation and storage [64, 131, 137].

Despite their outstanding antibacterial activity, the AgNPs can be used as drug/biomolecular delivery vehicles to the targeted cells and tissues with improved therapeutic efficacy [27, 43, 55]. Particularly, strategically functionalized AgNPs are suitable carriers for various therapeutic molecules such as anti-inflammatory, anticancer, antioxidant, and antimicrobials [27, 43, 55]. Literature reports that AgNPs exhibit dual nature as a drug carrier as well as a drug [4, 27, 43, 55]. Therefore, this intrinsic therapeutic activity of AgNPs enables maximum therapeutic efficiency of the drug at a lower concentration and, consequently, the toxic side effects [27, 43, 55]. For example, the inherent anticancer properties of AgNPs have intrinsic anticancer activity and are therefore evaluated as an efficient antitumor drug carrier [43].

3.3 Functionalized Zinc Oxide Nanoparticles for Targeted Drug Delivery

Zinc oxide nanoparticles are well known for their low toxicity a biodegradability [41, 57, 96]). It is an excellent choice as a metallic nanocarrier in treating cancer, diabetes, and inflammation. During delivery of various drugs such as paclitaxel (PTX), curcumin (CUR), DOX, baicalin, or DNA fragments using zinc oxide nanoparticles, better solubility, higher toxicity compared with individual agents, and effective delivery into cancer cells was observed [41, 57, 96]. ZnO itself represents cytotoxicity toward cancer cells without harming normal cells due to high ion concentration release in acidic media and increased ROS production which provides a favorable environment to eradicate cancer tumor [41, 55]. Green synthesized zinc ferrite magnetic NPs loaded with carfilzomib showed about 95% release in 6 h [41, 55]. According to the research findings, the ZnO-NPs-DNR combination promoted dramatically decreased anticancer drug cytotoxicity and a large increase in ROS-mediated cancer cell targeting in human hepatocarcinoma cells (SMMC-7721 cells) [88]. Along with cancer treatment, zinc oxide nanoparticles showed effective antibacterial activity because of their exceptional features, such as high specific surface area and high action to block a wide range of pathogenic agents (Jiang et al. 2020) [12, 114]. Currently, Gram-negative *Escherichia coli* and Gram-positive *Staphylococcus aureus* are mainly studied as model bacteria for evaluating the antibacterial activity of ZnO NPs [114]. Some other pathogens also showed inhibition zones, such as *Pseudomonas aeruginosa, Pseudomonas aeruginosa, Bacillus subtilis, and Pseudomonas aeruginosa.* Socotrina encapsulated in ZnO NPs depicted effective antibacterial effects against microbial infections [96, 114, 46].

3.4 Functionalized Magnetic Nanoparticles for Targeted Drug Delivery

Magnetic nanoparticles (MNPs) generally consist of two components: one is magnetic metal/metal oxide, and the second is functional molecules [84, 117]. Magnetic nanoparticles can be manipulated by applying an external magnetic field [84]. MNPs are actively studied by researchers as next-generation drug delivery nanosystems due to their distinct physical properties and application in magnetic resonance imaging [84]. MNPs act as colloidal mediators for cancer treatment through magnetic hyperthermia [11]. The main properties of magnetic nanoparticles are: (i) they can be visualized, (ii) hyperthermia treatment and (iii) Magnetic field control, and (iv) absorbance near-infrared, microwave, and ultrasound radiations [12,

11, 84]. The control of the external magnetic field over MNPs allows the transportation of particles at the targeted site, which eliminates the side effects and reduces the amount of drug required. These MNPs are further functionalized with silica, amino group, numerous surfactants compound, or organic functional molecules and with a novel metal coating [84, 95].

Iron oxide (Fe_3O_4) is a widely studied MNPs for biomedical applications [96]. Fe_3O_4 nanoparticles possess significant properties such as high saturation field, extra anisotropy contributions, or shifted loops after field cooling [96]. Delivery of doxorubicin using superparamagnetic Fe_3O_4 nanoparticles and release at the targeted site is achieved by pH variation. Iron oxide nanoparticles (NPs) are magnetic materials that may be used for site-specific medication administration, diagnostics, and magnetic separation of biological products and cells [11, 84]. Iron oxide nanoparticles are thought to kill cancer cells without damaging normal cells. However, the vascularization and permeability of iron NPs cause them to be recognized as macrophages in the reticular endothelial system [11, 84].

4 Functionalization of Ceramics for Targeted Drug Delivery

A ceramic is an inorganic nonmetallic solid composed of either metal or nonmetal compounds which provide a wide range of attractive features, including ease of preparation, adaptability in terms of size and shape, surface area to volume ratio, desired stability under physiological conditions, and good biodegradability, biological stability, and biocompatibility [83]. As a result, they have gained tremendous attention for their use as drug carriers for many years in the field of drug delivery [83, 133]. They are mostly made up of oxides and carbides, including popular bioceramics zirconia, hydroxyapatite, mesoporous silica, beta-tricalcium phosphate (β-TCP), hydroxyapatite, among others [83, 133]. Moreover, certain inorganic-organic composites employed as drug carriers incorporate bioceramics as an integral component. Among these, the prominent bioceramics including zirconia, zeolite, silica, hydroxyapatite, tricalcium phosphate, and more are often favored as ideal drug reservoirs, as depicted in Fig. 5. By exploring desirable drug reservoirs based on a tailor-made system, we can enhance the targeting ability and minimize the side effect [86, 132].

4.1 *Functionalized Hydroxyapatite for Targeted Drug Delivery*

The chief inorganic component of tough tissue, like teeth of vertebrates and bone, is hydroxyapatite (HAP). A bone in the human body contains about 70% mineral, which

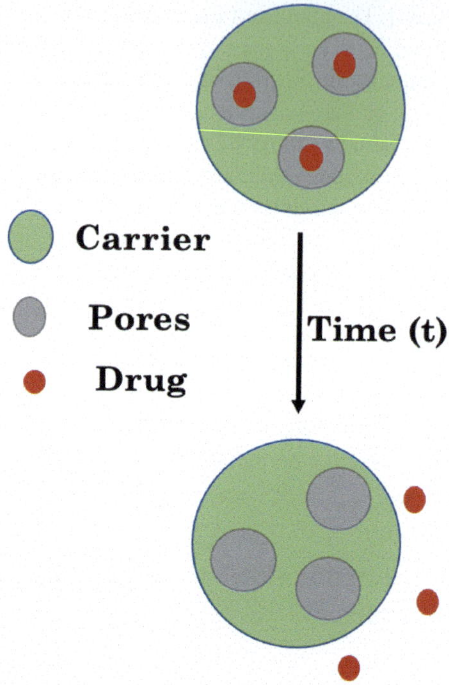

Fig. 5 Principle of ceramic-based desirable drug reservoirs

is nothing but HAP [62]. The HAP-based nanostructures have advanced properties such as hollow mesoporous structure, easy metabolism in vivo, osteoconductivity, biodegradability, excellent biocompatibility, and biological activity) [62]. It can be normally designed in the living organic body, and it can resist moisture, prevent shrinking, and have changeable hardness [62]. It has been a sought-after contender for the regulated and prolonged release of drugs, proteins, and genes, which are mentioned in Table 1.

It was discovered that the encapsulation of therapeutic molecules and the duration of sustained release are both significantly influenced by the characteristics of HAP [133]. Shi and group [111] produced nano-HAP and micro-HAP by combining lactoferrin (LF) with $Ca(NO_3)$ 24.H_2O to create HAP-LF [133]. The experimental profiles are summarized in Table 2.

4.2 Functionalized Tricalcium Phosphate for Targeted Drug Delivery

Tricalcium phosphate (TCP) comes in two varieties: high-temperature phase (α-TCP) and low-temperature phase (β-TCP). Aside from these two, β-TCP offers superior thermodynamic stability, biodegradability, and biocompatibility [21, 71]. As a result,

Table 1 Methods of synthesis and application of HAP in the field of drug delivery

Method and composition	Challenges	Outcomes	Key application	References
Coprecipitation method (*HAP and OLZ*)	Side effects of OZL, since OLZ does not release completely	After intramuscular injection, HAP-OLZ was able to release OLZ for a period of time that might last longer than 3 weeks in depressed rats. It was shown that this system improved locomotion, learning, and memory in depressed rats	HAP is employed as a carrier for OLZ's prolonged release, and HAP-OLZ had the potential to address the problem of nonadherent medication ingestion that arises after the administration of antidepressants through intramuscular injection	Shyong et al. [112]
Hydrothermal method (*HAP and IBU*)	Ibuprofen has side effects on the entire body. Ibuprofen did not release fully, possibly as a result of the formation of a hydrogen bond between the –COOH and –OH groups in IBU and HAP	Ibuprofen (IBU) has an optimum loading capacity in HAP at 413.65 mg/g, and the release process is pH-dependent. At pH values of 4.0, 5.6, and 7.4, respectively, cumulative release amounts of IBU have remained steady for more than 60 h at 38.55%, 40.42%, and 64.95%	As a tailored drug delivery system and in pH-sensitive DDS	Ma et al. [75]
Nano and micro-size (*HAP and Lactoferrin*)	Systemic side effects of ibuprofen	Due to the larger surface area, the maximum quantity of adsorbed drugs enhances the biocompatibility of HAP	They claimed that HAP-LF improved the biocompatibility of HAP. It was discovered that HAP may be used as a controlled release carrier of LF that may be used to improve bone regeneration and may potentially be a novel biomaterial	Shi et al. [111]

Table 2 Experimental profiles of two kinds of HAP [133]

Key properties of the material	Nano-size-based HAP material	Micro-size-based HAP material
Morphology	Rod-like	Microspheres
Scale	L-150 nm, and W-20 nm	D-15 μm
Size of pores (nm)	3–4	3–4 nm and 8–12 nm
Surface areas ($m^{2/}g$)	84.4	17.3
The amount of LF on the surface (mg/m^2)	2.9	1.0
The highest drug adsorbing capacity (μg/mg)	91.1	50.7

Table 3 Key application of β-TCP in the field of drug delivery

Method with composition	Challenge	Outcomes	Key application	References
Liquid porogen-based method (β-TCP and Ag_2O)	Biological effect of Ag_2O	Ag^+ was cumulatively released in amounts of 80–90 mol over the course of 60 days, enhancing osteoblast cell proliferation and osteoconductivity without being obviously harmful, not even for two wt% of Ag_2O	Smart release profile of Ag^+ during surgery, which prevents the bacterial infection	Hoover et al. [49]
Solution precipitation method (β-TCP/ CNT)	Optical concentration of SDBS	With 1% CNT, there is no discernible agglomeration	Bone graft biomaterials	Mirjalili et al. [87]
Ice bath extraction and freeze-drying (PLGA/ β-TCP Scaffolds and OIC-A006)	Biological effect of OIC-A006	OIC-A006 released in a biphasic rhythm, and its prolonged release significantly influenced cell adhesion and proliferation. Osteoinduction and bone growth were encouraged by the designed system	Osteoinduction and bone formation	Lin et al. [72]

TCP has been used as an alluring carrier in the drug delivery field due to its capacity for regulated, prolonged, and targeted drug molecule release, summarized in Table 3.

4.3 *Alginate-Brushite for Targeted Drug Delivery*

Dabiri and his group fabricated alginate-brushite-based composites and smart pH stimuli-responsive ibuprofen-loaded hydrogel using a unique in situ technique. The targeted application of composite was observed with significant benefits, such as brushite circumvented burst release of drug molecules [49].

4.4 *Functionalized Ceramics with Magnetic Nanoparticles for Targeted Drug Delivery*

Fe_3O_4 and Fe_2O_3-based nanoparticles have been employed as magnetic metal–organic frameworks (MOFs) and inductors for targeted delivery in medical applications [24]. Numerous investigations suggest that they could be used in smart drug delivery, which is explained in Table 4.

4.5 *Zeolitic Imidazole Framework (ZIFs) for Targeted Drug Delivery*

Conventional therapy systems made from organic or inorganic-based materials suffer from some major limitations associated with uncontrolled drug release, biocompatibility, cytotoxicity, etc. ZIFs have attracted significant interest as stimuli-responsive smart drug carriers because of their high drug loading and excellent biodegradability. The cytotoxicity and uncontrolled drug release of conventional materials for DDS, among other factors, limited their applicability [66, 121]. Due to their distinctive priority of structural design, higher area, and less toxicity, ZIFs have arisen quickly as a novel type of smart materials for drug delivery [66, 121]. Adhikari et al. [2] studied the activity of ZIF-7 and ZIF-8 for the release of DOX. Their research observed that the combination of DOX with ZIF-7 exhibited pH sensitivity and facilitated controlled drug delivery. Conversely, when DOX was paired with ZIF-8, it was determined to be non-pH-sensitive, and the drug release profile showed only a slight degree of control. In another work, Wu et al. [124]. fabricated biocompatible NIR-responsive chemodrug delivery platform using ZIFs and successfully controlling drug carrier degradation and its toxicity. Under the NIR and thermal irradiation, the drug was effectively released from the carrier and showed a biological response, which has an excellent effect on the inhibition of the tumor. Similarly, [128] delivered DOX with ZIF-8 to specific target sites, DOX was first modified with a pH-responsive linker containing two carboxyl groups to form the inactive prodrug CAD and subsequently seeded inside ZIF-8 by a 5 min mineralization process. The fabricated system shows efficient, biocompatible, and synergistic chemotherapy.

Table 4 Key application of ceramics with iron oxide NPs in drug delivery

Composition	Outcomes	Key application	References
Fe_3O_4/PLGA or HAP/ PLGA	With increasing PLGA concentration, the nanocomposites' hydrodynamic diameters increased without affecting loading capacity. In vitro cytotoxicity tests revealed their high biocompatibility	Targeted drug delivery	Bootdee et al. [24]
pH-sensitive Fe_3O_4@Fe-metal–organic Framework-HAP and DOX	The developed drug delivery system was highly biocompatible and released the drug in response to pH	Biocompatible drug carrier for targeted drug delivery	Yang et al. [130]
Fe_3O_4/SiO_2/HAP functionalized by APTES with Atenolol	The weight ratios of the constituent agents and the calcining temperature affect the size and morphology of nanocomposites	Smart release of drugs for hypertension	Mortazavi-Derazkola et al. [90]
HAP and Na-Alg coated on NPs surface loaded with curcumin and 6-gingerol	Drug release for curcumin and 6-gingerol was higher at pH 5.3 than at pH 7.4 The release time for both drugs was longer than seven days at 37 °C	pH-responsive smart drug delivery system for cancer treatment	Manatunga et al. [82]

4.6 *Silicon-Based Ceramics for Targeted Drug Delivery*

Mesoporous silicon-based nanoparticles (MSNs) are characterized by a mesoporous structure, high drug loading capacity, cheap, large area with pore volume, tunable pore size, nontoxicity, favorable biocompatibility, biodegradation, and their stability [15, 16]. As a result, they have enormous potential as drug delivery systems for targeted distribution and controlled, prolonged release of medicines [15, 16]. Tetra orthosilicate (TEOS, $Si(OC_2H_5)_4$) is typically hydrolyzed as part of another synthesis process [15, 16]. Numerous studies have shown their use in targeted, regulated, and sustained delivery of biomolecules which are explained in Table 5.

Table 5 Key uses of MSN in the field of drug delivery systems

Material	Outcomes	Key application	References
MSN and DOX	As-prepared MSNs measuring 150–200 nm in size. The study showed that in the therapy of the oral mucosa, MSNs/DOX systems have a better DOX absorption profile and steady release than free DOX. The ability to release drugs at the cellular physiological microenvironment (pH) was the most attractive feature of the system	DDS is a pH-sensitive and focused treatment for illnesses like cancer and brain ischemia–reperfusion damage	Bardhan et al. [20]
MSN/ZnO with DOX	pH-responsive drug release profile. Additionally, ZnO QDs' cytotoxicity may give composites the ability to treat tumors synergistically. As a result, this approach had great promise for cancer therapy and may increase the index of malignant therapy	Stimuli-responsive targeted drug delivery for tumor	Zhang et al. [134]
MSN/HA with DOX	Stronger cytotoxic effects and cellular absorption by cancer cells were associated with DOX release	pH and light-responsive targeted drug delivery	Lin et al. [72]
SiO_2/LDH/ Bevacizumab Composites with DOX	The approach improved the cellular uptake of DOX and increased its targeting of cancer and brain cells. The system demonstrated stronger effects against angiogenesis and neuroblastoma, according to both in vivo and in vitro tests	The composites may be used in targeted cancer therapies	Zhu et al. [139]

(continued)

Table 5 (continued)

Material	Outcomes	Key application	References
MSN/PG/SPION with chlorine6	MSNs and SPION linked by a PG layer could improve their aggregate stability In photodynamic treatment, Ce6 may be successfully conveyed into cells under the control of magnetism and increase its effectiveness	Photodynamic therapy enhances the biological; response	Yang et al. [129]
BMMs/ P(NIPAM-co-AA) with IBU	The nanocomposites were capable of reacting to pH and heat. The results of the pharmacokinetics test of P(NIPAM-co-AA) @BMMs showed that the power law model proposed by Korsmeyer and Peppas and non-Fickian diffusion were the two mechanisms through which ibuprofen was released	Temperature and pH-responsive drug release profile	Jin et al. [59]

5 Conclusions

Targeted drug delivery, also recognized as smart drug delivery, is a way of administering medication to patients by raising the drug dosages in diseased areas of the body. The benefits of the targeted release formulation include reducing the drug dosage to the patients, uniform distribution of the medication action, reducing the drug's bad belongings, and less variability in circulating drug levels. To address the shortcomings of traditional medication administration, targeted drug delivery is mostly based on polymer-mediated drug delivery technologies. Polymers-based delivery systems easily attach to medications and are directed to particular body sections with only sick tissue, avoiding contact with healthy tissue. Other than polymers, ceramics and metals were also widely functionalized to deliver medications at the targeted site. To date, the substantial deployment of high-throughput Ag and Au nanoparticle-based biomaterials in the targeted drug delivery system has resulted in increased functioning. From now on, the major difficulty remains to be handled to view these systems' acceptability as first-line therapy methods. In many laboratories in India, pharmaceutical research on targeted drug delivery systems is being undertaken zealously. They are occasionally examined in vivo in animals for pharmacokinetics and in vitro for release patterns but seldom for efficacy. There is a scarcity of data on clinical research and the drug delivery systems' usefulness in patients. If the research has achieved a significant result (clinical usage), pharmacologists must be involved in the research of pharmacokinetics and pharmacodynamics of drug delivery systems.

References

1. Adepu S, Ramakrishna S (2021) Controlled drug delivery systems: Current status and future directions. Molecules 26
2. Adhikari C, Das A, Chakraborty A (2015) Zeolitic imidazole framework (ZIF) nanospheres for easy encapsulation and controlled release of an anticancer drug doxorubicin under different external stimuli: a way toward smart drug delivery system. Mol Pharm 12:3158–3166. https://doi.org/10.1021/ACS.MOLPHARMACEUT.5B00043
3. Alle M, Bandi R, Sharma G et al (2021) Current trends in engineered gold nanoparticles for cancer therapy. Nanotechnology in the life sciences. Springer, Cham, pp 1–40
4. Alle M, Park SC, Bandi R et al (2021) Rapid in-situ growth of gold nanoparticles on cationic cellulose nanofibrils: recyclable nanozyme for the colorimetric glucose detection. Carbohydr Polym 253:117239. https://doi.org/10.1016/j.carbpol.2020.117239
5. Alle M, Reddy GB, Kim TH et al (2020) Doxorubicin-carboxymethyl xanthan gum capped gold nanoparticles: microwave synthesis, characterization, and anti-cancer activity. Carbohydr Polym 229:115511. https://doi.org/10.1016/j.carbpol.2019.115511
6. Alle M, Sharma G, Lee S-H, Kim J-C (2022) Next-generation engineered nanogold for multimodal cancer therapy and imaging: a clinical perspectives. J Nanobiotechnology 20:222. https://doi.org/10.1186/s12951-022-01402-z
7. Amina SJ, Guo B (2020) A review on the synthesis and functionalization of gold nanoparticles as a drug delivery vehicle. Int J Nanomedicine 15:9823–9857. https://doi.org/10.2147/IJN.S279094
8. An J, Gou Y, Yang C et al (2013) Synthesis of a biocompatible gelatin functionalized graphene nanosheets and its application for drug delivery. Mater Sci Eng C 33:2827–2837. https://doi.org/10.1016/j.msec.2013.03.008
9. An B, Lin YS, Brodsky B (2016) Collagen interactions: drug design and delivery. Adv Drug Deliv Rev 97:69–84
10. Anandhakumar S, Krishnamoorthy G, Ramkumar KM, Raichur AM (2017) Preparation of collagen peptide functionalized chitosan nanoparticles by ionic gelation method: an effective carrier system for encapsulation and release of doxorubicin for cancer drug delivery. Mater Sci Eng C 70:378–385. https://doi.org/10.1016/j.msec.2016.09.003
11. Anderson SD, Gwenin VV, Gwenin CD (2019) Magnetic functionalized nanoparticles for biomedical, drug delivery and imaging applications. Nanoscale Res Lett 14:1–6
12. Angelov B, Salviati G, Al Awadh AA et al (2022) Sustainable synthesis and characterization of zinc oxide nanoparticles using raphanus sativus extract and its biomedical applications. mdpi.com. https://doi.org/10.3390/cryst12081142
13. Aoki K, Saito N (2020) Biocompatibility and carcinogenicity of carbon nanotubes as biomaterials. Nanomaterials 10
14. Arun A, Malrautu P, Laha A et al (2021) Collagen nanoparticles in drug delivery systems and tissue engineering. Appl Sci (Switzerland) 11
15. Ashique S, Mittal V, Khatun T et al (2021) Micro-needles as an effective drug delivery system and associated patents in pharmaceutical field: a Review. irrespub.com. https://doi.org/10.55006/biolsciences.2021.1106
16. Ashurbekova K, Ashurbekova K, Alonso-Lerma B et al (2022) Biocompatible silicon-based hybrid nanolayers for functionalization of complex surface morphologies. ACS Appl Nano Mater 5:2762–2768. https://doi.org/10.1021/ACSANM.1C04428
17. Bajpai SK, Bajpai M, Dengre R (2003) Chemically treated hard gelatin capsules for colon-targeted drug delivery: a novel approach
18. Bandi R, Tummala S, Dadigala R et al (2021) Role of metal-doped carbon dots in bioimaging and cancer therapy. Nanotechnology in the life sciences. Springer, Cham, pp 101–123
19. Barbucci R, Pasqui D, Giani G et al (2011) A novel strategy for engineering hydrogels with ferromagnetic nanoparticles as crosslinkers of the polymer chains. Potential applications as a targeted drug delivery system. Soft Matter 7:5558–5565. https://doi.org/10.1039/c1sm05174a

20. Bardhan M, Majumdar A, Jana S et al (2018) Mesoporous silica for drug delivery: interactions with model fluorescent lipid vesicles and live cells. J Photochem Photobiol B 178:19–26. https://doi.org/10.1016/J.JPHOTOBIOL.2017.10.023
21. Barkallah R, Taktak R, Guermazi N et al (2021) Mechanical properties and wear behaviour of alumina/tricalcium phosphate/titania ceramics as coating for orthopedic implant. Elsevier
22. Bobde Y, Biswas S, Ghosh B (2020) Current trends in the development of HPMA-based block copolymeric nanoparticles for their application in drug delivery. Eur Polym J 139:110018. https://doi.org/10.1016/J.EURPOLYMJ.2020.110018
23. Bojarová P, Tavares MR, Laaf D et al (2018) Biocompatible glyconanomaterials based on HPMA-copolymer for specific targeting of galectin-3. J Nanobiotechnology 16:1–16. https://doi.org/10.1186/S12951-018-0399-1/FIGURES/2
24. Bootdee K, Grady BP, Nithitanakul M (2017) Magnetite/poly(D, L-lactide-co-glycolide) and hydroxyapatite/poly(D, L-lactide-co-glycolide) prepared by w/o/w emulsion technique for drug carrier: physical characteristic of composite nanoparticles. Colloid Polym Sci 295:2031–2040. https://doi.org/10.1007/S00396-017-4185-7
25. Brown SD, Nativo P, Smith JA et al (2010) Gold nanoparticles for the improved anticancer drug delivery of the active component of oxaliplatin. J Am Chem Soc 132:4678–4684. https://doi.org/10.1021/JA908117A/ASSET/IMAGES/LARGE/JA-2009-08117A_0006.JPEG
26. Brown AN, Smith K, Samuels TA et al (2012) Nanoparticles functionalized with ampicillin destroy multiple-antibiotic-resistant isolates of Pseudomonas aeruginosa and Enterobacter aerogenes and methicillin-resistant Staphylococcus aureus. Appl Environ Microbiol 78:2768–2774. https://doi.org/10.1128/AEM.06513-11
27. Bruna T, Maldonado-Bravo F (2021) PJ-IJ Silver nanoparticles and their antibacterial applications. Int J Mol Sci 22(13):7202. https://doi.org/10.3390/ijms22137202
28. Calderón-Jiménez B, Johnson ME, Montoro Bustos AR et al (2017) Silver nanoparticles: Technological advances, societal impacts, and metrological challenges. Front Chem 5.https://doi.org/10.3389/FCHEM.2017.00006/FULL
29. Chandrakala V, Aruna V, Angajala G (2022) Review on metal nanoparticles as nanocarriers: current challenges and perspectives in drug delivery systems. Emergent Mater 5:1593–1615
30. Chang J, Yuan X, Betbeder D et al (2010) Development of transferrin functionalized poly (ethylene glycol)/poly (lactic acid) amphiphilic block copolymeric micelles as a potential delivery system targeting brain. Springer 21:2673–2681. https://doi.org/10.1007/s10856-010-4106-5
31. Chytil P, Kostka L, Etrych T (2021) Hpma copolymer-based nanomedicines in controlled drug delivery. J Pers Med 11:1–22
32. Cui J, Alt K, Ju Y et al (2019) Ligand-functionalized poly(ethylene glycol) particles for tumor targeting and intracellular uptake. Biomacromol 20:3592–3600. https://doi.org/10.1021/acs.biomac.9b00925
33. Deb A, Vimala R (2018) Camptothecin loaded graphene oxide nanoparticle functionalized with polyethylene glycol and folic acid for anticancer drug delivery. J Drug Deliv Sci Technol 43:333–342. https://doi.org/10.1016/j.jddst.2017.10.025
34. Decuzzi P, Cook AB (2021) Harnessing endogenous stimuli for responsive materials in theranostics. ACS Nano 15:2068–2098. https://doi.org/10.1021/ACSNANO.0C09115
35. Delfi M, Ghomi M, Zarrabi A et al (2020) Functionalization of polymers and nanomaterials for biomedical applications: antimicrobial platforms and drug carriers. Prosthesis 2:117–139
36. Desai N, Momin M, Khan T et al (2021) Metallic nanoparticles as drug delivery system for the treatment of cancer. Expert Opin Drug Deliv 18:1261–1290. https://doi.org/10.1080/17425247.2021.1912008
37. Dutta SD, Ganguly K, Bandi R, Alle M (2021) A new era of cancer treatment: carbon nanotubes as drug delivery tools. Nanotechnology in the life sciences. Springer, Cham, pp 155–171
38. Elhissi AMA, Ahmed W, Hassan IU et al (2012) Carbon nanotubes in cancer therapy and drug delivery. J Drug Deliv 2012:1–10. https://doi.org/10.1155/2012/837327

39. Elias PZ, Liu GW, Wei H et al (2015) A functionalized, injectable hydrogel for localized drug delivery with tunable thermosensitivity: synthesis and characterization of physical and toxicological properties. J Control Release 208:76–84. https://doi.org/10.1016/j.jconrel.2015.03.003
40. Foroozandeh P, Aziz AA (2018) Insight into cellular uptake and intracellular trafficking of nanoparticles. Nanoscale Res Lett 13.https://doi.org/10.1186/S11671-018-2728-6
41. García-Hevia L, Valiente R (2016) Nano-ZnO leads to tubulin macrotube assembly and actin bundling, triggering cytoskeletal catastrophe and cell necrosis. Nanoscale 8(21):10963–73
42. Gattás-Asfura KM, Weisman E, Andreopoulos FM et al (2005) Nitrocinnamate-functionalized gelatin: synthesis and "smart" hydrogel formation via photo-cross-linking. Biomacromol 6:1503–1509. https://doi.org/10.1021/bm049238w
43. Gomes HIO, Martins CSM, Prior JA V, Taglietti M (2021) Silver nanoparticles as carriers of anticancer drugs for efficient target treatment of cancer cells. Nanomaterials 11:964. mdpi.com. https://doi.org/10.3390/nano11040964
44. Gowda B, Ahmed M, Chinnam S et al (2022) Current trends in bio-waste mediated metal/metal oxide nanoparticles for drug delivery. Elsevier
45. Guadie Assefa A, Adugna Mesfin A, Legesse Akele M et al (2017) Microwave-assisted green synthesis of gold nanoparticles using olibanum gum (boswellia serrate) and its catalytic reduction of 4-nitrophenol and hexacyanoferrate (III) by sodium borohydride. J Clust Sci. https://doi.org/10.1007/s10876-016-1078-8
46. Gudkov SV, Burmistrov DE, Serov DA et al (2021) A mini review of antibacterial properties of ZnO nanoparticles. Front Phys 9.https://doi.org/10.3389/FPHY.2021.641481/FULL
47. Gupta S, Singh R, Kumar V et al (2018) Ornamentation of triskelion peptide nanotori to produce gold nanoparticle (AuNP)-embedded peptide nanobangles. Chem Asian J 13:3285–3295. https://doi.org/10.1002/ASIA.201801270
48. Hoang Thi TT, Pilkington EH, Nguyen DH et al (2020) The importance of Poly(ethylene glycol) alternatives for overcoming PEG immunogenicity in drug delivery and bioconjugation. Polym (Basel) 12(2):298. https://doi.org/10.3390/polym12020298
49. Hoover S, Tarafder S, Bandyopadhyay A, Bose S (2017) Silver doped resorbable tricalcium phosphate scaffolds for bone graft applications. Mater Sci Eng C 79:763–769. https://doi.org/10.1016/J.MSEC.2017.04.132
50. Hu X, Tan H, Wang X, Chen P (2016) Surface functionalization of hydrogel by thiol-yne click chemistry for drug delivery. Colloids Surf A Physicochem Eng Asp 489:297–304. https://doi.org/10.1016/j.colsurfa.2015.11.007
51. Husen A, Siddiqi KS (2023) Advances in smart nanomaterials and their applications. Elsevier Inc., 50 Hampshire St., 5th Floor, Cambridge, MA 02139, USA https://doi.org/10.1016/C2021-0-02202-1
52. Hussain A, Hasan A, Babadaei MMN et al (2020) Application of gelatin nanoconjugates as potential internal stimuli-responsive platforms for cancer drug delivery. J Mol Liq 318
53. Imani R, Prakash S, Vali H, Faghihi S (2018) Polyethylene glycol and octa-arginine dual-functionalized nanographene oxide: an optimization for efficient nucleic acid delivery †. Cite Biomater Sci 6:1636. https://doi.org/10.1039/c8bm00058a
54. Ishii T, Otsuka H, Kataoka K, Nagasaki Y (2004) Preparation of functionally PEGylated gold nanoparticles with narrow distribution through autoreduction of auric cation by α-Biotinyl-PEG-block-[poly(2-(N, N-dimethylamino)ethyl methacrylate)]. Langmuir 20:561–564. https://doi.org/10.1021/LA035653I
55. Ivanova N, Gugleva V, Dobreva M, Pehlivanov I (2018) Silver nanoparticles as multi-functional drug delivery systems. IntechOpen, 2019. https://doi.org/10.5772/intechopen
56. Janisova L, Gruzinov A, Zaborova O et al (2020) Molecular mechanisms of the interactions of N-(2-Hydroxypropyl)methacrylamide copolymers designed for cancer therapy with blood plasma proteins. mdpi.com
57. Jiang J, Pi J (2018) applications JC-B chemistry and, 2018 undefined The advancing of zinc oxide nanoparticles for biomedical applications. Bioinorg Chem Appl 1062562. https://doi.org/10.1155/2018/1062562

58. Jin-Chul K, Madhusudhan A, Husen A (2021) Smart nanomaterials in biomedical applications. Springer Nature Switzerland AG, Gewerbestrasse 11, 6330 Cham, Switzerland. https://doi.org/10.1007/978-3-030-84262-8
59. Jin X, Wang Q, Sun J et al (2017) P(NIPAM-co-AA)@BMMs with mesoporous silica core and controlled copolymer shell and its fractal characteristics for dual pH- and temperature-responsive performance of ibuprofen release. 67:131–142. https://doi.org/10.1080/00914037.2017.1309544
60. Joshi P, Chakraborti S (2012) The anticancer activity of chloroquine-gold nanoparticles against MCF-7 breast cancer cells. Elsevier 95:195–200. https://doi.org/10.1016/j.colsurfb.2012.02.039
61. Kalimuthu K, Lubin BC, Bazylevich A, et al (2018) Gold nanoparticles stabilize peptide-drug-conjugates for sustained targeted drug delivery to cancer cells. J Nanobiotechnology 16.https://doi.org/10.1186/s12951-018-0362-1
62. Kebiroglu MH, Orek C, Bulut N et al (2017) Temperature dependent structural and vibrational properties of hydroxyapatite: a theoretical and experimental study. Ceram Int 43:15899–15904. https://doi.org/10.1016/j.ceramint.2017.08.164
63. Kesharwani K, Singh R, Khan MJ et al (2021) Hydrophobized short peptide amphiphile functionalized gold nanoparticles as antibacterial biomaterials. ChemistrySelect 6:6827–6833. https://doi.org/10.1002/SLCT.202102204
64. Kesharwani K, Singh R, Tripathi SK et al (2022) Antimicrobial activity of silver nanoparticles loaded biomimetic isomeric short lipopeptide nanostructures. ChemistrySelect 7.https://doi.org/10.1002/SLCT.202202234
65. Kim A, Kim, (2019) Oxidation- and temperature-responsive poly(hydroxyethyl acrylate-co-phenyl vinyl sulfide) micelle as a potential anticancer drug carrier. Pharmaceutics 11:462. https://doi.org/10.3390/pharmaceutics11090462
66. Kouser S, Hezam A, Khadri MJN, Khanum SA (2022) A review on zeolite imidazole frameworks: synthesis, properties, and applications. J Porous Mater 29:663–681. https://doi.org/10.1007/S10934-021-01184-Z
67. Koziolová E, Kostka L, Kotrchová L et al (2018) N-(2-Hydroxypropyl)methacrylamide-based linear, diblock, and starlike polymer drug carriers: advanced process for their simple production. Biomacromol 19:4003–4013. https://doi.org/10.1021/ACS.BIOMAC.8B00973/ASSET/IMAGES/LARGE/BM-2018-00973K_0006.JPEG
68. Kumar V, Gupta S, Mishra N et al (2017) Laser-induced fabrication of gold nanoparticles on shellac-driven peptide nanostructures. Mater Res Express 4(3):035036
69. Kumari S, Tiyyagura HR, Pottathara YB et al (2021) Surface functionalization of chitosan as a coating material for orthopaedic applications: a comprehensive review. Carbohydr Polym 255
70. Lei J, Li X, Wang S et al (2019) Facile fabrication of biocompatible gelatin-based self-healing hydrogels. ACS Appl Polym Mater 1:1350–1358. https://doi.org/10.1021/acsapm.9b00143
71. Li H, Xue F, Bai J et al (2015) Effect of sodium stearate on the surface properties of beta-calcium phosphate. Elsevier
72. Lin L, Dong Y, Zhou Q (2017) Sustained release of OIC-A006 from PLGA microspheres to induce osteogenesis of composite PLGA/β-TCP scaffolds. Sci Eng Compos Mater 24:721–730. https://doi.org/10.1515/SECM-2015-0372/HTML
73. Liu Z, Chen K, Davis C et al (2008) Drug delivery with carbon nanotubes for in vivo cancer treatment. Cancer Res 68:6652–6660. https://doi.org/10.1158/0008-5472.CAN-08-1468
74. Luo T, David MA, Dunshee LC et al (2017) Thermoresponsive elastin-b-collagen-like peptide bioconjugate nanovesicles for targeted drug delivery to collagen-containing matrices. Biomacromol 18:2539–2551. https://doi.org/10.1021/acs.biomac.7b00686
75. Ma F, Cui MF, Zhu JH, Li YL (2017) Porous hydroxyapatite microspheres prepared by using poly (allylamine hydrochloride) and its application in drug delivery. Wuji Cailiao Xuebao/J Inorganic Mater 32:1215–1222. https://doi.org/10.15541/jim20170041
76. Madhusudhan A, Reddy GB, Krishana IM (2019) Green synthesis of gold nanoparticles by using natural gums. Nanomaterials and plant potential. Springer International Publishing, Cham, pp 111–134

77. Madhusudhan A, Reddy G, Venkatesham M et al (2014) Efficient pH dependent drug delivery to target cancer cells by gold nanoparticles capped with carboxymethyl chitosan. Int J Mol Sci 15:8216–8234. https://doi.org/10.3390/ijms15058216
78. Madhusudhan A, Reddy GB, Venkatesham M, Veerabhadram G (2012) Design and evaluation of efavirenz loaded solid lipid nanoparticles to improve the oral bioavailability. Int J Pharm Pharmaceutical Sci Res 2:84–89
79. Madhusudhan A, Reddy GB, Ambikar DB (2018) The effect of colon targeted delivery of celecoxib loaded microspheres on experimental colitis induced by acetic acid in rats
80. Mahato K, Nagpal S, Shah MA et al (2019) Gold nanoparticle surface engineering strategies and their applications in biomedicine and diagnostics. 3 Biotech 9. https://doi.org/10.1007/S13205-019-1577-Z
81. Mahmoudi Saber M (2019) Strategies for surface modification of gelatin-based nanoparticles. Colloids Surf B Biointerfaces 183
82. Manatunga D, de Silva R (2017) pH responsive controlled release of anti-cancer hydrophobic drugs from sodium alginate and hydroxyapatite bi-coated iron oxide nanoparticles. Eur J Pharm and Biopharm 117:29–38. https://doi.org/10.1016/j.ejpb.2017.03.014
83. Manzano M (2014) Ceramics for drug delivery. Bio Ceram Clin Appl 9781118406755:343–382. https://doi.org/10.1002/9781118406748.CH12
84. Materón EM, Miyazaki CM, Carr O et al (2021) Magnetic nanoparticles in biomedical applications: a review. Appl Surf Sci Adv 6.https://doi.org/10.1016/j.apsadv.2021.100163
85. Mauri E, Rossi F, Sacchetti A (2016) Tunable drug delivery using chemoselective functionalization of hydrogels. Mater Sci Eng C 61:851–857. https://doi.org/10.1016/j.msec.2016.01.022
86. McNeel K, Siraj N, Negulescu I et al (2018) Sodium deoxycholate/TRIS-based hydrogels for multipurpose solute delivery vehicles: ambient release, drug release, and enantiopreferential release. Elsevier
87. Mirjalili F, Mohammadi H, Azimi M et al (2017) Synthesis and characterization of β-TCP/CNT nanocomposite: morphology, microstructure and in vitro bioactivity. Elsevier
88. Moghaddam AB, Moniri M, Azizi S et al (2017) Eco-friendly formulated zinc oxide nanoparticles: induction of cell cycle arrest and apoptosis in the MCF-7 cancer cell line. mdpi.com. https://doi.org/10.3390/genes8100281
89. Mohapatra A, Uthaman S, Park IK (2021) External and internal stimuli-responsive metallic nanotherapeutics for enhanced anticancer therapy. Front Mol Biosci 7.https://doi.org/10.3389/FMOLB.2020.597634/FULL
90. Mortazavi-Derazkola S (2017), Green synthesis of magnetic Fe3O4/SiO2/HAp nanocomposite for atenolol delivery and in vivo toxicity study. J Cleaner Prod 168:39–50. https://doi.org/10.1016/j.jclepro.2017.08.235
91. Murali Krishna I, Bhagavanth Reddy G, Veerabhadram G, Madhusudhan A (2016) Eco-friendly green synthesis of silver nanoparticles using salmalia malabarica: synthesis, characterization, antimicrobial, and catalytic activity studies. Appl Nanosci. https://doi.org/10.1007/s13204-015-0479-6
92. Nagy G, Király G, Veres P et al (2019) Controlled release of methotrexate from functionalized silica-gelatin aerogel microparticles applied against tumor cell growth. Int J Pharm 558:396–403. https://doi.org/10.1016/j.ijpharm.2019.01.024
93. Nahar M, Dubey V, Mishra D et al (2010) In vitro evaluation of surface functionalized gelatin nanoparticles for macrophage targeting in the therapy of visceral leishmaniasis. J Drug Target 18:93–105. https://doi.org/10.3109/10611860903115290
94. Nan A, Nanayakkara NPD, Walker LA et al (2001) N-(2-hydroxypropyl)methacrylamide (HPMA) copolymers for targeted delivery of 8-aminoquinoline antileishmanial drugs. J Control Release 77:233–243. https://doi.org/10.1016/S0168-3659(01)00514-4
95. Niederberger M, Pinna N (2009) Metal oxide nanoparticles in organic solvents: synthesis, formation, assembly and application. Springer Science & Business Media 17
96. Nikolova M, Biomimetics MC (2020) Metal oxide nanoparticles as biomedical materials. mdpi.com 201 https://doi.org/10.3390/biomimetics5020027

97. Novy Z, Lobaz V, Vlk M et al (2020) Head-To-head comparison of biological behavior of biocompatible polymers poly (ethylene oxide), poly (2-ethyl-2-oxazoline) and poly [N-(2-hydroxypropyl). mdpi.com
98. Park C, Youn H, Kim H et al (2009) Cyclodextrin-covered gold nanoparticles for targeted delivery of an anti-cancer drug. pubs.rsc.org 19:2261–2440. https://doi.org/10.1039/B816209C
99. Patil S V., Shelake SS, Patil SS (2018) Polymeric materials for targeted delivery of bioactive agents and drugs. In: Fundamental biomaterials: polymers. Elsevier Inc., pp 249–266
100. Prakash S, Malhotra M, Shao W et al (2011) Polymeric nanohybrids and functionalized carbon nanotubes as drug delivery carriers for cancer therapy. Adv Drug Deliv Rev 63:1340–1351
101. Prieto G, Barbosa S, Buyana B, et al (2022) Nanoparticles loaded with platinum drugs for colorectal cancer therapy. mdpi.com. https://doi.org/10.3390/ijms231911261
102. Qiu H, Guo H, Li D et al (2020) Intravesical hydrogels as drug reservoirs. Trends Biotechnol 38:579–583. https://doi.org/10.1016/J.TIBTECH.2019.12.012
103. Radaic A, Joo NE, Jeong S-H et al (2021) Phosphatidylserine-gold nanoparticles (PS-AuNP) induce prostate and breast cancer cell apoptosis. mdpi.com. https://doi.org/10.3390/pharmaceutics13071094
104. Saberinasab A, Raissi H, Hashemzadeh H (2019) Understanding the effect of vitamin B6 and PEG functionalization on improving the performance of carbon nanotubes in temozolomide anticancer drug transportation. J Phys D Appl Phys 52.https://doi.org/10.1088/1361-6463/ab2abf
105. Saikia N (2018) Functionalized carbon nanomaterials in drug delivery: emergent perspectives from application. In: Novel nanomaterials-synthesis and applications. InTech
106. Saito N, Haniu H, Usui Y et al (2014) Safe clinical use of carbon nanotubes as innovative biomaterials. Chem Rev 114:6040–6079
107. Sanginario A, Miccoli B, Demarchi D (2017) Carbon nanotubes as an effective opportunity for cancer diagnosis and treatment. Biosensors (Basel) 7
108. Santoro M, Tatara AM, Mikos AG (2014) Gelatin carriers for drug and cell delivery in tissue engineering. J Control Release 190:210–218
109. Selvakannan PR, Mandal S, Phadtare S et al (2003) Capping of gold nanoparticles by the amino acid lysine renders them water-dispersible. Langmuir 19:3545–3549. https://doi.org/10.1021/LA026906V
110. Shi L, Zhang J, Zhao M et al (2021) Effects of polyethylene glycol on the surface of nanoparticles for targeted drug delivery 13:10748. https://doi.org/10.1039/d1nr02065j
111. Shi P, Wang Q, Yu C et al (2017) Hydroxyapatite nanorod and microsphere functionalized with bioactive lactoferrin as a new biomaterial for enhancement bone regeneration. Elsevier
112. Shyong Y, Wang M, Kuo L et al (2017) Mesoporous hydroxyapatite as a carrier of olanzapine for long-acting antidepression treatment in rats with induced depression. Elsevier
113. Singh R, Suryavashi V, Vinayak V, Joshi KB (2019) Gold-ions-mediated diproline peptide nanocarpets and their inhibition of bacterial growth. ChemistrySelect 4:5810–5816. https://doi.org/10.1002/SLCT.201900847
114. Sirelkhatim A, Mahmud S, Seeni A et al (2015) Review on zinc oxide nanoparticles: antibacterial activity and toxicity mechanism. Nanomicro Lett 7:219–242. https://doi.org/10.1007/S40820 015-0040-X
115. Sreedharan S, Nanomedicine (2019) RS-IJ of, 2019 Ciprofloxacin functionalized biogenic gold nanoflowers as nanoantibiotics against pathogenic bacterial strains. Taylor & Francis 14:9905–9916. https://doi.org/10.2147/IJN.S224488
116. Sun L, Liu D, Wang Z (2008) Functional gold nanoparticle-peptide complexes as cell-targeting agents. Langmuir 24:10293–10297. https://doi.org/10.1021/LA8015063
117. Sun R, Lv Y, Niu Y et al (2017) Physicochemical and biological properties of bovine-derived porous hydroxyapatite/collagen composite and its hydroxyapatite powders. Ceram Int 43(18):16792–8
118. Sur S, Rathore A, Dave V et al (2019) Recent developments in functionalized polymer nanoparticles for efficient drug delivery system. Nano Struct Nano Objects 20

119. Tripathi SK, Kesharwani K, Kaul G et al (2022) Amyloid-β inspired short peptide amphiphile facilitates synthesis of silver nanoparticles as potential antibacterial agents. ChemMedChem 17.https://doi.org/10.1002/CMDC.202200251
120. Venkatasubbu G, Ramasamy S (2013) Surface modification and paclitaxel drug delivery of folic acid modified polyethylene glycol functionalized hydroxyapatite nanoparticles. Elsevier 235:437–42. https://doi.org/10.1016/j.powtec.2012.11.003
121. Wang Y, Cao Y, Jiang Z et al (2021) The neuropeptide Y1receptor ligand-modified cell membrane promotes targeted photodynamic therapy of zeolitic imidazolate frameworks for breast cancer. J Phys Chem Lett 12:11280–11287. https://doi.org/10.1021/ACS.JPCLETT.1C03562
122. Wang Y, Yang T (2020) Strategies for engineering advanced nanomedicines for gas therapy of cancer. National Science Review 7(9):1485-512.
123. Woehrle GH, Brown LO, Hutchison JE (2005) Thiol-functionalized, 1.5-nm gold nanoparticles through ligand exchange reactions: scope and mechanism of ligand exchange. J Am Chem Soc 127:2172–2183. https://doi.org/10.1021/JA0457718/SUPPL_FILE/JA0457718SI20041111_124011.PDF
124. Wu Q, Niu M, Chen X et al (2018) Biocompatible and biodegradable zeolitic imidazolate framework/polydopamine nanocarriers for dual stimulus triggered tumor thermo-chemotherapy. Biomaterials 162:132–143. https://doi.org/10.1016/J.BIOMATERIALS.2018.02.022
125. Xu Z, Wang S, Li Y et al (2014) Covalent functionalization of graphene oxide with biocompatible poly(ethylene glycol) for delivery of paclitaxel. ACS Appl Mater Interfaces 6:17268–17276. https://doi.org/10.1021/am505308f
126. Xu J-J, Zhang W-C, Guo Y-W et al (2022). Metal nanoparticles as a promising technology in targeted cancer treatment. https://doi.org/10.1080/10717544.2022.2039804
127. Yadav D, Dewangan HK (2020) PEGYLATION: an important approach for novel drug delivery system. 32:266–280. https://doi.org/10.1080/09205063.2020.1825304
128. Yan J, Liu C, Wu Q et al (2020) Mineralization of pH-sensitive doxorubicin prodrug in ZIF-8 to enable targeted delivery to solid tumors. Anal Chem 92:11453–11461. https://doi.org/10.1021/ACS.ANALCHEM.0C02599
129. Yang X, Wen Y, Wu A et al (2017) Polyglycerol mediated covalent construction of magnetic mesoporous silica nanohybrid with aqueous dispersibility for drug delivery. Mater Sci Eng C 80:517–525. https://doi.org/10.1016/J.MSEC.2017.06.022
130. Yang Y, Xia F, Yang Y et al (2017) Litchi-like Fe 3 O 4 @Fe-MOF capped with HAp gatekeepers for pH-triggered drug release and anticancer effect. J Mater Chem B 5:8600–8606. https://doi.org/10.1039/C7TB01680H
131. Yin IX, Zhang J, Zhao IS et al (2020) The antibacterial mechanism of silver nanoparticles and its application in dentistry. Taylor & Francis 15:2555–2562. https://doi.org/10.2147/IJN.S246764
132. Zamani M, Rostami M, Aghajanzadeh M et al (2018) Mesoporous titanium dioxide@ zinc oxide–graphene oxide nanocarriers for colon-specific drug delivery. J Mater Sci 53:1634–1645. https://doi.org/10.1007/S10853-017-1673-6
133. Zang S, Chang S, Shahzad MB et al (2019) Ceramics-based drug delivery system: a review and outlook. Rev Adv Mater Sci 58:82–97. https://doi.org/10.1515/RAMS-2019-0010/HTML
134. Zhang J, Wu D, Li M et al (2015) Multifunctional mesoporous silica nanoparticles based on charge-reversal plug-gate nanovalves and acid-decomposable ZnO quantum dots for intracellular drug. ACS Publications 7:26666–26673. https://doi.org/10.1021/acsami.5b08460
135. Zhang X, Liu Z, Shen W et al (2016) Silver nanoparticles: synthesis, characterization, properties, applications, and therapeutic approaches. mdpi.com. https://doi.org/10.3390/ijms17091534
136. Zhao Y, Tian Y, Cui Y et al (2010) Small molecule-capped gold nanoparticles as potent antibacterial agents that target gram-negative bacteria. J Am Chem Soc 132:12349–12356. https://doi.org/10.1021/JA1028843

137. Zharkova MS, Golubeva OY, Orlov DS et al (2021) Silver nanoparticles functionalized with antimicrobial polypeptides: benefits and possible pitfalls of a novel anti-infective tool. Front Microbiol 12.https://doi.org/10.3389/FMICB.2021.750556/FULL
138. Zhu ZJ, Ghosh PS, Miranda OR et al (2008) Multiplexed screening of cellular uptake of gold nanoparticles using laser desorption/ionization mass spectrometry. J Am Chem Soc 130:14139–14143. https://doi.org/10.1021/JA805392F
139. Zhu R, Wang Z, Liang P et al (2017) Efficient VEGF targeting delivery of DOX using Bevacizumab conjugated SiO2@ LDH for anti-neuroblastoma therapy. Elsevier 63:163–80. https://doi.org/10.1016/j.actbio.2017.09.009

Metallic Nanoparticles for Imaging and Therapy

Ibraq Khurshid, Hemant Singh, Alia Khan, Muzafar Ahmed Mir, Bilkees Farooq, Asif Iqbal Shawl, Shabir Hassan, Syed Salman Ashraf, Yarjan Abdul Samad, and Showkeen Muzamil

Abstract Metallic nanoparticles have captivated scientists due to their ability to be altered with different chemical functional groups, which enable them to conjugate with ligands, small molecules and medicines, thus creating a vast array of prospective implications in imaging and therapeutics. Additionally, several imaging techniques (e.g. CT scan, positron emission tomography, magnetic resonance imaging and ultrasound) have been invented for imaging various illnesses. These imaging techniques require contrast materials to highlight the body areas being examined. As a result, several metallic nanoparticles have been synthesized, including gold nanoparticles, magnetic nanoparticles, nanocages, nanoshells and silver nanoparticles, to be used as a contrast material for imaging techniques and maybe act as therapeutics carriers. Therefore, this chapter systematically discusses the various metallic nanoparticles and their application in imaging and therapy.

Keywords Metallic nanoparticles · Diagnostics · Drug drug delivery · Imaging · Therapeutics

I. Khurshid · A. Khan · M. Ahmed Mir · B. Farooq
Department of Zoology, Central University of Kashmir, Jammu and Kashmir, India

H. Singh
Biological Engineering, Indian Institute of Technology Gandhinagar, Gujarat, India

H. Singh · A. I. Shawl · S. Hassan (✉) · S. S. Ashraf
Department of Biology, College of Arts and Sciences, Khalifa University of Science and Technology, Abu Dhabi, United Arab Emirates
e-mail: shabir.hassan@ku.ac.ae

Y. A. Samad
Department of Aerospace Engineering, Khalifa University of Science and Technology, Abu Dhabi, United Arab Emirates

S. Muzamil (✉)
Laboratory of Molecular Biochemistry, Division of Veterinary Biochemistry, Faculty of Veterinary Sciences and Animal Husbandry, Shuhama, Alusteng, Jammu and Kashmir 190 006, Srinagar, India
e-mail: showkeen@skuastkashmir.ac.in

A. Madhusudhan et al. (eds.), *Functional Smart Nanomaterials and Their Theranostics Approaches*, Smart Nanomaterials Technology,
https://doi.org/10.1007/978-981-99-6597-7_3

1 Introduction

Nanotechnology is the field of science and engineering that has applications in many different areas and is still developing [113]. Some of the fields where nanotechnology is used are optics, electronics, catalysis, biomedicine, agriculture, magnetics and energy research [6, 38–43, 49, 52]. It focuses on materials with dimension range of nanometre till less than one-hundredth of a nanometre [107]. The technological advance of regulating materials at the nanoscale enabled a great revolution in medical and healthcare treatments and therapies [27]. This technology involves materials used to have atomic structures and arrangements so tiny that quantum-level phenomena increase their physical properties [73]. Metallic units known as nanoparticles can be of different shapes including spherical, triangular or rod-shaped [105]. Nanobiotechnology is a collaborative field that includes technological research and development in several fields, including physics, chemistry, material science and nanoscience which can be of greater use in biological sciences [91]. Due to its use in the fields of targeted drug delivery [20], magnetic separation and preconcentration of target analytes, biotechnology [92], electronic storage systems [51], vehicles for gene and drug delivery [106], nanotechnology has recently gained the support of various industrial sectors (Appenzeller 1991). Because of the numerous applications offered, such particles have the potential to have a big impact on civilization. Nanomaterials' distinct physiochemical characteristics can be attributed to specifically their high surface area. These special qualities make them a strong contender for biomedical applications because a wide range of biological activities occurs at the nanoscale [82]. In the period of nanotechnology, metal nanoparticles have been crucial in creating new biosensors and/or improving the current bio-sensing methods to meet the demand for more specialized and highly sensitive biomolecular diagnostics [46]. Due to these metals distinctive nanoscale physiochemical features, a wide range of biosensors, including: nanobiosensors for disease diagnostics at the point of care, nanoprobes for cell tracking disease etiology, in vivo sensing/imaging or therapeutic monitoring, more nanotechnology-based tools for basic biological research have been created [7]. Most of the metallic nanoparticles show features related to size that are notably different from those seen in micro-particles or bulk materials. We can see odd features such as surface plasmon resonance (SPR) in some metal NPs, quantum confinement in semiconductor nanocrystals and superparamagnetism in magnetic materials, depending upon the composition of their size [5]. Noble metal nanoparticles, especially gold and silver nanoparticles, are among the most thoroughly investigated nanomaterials and have sparked the creation of a plethora of techniques and procedures for molecular imaging, drug delivery and therapies [67]. Because they can act as agents with specific structures and functions suitable for cancer imaging, diagnosis and therapy. All of inorganic nanoparticles (for example, metallic nanoparticles, quantum dots, Fe_3O_4 nanoparticles) and polymeric nanoparticles (for example, dendrimers and micelles) are used in cancer nanomedicine [27]. There are significant connections between how nanoparticles are created and what they might be used for. Numerous investigations have revealed that silver nanoparticles have antimicrobial capabilities [32]. Gram-positive and Gram-negative bacteria's growth has been found to be effectively inhibited by nanoparticles made of silver and gold [32]. The investigation has started concentrating on antimicrobial nanoparticles as possible innovative therapeutic tools due to the growth in antibiotic resistance recently and the decrease in the discovery of new antibiotics. The production of small molecule adsorbates has also been detected optically using silver nanoparticles [77]. While Pt nanoparticle-based catalysts have been demonstrated to have high activity for the electro-oxidation of formic acid [121]. Almost all of these nanoparticles are typically produced using wet-chemical procedures that are usually low-cost and high-volume processes. Their potential usage in biological applications is however constrained by the necessity for hazardous solvents and chemical contamination during nanoparticle manufacturing [70]. Therefore, a "green" non-toxic method of creating metallic nanoparticles is required to enable their application in vast industrial range. Potentially, biological techniques c silver nanoparticles ould be used to accomplish this. There are two types of "natural" biogenic metallic nanoparticle production. The first is bioreduction, which involves physiologically reducing metal ions into more stable chemical forms. Dissimilatory metal reduction, in which the reduction of a metal ion is combined with the oxidation of an enzyme, can be used by many species [19]. Metal NPs must be non-toxic, biocompatible and stable in biological media in order to be employed in therapy and diagnosis. Additionally, they must specifically address the intended target. A biological target can be "tagged" or targeted by coating NPs with biological molecules that cause them to interact with or bind to the target. It is crucial to attach a vector that recognizes the target specifically in order to make sure the NP gets to the intended target [21]. Due to its potential to address issues with imaging and therapy of degenerative diseases, including cancer, this field is currently receiving a lot of press attention [120]. Some of the metallic nanoparticles with great properties in imaging and therapy are discussed as shown in Fig. 1.

2 Silver Nanoparticles in Imaging and Therapy

Silver nanoparticles (AgNPs) possess unique chemical and physical properties like catalytic activity, chemical stability, conductivity, and, most important antiviral, anti-inflammatory, antibacterial and antifungal properties have attracted the attention of researchers and are being used in different fields, including health, medical, food and industrial purposes [31, 72, 86]. Silver is an efficient antimicrobial agent and shows low toxicity [48 with both in vitro and in vivo applications [33, 77]. Silver nanoparticles due to their colloidal nature are being used as substrate for enhanced spectroscopy as it partially wants an electrically conducting surface [104, 119]. Silver nanoparticles may be fabricated by numerous physical, chemical and biological methods [111]. Physical methods like ball milling, laser ablation involves top-down approach of nanoparticle synthesis which are costlier. Chemical methods are mostly used to synthesize the AuNPs as stable colloidal dispersion in organic solvents or

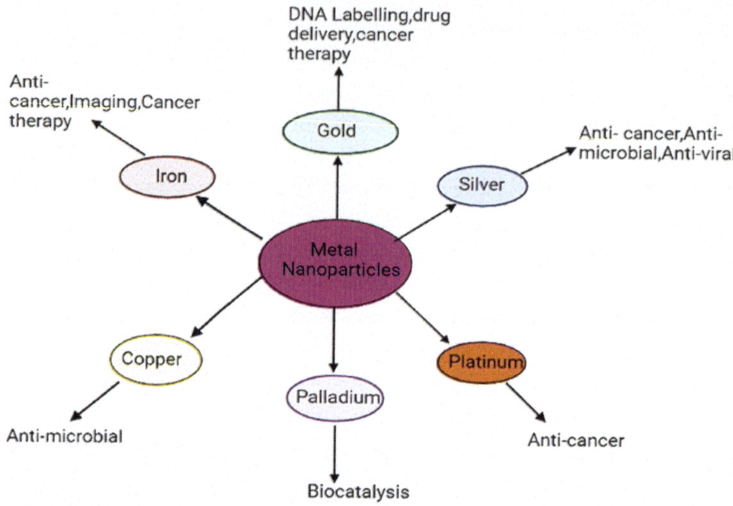

Fig. 1 Metallic nanoparticles with different properties

water [122] with borohydride, ascorbate, citrate and hydrogen as reductants which reduces silver ions to silver nanoparticles [69]. Chemical methods however involve the use of harmful chemicals. Therefore, biological method of nanoparticles synthesis involving use of eco-friendly and economical reductants is much being preferred nowadays. Silver nanoparticles due to their extraordinary optical properties are being utilized for development of imaging labels and optical sensors [13]. Magneto-silver core shell nanohybrids are currently being developed for use as contrast agents in imaging techniques like in magnetic resonance imaging and in CT contrasting applications [36].

2.1 Imaging

For detection and diagnosis of various diseases, non-invasive high-resolution imaging techniques like magnetic resonance imaging (MRI), computed tomography (CT), positron emission tomography (PET) and ultrasound are used. Physicians need good contrast of imaging for the clear visualization of structures for correct treatment [85], but contrast agents used in these techniques are toxic, with less imaging time and retention time. It has been found that core shell nanoparticles can increase retention time and biocompatibility when used as contrasting agents [8, 14, 119]. Because of these unique properties that noble metal nanoparticles display like SPR they have been extensively used in several types of sensors as contrasting agents because of their magnetic properties, low toxicity and good colloidal stability.

2.1.1 Magnetic Resonance Imaging (MRI)

MRI is an in vivo imaging technique for detailed tissue analysis, both for temporal and spatial analysis [98]. MRI technique is more sensitive to tissue composition because the difference between different tissues is based on local biochemical and metabolic activities in tissues and physiochemical properties of water. The reduction in magnetic relaxation time by MRI contrast agent provides the physical basis for MRI technique. MRI contrast agents like hydrogen protons from water tend to align themselves with an external magnetic field. A magnetic moment is developed onto the particle when another radio frequency pulse sequence is added to the external magnetic field. The protons align themselves again to the earlier magnetic field when the pulse sequence terminates. This phenomenon of returning to the actual position is called magnetic relaxation, which occurs either by longitudinal or transverse relaxation. Magnetic nanoparticles reduce relaxation time efficiently in water at a concentration of nanomolar even and the manipulation of physical and chemical properties of magnetic nanoparticles allows customization of their characters as per their applications [98].

2.1.2 Cancer Imaging

The most widely used metallic nanoparticle for treating cancer is silver [93]. The ability of nanoparticulated silver to cause cell death in human cells depends on its particular mechanism of action. Their method of action to cause cancer cell demise is rather dogmatic, regardless of their physical and chemical characteristics, such as heterogeneity in size, form and capping material [63]. Folic acid can be used to functionalize silver nanoparticles, enhancing their affinity for tumours that overexpress the folic acid receptor [117]. Silver nanoparticles possess the specific fluorescence properties which make them appropriate for the diagnostic purposes. Silver nanoparticles have plasmonic structure hence are in position to scatter and absorb light impinging certain regions. Selectively silver nanoparticles are taken up by cancerous cells, and after uptake, silver nanoparticles derived scattered light can be used for imaging and hence help in appropriate diagnosis and treatment of cancers. For the rat basophilic leukaemia imaging, luminescent nanosilver particles were used, and it was found that these silver nanoparticles possess toxic effect against basophilic leukaemia at the concentration of 10–100 μM [64].

2.1.3 Swept-Source Optical Coherence Tomography (SSOCT)

SSOCT is a unique and novel form of the optical coherence tomography (OCT) method with benefits in reducing motion artefacts and sensitivity. For enhancing the contrast of an OCT image, silver nanoparticles are exploited as exogenous contrast agents. A swept-source laser light source enables an imaging rate of 100 kHz. Meanwhile, OCT detects the light that is backscattered by tissue microstructures, and silver nanoparticles are appropriate for this role due to their exceptional localized SPR property [34]. The in vitro imaging of chicken tissue using silver nanoparticles as contrast agents revealed that the nanoparticles effectively backscattered near-infrared light and entered the tissue layers time-dependent, enhancing contrast and increasing imaging depth.

2.2 *Silver Nanoparticles in Therapy*

2.2.1 Antibacterial Therapy

Silver nanoparticles because of their extensive antibacterial property have attracted researcher as they appear alternative to antibiotics which have developed resistance. In broad-spectrum vancomycin-resistant strains of gram-positive and gram-negative bacteria, gold nanoparticles (AuNPs) act as efficient antibacterial representatives [124] Silver nanoparticles have a more superficial area-to-volume ratio than their bulk counterparts which allows them to interact in better way with biological membranes. For silver ions to have an antimicrobial property, they must exist in ionized form. The positive charge of the silver ions is believed to be responsible for its antimicrobial property [2]. As per the literature, silver nanoparticles interact with the sulphur and phosphorus groups of proteins and attack cell division and respiratory chain, thereby killing the cells by radicle oxygen formation [30, 100]. Some reports also show that electrostatic interaction do exist between positively charged silver nanoparticles and negatively charged cell membrane [12]. These silver nanoparticles accumulate in the membrane penetrating the cell, causing damage to cell. Gram-negative bacteria are extra prone to silver nanoparticles then gram positive due to presence of less peptidoglycan and thin cell wall in gram-negative bacteria then in gram positive, allowing more silver ions to pass the membrane and perform their function [4]. Synergic effect of gold nanoparticles with antibiotics have been examined which led to bacteria death by different mechanisms [26, 44, 75]. In this case, nanosilver makes it possible for antibiotics to reach the cell surface, making cells more sensitive to antibiotics.

2.2.2 Antiviral Therapy

Silver nanoparticles possess antiviral activity as reported by various studies. In antiviral therapy, the mechanism of silver nanoparticles antiviral action is crucial. It has been suggested that ind to viral particles and prevent viral nucleic acid replication in plant cells [88]. The size of silver nanoparticles plays an essential role in efficiency, subsequently, the direct interaction between the viral genome and proteins is an important factor for antiviral action [25]. One of the mechanisms of action of silver nanoparticles is the increased membrane hydrophilicity which was evaluated because silver nanoparticles were integrated into polysulfone ultrafiltration membrane against MS2 bacteriophage [129]. To investigate antiviral action of silver nanoparticles, a study was carried out and the results showed the sensitivity to the silver nanoparticle coated polyurethane condom by both macrophage (M)-tropic strain and T-lymphocyte (T)-tropic strain of HIV-1 [83]. Sulforhodamine assay [116] and hemagglutination assay [126, 127] demonstrated that silver nanoparticles could successfully inhibit the growth of influenza virus.

2.2.3 Antifungal

Silver nanoparticles show potent role in the inhibition of fungal pathogens. Fungus infections more commonly occur in immunosuppressed individuals. Antifungal drugs being limited therefore to overcome fungal diseases is a tedious process. Biologically synthesized silver nanoparticles show strong antifungal activity against various fungal pathogens like *Candida* species, *Alternaria, Fusarium oxysporium* and *Aspergillus* species. Silver nanoparticles prevent the growth of fungal species by inhibiting the enzyme involved in the synthesis of fungal cell wall. They also alter the structures of cell membrane by penetrating it, thereby changing the permeability of cell wall, causing the ultimate death of fungus [97]. Another possible mechanism of action of silver nanoparticles suggested is that they harm the surface proteins and nucleic acids by generating free radicals, thereby damaging intracellular organelles. Damage to the lipids, cellular content, nucleic acids and reduced mitochondrial activity are brought about by an increase in the intracellular ROS [17, 66].

2.2.4 Anticancer Therapy

Silver nanoparticles are being utilized in targeting cancer cells like breast cancer as reported by trypan blue exclusion method [23]. Since the chemotherapeutic agents are being utilized for cancer treatment has wide side effects. The photodynamic cancer therapy which is cytotoxic involves the use of specific dye to generated oxygen atoms in order to destroy the cancer cells, but the remaining dye molecules in the patient make them sensitive to the daylight exposure creating problems for them [107]. So, the focus of researchers was to develop the technique to target cancer cells specifically, for which target specific drug delivery by nanoparticles appears to be the best solution which will not only help in specific drug delivery but also in delivering more quantity of drug and prevent its spread to other parts of the body [79, 107]. Multidrug-resistant (MDR) cancer poses major challenge to researchers as they survive the chemotherapy. However, it has been shown that use of modified silver nanoparticles can treat MDR cancers also [128].

3 Gold Nanoparticles in Imaging and Therapy

Gold nanoparticles are widely employed in the medical profession owing to their remarkable biocompatibility, resulting from their great physical and chemical stability, ease of functionalization with physiologically active organic molecules [95]. Research is being done on gold nanoparticles for use as catalysts, biological sensors, optoelectronic components or ultrasensitive compounds [18]. Proteins, nucleic acids, enzymes, antibodies, drugs and fluorescent dyes are few of the many molecules that gold nanoparticles can directly bind and interact on their surfaces for several medical uses and biological processes [102, 115]. Despite the fact that gold nanoparticles have a lengthy history, creating compact, stable gold nanoparticles is a challenging task in nanotechnology. Based on these traits, the medical industry has identified applications for delivery carriers, treatments (PDT and RT), imaging, diagnostics and other biological processes as shown in Fig. 2.

3.1 Therapeutics

Therapeutic efficacy of gold nanoparticles depends on their capacity to bind with tumours and cancer cells leading to their destruction [62, 81]. Since gold nanoparticles are biocompatible, they are employed to treat epithelial ovarian cancer [94]. By

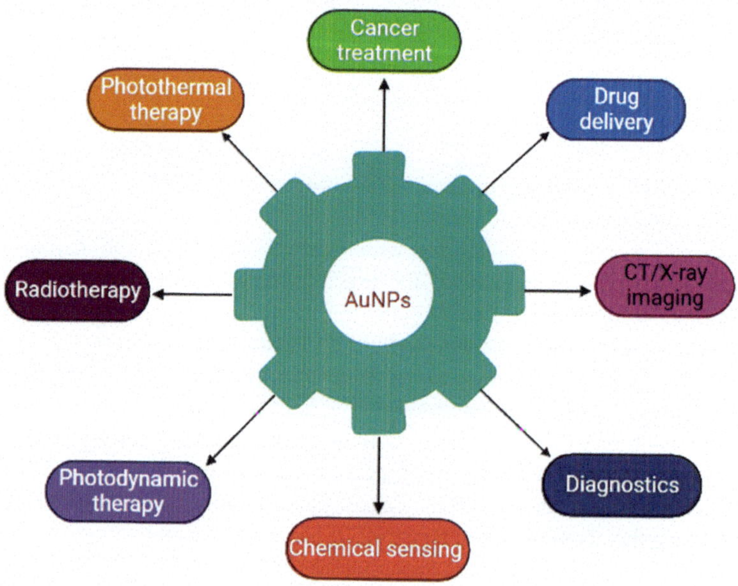

Fig. 2 Graphical representation of biomedical applications of gold nanoparticles (AuNPs)

identifying the gold nanoparticles-triggered processes at the subcellular and molecular levels, it may be possible to uncover novel targets and tactics to enhance gold nanoparticles-dependent treatment, such as photothermal agents, contrast agents and radio sensitizers [62]. The basis for photothermal therapy is the ability of photons to absorb light and transform it into thermal energy and is employed primarily in cancer treatments. Gold nanoparticles are ideally suited for the thermal eradication of cancer due to their ease in surface functionalization and aptitude for photothermal heating [56–58]. Photothermal treatment (PTT), often referred to as thermal ablation or optical hyperthermia, a popular non-invasive technique of treating cancer because it offers the benefits of real-time tumour site monitoring and photoinduced cell or tissue destruction [114]. The therapy of infectious disorders, which has been vigorously developed, and tumour therapy both point in the direction of photothermal cell injury as a potential route [37]. Targeting recognition technology is used in PTT to inject compounds into the body that congregate close to the tumour tissues [84, 87]. Photothermal materials (metal nanoparticles) may transform light energy into heat energy (photothermal conversion) when exposed to external light sources, often visible or near-infrared (NIR), which destroys tumour tissue and causes death of cancer cells [84, 87]. The depth of effective PTT penetration can also be changed by modifying the geometrical and physical properties of gold nanoparticles, such as size and shape [9, 10, 89]. Although the majority of gold nanoparticles used in PTT are nanorods or shells, cellular absorption may be limited when employed in biological environments [54]. Photodynamic therapy (PDT) is a different type of light treatment that was created recently and is used to kill harmful germs and cancer cells [1]. PDT uses visible light, photosensitizer (PS), and tissue-derived molecular oxygen (O_2). PDT directly results in tumour cell death by apoptosis, necrosis and autophagy which can greatly increase survival time and quality of life [68]. The availability of oxygen in tissues is necessary for PDT. In the PDT procedure, laser light of a certain wavelength stimulates the PS that has been absorbed by the tissue. The excited PS will transmit energy to the surrounding oxygen, producing reactive oxygen species (ROS) and increasing the level of ROS at the target areas. When ROS interact with adjacent biological macromolecules, they can significantly increase cytotoxicity, cell damage and even death or apoptosis in the affected cells [22, 45, 114]. Gold nanoparticles can gather in the vicinity of the tumour, absorb near-infrared (NIR) light, increase warmth and elevate amounts of ROS, which can inhibit tumour growth and promote cancer cell death (Jing et al. 2014). Radiation therapy (RT), in conjugation to PTT and PDT, is one of the least invasive and often used therapies in the treatment of a range of cancers [118]. RT entails delivering high-intensity ionizing radiation to all tumour tissues while the surrounding healthy cells, tissues and organs must be protected. This causes the tumour cells to die [60, 103]. Typically, water and biological components are ionized using X-rays and γ-rays (such as organelles). Water is the primary element of a cell and the primary target of ionizing radiation, which causes the lysis of water molecules. The creation of charged species and free radicals is the result of this lysis, known as radiolysis. The interplay of biologicals and membrane structure may potentially result in structural damage and cell death [65].

Due to high atomic number of gold in gold nanoparticles, there have been multiple studies of radio sensitization using them in RT recently [47, 78].

3.2 *Imaging*

Imaging is essential for diagnosis as well as for managing diseases and assessing how well treatments are working. Computed tomography (CT), magnetic resonance imaging (MRI), X-rays and ultrasounds are examples of common clinical imaging modalities that fall within the category of structural imaging modalities. One of the most significant and developed tissue imaging methods is X-ray computed tomography (CT), which is widely available and inexpensive [55]. CT is one of the most exciting imaging/diagnostic methods now employed in hospitals due to its availability, efficacy, and affordability. In instance, CT is a non-invasive clinical diagnostic technology that can separate tissues and rebuild images in three dimensions [74]. The cross-sectional 3D image created by a CT scan is made up of X-ray images obtained at various angles while spinning around an object [24, 74]. To distinguish between normal and abnormal cells using contrast agents (as iodinated molecules), CT imaging depends on the fact that healthy and sick tissues or cells have different densities [16]. Since they have the ability to transform light energy into heat energy via the SPR effect and substantially absorb ionizing radiation to enhance the coefficient of X-ray absorption, gold nanoparticles are gaining interest in imaging as an X-ray contrast agent [99]. Gold nanoparticles migration and concentration at target locations and prolonged vascular retention duration are crucial for their possible use in improved X-ray CT imaging because they enable non-invasive tracking and visualization of the therapeutic cells [80, 125]. Recently, numerous other nanoprobes, including gold nanoprobes and nanotags have been created as blood pool CT contrast agents [96]. According to research by [35] mice with millimetre-sized human breast tumours' (1.5 mm) are more easily seen when gold nanoparticles are used in conjunction with active tumour targeting (using anti-Her2 antibodies) as opposed to passive targeting as seen in Fig. 3. Compared to conventional molecules, gold nanoparticles have a longer vascular retention time, which could expand the imaging window [3]. According to Cai et al. (2007), in CT scans, 14 nm gold nanoparticles functionalized with PEG can be used, for achieving a longer imaging window than iodine compounds and enabling more detailed imaging.

4 Magnetic Nanoparticles in Therapeutics and Imaging

Magnetic nanoparticles are being used in biomedical applications because of their biocompatibily, small size, capacity for surface functionalization, superparamagnetism behaviour and targeting capabilities. Iron oxide nanoparticles can be directed on the target with limited negative effects on tissues other than the target. Both for

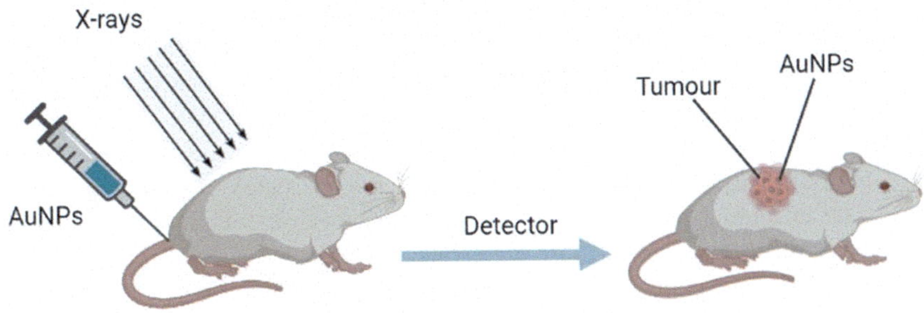

Fig. 3 Simple scheme for X-ray imaging

Fig. 4 Structure of a typical magnetic nanoparticle

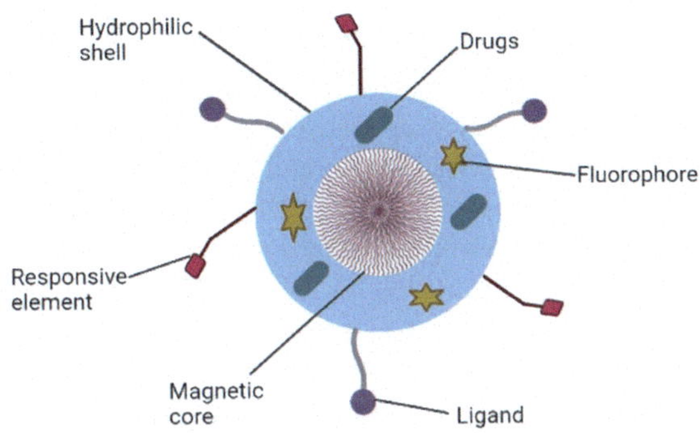

chemotherapeutic and photosensitizer drugs iron oxide magnetic nanoparticles have successfully been utilized as medication carriers. Magnetite (Fe_3O_4) and other magnetic nanoparticles have intriguing features that make them perfect nanocarriers in a variety of biomedical applications. These characteristics can be summed up as follows: ability to target a desired treatment site using an external magnetic field (magnetic targeting), more surface area, biocompatibility, chemical stability, small size, low production costs and multiple preparation routes including biogenic chemical and physical synthesis [108–110]. Iron oxide magnetic nanoparticles have thus been used in a number of biological applications, including magnetic hyperthermia, bio-sensing, immunoassay, magnetic resonance imaging and targeted drug delivery systems as shown in Fig. 4.

4.1 Photodynamic Therapy

PDT is a medical procedure which involves injecting a photosensitizer into the tumour site. When exposed to visible near-infrared light, the photosensitizer produces reactive oxygen species (ROS) that cause a deleterious effect on the tumour site. PDT's fundamental drawback is its nonexistence of selectivity, the combination of magnetic iron oxide nanoparticles with photosensitizer offers a novel and promising PDT strategy. The passive accumulation of nanoparticles in tumour locations, known as the EPR effect, is one of the main benefits of utilizing nanoparticles in PDT [56]. This effect can be understood in the light of the changed anatomy of tumour tissues, which exhibit increased angiogenesis, leaky blood vessels and a compromised lymphatic drainage system. As a result, nanoparticles can spontaneously concentrate in tumour areas and extravasate from blood vessels. Due to impaired lymphatic outflow, nanoparticles are maintained in tumour tissues. Once they have accumulated due to compromised lymphatic drainage, nanoparticles are retained in tumour tissues [56–58]. According to an intriguing finding, the EPR effect causes therapeutic levels of photosensitizer-containing nanoparticles to accumulate in solid tumours [76]. This makes nanoparticles promising carriers for singlet oxygen generators in PDT. PDT and magnetic hyperthermia can be combined to dramatically increase the cytotoxicity against cancerous tissues. Once superparamagnetic iron oxide nanoparticles have been properly targeted using magnetic targeting, an alternate magnetic field is applied for a predetermined amount of time. This causes a local increase in tissue temperature (42–45 °C), which targets cancer cell death by apoptosis. In order to increase the effectiveness of several cancer treatments, hyperthermia therapy can be used. Due to its superparamagnetism, magnetite is the material most frequently employed for the manufacture of iron oxide magnetic nanoparticles because it enables quick nanoparticle magnetization in response to a magnetic field. Based on a chromophore and its capacity to direct the photosensitizer to the intended site of application with a minimum amount of adverse effects on healthy tissues, PDT with magnetic iron oxide nanoparticles is used [50].

4.2 Photo Thermal Therapy (PTH)

PTT is a non-invasive cancer therapy strategy that involves use of visible light. Laser energy transforms into thermal energy, which causes the cancer to be thermally abated, in the presence of photo-absorbers like NPs. PTT is being used for clinical investigations. On the basis of the indication that tumour cells are more sensitive to temperature fluctuations than normal tissues, it is anticipated to be a potential substitute for cancer therapy that can efficiently eradicate the tumour. Damage to cell membrane and subcellular levels results from thermal ablation. According to the parameters that have been used, such as the laser power and exposure time, the method of cell death in PTT can either be necrosis or apoptosis. An intense laser power may cause necrosis, whereas a low laser power can result in apoptosis. Future medicines that can successfully slow tumour growth include NPs-based therapies since they are less invasive, effective, controlled and promising. Functional theranostics has given multifunctional nanoparticles (NPs) a lot of attention. [112] looked into the effects of photothermal therapy on cancer cell lines KB (human oral squamous carcinoma cell line) and MCF-7 (human breast carcinoma cell line). CMCTS coated Fe_3O_4 with a diameter of 177 nm was used. A magnetic field from outside was used during in vivo tests to promote accumulation in the tumour. The temperature then increased by being exposed to an 808 nm laser with a 1.5 W cm^2 power density. Individual and clustered Fe_3O_4 NPs both had a positive effect on photothermal therapy, but clustered Fe_3O_4 NPs were superior [112]

4.3 *Magnetic Hyperthermia Therapy (MHT)*

MHT takes advantage of the heating outcome of magnetic nanoparticles used in the cancer therapy method in an external alternating magnetic field (AMF). It is a non-invasive technique for producing heat that permeates tissues and kills the tumour. The exploitation of a high-frequency magnetic field by Gordon et al. [29] illustrated the intracellular use of magnetic NPs. Gilchrist et al. [28], depicted the use of magnetic particles for hyperthermia. In order to explain how magnetic energy is converted into heat, Jordan et al. investigated the specific loss power/specific absorption rate of several magnetic nanoparticles. Later, successful in vitro and in vivo investigations were carried out by employing magnetic fluid hyperthermia. MHT has been used in brain tumour treatment, and clinical trials are being carried out for pancreatic and oesophageal cancers.

4.4 *Immunotherapy*

Cancer immunotherapy is a distinctive yet effective form of cancer therapy, in contrast to PTT, PDT and MHT. In immunotherapy, the immune system of the body is the target rather than the tumour cells or tissues. The goal of cancer immunotherapy is to eradicate tumour tissues while selectively killing cancer cells by turning on T cells, dendritic cells and macrophages [90]. For the subsequent cancer cells cooperative attack, it was a task to carry adequate quantities of antigen into dendritic cells (DCs), which can activate enough cytotoxic T cells and CD4 + helper T cells. Attributing to the large specific area to volume and viable targeting molecular alteration, NPs have been established as required transporters for transporting tumour-associated antigens into dendritic cells [59]. ZnO can monitor mature DCs in vitro using its green fluorescence produced at 470 nm, while Fe_3O_4 may trace the mature DCs in vivo using improved MR T2 imaging, according to research by [15]. The outcomes showed that the planned method might efficiently harvest DCs within an hour and activate tumour-specific T lymphocytes to kill cancer cells successfully. Iron oxide nanoparticles are also known to enhance the efficacy of immune-suppressive drugs by inhibiting the release of cytokines interleukins-2 and tumour necrosis factor-α at the time of transplantation [101].

4.5 *Imaging*

4.5.1 MRI

Magnetic resonance imaging (MRI) is a technique that is most frequently utilized in imaging. Iron oxide formulations are employed in this technique as contrast agents. The weighting of image contrast can be used to highlight certain diseases or anatomical structures. The hydrogen protons of water molecules inside imaging tissue revert to their equilibrium states after a magnetic field produces an orientation of their nuclear magnetic spins [98].

4.5.2 MPI

Myocardial perfusion imaging (MPI) utilized the magnetic particles themselves as a tracer and not as a contrast agent, unlike MRI which uses iron oxide nanoparticles as a contrast agent to produce the MR signal. It accomplishes this by creating a signal from the nonlinear magnetization of the magnetic nanoparticles using an oscillating magnetic field (Ferguson et al. 2015). Since, its initial introduction in 2005, the technology has only been used in pre-clinical settings [61]. Iron oxide nanoparticles (IONPs) were thoroughly researched in recent years to be adapted to this kind of imaging [53] and used in research projects such as stem cell tracking [11].

4.5.3 Dual Modality Imaging

While not only giving importance to the two main techniques of imaging, MRI and MPI, many of nanoparticles with multifunctional imaging activities exist. This technique is called dual-modality imaging or multifunctional imaging [71]. This technique is based on iron oxide platform. The multimodality technique combines MRI, MPI, positron emission tomography (PET) and computed tomography (CT), because iron oxide can be encapsulated, grafted and coated. The iron oxide nanoparticles are usually grafted with Infrared Florescence dye for optical imaging [123].

5 Conclusion and Future Perspectives

This book chapter mainly deals with materials with less than one-hundredth of nanometre dimensions. These materials are called nanomaterials, and the technology employed is nanotechnology. We have concluded that nanotechnology is being used in almost every field of science. It has revolutionized the fields of imaging and therapy by regulating materials at the nanoscale. Nobel metal nanoparticles have been crucial in creating new biosensors and have led to improving new bio-sensing methods to meet the requirements in highly specialized and sensitive biomolecular diagnostics. Functionalization of MNPs has been primarily credited with their use as potential tools in biomedical applications such as drug transport, magnetic hyperthermia and diagnostics. Many clinical trials have been conducted for MNPs over the last ten years. A significant connection occurs between the formation and utilization of nanoparticles. This chapter discusses the imaging and therapeutic properties of silver, gold (Au), iron and magnetic nanoparticles. Despite the fact that recent research has shown multivalent composite materials to have several benefits, it is still difficult to create a more specific target with excellent specificity. Providentially, the topic of nanotechnology endures piquing attention among scientists studying chemicals thanks to important discoveries and fresh scientific problems. Future research should focus on the biocompatibility and safety of these nanoparticles, particularly their long-term harmfulness. Further, scientific studies on human and animal models must be carried out with nanomaterials to support their usage, particularly in biomedical imaging using CT scans, MRI, ultrasound and other optical imaging.

Acknowledgements The authors gratefully acknowledge the BioRender for figure processing and Indian Institute of Technology Gandhinagar, Indian Institute of Technology Roorkee and Sher-e-Kashmir University of Agricultural Sciences and Technology of Jammu (SKUASTJammu), India. Authors also acknowledge the Khalifa University, Abu Dhabi, United Arab Emirates.

References

1. Abrahamse H, Hamblin MR (2016) New photosensitizers for photodynamic therapy. Biochem J 473(4):347–364. https://doi.org/10.1042/BJ20150942
2. Ahmed S, Ahmad M, Swami BL, Ikram S (2016) A review on plants extract mediated synthesis of silver nanoparticles for antimicrobial applications: a green expertise. J Adv Res 7(1):17–28. https://doi.org/10.1016/j.jare.2015.02.007
3. Ahn S, Jung SY, Lee SJ (2013) Gold nanoparticle contrast agents in advanced X-ray imagingtechnologies. Molecules 18(5):58585890. https://doi.org/10.3390/molecules18055858
4. Ankanna STNVKVP, Prasad TNVKV, Elumalai EK, Savithramma N (2010) Production of biogenic silver nanoparticles using Boswellia ovalifoliolata stem bark. Dig J Nanomater Biostruct 5(2):369–372
5. Azzazy HM, Mansour MM (2009) In vitro diagnostic prospects of nanoparticles. Clin Chim Acta 403(1–2):1–8. https://doi.org/10.1016/j.cca.2009.01.016
6. Bachheti RK, Bachheti A, Husen A (2023) Nanomaterials for environmental and agricultural sectors. Springer Nature Singapore Pte Ltd., 152 Beach Road, #21-01/04 Gateway East, Singapore 189721, Singapore. https://doi.org/10.1007/978-981-99-2874-3
7. Baptista PV, Doria G, Quaresma P, Cavadas M, Neves CS, Gomes I, Franco R et al (2011) Nanoparticles in molecular diagnostics. Prog Mol Biol Transl Sci 104:427–488. https://doi.org/10.1016/B978-0-12-416020-0.00011-5
8. Berry CC, Curtis AS (2003) Functionalisation of magnetic nanoparticles for applications in biomedicine. J Phys D Appl Phys 36(13):R198. https://doi.org/10.1088/0022-3727/36/13/203
9. Bibikova O, Singh P, Popov A, Akchurin G, Skaptsov A, Skovorodkin I, Tuchin V et al (2017) Shape-dependent interaction of gold nanoparticles with cultured cells at laser exposure. Laser Phys Lett 14(5):055901. https://doi.org/10.1088/1612-202X/aa63ae

10. Boyer D, Tamarat P, Maali A, Lounis B, Orrit M (2002) Photothermal imaging of nanometer-sized metal particles among scatterers. Science 297(5584):1160–1163. https://doi.org/10.1126/science.1073765
11. Bulte JW, Walczak P, Janowski M, Krishnan KM, Arami H, Halkola A, Rahmer J et al (2015) Quantitative "hot-spot" imaging of transplanted stem cells using superparamagnetic tracers and magnetic particle imaging. Tomography 1(2):91–97. https://doi.org/10.18383/j.tom.2015.00172
12. Cao YC, Jin R, Mirkin CA (2002) Nanoparticles with Raman spectroscopic fingerprints for DNA and RNA detection. Science 297(5586):1536–1540. https://doi.org/10.1126/science.297.5586.1536
13. Caro C, Castillo PM, Klippstein R, Pozo D, Zaderenko AP (2010) Silver nanoparticles: sensing and imaging applications. Silver Nanoparticles:201
14. Chatterjee K, Sarkar S, Rao KJ, Paria S (2014) Core/shell nanoparticles in biomedical applications. Adv Colloid Interface Sci 209:8–39. https://doi.org/10.1016/j.cis.2013.12.008
15. Cho NH, Cheong TC, Min JH, Wu JH, Lee SJ, Kim D, Seong SY et al (2011) A multifunctional core–shell nanoparticle for dendritic cell-based cancer immunotherapy. Nat Nanotechnol 6(10):675–682. https://doi.org/10.1038/nnano.2011.149
16. Cormode DP, Naha PC, Fayad ZA (2014) Nanoparticle contrast agents for computed tomography: a focus on micelles. Contrast Media Mol Imaging 9(1):37–52. https://doi.org/10.1002/cmmi.1551
17. Costa V, Moradas-Ferreira P (2001) Oxidative stress and signal transduction in Saccharomyces cerevisiae: insights into ageing, apoptosis and diseases. Mol Aspects Med 22(4–5):217–246. https://doi.org/10.1016/S0098-2997(01)00012-7
18. Deepa C, Rajeshkumar L, Ramesh M (2022) Preparation, synthesis, properties and characterization of graphene-based 2D nano-materials for biosensors and bioelectronics. J Market Res. https://doi.org/10.1016/j.jmrt.2022.06.023
19. Deplanche K, Caldelari I, Mikheenko IP, Sargent F, Macaskie LE (2010) Involvement of hydrogenases in the formation of highly catalytic Pd (0) nanoparticles by bioreduction of Pd (II) using Escherichia coli mutant strains. Microbiology 156(9):2630–2640. https://doi.org/10.1099/mic.0.036681-0
20. Dobson J (2006) Gene therapy progress and prospects: magnetic nanoparticle-based gene delivery. Gene Ther 13(4):283–287. https://doi.org/10.1038/sj.gt.3302720
21. Efremov RG, Chugunov AO, Pyrkov TV, Priestle JP, Arseniev AS, Jacoby E (2007) Molecular lipophilicity in protein modeling and drug design. Curr Med Chem 14(4):393–415. https://doi.org/10.2174/092986707779941050
22. Falahati M, Attar F, Sharifi M, Saboury AA, Salihi A, Aziz FM, El-Sayed MA et al (2020) Gold nanomaterials as key suppliers in biological and chemical sensing, catalysis, and medicine. Biochimica et Biophysica Acta (BBA)-Gen Subj 1864(1):129435. https://doi.org/10.1016/j.bbagen.2019.129435
23. Franco-Molina MA, Mendoza-Gamboa E, Sierra-Rivera CA, Gómez-Flores RA, Zapata-Benavides P, Castillo-Tello P, Rodríguez-Padilla C et al (2010) Antitumor activity of colloidal silver on MCF-7 human breast cancer cells. J Exp Clin Cancer Res 29(1):1–7. https://doi.org/10.1186/1756-9966-29-148
24. Fuller MA, Köper I (2019) Biomedical applications of polyelectrolyte coated spherical gold nanoparticles. Nano convergence 6:1–10. https://doi.org/10.1186/s40580-019-0183-4
25. Galdiero S, Falanga A, Vitiello M, Cantisani M, Marra V, Galdiero M (2011) Silver nanoparticles as potential antiviral agents. Molecules 16(10):8894–8918. https://doi.org/10.3390/molecules16108894
26. Ghosh IN, Patil SD, Sharma TK, Srivastava SK, Pathania R, Navani NK (2013) Synergistic action of cinnamaldehyde with silver nanoparticles against spore-forming bacteria: a case for judicious use of silver nanoparticles for antibacterial applications. Int J Nanomed:4721–4731. https://doi.org/10.2147/IJN.S49649
27. Gil PR, Parak WJ (2008) Composite nanoparticles take aim at cancer. ACS Nano 2(11):2200–2205. https://doi.org/10.1021/nn800716j

28. Gilchrist RK, Medal R, Shorey WD, Hanselman RC, Parrott JC, Taylor CB (1957) Selective inductive heating of lymph nodes. Ann Surg 146(4):596
29. Gordon RT, Hines JR, Gordon D (1979) Intracellular hyperthermia a biophysical approach to cancer treatment via intracellular temperature and biophysical alterations. Med Hypotheses 5(1):83–102. https://doi.org/10.1016/0306-9877(79)90063-x
30. Gordon O, Vig Slenters T, Brunetto PS, Villaruz AE, Sturdevant DE, Otto M, Fromm KM et al (2010) Silver coordination polymers for prevention of implant infection: thiol interaction, impact on respiratory chain enzymes, and hydroxyl radical induction. Antimicrob Agents Chemother 54(10):4208–4218. https://doi.org/10.1128/AAC.01830-09
31. Gurunathan S, Park JH, Han JW, Kim JH (2015) Comparative assessment of the apoptotic potential of silver nanoparticles synthesized by Bacillus tequilensis and Calocybe indica in MDA-MB-231 human breast cancer cells: targeting p53 for anticancer therapy. Int J Nanomed 10:4203. https://doi.org/10.2147/IJN.S83953
32. Guzman M, Dille J, Godet S (2012) Synthesis and antibacterial activity of silver nanoparticles against gram-positive and gram-negative bacteria. Nanomed Nanotechnol Biol Med 8(1):37–45. https://doi.org/10.1016/j.nano.2011.05.007
33. Haes AJ, Van Duyne RP (2002) A nanoscale optical biosensor: sensitivity and selectivity of an approach based on the localized surface plasmon resonance spectroscopy of triangular silver nanoparticles. J Am Chem Soc 124(35):10596–10604. https://doi.org/10.1021/ja020393x
34. Haes AJ, Zou S, Schatz GC, Van Duyne RP (2004) A nanoscale optical biosensor: the long range distance dependence of the localized surface plasmon resonance of noble metal nanoparticles. J Phys Chem B 108(1):109–116. https://doi.org/10.1021/jp0361327
35. Hainfeld JF, O'Connor MJ, Dilmanian FA, Slatkin DN, Adams DJ, Smilowitz HM (2011) Micro-CT enables microlocalisation and quantification of Her2-targeted gold nanoparticles within tumour regions. Br J Radiol 84(1002):526–533. https://doi.org/10.1259/bjr/42612922
36. Haribabu V, Girigoswami K, Girigoswami A (2021) Magneto-silver core–shell nanohybrids for theragnosis. Nano-Struct Nano-Objects 25:100636. https://doi.org/10.1016/j.nanoso.2020.100636
37. Huang X, Jain PK, El-Sayed IH, El-Sayed MA (2008) Plasmonic photothermal therapy (PPTT) using gold nanoparticles. Lasers Med Sci 23:217–228. https://doi.org/10.1007/s10103-007-0470-x
38. Husen A (2022) Engineered nanomaterials for sustainable agricultural production, soil improvement and stress management. Elsevier Inc., 50 Hampshire St., 5th Floor, Cambridge, MA 02139, USA. ISBN 978-0-323-91933-3 https://doi.org/10.1016/C2021-0-00054-7
39. Husen A (2023a) Nanomaterials from agricultural and horticultural products. Springer Nature Singapore Pte Ltd., 152 Beach Road, #21-01/04 Gateway East, Singapore 189721, Singapore https://doi.org/10.1007/978-981-99-3435-5
40. Husen A (2023b) Nanomaterials and nanocomposites exposures to plants (Response, interaction, phytotoxicity and defense mechanisms). Springer Nature Singapore Pte Ltd., 152 Beach Road, #21-01/04 Gateway East, Singapore 189721, Singapore https://doi.org/10.1007/978-981-99-2419-6
41. Husen A, Bachheti RK, Bachheti A (2023) Current trends in green nano-emulsions (Food, agriculture and biomedical sectors). Springer Nature Singapore Pte Ltd., 152 Beach Road, #21-01/04 Gateway East, Singapore 189721, Singapore
42. Husen A, Iqbal M (2019) Nanomaterials and plant potential. Springer International Publishing AG, Gewerbestrasse 11, 6330 Cham, Switzerland. https://doi.org/10.1007/978-3-030-05569-1
43. Husen A, Siddiqi KS (2023) Advances in smart nanomaterials and their applications. Elsevier Inc., 50 Hampshire St., 5th Floor, Cambridge, MA 02139, USA. https://doi.org/10.1016/C2021-0-02202-1
44. Hwang IS, Lee J, Hwang JH, Kim KJ, Lee DG (2012) Silver nanoparticles induce apoptotic cell death in Candida albicans through the increase of hydroxyl radicals. FEBS J 279(7):1327–1338. https://doi.org/10.1111/j.1742-4658.2012.08527.x

45. Imanparast A, Bakhshizadeh M, Salek R, Sazgarnia A (2018) Pegylated hollow gold-mitoxantrone nanoparticles combining photodynamic therapy and chemotherapy of cancer cells. Photodiagn Photodyn Ther 23:295–305. https://doi.org/10.1016/j.pdpdt.2018.07.011
46. Jain KK (2007) Applications of nanobiotechnology in clinical diagnostics. Clin Chem 53(11):2002–2009
47. Jain N, Bhargava A, Majumdar S, Tarafdar JC, Panwar J (2011) Extracellular biosynthesis and characterization of silver nanoparticles using aspergillus flavus NJP08: a mechanism perspective. Nanoscale 3(2):635–641. https://doi.org/10.1039/C0NR00656D
48. Jain D, Kachhwaha S, Jain R, Srivastava G, Kothari SL (2010) Novel microbial route to synthesize silver nanoparticles using spore crystal mixture of Bacillus thuringiensis. http://nopr.niscpr.res.in/handle/123456789/10471
49. Jin-Chul K, Madhusudhan A, Husen A (2021) Smart nanomaterials in biomedical applications. Springer Nature Switzerland AG, Gewerbestrasse 11, 6330 Cham, Switzerland https://doi.org/10.1007/978-3-030-84262-8
50. Kandasamy G, Maity D (2015) Recent advances in superparamagnetic iron oxide nanoparticles (SPIONs) for in vitro and in vivo cancer nanotheranostics. Int J Pharm 496(2):191–218. https://doi.org/10.1016/j.ijpharm.2015.10.058
51. Kang YS, Risbud S, Rabolt JF, Stroeve P (1996) Synthesis and characterization of nanometer-size Fe_3O_4 and γ-Fe_2O_3 particles. Chem Mater 8(9):2209–2211. https://doi.org/10.1021/cm960157j
52. Khan A, Roy A, Bhasin S, Emran TB, Khusro A, Eftekhari A, Karimi F et al (2022) Nanomaterials: an alternative source for biodegradation of toxic dyes. Food Chem Toxicol:112996. https://doi.org/10.1016/j.fct.2022.112996
53. Khandhar AP, Keselman P, Kemp SJ, Ferguson RM, Goodwill PW, Conolly SM, Krishnan KM (2017) Evaluation of PEG-coated iron oxide nanoparticles as blood pool tracers for preclinical magnetic particle imaging. Nanoscale 9(3):1299–1306. https://doi.org/10.1039/C6NR08468K
54. Kim HS, Lee DY (2018) Near-infrared-responsive cancer photothermal and photodynamic therapy using gold nanoparticles. Polymers 10(9):961. https://doi.org/10.3390/polym10090961
55. Kim D, Park S, Lee JH, Jeong YY, Jon S (2007) Antibiofouling polymer-coated gold nanoparticles as a contrast agent for in vivo X-ray computed tomography imaging. J Am Chem Soc 129(24):7661–7665
56. Kim H, Chung K, Lee S, Kim DH, Lee H (2016a) Near-infrared light-responsive nanomaterials for cancer theranostics. Wiley Interdisc Rev Nanomed Nanobiotechnol 8:23–45
57. Kim J, Kim J, Jeong C, Kim WJ (2016b) Synergistic nanomedicine by combined gene and photothermal therapy. Adv Drug Deliv Rev 98:99–112. https://doi.org/10.1016/j.addr.2015.12.018
58. Kim KS, Kim J, Lee JY, Matsuda S, Hideshima S, Mori Y, Na K et al (2016c) Stimuli-responsive magnetic nanoparticles for tumor-targeted bimodal imaging and photodynamic/hyperthermia combination therapy. Nanoscale 8(22):11625–11634. https://doi.org/10.1039/C6NR02273A
59. Klippstein R, Pozo D (2010) Nanotechnology-based manipulation of dendritic cells for enhanced immunotherapy strategies. Nanomed Nanotechnol Biol Med 6(4):523–529. https://doi.org/10.1016/j.nano.2010.01.001
60. Klębowski B, Depciuch J, Parlińska-Wojtan M, Baran J (2018) Applications of noble metal-based nanoparticles in medicine. Int J Mol Sci 19(12):4031. https://doi.org/10.3390/ijms19124031
61. Knopp T, Gdaniec N, Möddel M (2017) Magnetic particle imaging: from proof of principle to preclinical applications. Phys Med Biol 62(14):R124. https://doi.org/10.1088/1361-6560/aa6c99
62. Kodiha M, Wang YM, Hutter E, Maysinger D, Stochaj U (2015) Off to the organelles-killing cancer cells with targeted gold nanoparticles. Theranostics 5(4):357. 10.7150%2Fthno.10657

63. Kovács D, Igaz N, Gopisetty MK, Kiricsi M (2022) Cancer therapy by silver nanoparticles: fiction or reality? Int J Mol Sci 23(2):839. https://doi.org/10.3390/ijms23020839
64. Kravets V, Almemar Z, Jiang K, Culhane K, Machado R, Hagen G, Pinchuk A et al (2016) Imaging of biological cells using luminescent silver nanoparticles. Nanoscale Res Lett 11(1):1–9
65. Kwatra D, Venugopal A, Anant S (2013) Nanoparticles in radiation therapy: a summary of various approaches to enhance radio sensitization in cancer. Transl Cancer Res 2(4):330–342. https://doi.org/10.3978/j.issn.2218-676X.2013.08.06
66. Landolfo S, Politi H, Angelozzi D, Mannazzu I (2008) ROS accumulation and oxidative damage to cell structures in Saccharomyces cerevisiae wine strains during fermentation of high-sugar-containing medium. Biochimica et Biophysica Acta (BBA) Gen Subj 1780(6):892–898. https://doi.org/10.1016/j.bbagen.2008.03.008
67. Larguinho M, Baptista PV (2012) Gold and silver nanoparticles for clinical diagnostics—from genomics to proteomics. J Proteomics 75(10):2811–2823. https://doi.org/10.1016/j.jprot.2011.11.007
68. Lee CN, Hsu R, Chen H, Wong TW (2020) Daylight photodynamic therapy: an update. Molecules 25(21):5195. https://doi.org/10.3390/molecules25215195
69. Lee JY, Shou Z, Balazs AC (2003) Modeling the self-assembly of copolymer-nanoparticle mixtures confined between solid surfaces. Phys Rev Lett 91(13):136103. https://doi.org/10.1103/PhysRevLett.91.136103
70. Li X, Xu H, Chen ZS, Chen G (2011) Biosynthesis of nanoparticles by microorganisms and their applications. J Nanomater 2011:1–16. https://doi.org/10.1155/2011/270974
71. Li S, Zhang L, Wang T, Li L, Wang C, Su Z (2015) The facile synthesis of hollow Au nanoflowers for synergistic chemo-photothermal cancer therapy. Chem Commun 51(76):14338–14341. https://doi.org/10.1039/C5CC05676D
72. Li JF, Huang YF, Ding Y, Yang ZL, Li SB, Zhou XS, Tian ZQ et al (2010) Shell-isolated nanoparticle-enhanced Raman spectroscopy. Nature 464(7287):392–395. https://doi.org/10.1038/nature08907
73. Link S, El-Sayed MA (2003) Optical properties and ultrafast dynamics of metallic nanocrystals. Annu Rev Phys Chem 54(1):331–366. https://doi.org/10.1146/annurev.physchem.54.011002.103759
74. Lusic H, Grinstaff MW (2013) X-ray-computed tomography contrast agents. Chem Rev 113(3):1641–1666. https://doi.org/10.1021/cr200358s
75. Markowska K, Grudniak AM, Krawczyk K, Wrobel I, Wolska KI (2014) Modulation of antibiotic resistance and induction of a stress response in Pseudomonas aeruginosa by silver nanoparticles. J Med Microbiol 63(6):849–854. https://doi.org/10.1099/jmm.0.068833-0
76. Mbakidi JP, Drogat N, Granet R, Ouk TS, Ratinaud MH, Rivière E, Sol V et al (2013) Hydrophilic chlorin-conjugated magnetic nanoparticles—potential anticancer agent for the treatment of melanoma by PDT. Bioorg Med Chem Lett 23(9):2486–2490. https://doi.org/10.1016/j.bmcl.2013.03.039
77. McFarland AD, Van Duyne RP (2003) Single silver nanoparticles as real-time optical sensors with zeptomole sensitivity. Nano Lett 3(8):1057–1062. https://doi.org/10.1021/nl034372s
78. McMahon SJ, Hyland WB, Muir MF, Coulter JA, Jain S, Butterworth KT, Currell FJ et al (2011) Nanodosimetric effects of gold nanoparticles in megavoltage radiation therapy. Radiother Oncol 100(3):412–416. https://doi.org/10.1016/j.radonc.2011.08.026
79. McNamara K, Tofail SA (2017) Nanoparticles in biomedical applications. Adv Phys X 2(1):54–88. https://doi.org/10.1080/23746149.2016.1254570
80. Meir R, Popovtzer R (2018) Cell tracking using gold nanoparticles and computed tomography imaging. Wiley Interdisc Rev Nanomed Nanobiotechnol 10(2):e1480. https://doi.org/10.1002/wnan.1480
81. Mieszawska AJ, Mulder WJ, Fayad ZA, Cormode DP (2013) Multifunctional gold nanoparticles for diagnosis and therapy of disease. Mol Pharm 10(3):831–847. https://doi.org/10.1021/mp3005885

82. Mody VV, Nounou MI, Bikram M (2009) Novel nanomedicine-based MRI contrast agents for gynecological malignancies. Adv Drug Deliv Rev 61(10):795–807. https://doi.org/10.1016/j.addr.2009.04.020
83. Mohammed Fayaz A, Ao Z, Girilal M, Chen L, Xiao X, Kalaichelvan PT, Yao X (2012) Inactivation of microbial infectiousness by silver nanoparticles-coated condom: a new approach to inhibit HIV-and HSV-transmitted infection. Int J Nanomed:5007–5018. https://doi.org/10.2147/IJN.S34973
84. MubarakAli D, Thajuddin N, Jeganathan K, Gunasekaran M (2011) Plant extract mediated synthesis of silver and gold nanoparticles and its antibacterial activity against clinically isolated pathogens. Colloids Surf B 85(2):360–365. https://doi.org/10.1016/j.colsurfb.2011.03.009
85. Mujahid MH, Upadhyay TK, Khan F, Pandey P, Park MN, Sharangi AB, Kim B et al (2022) Metallic and metal oxide-derived nanohybrid as a tool for biomedical applications. Biomed Pharmacother 155:113791. https://doi.org/10.1016/j.biopha.2022.113791
86. Mukherjee P, Ahmad A, Mandal D, Senapati S, Sainkar SR, Khan MI, Sastry M et al (2001) Fungus-mediated synthesis of silver nanoparticles and their immobilization in the mycelial matrix: a novel biological approach to nanoparticle synthesis. Nano Lett 1(10):515–519. https://doi.org/10.1021/nl0155274
87. Murphy CJ, Thompson LB, Alkilany AM, Sisco PN, Boulos SP, Sivapalan ST, Huang J et al (2010) The many faces of gold nanorods. J Phys Chem Lett 1(19):2867–2875. https://doi.org/10.1021/jz100992x
88. Noha K, Bondok AM, El-Dougdoug KA (2018) Evaluation of silver nanoparticles as antiviral agent against ToMV and PVY in tomato plants. Sciences 8(01):100–111
89. Orendorff CJ, Sau TK, Murphy CJ (2006) Shape-dependent plasmon-resonant gold nanoparticles. Small 2(5):636–639. https://doi.org/10.1002/smll.200500299
90. Palucka AK, Ueno H, Fay JW, Banchereau J (2007) Taming cancer by inducing immunity via dendritic cells. Immunol Rev 220(1):129–150. https://doi.org/10.1111/j.1600-065X.2007.00575.x
91. Pandit C, Roy A, Ghotekar S, Khusro A, Islam MN, Emran TB, Bradley DA et al (2022) Biological agents for synthesis of nanoparticles and their applications. J King Saud Univ Sci 34(3):101869. https://doi.org/10.1016/j.jksus.2022.101869
92. Pankhurst QA, Connolly J, Jones SK, Dobson J (2003) Applications of magnetic nanoparticles in biomedicine. J Phys D Appl Phys 36(13):R167. https://doi.org/10.1088/0022-3727/36/13/201
93. Pannerselvam B, Thiyagarajan D, Pazhani A, Thangavelu KP, Kim HJ, Rangarajulu SK (2021) Copperpod plant synthesized AgNPs enhance cytotoxic and apoptotic effect in cancer cell lines. Processes 9(5):888. https://doi.org/10.3390/pr9050888
94. Piktel E, Ościłowska I, Suprewicz Ł (2021) Suprewicz-mediated apoptosis and autophagy in ovarian cancer cells treated with peanut-shaped gold nanoparticles. Int J Nanomed 16:1993. https://doi.org/10.1097/00000658-195710000-00007
95. Pissuwan D, Gazzana C, Mongkolsuk S, Cortie MB (2020) Single and multiple detections of foodborne pathogens by gold nanoparticle assays. Wiley Interdisc Rev Nanomed Nanobiotechnol 12(1):e1584. https://doi.org/10.1002/wnan.1584
96. Popovtzer R, Agrawal A, Kotov NA, Popovtzer A, Balter J, Carey TE, Kopelman R (2008) Targeted gold nanoparticles enable molecular CT imaging of cancer. Nano Lett 8(12):4593–4596. https://doi.org/10.1021/nl8029114
97. Prabhu S, Poulose EK (2012) Silver nanoparticles: mechanism of antimicrobial action, synthesis, medical applications, and toxicity effects. Int Nanolett 2:1–10. https://doi.org/10.1186/2228-5326-2-32
98. Qiao R, Yang C, Gao M (2009) Superparamagnetic iron oxide nanoparticles: from preparations to in vivo MRI applications. J Mater Chem 19(35):6274–6293. https://doi.org/10.1039/B902394A

99. Rahman WN, Corde S, Yagi N, Abdul Aziz SA, Annabell N, Geso M (2014) Optimal energy for cell radiosensitivity enhancement by gold nanoparticles using synchrotron-based monoenergetic photon beams. Int J Nanomed:2459–2467. https://doi.org/10.2147/IJN.S59471
100. Rai M, Yadav A, Gade A (2009) Silver nanoparticles as a new generation of antimicrobials. Biotechnol Adv 27(1):76–83. https://doi.org/10.1016/j.biotechadv.2008.09.002
101. Raj G, Rajendhran et al (2022) Iron oxide nanoparticles-a boon for therapeutic applications. World J Pharm Life Sci 8:54–61
102. Ramalingam V (2019) Multifunctionality of gold nanoparticles: plausible and convincing propertie. Adv Colloid Int Sci 271:101989
103. Retif P, Pinel S, Toussaint M, Frochot C, Chouikrat R, Bastogne T, Barberi-Heyob M (2015) Nanoparticles for radiation therapy enhancement: the key parameters. Theranostics 5(9):1030. 10.7150%2Fthno.11642
104. Rosi NL, Mirkin CA (2005) Nanostructures in biodiagnostics. Chem Rev 105(4):1547–1562. https://doi.org/10.1021/cr030067f
105. Roy A, Sharma A, Yadav S, Jule LT, Krishnaraj R (2021) Nanomaterials for remediation of environmental pollutants. Bioinorg Chem Appl 2021. https://doi.org/10.1155/2021/1764647
106. Rudge S, Peterson C, Vessely C, Koda J, Stevens S, Catterall L (2001) Adsorption and desorption of chemotherapeutic drugs from a magnetically targeted carrier (MTC). J Control Release 74(1–3):335–340. https://doi.org/10.1016/S0168-3659(01)00344-3
107. Salata O (2004) Applications of nanoparticles in biology and medicine. J Nanobiotechnol 2:3. https://doi.org/10.1186/1477-3155-2-3
108. Seabra AB, Durán N (2015) Nanotoxicology of metal oxide nanoparticles. Metals 5(2):934–975. https://doi.org/10.3390/met5020934
109. Seabra AB, Haddad P, Duran N (2013) Biogenic synthesis of nanostructured iron compounds: applications and perspectives. IET Nanobiotechnol 7(3):90–99. https://doi.org/10.1049/iet-nbt.2012.0047
110. Seabra AB, Pasquôto T, Ferrarini ACF, Santos MDC, Haddad PS, de Lima R (2014) Preparation, characterization, cytotoxicity, and genotoxicity evaluations of thiolated-and S-nitrosated superparamagnetic iron oxide nanoparticles: implications for cancer treatment. Chem Res Toxicol 27(7):1207–1218. https://doi.org/10.1021/tx500113u
111. Shankar SS, Ahmad A, Sastry M (2003) Geranium leaf assisted biosynthesis of silver nanoparticles. Biotechnol Prog 19(6):1627–1631. https://doi.org/10.1021/bp034070w
112. Shen S, Wang S, Zheng R, Zhu X, Jiang X, Fu D, Yang W (2015) Magnetic nanoparticle clusters for photothermal therapy with near-infrared irradiation. Biomaterials 39:67–74. https://doi.org/10.1016/j.biomaterials.2014.10.064
113. Sikiru S, Ayodele OA, Sanusi YK et al (2022) A comprehensive review on nanotechnology application in wastewater treatment a case study of metal-based using green synthesis. J Environ Chem Eng 10
114. Singh RK, Kurian AG, Patel KD, Mandakhbayar N, Lee NH, Knowles JC, Kim HW et al (2020) Label-free fluorescent mesoporous bioglass for drug delivery, optical triple-mode imaging, and photothermal/photodynamic synergistic cancer therapy. ACS Appl Bio Mater 3(4):2218–2229. https://doi.org/10.1021/acsabm.0c00050
115. Slocik JM, Stone MO, Naik RR (2005) Synthesis of gold nanoparticles using multifunctional peptides. Small 1(11):1048–1052. https://doi.org/10.1002/smll.200500172
116. Sreekanth TVM, Nagajyothi PC, Muthuraman P, Enkhtaivan G, Vattikuti SVP, Tettey CO, Yoo K et al (2018) Ultra-sonication-assisted silver nanoparticles using Panax ginseng root extract and their anti-cancer and antiviral activities. J Photochem Photobiol B 188:6–11. https://doi.org/10.1016/j.jphotobiol.2018.08.013
117. Srinivasan S, Bhardwaj V, Nagasetti A, Fernandez-Fernandez A, McGoron AJ (2016) Multifunctional surface-enhanced raman spectroscopy-detectable silver nanoparticles for combined photodynamic therapy and pH-triggered chemotherapy. J Biomed Nanotechnol 12(12):2202–2219. https://doi.org/10.1166/jbn.2016.2312

118. Sztandera K, Gorzkiewicz M, Klajnert-Maculewicz B (2018) Gold nanoparticles in cancer treatment. Mol Pharm 16(1):1–23. https://doi.org/10.1021/acs.molpharmaceut.8b00810
119. Tartaj P, del Puerto Morales M, Veintemillas-Verdaguer S, Gonzalez-Carreno T, Serna CJ (2003) The preparation of magnetic nanoparticles for applications in biomedicine. J Phys D Appl Phys 36(13):R182. https://doi.org/10.1088/0022-3727/36/13/202
120. Wagner V, Dullaart A, Bock AK, Zweck A (2006) The emerging nanomedicine landscape. Nat Biotechnol 24(10):1211–1217. https://doi.org/10.1038/nbt1006-1211
121. Waszczuk P, Barnard TM, Rice C, Masel RI, Wieckowski A (2002) A nanoparticle catalyst with superior activity for electrooxidation of formic acid. Electrochem Commun 4(7):599–603. https://doi.org/10.1016/S1388-2481(02)00386-7
122. Wiley BJ, Im SH, Li ZY, McLellan J, Siekkinen A, Xia Y (2006) Maneuvering the surface plasmon resonance of silver nanostructures through shape-controlled synthesis. J Phys Chem B 110(32):15666–15675. https://doi.org/10.1021/jp0608628
123. Wu W, Jiang CZ, Roy VA (2016) Designed synthesis and surface engineering strategies of magnetic iron oxide nanoparticles for biomedical applications. Nanoscale 8(47):19421–19474. https://doi.org/10.1039/C6NR07542H
124. Xu L, Yi-Yi W, Huang J, Chun-Yuan C, Zhen-Xing W, Xie H (2020) Silver nanoparticles: synthesis, medical applications and biosafety. Theranostics 10(20):8996. https://doi.org/10.7150/thno.45413
125. Yin D, Li X, Ma Y, Liu Z (2017) Targeted cancer imaging and photothermal therapy via monosaccharide-imprinted gold nanorods. Chem Commun 53(50):6716–6719. https://doi.org/10.1039/C7CC02247F
126. Zhang Q, Han L, Jing H, Blom DA, Lin Y, Xin HL, Wang H (2016a) Facet control of gold nanorods. ACS Nano 10(2):2960–2974. https://doi.org/10.1021/acsnano.6b00258
127. Zhang XF Liu ZG, Shen W, Gurunathan S (2016b) Silver nanoparticles: synthesis, characterization, properties, applications, and therapeutic approaches. Int J Mol Sci 17(9):1534. https://doi.org/10.3390/ijms17091534
128. Zhang M, Liu E, Cui Y, Huang Y (2017) Nanotechnology-based combination therapy for overcoming multidrug-resistant cancer. Cancer Biol Med 14(3):212. 10.20892%2Fj.issn.2095-3941.2017.0054
129. Zodrow K, Brunet L, Mahendra S, Li D, Zhang A, Li Q, Alvarez PJ (2009) Polysulfone ultrafiltration membranes impregnated with silver nanoparticles show improved biofouling resistance and virus removal. Water Res 43(3):715–723. https://doi.org/10.1016/j.watres.2008.11.014

Theranostic Applications of Functional Nanomaterials Using Microscopic and Spectroscopic Techniques

Sahil Tahiliani, Nishtha Lukhmana, and Shyam Aravamudhan

Abstract Theranostics encompasses the diagnostic and therapeutic capabilities concurrently offered by nanomaterials through imaging and spectral techniques. Several nanomaterials have been functionalized for specialized theranostic applications, including metal and metal-oxide nanoparticles, carbon-based nanomaterials, polymeric nanomaterials, lipid-based nanoparticles, quantum dots, and biomimetic nanomaterials. Some of the functionalization of these nanomaterials are steered to develop superior properties that are desirable for theranostic applications, including biocompatibility, nontoxicity, hydrophilicity, lipophilicity, and colloidal stability. This functionalization assists the nanomaterials in transferring and diffusing across various biological barriers, such as blood–brain barrier, extracellular matrix, mucous, and cell walls, in terms of acting as effective imaging, spectral and theranostic agents. Some of the innovative imaging techniques like in vivo and in vitro fluorescence microscopy, electron microscopy, 2D/3D-confocal microscopy, 3D-two-photon laser scanning microscopy, and spectral techniques like inductively coupled mass spectrometry (ICP-MS), Raman spectroscopy, Fourier transform infrared (FTIR) spectroscopy, impedance spectroscopy and recent hybrid techniques like hyperspectral imaging (HSI) and UV-Vis-NIR spectroscopy have been successfully developed as promising and feasible theranostic applications. By applying these cutting-edge functionalized nanomaterials, early diagnosis and prompt delivery of medicine can be made possible, allowing hospitals to include hybrid microscopic and spectral imaging modalities into routine patient care.

S. Tahiliani (✉) · S. Aravamudhan
Joint School of Nanoscience and Nanoengineering, NC A&T State University, Greensboro, NC, USA
e-mail: stahiliani22@gmail.com

S. Aravamudhan
e-mail: saravamu@ncat.edu

N. Lukhmana
Department of Food Science and Technology, University of Georgia, Athens, GA, USA
e-mail: nlukhmana@gmail.com

A. Madhusudhan et al. (eds.), *Functional Smart Nanomaterials and Their Theranostics Approaches*, Smart Nanomaterials Technology,
https://doi.org/10.1007/978-981-99-6597-7_4

Keywords Functionalized nanomaterials · Therapeutic · Diagnostic · Spectroscopy · Microscopy · Theranostics

1 Introduction

Over the past few decades, there has been a growing interest in the medical applications of nanotechnology. Liposomes, polymeric micelles, dendrimers, and inorganic nanocarriers are some of the most common types of functional nanocarriers used in biomedicine [1]. The inclusion of diagnostic and therapeutic compounds onto biocompatible nanoscale carriers has the potential to improve biodistribution while simultaneously lowering toxicity levels. There are cases in which the carrier itself possesses imaging and therapeutic qualities [2].

The term "theranostic," which was first used by Funkhouser in 2002, refers to a material that combines therapeutic and diagnostic imaging modalities [3]. The word "theranostics" is a combination of "therapeutics" and "diagnostics," and it refers to the practice of employing bioactive materials in the diagnostic process as well as in the administration of treatment for the management of diseases. For example, theranostics is an indispensable component of the treatment of human epidermal growth factor receptor 2 (HER 2) in breast cancer treatment with anti-HER 2 receptor antibodies and radioiodine therapy for differentiated thyroid tumors [4]. For the treatment of metastatic bone lesions, therapeutic radiopharmaceuticals such as 186Re and 188Re—hydroxyethylidene diphosphonic acid (HEDP) and 153Sm-ethylenediamine tetramethylene phosphonic acid (EDTMP) are utilized [5]. The majority of these theranostic medicines have been evaluated in the context of evidence-based medications and compared to the effects of other conventional treatments to determine their efficacy. Theranostics is fundamentally involved in identifying, diagnosing, and treating many diseases on the molecular level [6].

In contrast to a conventional small molecule, theranostic functional nanomaterials have the potential to improve the delivery of pharmacological agents and guide the nanoparticles after they have been administered. This would allow for the greatest possible therapeutic efficacy while minimizing the risk of adverse effects in unintended areas [2]. These theranostic systems provide the capability of monitoring passive and active targeting, triggered release, as well as other therapeutic activities, by merging pharmaceuticals and imaging contrast agents within a single particular customized platform [7].

In the context of this discussion, nanotechnology is a rapidly developing area of research for applications in the medical field [8–10]. As a molecular probe, nanoparticles are essential components of the technology that is being developed to diagnose and treat diseases. The nanoparticles have a size of ten to the ninth of a meter and are remarkably minuscule. In comparison to the features of bulk materials, nanoparticles possess exceptional chemical and physical characteristics [11]. Top-down and bottom-up methods are used to broadly categorize nanoparticle production. Mechanical milling, physical vapor deposition (PVD), lithography, and pyrolysis by thermal evaporation are all examples of top-down techniques that entail physical engagement [12]. Chemical and biological techniques are examples of bottom-up methods. The bottom-up chemical techniques include sol-gel, chemical vapor deposition (CVD), chemical co-precipitation, microemulsions, the hydrothermal method, the sonochemical technique, and the microwave technique [13, 14]. In addition, further approaches to the synthesis of nanoparticles include the utilization of plant extracts, enzymes, microbes, and actinomycetes. Nanoparticles that have been synthetically produced can be used to characterize the properties of nanomaterials. Synthesized nanoparticles have been used in a wide variety of biological applications, and there are many different kinds. Quantum dots, magnetic nanoparticles, optically active nanoparticles, carbon nanoparticles, emulsions, micelles, liposomes, microcapsules, microspheres, and thin films are only a few of the materials that fall under this category [15]. Each nanoparticle possesses its one-of-a-kind property and contributes to the sensing, detecting, and diagnosing of disease in any organism. When exposed to nanoparticles in a biofluid, the particles tend to aggregate, which results in the fluid losing its properties. In addition, applications in the biomedical field require nanoparticles to possess multiple functions simultaneously. It is anticipated that the scientific field would produce diverse findings from the same nanoparticles [16]. The dimensional structures of nanomaterials (Fig. 1), which are used for electron confinement, play a significant role in the classification of these materials. Nanomaterials in zero dimension (0D), one dimension (1D), two dimensions (2D), and three dimensions (3D) are all included in this category [17]. Nanotubes (such as carbon nanotubes, CNT), nanorods (such as gold nanorods (AuNRs), silver nanorods), and other similar structures are all examples of 1D nanomaterials [18]. The two-dimensional nanomaterials can be shaped into nanosheets (like graphene sheets) and nanoplates due to their malleability (such as gold nanoplates). It can be drawn into very thin nanowires because it is ductile. Examples are gold nanowires and silver nanowires [19].

Nanocages and nanocubes are examples of 3D nanomaterials. Nanomaterials with zero dimensions can also be created, and some examples of these are carbon nanoparticles and quantum dots (c-dot) [20]. It is possible to further classify it into two different categories based on its morphological nature. They represent nature in both its pure and amorphous states [21]. In various medicinal applications, these nanomaterials are often utilized. Every single nanomaterial that has been described possesses a one-of-a-kind property, both physically and chemically. In the field of biomedical sciences, a single nanomaterial would make it possible to develop multifunctional applications [22].

This chapter aims to elaborate on different microscopic and spectroscopic techniques involved in applying nanomaterials as a theranostics agent.

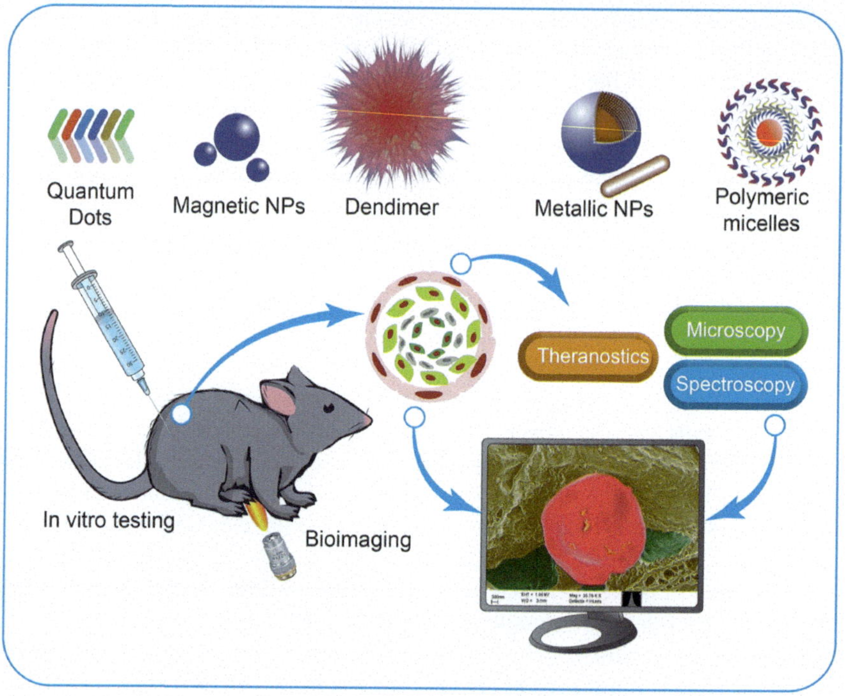

Fig. 1 Various types of nanomaterials used in theranostic applications in a model organism. (Figures and graphic illustrations created using Adobe Illustrator). The illustrations are from Adobe Illustrator, and the red blood cell image is an SEM image imaged by authors that were false-colored using Adobe Photoshop

2 Characteristics of Theranostic Nanomaterials

The idea of theranostic nanoparticles has been covered in several previous papers. Theranostic nanomaterials are divided into three components, according to Ferrari's classification which takes into account each component's function as well as its physical location: (a) biomedical payload: includes imaging agents such as organic dyes, MRI contrast agents, and CT contrast agents and therapeutic agents such as anticancer drugs, DNA, proteins, hyperthermia inducing nanoparticles; (b) carrier: includes providing physical protection for the biological payload during delivery to the specific target site under physiological conditions. In addition to organic-based carriers, such as dendrimers, polymers, and lipids, inorganic-based carriers have also been developed; (c) a surface modifier is a component that is attached to the surface of the carrier to provide additional properties for theranostic nanomaterials. These additional properties include a long duration time, the ability to barrier penetration, and a target-specific binding [23].

3 Application of Nanoparticles in Molecular Imaging and Theranostics

The past few decades have revealed that molecular imaging has a huge potential to encourage earlier and more accurate diagnosis of cancers. This is because molecular imaging is able to uncover molecular phenotypes that directly reflect the biological processes occurring in cancer [24]. Optical imaging is still demanding even though other molecular imaging modalities, such as positron emission tomography (PET), single photon emission computed tomography (SPECT), and magnetic resonance spectroscopy (MRS), have been utilized in the clinic for cancer detection. This is because optical imaging can provide images in real time with a high spatial resolution, and it does not require the use of ionizing radiation [25]. The fact that the molecular profiles of most malignancies differ substantially from patient to patient as well as from location to location, and time to time within a single tumor mass presents a difficulty for molecular imaging of tumors [26]. Because of this, many disease-related molecules, also known as biomarkers, should preferably be examined to improve the sensitivity and specificity of cancer detection, which in turn necessitates the use of multiplexed molecular imaging techniques [27].

Molecular imaging using exogenous contrast agents enables one to selectively image protein biomarkers of interest with enhanced signal, image contrast, spatial resolution, and speed. This is in contrast to label-free imaging methods, which are unable to provide information about the expression of large macromolecular targets [28]. For multiplexed molecular imaging, a wide variety of exogenous contrast agents, including traditional fluorescent dyes, quantum dots (QDs), and surface-enhanced Raman-scattering nanoparticles (SERS NPs), can be utilized. Fluorescent dyes that are used traditionally have been developed and marketed to a point where some of them have even been approved for use in clinical settings [29]. On the other hand, they suffer from several drawbacks when it comes to multiplexed imaging. As an illustration, its comparatively wide emission bandwidth of 50 nm (FWHM) [30] restricts the level of multiplexing that may be accomplished, particularly in the near-infrared range (650–800 nm), where tissue autofluorescence and water absorbance are both quite low [31]. To image numerous dyes, it is frequently necessary to use a variety of excitation wavelengths. Imaging probes, like QDs and SERS NPs, on the other hand, have narrower emission line widths, and they also have the added benefit of being able to have different "flavors" of probes excited by a single wavelength of light. This often allows for the imaging of five or more molecular targets using a single-excitation spectral imaging system [32]. An overview of microscopic applications is shown below (Fig. 2).

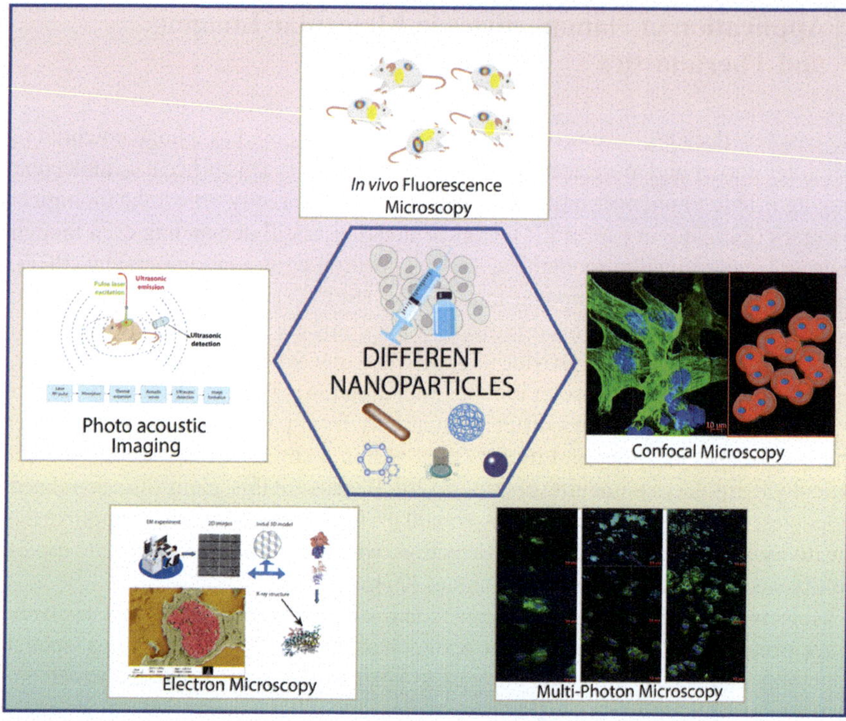

Fig. 2 Different types of applications of nanotheranostics using microscopy. (Figures and graphic illustrations created using Adobe Illustrator). The illustrations are from Adobe Illustrator, and the cell images were imaged by authors by the confocal microscope that was false-colored using Adobe Photoshop

3.1 Fluorescence Microscopy

Near-infrared fluorescence (NIRF) imaging using fluorescence imaging technologies can provide the finest spatial resolution for microscopic disease diagnostics. NIRF is preferable to visible light because it can reach deeper into tissues and produces less background autofluorescence [33]. However, practical utility is still hindered by factors such as limited penetration depth and autofluorescence and scattering qualities in different tissues. Additionally, low sensitivity for abnormality detection may result from limited fluorescence in the target lesion in addition to probable blink and photobleaching effects [34].

Both confocal fluorescence microscopy (CFM) and TPF are optical techniques that permit non-invasive imaging of untreated biological tissues in vivo at cellular resolution. Using a pinhole in front of the detector, CFM allows for high-resolution 3D viewing by eliminating signals outside the focus plane. Changes in pinhole aperture size allow for selective focus plane imaging of the specimen. [35, 36]. When compared to wide-field fluorescence spectroscopy, the lateral and axial resolution of an autofluorescence image acquired using CFM is significantly better. CFM is a micro-endoscopy technique that can be used in the GI tract, either ex vivo on untreated or treated tissue samples or in the body [37, 38].

Dextran-coated superparamagnetic nanoparticles (MN) were synthesized by Medarova et al., with siRNA targeting green fluorescent protein (siGFP; 5 per particle), a NIR emitting dye, Cy5.5 (NIRF; 3 per particle), and a membrane translocation peptide, myristoylated polyarginine peptide (MPAP; 4 per particle) covalently attached to promote nanoparticle trafficking into the cytoplasm of targeted cells [39]. Using fluorescence microscopy, the uptake of MN-NIRF-siGFP by 9 L-glioma cells was followed which stably expressed green fluorescent protein (GFP). For further analysis of the specificity of the siRNA target toward GFP, target mRNA levels were analyzed using quantitative RT-PCR and compared the results with those obtained from 9 L-glioma cells expressing red fluorescent protein (RFP). In vivo investigations using tumor-bearing mice implanted on both sides with glioma cells expressing 9 light-green fluorescent protein (L-GFP) and 9 light-red fluorescent protein (L-RFP) demonstrated considerable mRNA reduction in 9 L-GFP tumors compared to saline alone and mismatch controls [40]. As a result of the encouraging results of the GFP experiments, MNNIRF- MPAP-siSurvivin was designed to specifically target the inhibitor of apoptosis protein (IAP) survivin. A substantial drop in survivin mRNA levels was found after 2 weeks of MNNIRF-MPAP-siSurvivin injections into nude mice harboring subcutaneous human colorectal carcinoma tumors (LS174T) as measured by RT-PCR. In addition, in vivo trials with aspartate aminotransferase (AST) and alanine aminotransferase (ALT) showed a lack of cytotoxicity and an absence of an immunostimulatory response [41].

Using FRET, Schneider et al. investigated the intracellular transport of DNA/lipid complexes (lipoplexes) in rat smooth muscle cells (A10). Flow cytometry, confocal microscopy, and fluorometric techniques were used to examine Cy3- and Cy5-labeled DNA and lipids, respectively, to determine the efficiency of energy transfer from the donor to an acceptor (E_d) and acceptor to the donor (E_a). Flow cytometry (excitation = 488 nm) and microscopy (excitation = 523 nm) were utilized with different optical settings (donor excitation wavelengths) to examine the kinetics of lipoplex dissociation. According to a study that compared the effectiveness of donor- and acceptor-based energy transfer at three, five-, and nine-hours post-transfection, the majority of DNA was freed from the lipoplexes at the end of 24 h. This highlighted the usefulness of FRET to permit information capture with high spatial resolution inside the cell, and these findings are essential for understanding intracellular trafficking and the mechanism of vehicle unpackaging [42].

There are several drawbacks to fluorescence imaging, however, these can be addressed with the help of nanoparticles. More signals, for instance, can be generated by loading more fluorescent dye molecules onto nanoparticles [43]. Furthermore, the nanoparticles can be engineered (or reorganized) to avoid the potential for NIR fluorescence quenching. [44]. The concentration of fluorescent dye in a lesion can be increased by increasing the concentration of nanoparticles in the lesion, and both active and passive techniques can be used to achieve this goal. Further, the comparatively prolonged half-life allows for increased absorption in certain lesions. To mitigate the blink and photobleaching effects, nanoparticles can be engineered to boost the energy of incident light. Due to their advantageous characteristics, fluorescent nanoparticle platforms have been the subject of much research and development [45].

When monitoring molecular events in biological systems, organic fluorescent dyes like cyanine 5.5 (Cy5.5) and fluorescein isothiocyanate (FITC) are frequently utilized. Because the visible or ultraviolet (UV) light that is used to excite organic dyes cannot penetrate deeply into the tissue, the use of organic dyes in bioimaging is limited primarily to the examination of cells [46]. In addition, individual organic colors can be bleached off by the light and can be quite poisonous. As a result, methods have been developed to provide the required photostability and reduce loss during delivery, both of which would lead to a reduction in imaging ability as well as an increase in toxicity. These methods involve encasing organic dyes in a protective matrix made of SiO_2 or a polymer. In this context, the two-photon dye, which makes use of NIR light offers significant potential because NIR light is minimally absorbed by the tissues of the human body [47].

Chitosan-based nanoparticles (CNPs) were manufactured by Park and Kwon et al. to simultaneously diagnose and treat cancer [48]. For imaging, CNPs were tagged with the NIRF dye Cy5.5, and they were loaded with the anticancer medication paclitaxel (PTX). In addition, a water-soluble hyperbranched polyhydroxy polymer was synthesized with cytochrome c and amphiphilic fluorescent dye molecules (indocyanine green [ICG]), and it was then further conjugated with folic acid. This was done in a study that was conducted by Perez et al. [49].

3.2 Mass Spectrometry

In proteomics research, liquid chromatography–tandem mass spectrometry, or LC–MS/MS can be used to profile the protein corona surrounding magnetic nanoparticles. This is one of the applications of LC–MS/MS [50]. To conduct high-throughput screening, this method cannot be used because it calls for laborious and time-consuming sample preparation. As a result of this limitation, the method is inappropriate for application. Inductively coupled plasma-mass spectrometry (ICP-MS), which may function in a variety of high-resolution modes, is one of the alternative methods that can solve this problem. This method has a great deal of potential. This is a result of the method's shown capability for very sensitive and interference-free monitoring of iron (in the form of a variety of nano-sized structures) in a range of diverse biological settings [51]. Nonetheless, the creation of the protein corona and its potential to inhibit nanoparticle agglomeration is not addressed in the published ICP-MS-based studies. Some evidence suggests that disaggregation can occur because of steric stabilization by proteins, such as albumin. Circular dichroism spectroscopy and dynamic light scattering are used, however, they aren't well suited for usage with complicated protein combinations like a serum [52].

SP-ICP-MS/MS has only seldom been utilized to describe SPIONs, and the impact of biological media on their stability has not been studied. This research aims to overcome this knowledge gap by creating a reliable assay for investigating how SPIONs interact with serum proteins. This proof-of-principle investigation confirms the promise of SP-ICP-MS/MS as a practical tool for SPION-protein analysis. The proposed method makes it easy and quick to track how SPIONs' structures evolve in response to interactions with proteins in the blood. We think it could find a use in evaluating various nanoparticle-protein combinations, which would aid in the culling of potential nanomaterials for more rigorous preclinical testing [53].

In recent years, metallic and hybrid ultrasmall nanoparticles (USNPs) with diameters of less than 10 nm have received enormous interest. Because of how little they are, the kidneys can flush them out of the body very rapidly, which lowers the risk of toxicity. Not just for use in diagnostics but also in therapy, with the primary focus being on the treatment of cancer. Numerous uses for these USNPs have been proposed [54].

Due to its quickness and convenience, this technique has become the standard for size characterization in everyday applications. USNPs in a complex medium cannot be accurately sized because of the method's lack of specificity and its significant bias toward larger objects. Many other techniques relying on specific detection have been developed, with inductively coupled plasma-mass spectrometry (ICP-MS) being the most common. Single-particle ICP-MS (sp-ICP-MS) is a relatively new technique for determining the size of nanoparticles made of metal. To determine the optimal pH for sizing the AGuIX USNP, this procedure has been utilized [55]. Selecting a capillary with an internal diameter of 75 m, as suggested by Chamieh et al., enables the optimal size range for a USNP and the detection of probable deterioration or aggregation processes (hydrodynamic radii of 0.2–50 nm) [56].

The use of mass spectrometry imaging (MSI) in the study of biomolecules extracted from tissue samples has gained a lot of attention recently. Several other MSI techniques, including matrix-assisted laser desorption/ionization (MALDI), matrix-free laser desorption/ionization, laser ablation electrospray ionization and desorption electrospray ionization, and secondary ion bombardment, can be utilized [57]. Biomolecules in tissue samples can be imaged or mapped using MS. Many-sequence alignment (MSI) has the advantage of being able to identify multiple biomolecules at once; nevertheless, it necessitates a large database for molecular identification, and its sensitivity and resolution are poor for big proteins (molecular weight >80 kDa). To facilitate the desorption and ionization of biomolecules, MALDI-MSI experiments are typically performed in an organic matrix [57]. However, the identification of molecules with low molecular weights is sometimes hindered by excessive background signals from the organic substrate. Analysis of nanomaterial dispersion in cell and tissue samples may be performed using laser desorption/ionization mass spectrometry (LDI-MS) methods [58]. To improve resolution and decrease background noise in mass spectrometry (MS), nanomaterials and nanostructured substrates produced from a metal, metal oxide, silicon, and carbon materials have been frequently used as matrices in recent years [59].

For the investigation of small compounds such as lipid components, LDI-MS-based tissue imaging techniques have been widely used, employing a suitable organic matrix or nanoparticle substrates [60]. However, there are only a few reports of LDI-MS investigation of membrane-bound proteins in cells. These used AS1411-Au NPs/Au@PC NP-LDI-MS platform to analyze nucleolin expression in a conventional tissue microarray by tracking the [Au] + signal [57]. To target proteins, a self-assembled and label-free AS1411-Au NPs/Au@PC NP may be prepared with less effort than a previous aptamer-Au NPs modified graphene oxide nanocomposite. In addition, the LDI-MS results for the AS1411-Au NPs/Au@PC NP show a strong signal for [Au] + cluster ions. AS1411-Au NPs/Au@PC NPLDI-MS was also shown to be a sensitive platform for the detection of tumor cells and the imaging of tumor tissue [61].

3.3 Inductively Coupled Plasma (ICP) Mass Spectrometry

Mass spectrometry immunohistochemistry (MSIHC) using ICP-MS-based technologies like MS-based cytometry and laser ablation ICP-MS (LA-ICP-MS) imaging is becoming increasingly popular in molecular pathology. In addition to its multiplexing capability (i.e., the capacity to identify several targets within a single sample), this technology's large dynamic range and the lack of background effects due to autofluorescence show promise in the elucidation of molecular profiles. Non-ionizing lanthanide isotope functionalized receptor-targeting vectors (e.g., antibodies) are necessary for this [51].

Theranostics has helped advance the study of radioimmunotherapy and peptide-based systemic radionuclide therapy [62]. Peptides have many advantages over antibodies in the clinic, including reduced production cost, easier GMP manufacturing methods, and better pharmacokinetics [63]. Using two separate (radio)isotopes, such as In or Ga for diagnostic applications and Lu or Y for therapeutic applications, peptides can be connected between these settings without having to switch tracers. It may seem that certain radioisotopes are nearly made for this dual purpose. For diagnostic gamma-rays in scintigraphy and therapeutic alpha particles, ^{165}Ho (Holmium) is a good example [51].

CXCR4 is intimately associated with the aggressiveness and spreading character of cancer, making it an attractive oncologic target that has been pursued effectively in a peptide-based theranostic setup [64]. CXCR4 can be targeted with a variety of vectors, each optimized for a certain type of label. It is important to note that (currently) fluorescent or bimodal (or rather hybrid) labels appear to necessitate a (bigger) targeting vector, such as Ac-TZ14011 [65].

While the efficacy of theranostic concepts can be demonstrated in vivo using tumor models, microscopical analysis of the tracer uptake in cells and tissue can be a barrier to their early phase development. Successful integration of non-invasive diagnostic imaging (based on the radioisotope) with the examination of diseased tissue sections was made possible by the use of hybrid imaging tracers in a different scenario, namely image-guided surgery (based on fluorescence) [66]. Hybrid tracers, because of their fluorescent component, may theoretically bridge the gap between MS and radioisotope-based discoveries. At the same time, a hybrid label might be used to ascertain whether "big" multi-chelate labels are needed for MS-based diagnostics [67].

3.4 Electron Microscopy

Scanning electron microscopy (SEM) creates seemingly three-dimensional images by scanning a material with a focused beam of electrons. Surface imaging is all that can be achieved with high resolution (3–20 nm) [68, 69]. To characterize the spatial relationships between nanoparticles and the cell surface, particularly concerning the internalization process and cell shape modification, SEM has primarily been used in nanomedical research. However, field emission SEM (using a high-energy electron beam) has made it possible to see nanoparticles in the endosomal compartment [70].

Transmission electron microscopy (TEM) is a technique that allows for the precise observation of the inside of a sample by obtaining images from a beam of electrons that are transmitted through a thin specimen. This method of microscopy has seen extensive use in the field of nanomedical research, and due to the one-of-a-kind information that its high resolution makes possible, it can shed light on the intricate relationships that exist between nanoparticulates and the components of cells and tissues [71]. If the sample was processed in such a way as to preserve the nanoparticles' spatial relationship with the cell surface, transmission electron microscopy (TEM) can provide unequivocal information on the mechanism(s) allowing nanoparticles to enter the cells by crossing the plasma membrane at the cellular level. This information can be used to determine how nanoparticles can enter the cells [72]. Endocytosis is the most prevalent route of cellular uptake by transmission electron microscopy (TEM, nanocarriers have been shown to make contact with the plasma membrane, and they enter the cell encased in endosomes [73]. Endocytosis is a process that can take place for single nanoparticles as well as for tiny nanoparticle groups, however, phagocytosis and macropinocytosis are the processes that allow single massive nanoconstructs and huge clusters of nanoparticles to enter the cells, respectively [74]. All of these different forms of uptake methods might be made easier by a variety of receptors [75]. According to the findings of ultrastructural investigations, a few nanoconstructs can cross the biological membranes even in the absence of endosomal structures. This process has been mostly observed for lipid-based or lipid-coated nanocarriers, which are expected to fuse with the plasma membrane, even though conclusive ultrastructural evidence of this event has not been shown so far [76].

3.5 Confocal Microscopy

Using laser scanning on an optical platform, confocal laser scanning microscopy, also known as CLSM, is a tried-and-true method for producing images from samples of cells or tissue. When compared to more traditional methods of optical or fluorescent microscopy, the pictures acquired by this method have a higher resolution and greater depth selectivity [77]. The capability of CLSM to perform optical sectioning, in which the images are reconstructed based on point-by-point scanning, is the most important function of the instrument. The three-dimensional structure of the superficial interior can be created once the laser can pierce the skin and "resolve" the interior details of the nonopaque specimens. The surface of the samples can be photographed even when the specimens themselves are opaque. Because of the one-kind sectioning approach used by the laser, the overall quality of the images has been much improved, and the information from the field that is out of focus is not superimposed on the portion of the image that is in focus [78].

The concept of theranostics was developed because of research into the genetics of cancer, which led to the production of therapy routes through the application of bioinformatics tools. Because recent developments in molecular science have made it possible to "identify" an individual tumor by genomic and proteomic profiling, individualized theranostic medicines that target-specific compartments of the tumor's microenvironment can now be produced. Even though theranostic imaging opens up whole new doors for individualized cancer treatment by bringing together the fields of chemistry, molecular biology, and imaging, quantitative image analysis continues to be one of the field's greatest obstacles [57]. To visualize and target every element of the tumor microenvironment in conjunction with molecular medicines, more advanced image analysis approaches are required. The power of CLE technology to image at the cell level intraoperatively on the fly makes it possible for a possibly more personalized, precise, or customized approach to the surgical operation to remove an invasive brain tumor. This is because CLE technology can picture at cell resolution. Imaging and perhaps targeting cells is now possible because of fluorescent stains and markers; we are getting very close to being able to do "cell surgery" as particular stains and luminous markers continue to be created [79].

3.6 Two-Photon Laser Scanning Microscopy

Two-photon excitation microscopy, often known as TPEF or 2PEF, is a fluorescence imaging technique that can image living tissue with a spatial resolution of 0.64 m laterally and 3.35 m axially. This technique may be used to image living tissue up to around one millimeter in thickness. Two-photon excitation, in contrast to conventional fluorescence microscopy, in which the excitation wavelength is shorter than the emission wavelength, needs simultaneous excitation by two photons with a larger wavelength than the light that is emitted. Near-infrared (NIR) excitation light is often utilized in two-photon excitation microscopy since this type of light may also stimulate fluorescent dyes [80]. On the other hand, two photons of NIR light are taken in for every excitation. The use of infrared light helps to reduce the amount of scattering that occurs in the tissue. Because of the multiphoton absorption, there is a significant reduction in the signal from the background. These factors result in an improved penetration depth for this method. Because of its deeper tissue penetration, effective light detection, and reduced photobleaching, two-photon excitation can be advantageous compared to confocal microscopy [81].

In X-ray imaging applications of cancer, gold nanoparticles have typically been used as contrasting agents. However, the most recent investigations have revealed two-photon photoluminescence (2PPL) from gold nanoparticles for nonlinear bioimaging applications. Conventional organic dyes have poor photostability along with relatively small two-photon absorption cross-Sects. (1–100 GM) (GM = 10–50 cm^4 s $photon^{-1}$. The molecular two-photon absorption cross-section is usually quoted in the units of Goeppert-Mayer (GM)), which makes them unsuitable for use in nonlinear bioimaging applications. On the other hand, semiconductor quantum dots (QDs) are hampered by their photo-blinking behavior and high cytotoxicity, which makes them unsuitable for use [82]. The strong 2PPL of a range of gold nanoparticles, such as gold nanospheres, gold nanoshells, gold nanocages, and gold nanorods, has been brought to attention to overcome these limits. Because of the enormous two-photon action cross-section of 4.2104 GM that GNRs possess, they have been at the forefront of this study. The "lightning rod effect" and localized surface plasmon resonance modes can explain the enormous two-photon action cross-section seen by 11 GNR. As a direct consequence of this, GNRs have found use in two-photon imaging. In addition, two-photon excitation microscopy has been utilized in conjunction with GNRs to perform the photothermal treatment (PTT) on cancer patients [83].

Since the beginning of TPM, there has been an awareness of the possibility that it can be used to investigate tissue physiology. The physiology of many different types of tissue, such as the corneal structure of rabbit eyes, the light-induced calcium signals in salamander retina, the human and mouse dermal and subcutaneous structures, the toxin effect on the human intestinal mucosa, and the metabolic processes of pancreatic islets, have all been successfully studied using two-photon tissue imaging [84]. The usage of TPM is particularly widespread in the fields of neuroscience and embryology in the modern day. TPM has been utilized in the field of neurobiology to investigate topics such as neuronal structure and function in intact brain slices, the function of dendritic spines, the role of calcium signaling in dendritic spine function, neuronal plasticity and the associated cellular morphological changes, and hemodynamics in rat neocortex [85, 86]. In the field of embryology, two-photon imaging has been used to investigate the genesis of the bilateral axis in sea urchin embryos, cell fusion events in the hypodermis of C. elegans, and the development of hamster embryos. As more advanced and affordable commercial equipment becomes available, it is anticipated that two-photon imaging will be applied to an expanding range of tissue systems [87].

Clinical diagnosis and therapy are an area that has a great deal of promise for two-photon imaging of tissue. A paradigm shift in clinical diagnosis has been seen with the introduction of optical biopsy. A conventional biopsy requires the removal of tissues, their fixation, and subsequent imaging. The histological method is invasive, and the information regarding the biochemical makeup of the tissue is not well retained while these preparation procedures are being conducted. An optical biopsy based on TPE has been proposed, where three-dimensional images of the patient's tissues will be acquired without causing them any pain or discomfort. In the next steps, a pathological examination of the image stack will be performed. TPM has been utilized well to photograph the underlying structure of the skin of human volunteers down to a depth of 150 μm [88]. In most cases, it is possible to resolve four separate structural layers in both the epidermis and the dermis. The stratum corneum is the outermost layer of the epidermis, and it is composed of cornified cells that form a protective layer. Keratinocytes from the epidermis make up the second layer of the skin. A layer of germinative basal cells can be found at the intersection of the epidermis and the dermis. It is possible to observe collagen and elastin fiber structures in the dermal layer of the skin. It is also possible to identify pathological conditions such as abnormal alterations in cellular morphology or cellular hyperfoliation. However, as of right now, it is still unknown whether the quality of images acquired by two-photon optical biopsy is sufficient to produce pathological analysis results with accuracy comparable to that of standard histology. The creation of a two-photon endoscope may provide a mechanism to biopsy various organs in the body if it is possible to demonstrate that this technology is both feasible and useful. Confocal imaging is undergoing similar efforts at the moment as well [89].

4 Application of Nanoparticles in Spectroscopy and Theranostics

Functional nanomaterials are most compatible and adapted for theranostic applications, because of their unique characteristics. In that, they offer flexible surface chemistry, surface area that is large enough and ability to interact with target systems at the level of molecularity. Their interaction with biological systems has been investigated by utilizing spectroscopic methods, which are often beneficial for tracking the transportation of functional nanomaterials, and their uptake, in vivo, and offer an advantage where they can be employed to target-specific tissues or cells. Additionally, functional nanomaterials are also used to effectively enhance the specificity and sensitivity of imaging methods, thereby leading to early and precise detection of diagnosis for various illnesses. Therefore, to conclude, functional nanomaterials, combined with the utilization of spectroscopic methods, offer a high potential for theranostic treatments, thereby providing a unique means around detection and monitoring diseases with precision and in a timely manner.

4.1 Fluorescence Spectroscopy

The phenomenon of fluorescence plays a major role in the process of diagnosing cancer. It is often shown as a function of intensity vs wavelength and is derived from the examination of tissue emission spectra. Fluorescence spectroscopy can be carried out utilizing steady-state or time-resolved approaches, employing external fluorescent protein tagging, single or multiple wavelength excitations, and a variety of other methodologies [90]. The steady-state fluorescence technique is the one that is utilized for emission spectral analysis most of the time. Because autofluorescence can be produced without the addition of any external fluorescence markers, in vivo detection can take place without the need for any pretreatment of tissue samples. The clinical task, such as establishing an initial diagnosis for an unknown lesion or monitoring its development, evaluating tumor treatment, or guiding its surgical excision, determines the measurement mode to be used and the parameters of the fluorescence signal that is to be evaluated [35]. The morphological structure, molecular composition, and metabolic state of the lesion are all important factors that determine how effective fluorescence spectroscopy is at detecting tumors. When doing experiments in vivo, the fluorescence signal has a significant impact on the pH of the tissue, the temperature, and the ionic balance within the cells. A fluorophore that is sensitive to pH can be used to determine the pH of the tissue. For instance, a change in pH from neutral to strongly acidic causes a shift in the serotonin emission maximum from 330 to 550 nm without causing a change in the spectrum of the serotonin's absorption. The fluorescence of pH-sensitive molecules, which has the potential to be exploited as a diagnostic for cancer, is also affected when the environment of a tumor is more acidic [91].

Over the last few decades, a great deal of effort has been put to create and develop delivery systems to produce delivery vehicles that are both extremely efficient and effective. These approaches utilize both viral and non-viral delivery vehicles in their respective protocols. Virus-based therapeutic delivery requires the inactivation of viral genes that cause disease and the incorporation of a copy of the therapeutic gene of interest. This copy can be incorporated as RNA for retroviral and lentiviral vectors or as DNA for adenoviral or adeno-associated viral vectors, depending on the type of viral vector being used [92]. Transfection with viral vectors can be achieved with a high degree of effectiveness, and the target gene can be integrated into the chromosomal DNA of the recipient cell (s). Concerns have been raised following several clinical experiments involving viral vectors because of the possibility of undesired interference with the host genome as well as the ability to generate a strong immune response. To this end, many researchers have been concentrating their efforts over the past few years on the development of new synthetic, non-viral vehicles. These vehicles have the potential to be nontoxic and less immunostimulatory than viral vehicles, and they also have the potential to be more versatile in terms of modification and the types of drug payloads they can carry. Researchers have developed and extensively researched non-viral vehicles that are based on synthetic polymers, dendrimers, liposomes, cell-penetrating peptides, and inorganic nanoparticles [93].

These kinds of vehicles also have the advantageous capacity to be adjusted in a theranostic manner straightforwardly [94].

To build highly effective gene delivery vectors, it is essential to first identify the rate-limiting steps involved in the path that the delivery vehicle takes from the site of administration to the target of interest. Only then can highly efficient gene delivery vectors be designed. Recent research in which the delivery vehicle and DNA were both labeled with organic dyes has enabled the use of more advanced imaging techniques, such as time-lapse microscopy, fluorescence resonance energy transfer (FRET), and fluorescence correlation spectroscopy (FCS) [95]. These methods have the potential to provide more accurate information regarding the trafficking and dissociation of polyplex molecules, which would be a step toward the development of theranostic systems. For instance, Lucas et al. utilized dual-color fluorescence fluctuation spectroscopy (FFS) to monitor complexes that contained Cy5-labeled poly(L-lysine) (PLL) and PEGylated poly (dimethylamine)ethyl] methacrylate-co-aminoethyl methacrylate (pEG-pDMAEMA-co-AEMA). The great spatial and temporal precision that FFS possesses enables it to distinguish more clearly between polyplexes that relate to one another and those that have become separated from one another. The authors' utilization of FFS resulted in the discovery of novel and interesting findings, such as the finding that high molecular weight pDMAEMA (1700 kDa) dissociates from ON within the cytoplasm without entering the nucleus, whereas low molecular weight PLL (30 kDa) releases ON after nuclear entry. This discovery was made possible by the fact that the authors used FFS [96].

4.2 Near-Infrared Techniques

Cancer's complex nature, widespread drug resistance, and relapse tendency pose a serious risk to human health. Extensive efforts have been put into designing multifunctional nanoplatforms for the theranostics of tumors, such as liposomes, polymers, mesoporous silica nanoparticles, magnetic nanocrystals, and gold nanoparticles because real-time visualization of these therapeutic courses is very important for clinicians to rapidly assess whether a drug is effective in a particular patient or not [97]. However, improved reticuloendothelial system (RES) uptake, nonspecific drug release, and inadequate spatial resolution of in vivo imaging continue to present significant obstacles to achieving the theranostic goal. To evaluate and enhance therapeutic efficacies during tumor treatment, real-time feedback from a limited number of systems is necessary [98].

Quantum dots (QDs) emitting in the second near-infrared window (NIR-II, 1,000–1,400 nm) could be the optimal platform for providing real-time observation of the in vivo theranostic process. This could be accomplished by strategically integrating several functions [99]. Previous research by Feng et al. has shown that Ag2S QDs, a new type of NIR-II QDs, have desirable properties for in vivo imaging, such as a large tissue penetration depth (1.2 cm), (2) high spatial resolution (25 m), and temporal resolution (50 ms), a long circulation time ($t1/2 = 4.37$ h) after coating with polyethylene glycol (PEG), and high tumor uptake (10% ID/gram tumor accumulation. These merits of Ag2S QDs make them a promising probe for early tumor detection and imaging of the dynamic process of tumor treatment in vivo. Loading PEG-Ag2S QDs with common anticancer medication, doxorubicin (DOX), served as proof of their utility as novel theranostic nanoplatforms for the in vivo simultaneous diagnosis, drug delivery, and treatment monitoring of malignancies [97].

For real-time in vivo visualization of tumor theranostics, a smart nanoplatform of DOX@PEG-Ag2S has been developed. This platform successfully integrates drug administration, tumor microenvironment-response drug release, and NIR-II fluorescence imaging. Active tumor-targeting efficiency (8.9% ID/gram) and drug loading capability (DOX, 93 wt.% w.r.t Ag2S QDs) were both high in as-prepared DOX@PEG-Ag2S in living mice. Additionally, the acidic milieu present in a tumor might cause a rapid and selective release of DOX from the Ag2S payloads into cancer cells, leading to substantial tumor inhibition. Because Ag2S QDs can be imaged using NIR-II fluorescence imaging at a high spatial and temporal resolution, DOX@PEG-Ag2S allowed for real-time monitoring of their therapeutic activity in vivo [100].

An innovative synthesis process for biocompatible methionine-capped superparamagnetic cobalt ferrite NPs ($CoFe_2O_4$@Met) has been reported by Feng Hu et al., which can decrease and attach gold species. First-of-its-kind hybrid magnetoplasmonic cobalt ferrite NPs adorned with Au0/Au1 + quantum dots (QDs) were produced in this method. HRTEM, AFM, FTIR, XPS, and chemical analyses all confirmed the creation of plasmonic gold QDs on the surface of iron oxide-based NPs [97].

When compared to visible light and UV light, NIR light poses a lower risk of damage to cellular DNA. For instance, Cho et al. just recently produced a series of two-photon dyes and uncovered certain essential signaling events that take place in mammalian cells. Comparable to two-photon dyes are UCNPs, which are made of inorganic materials rather than organic ones. UCNPs with organic two-photon dyes, absorb the energy of NIR light and then release it as light in the visible or ultraviolet spectrum. Even though UCNPs are far larger than organic dyes, they can be utilized to shed light on significant events that occur within cells. The action of a motor protein was recently seen by Lee et al. by employing UCNPs as imaging probes [49].

Park and Choi et al. revealed the use of tumor-targeting hyaluronic acid nanoparticles (HANPs) as a carrier of a hydrophobic photosensitizer known as chlorin e6 (Ce6) for simultaneous photodynamic imaging and therapy. Chemical conjugation of aminated 5′-cholanic acid, polyethylene glycol (PEG), and black hole quencher-3 (BHQ-3) to the hyaluronic acid polymers allowed for the synthesis of self-assembled HANPs. The HANPs had Ce6 injected into them using a dialysis technique. Fluorescence for optical imaging and singlet oxygen for photodynamic treatment (PDT) was produced when NIR laser irradiation dequenched the released Ce6 [101].

Matrix metalloproteinases (MMPs) on cancer cells can be useful targets for cancer imaging and treatment. Usually, optical imaging probes are coupled onto the therapeutic nanoparticle using MMPs-cleavable peptide linkers, and the optical probes are "turned off" by a FRET process. When the nanoparticle reaches the cancer cells that are to be imaged, the optical imaging probes inside of it are released, which causes the fluorescence to be "switched on." [49]. Cathepsin B and MMP imaging polymeric nanoprobes were produced by Yhee and Kim et al. by conjugating near-infrared fluorescence (NIRF) dye (Cy5.5) and dark-quencher (BHQ-3) to the substrate peptide of MMPs and cathepsin B, respectively. It was interesting to see that the fluorescence recovery of the cathepsin B probe occurred after the probe had entered the cytoplasm, which helped determine how well the drug was delivered [102].

4.3 Raman Spectroscopy

Photons can be scattered in either an elastic or inelastic manner depending on the nature of the interaction between light and matter. In a process referred to as Rayleigh scattering, the vast majority of photons are dispersed in an elastic manner, which means that there is no transfer of energy. However, a small population of photons, approximately one in every 107 photons, are scattered inelastically. This means that the photons exchange energy with the material that is scattering them via vibrational transitions within the molecule. This phenomenon is referred to as the Raman effect, or just Raman scattering for short. Raman scattered photons have the potential to either lose or gain energy (i.e., they undergo Stokes or anti-Stokes Raman scattering, respectively) [103].

Raman imaging, also known as label-free or intrinsic imaging, is a type of optical spectroscopy that does not involve the use of nanoparticles as a contrast medium. It has been investigated for clinical use to distinguish cancerous tissues from healthy tissues based on differences in their Raman spectral fingerprints. On the other hand, to do intrinsic Raman imaging, a longer acquisition period is required. This is because tissues have relatively little Raman scattering, which results in a faint Raman signal. Because of this, the intrinsic Raman spectrum imaging technique cannot be used in most biomedical contexts, which restricts its application in clinical research. Real-time in vivo imaging requires the Raman signal intensity to be greatly enhanced. This can be accomplished by utilizing exogenous contrast agents, which generate spectra that are more intense than the intrinsic Raman signal [104]. Typical components of these contrast agents include cores made of plasmonic nanoparticles, Raman reporter molecules, dielectric shells, ligands against specific targets, and other functional moieties. To improve the SERS signal of exogenous contrast agents, it is necessary to optimize several parameters. These include the type, shape, and size of the metal nanoparticle core; the quantity and optical absorbance of Raman reporter molecules attached to the nanoparticle; the composition of the shell layer; its thickness; and the choice of excitation laser wavelength. The use of SERS nanoparticles in Raman imaging has demonstrated outstanding sensitivity, high specificity, low background, and the possibility of multiplexing [105].

The development of portable fiber-optic Raman endoscopic probes was driven because of the requirement for Raman-based detection and imaging of malignancies on the internal epithelial surfaces of the body cavities. A clinically translatable

Raman endoscope has been developed for the detection of topically applied SERS nanoparticles and, more recently, for the detection of pre-malignant lesions of the gastrointestinal (GI) tract [106, 107]. Harmsen and colleagues demonstrated that SERRS contrast agents can detect pre-malignant lesions along the GI tract as small as 0.5 mm using a commercially available Raman probe that is capable of being inserted into the accessory channel of a clinical white-light endoscope to enable simultaneous dual-modal white-light/Raman imaging. The revolving mirror that was incorporated into the scanning Raman endoscope was what made it possible to achieve circumferential distribution of the laser along the intestinal epithelium. It has also been reported that a preclinical endoscope with a dual mode (Raman and fluorescence) has been developed for the detection of nanoparticles with fluorescence and surface-enhanced Raman-scattering contrast (F-SERS dots). This endoscope is intended to be used in the examination of patients. It has been reported that handheld fiber-optic probes can be used to detect SERS contrast agents in vivo for imaging of immune response as well as image-guided resection of glioblastoma multiforme (GBM) [108].

The surface-enhanced Raman-scattering (SERS) effect is a potential and complementary optical imaging approach that can significantly improve Raman imaging. Raman imaging is a promising optical imaging technique. Raman is capable of very specific and sensitive detection of SERS contrast agents, as well as the multiplexing of numerous agents in living subjects [109]. When compared to the nanomolar sensitivity that can be achieved by fluorescence imaging of quantum dots, the ultra-high sensitivity that can be achieved with Raman imaging in conjunction with MPRs can be measured in the picomolar range. In contrast to conventional optical imaging techniques, Raman imaging of MPRs does not suffer from the effects of autofluorescence or background signal. This is because the spectral signature of an MPR is greatly amplified and distinct (also known as its "fingerprint") [110]. Even though the most significant drawback of Raman imaging is that it only has a limited depth of penetration, this study was successful in visualizing tumors through the undamaged skin and skulls of living mice (to a depth of 2–5 mm). This outcome is due to a combination of factors, including the design of the nanoparticle, which features a gold core that generates surface plasmon resonance for enhanced Raman signaling, the Raman substrate that was utilized, and the number of nanoparticles that accumulated within the tumor [111]. In contrast to organic fluorochromes, Raman nanoparticles are not susceptible to the photodegradation that is characteristic of organic fluorochromes. In contrast to quantum dots, which are known to be cytotoxic, MPR nanoparticles are composed of inert gold and silica, and as a result, they may have a greater potential for application in therapeutic settings [110].

Raman imaging using SERS nanoparticles holds great promise as a cancer imaging tool. Recently, the approach has experienced a great deal of development, particularly in the area of in vivo imaging for use in preclinical studies. Since the contrast agent is a nanoparticle, this method has many advantages over traditional medical imaging methods. As a result of the enhanced extravasation and uptake of nanoparticles by malignant and pre-malignant tumors, SERS nanoparticles can detect these diseases. In addition, functional compounds can be added to nanoparticles for targeted delivery or detection with alternative imaging techniques [108].

4.4 FTIR Spectroscopy

For the characterization of biological materials, FTIR spectroscopic imaging has several benefits over many other imaging techniques since it is based on the characteristic absorbance of corresponding molecular vibrations in the sample Consequently, FTIR imaging eliminates the need for the use of dyes or labeling techniques to enable the viewing of various chemical components in the sample [112, 113]. There have been several publications that have discussed the use of FTIR imaging on medical samples, with the most recent examples being included in this special issue. The investigation of the compositional changes in genetically engineered bovine chondrocytes is another example. Other examples include the ratio of protein and lipid in the cerebellum and skin, as well as the imaging of bone [114, 115]. In their research on human colorectal cancer, Lasch et al. were able to use FTIR imaging with great success [116]. Choo et al. used FTIR spectroscopic imaging to evaluate the amount of beta-amyloid protein that was deposited in a slice of human brain tissue that had Alzheimer's disease. All of these recent applications have proved the potency and usability of this imaging approach in the biomedical area using the focal plane array (FPA) detector [117].

Another study has used an extract from the leaf of the Olax scandens plant to biosynthesize silver nanoparticles. This extract has phytochemicals thought to be responsible for the nanoparticles' synthesis and stability. FTIR spectroscopy was performed on both Olax leaf extract and b-AgNPs-500 to determine the precise function of the phytochemicals present in the extract. Confirming the significance of proteins in the creation and stabilization of b-AgNPs-500, the major stretching frequencies observed at $v = 1512.18$ and $= 1258.79\ cm^{-1}$ due to the presence of amide II and amide III in proteins of Olax scandens have nearly gone [118].

4.5 Impedance Spectroscopy

Over the past few decades, impedance spectroscopy (IS) measurements have steadily expanded into new fields of study as shown in Fig. 3. The technique was first used to characterize liquid-phase electrochemical cells, but since then it has been increasingly employed to characterize the resistive and capacitive properties of solids [119]. Impedance spectroscopy data obtained from macroscopic bulk materials are frequently analyzed, and equivalent circuit modeling of that data is a topic that has reached a level of understanding where it can be conducted accurately. On the other hand, the use of impedance spectroscopy with nanomaterials has only lately started to gain attention and its primary goal is to obtain local impedance characterization on small length scales that are as small as the nanoscale (126).

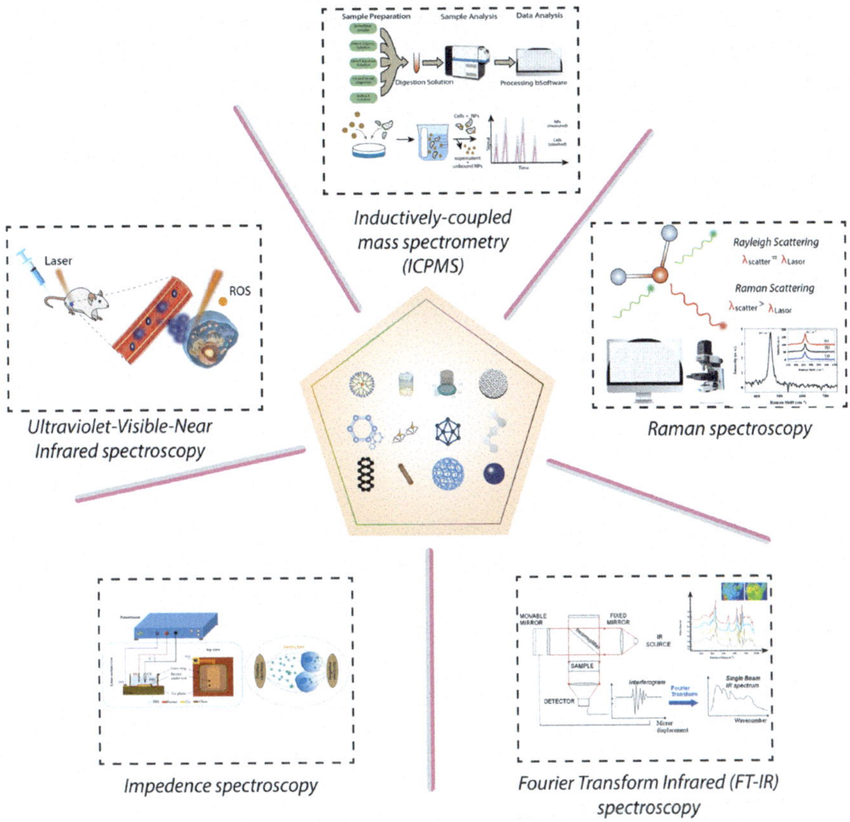

Fig. 3 Different types of spectroscopic techniques used for theranostics using functional nanomaterials. (Figures and all graphic illustrations were created using Adobe Illustrator)

The interaction of cells with nanoparticles is a topic of great interest in the field of cancer research at present because nanoparticles play a significant role. In 2005, De Nardo et al. conducted experiments to assess a method for the design of bio-reagent probes. These probes were generated by the coupling of a ferromagnetic nanoparticle complex and a chimeric monoclonal ChL6 [120]. Another work published in 2011 demonstrates the scientific and practical viability of labeling MCF-7 breast cancer cells with ferromagnetic nanoparticles bioconjugated with monoclonal anti-c-erbB-2 antibodies [120]. Recently, this research was extended to include three other cancer cell lines, each of which uniquely expresses the antigen c-erbB-2. The findings suggest that it may be possible to use magnetic nanoparticles to differently label cancer cells in response to an antigen–antibody reaction [121]. The isolation and labeling of BC cells can be conducted with the help of a versatile technique that consists of magnetic nanoparticles that have been linked to antibodies or other ligands. To determine whether additional treatment to prevent cancer dissemination is required, ideal BC adjuvant therapeutic strategies in underdeveloped regions require monitoring of patients with regular blood tests to detect circulating tumor cells (CTCs). This requires simple and inexpensive CTC biosensors in whole blood that are applicable for use in underdeveloped countries [122]. Measurements of relative electrical impedance spectroscopy with the use of magnetic nanoparticles were proposed as a viable method of monitoring by JG Silva et al. It is possible to fix a tumor cell interphase near gold electrodes by using a magnet to split cells in suspension coupled to magnetic nanoparticles. Below 20 kHz, relative impedance variations given by the differences of the interphase of tumor cells-homogenous condition demonstrated their potential value as a biosensor approach for evaluating BC cells in suspension, with particular sensitivity observed in the whole blood (WB) condition [123].

4.6 Hyperspectral Imaging

Spectral imaging is a new technology that combines imaging and spectrum to acquire spatial and spectral information from a scene simultaneously. It is also known as imaging spectroscopy and goes by the name imaging spectral imaging. In the field of spectral imaging, HSI is a subcategory that continually collects tens to hundreds of spectral bands. It gathers data in three dimensions, two of which are spatial (x and y), and one of which is spectral. This produces a dataset with three dimensions (x, y, and z), which is more often known as a data cube [124]. Briefly, HSI can capture a group of spatially resolved spectra or spectrally resolved images. Despite its roots in remote sensing, HSI has found widespread use in other fields such as biomedical imaging, the evaluation of food quality, surveillance, and many more. There are numerous ways to design an HSI system because of the wide variety of possible pairings of imaging concepts (such as spatial scanning, spectrum scanning, and snapshot) and light splitting techniques (such as dispersion, filtering, and modulation) [125].

Tumor identification using HSI can be classified into three distinct types of use: in vitro imaging of diseased samples, in vivo imaging of shallow surfaces, and intra-operative use. A histopathological analysis is still the preferred method of diagnosis nowadays. It takes a lot of effort and relies heavily on prior expertise [126].

The morphology of histiocytes and the behavior of malignancies have both been the subject of research using HSI detection of tissue samples as shown in an illustration in Fig. 4. The samples of tissue can either be fresh or processed in some way, for as by being frozen, fixed with formalin, embedded in paraffin, or dyed. In addition to this, they come in a variety of forms such as tissue biopsy sections, tissue microarrays (TMA), and cells [127, 128].

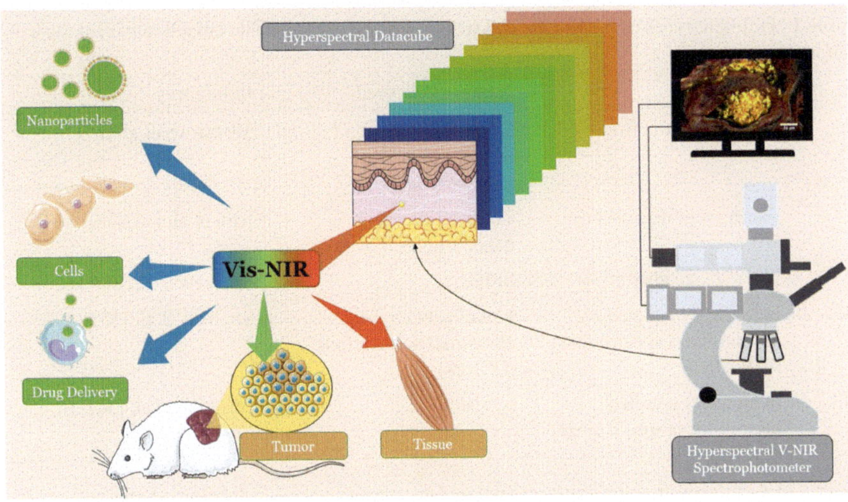

Fig. 4 Schematic illustration of hyperspectral imaging application in nanotheranostics. (Figures and graphic illustrations created using Adobe Illustrator). The cell and tissue illustrations are from Adobe Illustrator, and the blood vessel image is an enhanced dark field microscope image taken by authors that were false-colored using Adobe Photoshop

The majority of the histological characteristics are directly connected to the development of tumors; hence, benign, normal, and malignant tumors can be categorized following these characteristics [129]. This is done by looking at the nucleus and stroma of cells under a microscope and making a direct morphological observation. Microscopical examination of treated samples is the standard procedure, with illness stages being assigned based on the degree of alteration to tissue and cell structures and distributions. HSI can directly identify cell types and tissue patterns without the use of labels by taking advantage of composition-specific spectral markers. In addition, the increased information contained in hyperspectral images above standard red, green, and blue (RGB) images aids in a more precise diagnosis [130].

Molecular alterations in certain proteins, nucleic acids, and glycogens in the microenvironment can indicate distinct stages of cancers for non-invasive diagnosis and targeted therapy [125]. The tissue blood volume and the blood oxygen level are both essential markers that can be found in the microenvironment. The consumption of oxygen is significantly higher than the supply during the development of tumors, which results in hypoxia within the tumors themselves. Additionally, HSI can detect these variations of change in the microenvironment [131]. Although it is still in its development, the area of theranostics is progressing at a breakneck speed as summarized in Table 1.

Table 1 Different types of microcopy and spectroscopy techniques that employ nanomaterials for theranostic applications

S. No	Technique	Nanoparticle used	References
1	Fluorescent microscopy	Dextran-coated superparamagnetic nanoparticle	Medarova et al. [39]
2	Fluorescent microscopy	Chitosan-based nanoparticles (CNPs)	Kim et al. [48]
3	Mass spectrometry	SPIONs	Kruszewska et al. [53]
4	Inductively coupled plasma-mass spectrometry (ICP-MS)	Metallic and hybrid ultrasmall nanoparticles (USNPs)	Kuchma et al. [55]
5	Inductively coupled plasma-mass spectrometry (ICP-MS)	(USNPs)	Chamieh et al. [56]
6	Laser desorption/ionization mass spectrometry (LDI-MS) method	Silicon, and carbon nanomaterial	Mohammadpour and Majidzadeh [59]
7	LDI-MS	Aptamer-Au NPs modified graphene oxide nanocomposite	Lv et al. [61]
8	Electron microscopy	Lipid-based or lipid-coated nanocarriers	Yang et al. [76]
9	Two-photon laser scanning microscopies	Gold nanoparticles	Vickers [83]
10	Fluorescent spectroscopy	Inorganic nanoparticles	Dufes [93]
11	Near-infrared	Mesoporous silica nanoparticles	Hu et al. [97]
		Silver nanoparticle	Chen et al. [98]
		Superparamagnetic cobalt ferrite NPs	Hu et al. [97]
12	Raman spectroscopy	SERS nanoparticles	Kenry et al. [108, Cialla-May et al. [109], De La Zerda et al. [110]
13	FTIR	Silver nanoparticles	Mukherjee et al. [118]
14	Impedance spectroscopy	Ferromagnetic nanoparticle	DeNardo et al. [120]
15	Impedance spectroscopy	Magnetic nanoparticles	Silva et al. [123]

5 Conclusions

The field of theranostics has been around for a few years but the mounting amount of novel research and innovative results are fueling the fast-paced rise of cutting-edge applications. Researchers have developed a wide variety of systems based on nanoparticles that have the potential to be used for treatment, such as in the case of hyperthermia or radiation therapy, and/or for the administration of chemotherapeutic medications and diagnostic imaging agents. There are a few different techniques, one of which is the utilization of multiple imaging modalities, which can help with the tracking of nanoparticles in vitro as well as in vivo. The use of fluorescent labeling agents, such as organic fluorescent dyes or quantum dots, can make it possible to achieve high-resolution images that differentiate between the various organelles found within a cell. This allows for the therapy and the patient's response to the therapy to be monitored on a cellular level. However, to overcome issues with background fluorescence and photobleaching, microscopic and spectroscopic techniques are the ones that are recommended.

Theranostic compounds have the potential to play significant roles not only in the early stages of drug development but also as therapeutic-containing drug candidates in later clinical stages. During the early stages of drug development, in vitro imaging agents are quite useful, particularly for the modification of chemical structure and architecture to fine-tune therapeutic qualities.

References

1. Germain M, Caputo F, Metcalfe S, Tosi G, Spring K, Aslund AKO, Pottier A, Schiffelers R, Ceccaldi A, Schmid R (2020) Delivering the power of nanomedicine to patients today. J Control Release 326:164–171. https://doi.org/10.1016/j.jconrel.2020.07.007
2. Huang H, Lovell JF (2017) Advanced functional nanomaterials for theranostics. Adv Funct Mater 27(2):1603524. https://doi.org/10.1002/adfm.201603524
3. Funkhouser J (2002) Reinventing pharma: the theranostic revolution. Curr Drug Discov 2:17–19
4. McCarthy JR (2009) The future of theranostic nanoagents. Nanomedicine (Lond) 4(7):693–695. https://doi.org/10.2217/nnm.09.58
5. DeNardo GL, DeNardo SJ (2012) Concepts, consequences, and implications of theranosis. In: Seminars in nuclear medicine, vol 3. Elsevier, pp 147–150
6. ESR (2015) Medical imaging in personalised medicine: a white paper of the research committee of the European Society of Radiology (ESR). Insights Imaging 6:141–155
7. Caldorera-Moore ME, Liechty WB, Peppas NA (2011) Responsive theranostic systems: integration of diagnostic imaging agents and responsive controlled release drug delivery carriers. Acc Chem Res 44(10):1061–1070. https://doi.org/10.1021/ar2001777
8. Husen A, Bachheti RK, Bachheti A (2023) Current trends in green nano-emulsions (Food, Agriculture and Biomedical Sectors). Springer Nature Singapore Pte Ltd., 152 Beach Road, #21–01/04 Gateway East, Singapore 189721, Singapore
9. Husen A, Siddiqi KS (2023) Advances in smart nanomaterials and their applications. Elsevier Inc., 50 Hampshire St., 5th Floor, Cambridge, MA 02139, USA. https://doi.org/10.1016/C2021-0-02202-1
10. Jin-Chul K, Madhusudhan A, Husen A (2021) Smart nanomaterials in biomedical applications. Springer Nature Switzerland AG, Gewerbestrasse 11, 6330 Cham, Switzerland. https://doi.org/10.1007/978-3-030-84262-8
11. Bayda S, Adeel M, Tuccinardi T, Cordani M, Rizzolio F (2019) The history of nanoscience and nanotechnology: from chemical-physical applications to nanomedicine. Molecules 25(1):112. https://doi.org/10.3390/molecules25010112
12. Baptista A, Silva F, Porteiro J, Míguez J, Pinto G (2018) Sputtering physical vapour deposition (PVD) coatings: a critical review on process improvement and market trend demands. Coatings 8(11):402

13. Patel J, Patel A, Patel M, Vyas G (2022) Introduction to nanoparticulate drug delivery systems. In: Pharmacokinetics and pharmacodynamics of nanoparticulate drug delivery systems. Springer, pp 3–23
14. Saxena J, Jyoti A (2020) Nanomaterials: novel preparation routes, characterizations, and applications. In: Nanobiotechnology. Apple Academic Press, pp 23–33
15. Shaban M, Hasanzadeh M (2020) Biomedical applications of dendritic fibrous nanosilica (DFNS): recent progress and challenges. RSC Adv 10(61):37116–37133. https://doi.org/10.1039/d0ra04388e
16. Raliya R, Singh Chadha T, Haddad K, Biswas P (2016) Perspective on nanoparticle technology for biomedical use. Curr Pharm Des 22(17):2481–2490. https://doi.org/10.2174/1381612822666160307151409
17. Jeevanandam J, Barhoum A, Chan YS, Dufresne A, Danquah MK (2018) Review on nanoparticles and nanostructured materials: history, sources, toxicity and regulations. Beilstein J Nanotechnol 9(1):1050–1074. https://doi.org/10.3762/bjnano.9.98
18. Huo D, Kim MJ, Lyu Z, Shi Y, Wiley BJ, Xia Y (2019) One-dimensional metal nanostructures: from colloidal syntheses to applications. Chem Rev 119(15):8972–9073. https://doi.org/10.1021/acs.chemrev.8b00745
19. He Z, Yang Y, Liang HW, Liu JW, Yu SH (2019) Nanowire genome: a magic toolbox for 1D nanostructures. Adv Mater 31(51):e1902807. https://doi.org/10.1002/adma.201902807
20. Mauro N, Utzeri MA, Varvara P, Cavallaro G (2021) Functionalization of metal and carbon nanoparticles with potential in cancer theranostics. Molecules 26(11):3085. https://doi.org/10.3390/molecules26113085
21. Pascui OF, Lohwasser R, Sommer M, Thelakkat M, Thurn-Albrecht T, Saalwachter K (2010) High crystallinity and nature of crystal– crystal phase transformations in regioregular poly (3-hexylthiophene). Macromolecules 43(22):9401–9410
22. Salata O (2004) Applications of nanoparticles in biology and medicine. J Nanobiotechnology 2(1):3. https://doi.org/10.1186/1477-3155-2-3
23. Wolfram J, Ferrari M (2019) Clinical cancer nanomedicine. Nano Today 25:85–98. https://doi.org/10.1016/j.nantod.2019.02.005
24. Hussain T, Nguyen QT (2014) Molecular imaging for cancer diagnosis and surgery. Adv Drug Deliv Rev 66:90–100. https://doi.org/10.1016/j.addr.2013.09.007
25. Cai W, Chen X (2007) Nanoplatforms for targeted molecular imaging in living subjects. Small 3(11):1840–1854. https://doi.org/10.1002/smll.200700351
26. Wang YW, Reder NP, Kang S, Glaser AK, Liu JTC (2017) Multiplexed optical imaging of tumor-directed nanoparticles: a review of imaging systems and approaches. Nanotheranostics 1(4):369–388. https://doi.org/10.7150/ntno.21136
27. Wu L, Qu X (2015) Cancer biomarker detection: recent achievements and challenges. Chem Soc Rev 44(10):2963–2997. https://doi.org/10.1039/c4cs00370e
28. Pierce MC, Javier DJ, Richards-Kortum R (2008) Optical contrast agents and imaging systems for detection and diagnosis of cancer. Int J Cancer 123(9):1979–1990. https://doi.org/10.1002/ijc.23858
29. Wojtynek NE, Mohs AM (2020) Image-guided tumor surgery: the emerging role of nanotechnology. Wiley Interdiscip Rev Nanomed Nanobiotechnol 12(4):e1624. https://doi.org/10.1002/wnan.1624
30. Resch-Genger U, Grabolle M, Cavaliere-Jaricot S, Nitschke R, Nann T (2008) Quantum dots versus organic dyes as fluorescent labels. Nat Methods 5(9):763–775
31. Ntziachristos V, Bremer C, Weissleder R (2003) Fluorescence imaging with near-infrared light: new technological advances that enable in vivo molecular imaging. Eur Radiol 13(1):195–208
32. Kobayashi H, Ogawa M, Alford R, Choyke PL, Urano Y (2010) New strategies for fluorescent probe design in medical diagnostic imaging. Chem Rev 110(5):2620–2640. https://doi.org/10.1021/cr900263j
33. He X, Gao J, Gambhir SS, Cheng Z (2010) Near-infrared fluorescent nanoprobes for cancer molecular imaging: status and challenges. Trends Mol Med 16(12):574–583. https://doi.org/10.1016/j.molmed.2010.08.006

34. Ding F, Zhan Y, Lu X, Sun Y (2018) Recent advances in near-infrared II fluorophores for multifunctional biomedical imaging. Chem Sci 9(19):4370–4380. https://doi.org/10.1039/c8sc01153b
35. Kaniyala Melanthota S, Kistenev YV, Borisova E, Ivanov D, Zakharova O, Boyko A, Vrazhnov D, Gopal D, Chakrabarti S, K SP, Mazumder N (2022) Types of spectroscopy and microscopy techniques for cancer diagnosis: a review. Lasers Med Sci 37(8):3067–3084. https://doi.org/10.1007/s10103-022-03610-3
36. Wachsmann-Hogiu S, Annala AJ, Farkas DL (2021) Laser applications in biology and biotechnology. In: Handbook of laser technology and applications. CRC Press, pp 321–344
37. Chen J, Zhuo S, Chen G, Yan J, Yang H, Liu N, Zheng L, Jiang X, Xie S (2011) Establishing diagnostic features for identifying the mucosa and submucosa of normal and cancerous gastric tissues by multiphoton microscopy. Gastrointest Endosc 73(4):802–807. https://doi.org/10.1016/j.gie.2010.12.016
38. Hwang JY, Park J, Kang BJ, Lubow DJ, Chu D, Farkas DL, Shung KK, Medina-Kauwe LK (2012) Multimodality imaging in vivo for preclinical assessment of tumor-targeted doxorubicin nanoparticles. PLoS ONE 7(4):e34463. https://doi.org/10.1371/journal.pone.0034463
39. Medarova Z, Pham W, Farrar C, Petkova V, Moore A (2007) In vivo imaging of siRNA delivery and silencing in tumors. Nat Med 13(3):372–377. https://doi.org/10.1038/nm1486
40. Kelkar SS, Reineke TM (2011) Theranostics: combining imaging and therapy. Bioconjug Chem 22(10):1879–1903. https://doi.org/10.1021/bc200151q
41. Fernandez-Fernandez A, Manchanda R, McGoron AJ (2011) Theranostic applications of nanomaterials in cancer: drug delivery, image-guided therapy, and multifunctional platforms. Appl Biochem Biotechnol 165(7–8):1628–1651. https://doi.org/10.1007/s12010-011-9383-z
42. Schneider S, Lenz D, Holzer M, Palme K, Suss R (2010) Intracellular FRET analysis of lipid/DNA complexes using flow cytometry and fluorescence imaging techniques. J Control Release 145(3):289–296. https://doi.org/10.1016/j.jconrel.2010.04.016
43. Chatterjee DK, Gnanasammandhan MK, Zhang Y (2010) Small upconverting fluorescent nanoparticles for biomedical applications. Small 6(24):2781–2795. https://doi.org/10.1002/smll.201000418
44. Han X, Xu K, Taratula O, Farsad K (2019) Applications of nanoparticles in biomedical imaging. Nanoscale 11(3):799–819. https://doi.org/10.1039/c8nr07769j
45. He J, Li C, Ding L, Huang Y, Yin X, Zhang J, Zhang J, Yao C, Liang M, Pirraco RP (2019) Tumor targeting strategies of smart fluorescent nanoparticles and their applications in cancer diagnosis and treatment. Adv Mater 31(40):1902409
46. Zhou T, Liang X, Wang P, Hu Y, Qi Y, Jin Y, Du Y, Fang C, Tian J (2020) A hepatocellular carcinoma targeting nanostrategy with hypoxia-ameliorating and photothermal abilities that, combined with immunotherapy, inhibits metastasis and recurrence. ACS Nano 14(10):12679–12696
47. Zavaleta C, Ho D, Chung EJ (2018) Theranostic nanoparticles for tracking and monitoring disease state. SLAS Technol 23(3):281–293. https://doi.org/10.1177/2472630317738699
48. Kim K, Kim JH, Park H, Kim YS, Park K, Nam H, Lee S, Park JH, Park RW, Kim IS, Choi K, Kim SY, Park K, Kwon IC (2010) Tumor-homing multifunctional nanoparticles for cancer theragnosis: simultaneous diagnosis, drug delivery, and therapeutic monitoring. J Control Release 146(2):219–227. https://doi.org/10.1016/j.jconrel.2010.04.004
49. Lim EK, Kim T, Paik S, Haam S, Huh YM, Lee K (2015) Nanomaterials for theranostics: recent advances and future challenges. Chem Rev 115(1):327–394. https://doi.org/10.1021/cr300213b
50. Blume JE, Manning WC, Troiano G, Hornburg D, Figa M, Hesterberg L, Platt TL, Zhao X, Cuaresma RA, Everley PA, Ko M, Liou H, Mahoney M, Ferdosi S, Elgierari EM, Stolarczyk C, Tangeysh B, Xia H, Benz R, Siddiqui A, Carr SA, Ma P, Langer R, Farias V, Farokhzad OC (2020) Rapid, deep and precise profiling of the plasma proteome with multi-nanoparticle protein corona. Nat Commun 11(1):3662. https://doi.org/10.1038/s41467-020-17033-7

51. Buckle T, Van Der Wal S, Van Malderen SJ, Müller L, Kuil J, Van Unen V, Peters RJ, van Bemmel ME, McDonnell LA, Velders AH (2017) Hybrid imaging labels: providing the link between mass spectrometry-based molecular pathology and theranostics. Theranostics 7(3):624
52. Sikorski J, Matczuk M, Kaminska A, Kruszewska J, Trzaskowski M, Timerbaev AR, Jarosz M (2022) Protein-mediated transformations of superparamagnetic nanoparticles evidenced by single-particle inductively coupled plasma tandem mass spectrometry: a disaggregation phenomenon. Int J Mol Sci 23(3):1088. https://doi.org/10.3390/ijms23031088
53. Kruszewska J, Sikorski J, Samsonowicz-Gorski J, Matczuk M (2020) A CE-ICP-MS/MS method for the determination of superparamagnetic iron oxide nanoparticles under simulated physiological conditions. Anal Bioanal Chem 412(29):8145–8153. https://doi.org/10.1007/s00216-020-02948-3
54. Labied L, Rocchi P, Doussineau T, Randon J, Tillement O, Lux F, Hagege A (2021) Taylor dispersion analysis coupled to inductively coupled plasma-mass spectrometry for ultrasmall nanoparticle size measurement: from drug product to biological media studies. Anal Chem 93(3):1254–1259. https://doi.org/10.1021/acs.analchem.0c03988
55. Kuchma EA, Zolotukhin PV, Belanova AA, Soldatov MA, Lastovina TA, Kubrin SP, Nikolsky AV, Mirmikova LI, Soldatov AV (2017) Low toxic maghemite nanoparticles for theranostic applications. Int J Nanomedicine 12:6365–6371. https://doi.org/10.2147/IJN.S140368
56. Chamieh J, Leclercq L, Martin M, Slaoui S, Jensen H, Ostergaard J, Cottet H (2017) Limits in size of Taylor dispersion analysis: representation of the different hydrodynamic regimes and application to the size-characterization of cubosomes. Anal Chem 89(24):13487–13493. https://doi.org/10.1021/acs.analchem.7b03806
57. Tseng Y, Harroun S, Wu C, Mao J, Chang H, Huang C (2017) Satellite-like gold nanocomposites for targeted mass spectrometry imaging of tumor tissues. Nanotheranostics 1(2):141–153. https://doi.org/10.7150/ntno. 18897
58. Zhang W, Wang X, Xia Y, Ouyang Z (2017) Ambient ionization and miniature mass spectrometry systems for disease diagnosis and therapeutic monitoring. Theranostics 7(12):2968–2981. https://doi.org/10.7150/thno.19410
59. Mohammadpour Z, Majidzadeh AK (2020) Applications of two-dimensional nanomaterials in breast cancer theranostics. ACS Biomater Sci Eng 6(4):1852–1873. https://doi.org/10.1021/acsbiomaterials.9b01894
60. Chu HW, Unnikrishnan B, Anand A, Mao JY, Huang CC (2018) Nanoparticle-based laser desorption/ionization mass spectrometric analysis of drugs and metabolites. J Food Drug Anal 26(4):1215–1228. https://doi.org/10.1016/j.jfda.2018.07.001
61. Lv Q, Cui H, Song X (2021) Aptamers for the diagnosis of malign tumors. In: Aptamers for medical applications. Springer, pp 239–277
62. Mesguich C, Zanotti-Fregonara P, Hindie E (2016) New perspectives offered by nuclear medicine for the imaging and therapy of multiple myeloma. Theranostics 6(2):287–290. https://doi.org/10.7150/thno.14400
63. van Leeuwen FW, van der Poel HG (2016) Surgical guidance in prostate cancer: "From Molecule to Man" translationsfluorescence-guided surgery for prostate cancer patients. Clin Cancer Res 22(6):1304–1306
64. Mukherjee D, Zhao J (2013) The role of chemokine receptor CXCR4 in breast cancer metastasis. Am J Cancer Res 3(1):46–57
65. Kuil J, Buckle T, van Leeuwen FW (2012) Imaging agents for the chemokine receptor 4 (CXCR4). Chem Soc Rev 41(15):5239–5261. https://doi.org/10.1039/c2cs35085h
66. Buckle T, Kuil J, van den Berg NS, Bunschoten A, Lamb HJ, Yuan H, Josephson L, Jonkers J, Borowsky AD, van Leeuwen FW (2013) Use of a single hybrid imaging agent for integration of target validation with in vivo and ex vivo imaging of mouse tumor lesions resembling human DCIS. PLoS ONE 8(1):e48324. https://doi.org/10.1371/journal.pone.0048324
67. van Leeuwen FW, Valdes-Olmos R, Buckle T, Vidal-Sicart S (2016) Hybrid surgical guidance based on the integration of radionuclear and optical technologies. Br J Radiol 89(1062):20150797. https://doi.org/10.1259/bjr.20150797

68. Goldstein A, Soroka Y, Frusic-Zlotkin M, Popov I, Kohen R (2014) High resolution SEM imaging of gold nanoparticles in cells and tissues. J Microsc 256(3):237–247. https://doi.org/10.1111/jmi.12179
69. Li L, Mak KY, Shi J, Koon HK, Leung CH, Wong CM, Leung CW, Mak CS, Chan NM, Zhong W, Lin KW, Wu EX, Pong PW (2012) Comparative in vitro cytotoxicity study on uncoated magnetic nanoparticles: effects on cell viability, cell morphology, and cellular uptake. J Nanosci Nanotechnol 12(12):9010–9017. https://doi.org/10.1166/jnn.2012.6755
70. Havrdova M, Polakova K, Skopalik J, Vujtek M, Mokdad A, Homolkova M, Tucek J, Nebesarova J, Zboril R (2014) Field emission scanning electron microscopy (FE-SEM) as an approach for nanoparticle detection inside cells. Micron 67:149–154
71. Malatesta M (2021) Transmission electron microscopy as a powerful tool to investigate the interaction of nanoparticles with subcellular structures. Int J Mol Sci 22(23):12789
72. Costanzo M, Malatesta M (2019) Embedding cell monolayers to investigate nanoparticleplasmalemma interactions at transmission electron microscopy. Eur J Histochem EJH 63(1)
73. Lerch S, Ritz S, Bley K, Messerschmidt C, Weiss CK, Musyanovych A, Landfester K, Mailander V (2015) Nanoprobing the acidification process during intracellular uptake and trafficking. Nanomedicine 11(6):1585–1596. https://doi.org/10.1016/j.nano.2015.04.010
74. Costanzo M, Carton F, Marengo A, Berlier G, Stella B, Arpicco S, Malatesta M (2016) Fluorescence and electron microscopy to visualize the intracellular fate of nanoparticles for drug delivery. Eur J Histochem 60(2):2640. https://doi.org/10.4081/ejh.2016.2640
75. Cabezón I, Manich G, Martín-Venegas R, Camins A, Pelegrí C, Vilaplana J (2015) Trafficking of gold nanoparticles coated with the 8D3 anti-transferrin receptor antibody at the mouse blood–brain barrier. Mol Pharm 12(11):4137–4145
76. Yang J, Tu J, Lamers GEM, Olsthoorn RCL, Kros A (2017) Membrane fusion mediated intracellular delivery of lipid bilayer coated mesoporous silica nanoparticles. Adv Healthc Mater 6(20):1700759. https://doi.org/10.1002/adhm.201700759
77. Heo DN, Yang DH, Moon H-J, Lee JB, Bae MS, Lee SC, Lee WJ, Sun I-C, Kwon IK (2012) Gold nanoparticles surface-functionalized with paclitaxel drug and biotin receptor as theranostic agents for cancer therapy. Biomaterials 33(3):856–866
78. Zhang LW, Monteiro-Riviere NA (2012) Use of confocal microscopy for nanoparticle drug delivery through skin. J Biomed Opt 18(6):061214
79. Du Y, Fan K, Zhang H, Li L, Wang P, He J, Ding S, Yan X, Tian J (2018) Endoscopic molecular imaging of early gastric cancer using fluorescently labeled human H-ferritin nanoparticle. Nanomedicine 14(7):2259–2270. https://doi.org/10.1016/j.nano.2018.07.007
80. Xu L, Zhang J, Yin L, Long X, Zhang W, Zhang Q (2020) Recent progress in efficient organic two-photon dyes for fluorescence imaging and photodynamic therapy. J Mater Chem C 8(19):6342–6349
81. Wang H, Zhou S (2016) Magnetic and fluorescent carbon-based nanohybrids for multi-modal imaging and magnetic field/NIR light responsive drug carriers. Biomater Sci 4(7):1062–1073. https://doi.org/10.1039/c6bm00262e
82. Vickers ET, Garai M, Bonabi Naghadeh S, Lindley S, Hibbs J, Xu Q-H, Zhang JZ (2017) Two-photon photoluminescence and photothermal properties of hollow gold nanospheres for efficient theranostic applications. J Phys Chem C 122(25):13304–13313
83. Vickers ET (2020) Synthesis, characterization, surface tailoring, and applications of hollow gold nanospheres and perovskite nanomaterial. University of California, Santa Cruz
84. So PT, Dong CY, Masters BR, Berland KM (2000) Two-photon excitation fluorescence microscopy. Annu Rev Biomed Eng 2(1):399–429. https://doi.org/10.1146/annurev.bioeng.2.1.399
85. Kleinfeld D, Mitra PP, Helmchen F, Denk W (1998) Fluctuations and stimulus-induced changes in blood flow observed in individual capillaries in layers 2 through 4 of rat neocortex. Proc Natl Acad Sci 95(26):15741–15746
86. Shi SH, Hayashi Y, Petralia RS, Zaman SH, Wenthold RJ, Svoboda K, Malinow R (1999) Rapid spine delivery and redistribution of AMPA receptors after synaptic NMDA receptor activation. Science 284(5421):1811–1816. https://doi.org/10.1126/science.284.5421.1811

87. Squirrell JM, Wokosin DL, White JG, Bavister BD (1999) Long-term two-photon fluorescence imaging of mammalian embryos without compromising viability. Nat Biotechnol 17(8):763–767
88. Masters BR, So PT, Gratton E (1997) Multiphoton excitation fluorescence microscopy and spectroscopy of in vivo human skin. Biophys J 72(6):2405–2412. https://doi.org/10.1016/S0006-3495(97)78886-6
89. Giniunas L, Juskaitis R, Shatalin SV (1993) Endoscope with optical sectioning capability. Appl Opt 32(16):2888–2890. https://doi.org/10.1364/AO.32.002888
90. Palsson S, Stenram U, Thompson MS, Vaitkuviene A, Poskiene V, Ziobakiene R, Oyama J, DeWeert MJ, Gustafsson U, Bendsoe N, Andersson-Engels S, Svanberg S, Svanberg K (2006) Methods for detailed histopathological investigation and localization of biopsies from cervix uteri to improve the interpretation of autofluorescence data. J Environ Pathol Toxicol Oncol 25(1–2):321–340. https://doi.org/10.1615/jenvironpatholtoxicoloncol.v25.i1-2.210
91. Berland K (2001) Basics of fluorescence. In: Methods in cellular imaging. Springer, pp 5–19
92. Ramanathan S, Archunan G, Sivakumar M, Tamil Selvan S, Fred AL, Kumar S, Gulyas B, Padmanabhan P (2018) Theranostic applications of nanoparticles in neurodegenerative disorders. Int J Nanomedicine 13:5561–5576. https://doi.org/10.2147/IJN.S149022
93. Dufes C (2006) Beyond delivery. Gene Ther 13:739–740
94. Medintz IL, Uyeda HT, Goldman ER, Mattoussi H (2005) Quantum dot bioconjugates for imaging, labelling and sensing. Nat Mater 4(6):435–446. https://doi.org/10.1038/nmat1390
95. Schwille P (2001) Fluorescence correlation spectroscopy and its potential for intracellular applications. Cell Biochem Biophys 34(3):383–408. https://doi.org/10.1385/CBB:34:3:383
96. Lucas B, Remaut K, Sanders NN, Braeckmans K, De Smedt SC, Demeester J (2005) Studying the intracellular dissociation of polymer-oligonucleotide complexes by dual color fluorescence fluctuation spectroscopy and confocal imaging. Biochemistry 44(29):9905–9912. https://doi.org/10.1021/bi0476883
97. Hu F, Li C, Zhang Y, Wang M, Wu D, Wang Q (2015) Real-time in vivo visualization of tumor therapy by a near-infrared-II Ag2S quantum dot-based theranostic nanoplatform. Nano Res 8(5):1637–1647
98. Chen H, Li B, Zhang M, Sun K, Wang Y, Peng K, Ao M, Guo Y, Gu Y (2014) Characterization of tumor-targeting Ag 2 S quantum dots for cancer imaging and therapy in vivo. Nanoscale 6(21):12580–12590
99. Hong G, Diao S, Chang J, Antaris AL, Chen C, Zhang B, Zhao S, Atochin DN, Huang PL, Andreasson KI, Kuo CJ, Dai H (2014) Through-skull fluorescence imaging of the brain in a new near-infrared window. Nat Photonics 8(9):723–730. https://doi.org/10.1038/nphoton.2014.166
100. Du Y, Xu B, Fu T, Cai M, Li F, Zhang Y, Wang Q (2010) Near-infrared photoluminescent Ag2S quantum dots from a single source precursor. J Am Chem Soc 132(5):1470–1471. https://doi.org/10.1021/ja909490r
101. Park W, Shin H, Choi B, Rhim W-K, Na K, Han DK (2020) Advanced hybrid nanomaterials for biomedical applications. Prog Mater Sci 114:100686
102. Yhee JY, Kim SA, Koo H, Son S, Ryu JH, Youn IC, Choi K, Kwon IC, Kim K (2012) Optical imaging of cancer-related proteases using near-infrared fluorescence matrix metalloproteinase-sensitive and cathepsin B-sensitive probes. Theranostics 2(2):179–189. https://doi.org/10.7150/thno.3716
103. Jones RR, Hooper DC, Zhang L, Wolverson D, Valev VK (2019) Raman techniques: fundamentals and Frontiers. Nanoscale Res Lett 14(1):231. https://doi.org/10.1186/s11671-019-3039-2
104. Kong K, Kendall C, Stone N, Notingher I (2015) Raman spectroscopy for medical diagnostics–from in-vitro biofluid assays to in-vivo cancer detection. Adv Drug Deliv Rev 89:121–134. https://doi.org/10.1016/j.addr.2015.03.009
105. Nicolson F, Clark L, Panikkanvalappil SR, Andreiuk B, Andreou C (2022) Advances in surface enhanced Raman spectroscopy for in vivo imaging in oncology. Nanotheranostics 6(1):31

106. Ko KH, Kown CI, Park JM, Lee HG, Han NY, Hahm KB (2014) Molecular imaging for theranostics in gastroenterology: one stone to kill two birds. Clin Endosc 47(5):383–388. https://doi.org/10.5946/ce.2014.47.5.383
107. Krafft C, Dochow S, Latka I, Dietzek B, Popp J (2012) Diagnosis and screening of cancer tissues by fiber-optic probe Raman spectroscopy. Biomed Spectrosc Imaging 1(1):39–55
108. Kenry NF, Clark L, Panikkanvalappil S, Andreiuk B, Andreou C (2022) Advances in surface enhanced raman spectroscopy for in vivo imaging in oncology. Nanotheranostics 6(1):31–49. https://doi.org/10.7150/ntno. 62970
109. Cialla-May D, Zheng XS, Weber K, Popp J (2017) Recent progress in surface-enhanced Raman spectroscopy for biological and biomedical applications: from cells to clinics. Chem Soc Rev 46(13):3945–3961. https://doi.org/10.1039/c7cs00172j
110. De La Zerda A, Kircher MF, Jokerst JV, Zavaleta CL, Kempen PJ, Mittra E, Pitter K, Huang R, Campos C, Habte F (2013) A brain tumor molecular imaging strategy using a new triple-modality MRI-photoacoustic-Raman nanoparticle. In: Photons plus ultrasound: imaging and sensing. SPIE, pp 63–74
111. Kircher MF, de la Zerda A, Jokerst JV, Zavaleta CL, Kempen PJ, Mittra E, Pitter K, Huang R, Campos C, Habte F, Sinclair R, Brennan CW, Mellinghoff IK, Holland EC, Gambhir SS (2012) A brain tumor molecular imaging strategy using a new triple-modality MRI-photoacoustic-Raman nanoparticle. Nat Med 18(5):829–834. https://doi.org/10.1038/nm.2721
112. Lasch P, Haensch W, Naumann D, Diem M (2004) Imaging of colorectal adenocarcinoma using FT-IR microspectroscopy and cluster analysis. Biochim Biophys Acta 1688(2):176–186. https://doi.org/10.1016/j.bbadis.2003.12.006
113. Tang XC, Pikal MJ, Taylor LS (2002) A spectroscopic investigation of hydrogen bond patterns in crystalline and amorphous phases in dihydropyridine calcium channel blockers. Pharm Res 19(4):477–483. https://doi.org/10.1023/a:1015147729564
114. Boskey AL, Mendelsohn R (2005) Infrared spectroscopic characterization of mineralized tissues. Vib Spectrosc 38(1–2):107–114. https://doi.org/10.1016/j.vibspec.2005.02.015
115. Kidder LH, Colarusso P, Stewart SA, Levin IW, Appel NM, Lester DS, Pentchev PG, Lewis EN (1999) Infrared spectroscopic imaging of the biochemical modifications induced in the cerebellum of the Niemann-pick type C mouse. J Biomed Opt 4(1):7–13. https://doi.org/10.1117/1.429915
116. Lasch P, Boese M, Pacifico A, Diem M (2002) FT-IR spectroscopic investigations of single cells on the subcellular level. Vib Spectrosc 28(1):147–157
117. Choo LP, Wetzel DL, Halliday WC, Jackson M, LeVine SM, Mantsch HH (1996) In situ characterization of beta-amyloid in Alzheimer's diseased tissue by synchrotron Fourier transform infrared microspectroscopy. Biophys J 71(4):1672–1679. https://doi.org/10.1016/S0006-3495(96)79411-0
118. Mukherjee S, Chowdhury D, Kotcherlakota R, Patra S, B V, Bhadra MP, Sreedhar B, Patra CR (2014) Potential theranostics application of bio-synthesized silver nanoparticles (4-in-1 system). Theranostics 4(3):316–335. https://doi.org/10.7150/thno.7819
119. Barsukov Y, Macdonald JR (2012) Electrochemical impedance spectroscopy. Charact Mater 2:898-913
120. DeNardo SJ, DeNardo GL, Miers LA, Natarajan A, Foreman AR, Gruettner C, Adamson GN, Ivkov R (2005) Development of tumor targeting bioprobes (111In-chimeric L6 monoclonal antibody nanoparticles) for alternating magnetic field cancer therapy. Clin Cancer Res 11(19):7087s–7092s
121. Silva JG, Sánchez V, Polo SM, González CA (2013) Expression of c-erbB-2 in breast cancer cell lines as experimental receptor of magnetic nanoparticles. In: 2013 35th annual international conference of the IEEE engineering in medicine and biology society (EMBC). IEEE, pp 4498–4501
122. Malagrino TRS, Godoy AP, Barbosa JM, Lima AGT, Sousa NCO, Pedrotti JJ, Garcia PS, Paniago RM, Andrade LM, Domingues SH, Silva WM, Ribeiro H, Taha-Tijerina J (2022) Multifunctional hybrid MoS(2)-PEGylated/Au nanostructures with potential theranostic applications in biomedicine. Nanomaterials (Basel) 12(12):2053. https://doi.org/10.3390/nano12122053

123. Silva JG, Cárdenas RA, Quiróz AR, Sánchez V, Lozano LM, Pérez NM, López J, Villanueva C, González CA (2014) Impedance spectroscopy assisted by magnetic nanoparticles as a potential biosensor principle for breast cancer cells in suspension. Physiol Meas 35(6):931
124. Gao L, Smith RT (2015) Optical hyperspectral imaging in microscopy and spectroscopy—a review of data acquisition. J Biophotonics 8(6):441–456. https://doi.org/10.1002/jbio.201400051
125. Zhang Y, Wu X, He L, Meng C, Du S, Bao J, Zheng Y (2020) Applications of hyperspectral imaging in the detection and diagnosis of solid tumors. Transl Cancer Res 9(2):1265–1277. https://doi.org/10.21037/tcr.2019.12.53
126. Li Q, He X, Wang Y, Liu H, Xu D, Guo F (2013) Review of spectral imaging technology in biomedical engineering: achievements and challenges. J Biomed Opt 18(10):100901. https://doi.org/10.1117/1.JBO.18.10.100901
127. Bird B, Bedrossian K, Laver N, Miljkovic M, Romeo MJ, Diem M (2009) Detection of breast micro-metastases in axillary lymph nodes by infrared micro-spectral imaging. Analyst 134(6):1067–1076. https://doi.org/10.1039/b821166c
128. Verdonck M, Denayer A, Delvaux B, Garaud S, De Wind R, Desmedt C, Sotiriou C, Willard-Gallo K, Goormaghtigh E (2016) Characterization of human breast cancer tissues by infrared imaging. Analyst 141(2):606–619. https://doi.org/10.1039/c5an01512j
129. Boucheron LE, Bi Z, Harvey NR, Manjunath B, Rimm DL (2007) Utility of multispectral imaging for nuclear classification of routine clinical histopathology imagery. BMC Cell Biol 8(Suppl 1):S8. https://doi.org/10.1186/1471-2121-8-S1-S8
130. Berisha S, Lotfollahi M, Jahanipour J, Gurcan I, Walsh M, Bhargava R, Van Nguyen H, Mayerich D (2019) Deep learning for FTIR histology: leveraging spatial and spectral features with convolutional neural networks. Analyst 144(5):1642–1653. https://doi.org/10.1039/c8an01495g
131. Perou CM, Sorlie T, Eisen MB, van de Rijn M, Jeffrey SS, Rees CA, Pollack JR, Ross DT, Johnsen H, Akslen LA, Fluge O, Pergamenschikov A, Williams C, Zhu SX, Lonning PE, Borresen-Dale AL, Brown PO, Botstein D (2000) Molecular portraits of human breast tumours. Nature 406(6797):747–752. https://doi.org/10.1038/35021093
132. Tseng YT, Harroun SG, Wu CW, Mao JY, Chang HT, Huang CC (2017) Satellite-like gold nanocomposites for targeted mass spectrometry imaging of tumor tissues. Nanotheranostics 1(2):141–153. https://doi.org/10.7150/ntno.18897

Next-Generation Therapies for Breast Cancer

Anindita De, Sonam Patel, and K. Gowthamarajan

Abstract Breast cancer (BC) is predominantly diagnosed and is a major cause of malignancy-related death among women globally. Traditional BC therapies include surgical excision of damaged tissue, radiation therapy, systemic chemotherapy, and endocrine therapy. BC typically targets many important pathways, including the hormonal receptors, HER2, and PI3K/AKT/mTOR signaling pathways. Other therapy strategies include blocking immunological checkpoints, addressing DNA repair, and developing antibody–drug conjugates. As technology advances, the approach to BC therapy has changed from using toxic, high dosages of drugs to using some novel targeted therapies like biomarker-specific, line-agnostic, and receptor-agnostic. The success of Trastuzumab (Herceptin) in BC therapy has prepared the path for a next-generation strategy to target the tumor's molecular mechanism. Research is in progress to identify novel blood-based and non-blood-based biomarkers for early detection and therapy for BC. Tumor heterogeneity, on the other hand, remains a major challenge. Drug repositioning and advanced formulation strategies are also on the list of next-generation therapies for BC. It is necessary to thoroughly analyze the patient's mindset, financial capacity, and utility of innovative medical attention to effectively develop next-generation cancer medications. This chapter focuses on next-generation therapeutic strategies for the treatment of BC.

Keywords Breast cancer · Systemic therapy · Biomarker-specific therapy · Next-generation formulation

A. De (✉)
College of Pharmacy, Department of Pharmaceutics, Ajou University, Suwon 16502, South Korea
e-mail: aninditanirupa@gmail.com

S. Patel
College of Pharmacy, Department of Pharmaceutical Analysis, JSS Academy of Technical Education, Noida, India

K. Gowthamarajan
Departmnet of Pharmaceutics, JSS College of Pharmacy, JSS Academy of Higher Education and Research, Ooty, Nilgiris, Tamil Nadu 643001, India

A. Madhusudhan et al. (eds.), *Functional Smart Nanomaterials and Their Theranostics Approaches*, Smart Nanomaterials Technology,
https://doi.org/10.1007/978-981-99-6597-7_5

1 Introduction

Breast cancer (BC) is one of the most common invasive tumors among women globally and is one of the prime causes of malignancy-related death. Men account for less than 1% of all diagnosed BC cases [40, 76]. BC is an etiologically and clinically complex disorder. With the development of screening mammography and effective therapy over the last few decades, the rate of survival has improved. However, overall success is still far away due to the heterogenic nature of the BC, drug resistance, and lack of novel targeted biomarkers. The success of BC management depends largely on the effective targeting of the biomarkers or the early diagnostic biomarkers with an effective therapeutic formulation strategy. The pathway-specified biomarkers open up new strategies to target BC, and early detection biomarkers improve the life span of BC patients. Still, the success path for BC management is long and challenging. Research needs more biomarker-specific chemotherapeutics or advanced cancer cell targeting formulations. This chapter focuses on next-generation therapeutic strategies for the treatment of BC [40].

2 Current Status of Global Breast Cancer Count

2.1 Descriptive Epidemiology

BC accounts for almost 30% of newly diagnosed malignancy cases among females worldwide, with an estimated 2,188,849 BC cases in 2020 [37]. Statistics show that the rates of cancer cases are comparatively higher in Australia, New Zealand, most of Europe, and the northern part of America compared to Asia and Africa in general [3, 37]. The cases reported are moderate in the southern part of America and Eastern Europe. Over their lifetime, one in every ten women is affected by BC [76]. The diagnosis probability is less than 1% before the age of 40 years, but afterward, the rate of BC increases until about the age of 70 years [28]. According to the historical data, the rate of BC cases increased from the years of the 1980s to the 1990s. The reasons may be the alterations in reproduction nature, increased use of hormonal therapy, and advanced mammographic screening technology [3, 5, 40]. The rate of BC decreased at the beginning of the 2000s, notably among females over 45 years old, as a result of the Women's Health Initiative Research [3, 5, 28, 40]. Since 2004, incidence rates have been slowly increasing (0.3%/year), perhaps as a result of high obesity rates and low reproductive rates [5, 37]. According to statistical models, increases in estrogen receptor + (ER) malignancies, especially in situ, and persistent falls in ER cancers are expected to continue. However, in several other westernized nations (Canada, the UK, France, and Australia), occurrence rates have continued to fall, while rates in formerly low-risk countries (e.g., Latin America, Africa, and Asia) have been quickly rising. The potential causes might be higher life expectancy due to infectious disease reductions, rising obesity, changes in reproductive ability, and enhanced BC screening [37, 76].

2.2 *Etiologic Heterogeneity*

BC is a complicated disorder with high genomic and clinical heterogeneity [20, 21]. BC is classified based on invasiveness (i.e., in situ or invasive), morphology, immunohistochemistry (IHC) marker expressibility, and genomic panels. These characteristics have been connected to differential therapeutic response and prognosis [1]. In situ BC is restricted to the ducts or lobules [9, 87]. Ductal carcinoma *in situ* (DCIS) was detected more frequently than lobular carcinoma in situ (LCIS). DCIS and LCIS both show a high risk of invasiveness. But the exact etiology and history of in situ malignancies are still poorly understood. Invasive BC is classified histopathologically as invasive-ductal (75%–80% of BCs), invasive-lobular (5–15% of BCs), and papillary tumors (less common) [22, 40]. IHC staining is the direct-targeted approach for the identification of hormone-specific and human epidermal growth factor receptor 2 (HER2). Other assays, like gene array profiling or biomarker-based IHC positive markers, are not mostly used therapeutically [79]. Many significant intrinsic molecular subtypes were found using gene microarrays, including Luminal A, B, HER2, basal-like, and normal-like [40, 79, 81, 82]. Claudin-low and six triple-negative cancers' (TNBC) subtypes have been established in further categories of primarily TNBC (16, 17) (Fig. 1).

The BC has been categorized predominantly into four molecular subtypes based on the presence of hormone receptors (HRs), increased HER2 protein level, or HER2 gene: 1. Luminal A (HR^+/HER2-); 2. Luminal B ($HR^+/HER2^+$); 3. HER2+ ; and 4. triple-negative (TNBC; HR^-/HER^-) [15, 95]. Each of the subtypes has its characteristics of toxicity, therapeutic action, disease progression, and preferred organ metastases.

BC has HR positivity [ER and PR subtypes]. It is sub-categorized into Luminal A and B. Luminal A ($HR^+/HER2^-$) subtypes are often slow to develop and less aggressive. They respond well to hormonal treatments [36]. On the other hand, Luminal B ($HR^+/HER2^+$) is distinguished by the presence of high levels of Ki67 (an overexpressed biomarker) and HER2. Luminal B has a lower prediction rate and medical prognosis than Luminal A [26]. $HER2^+$ BC overexpresses the HER2/ERBB2 oncogene and is mostly treated with anti-HER2 chemotherapeutics [21, 90]. Because basal-like BC lacks HR and HER2 receptors, it is often referred to as TNBC. The majority of BC-affected patients (85%) have HR^+ conditions, with 71% suffering from Luminal A and 12% suffering from Luminal B. Only 6% of BC cases have HER2+ but HR−. TNBC accounts for the rest (11%) of the total cases.

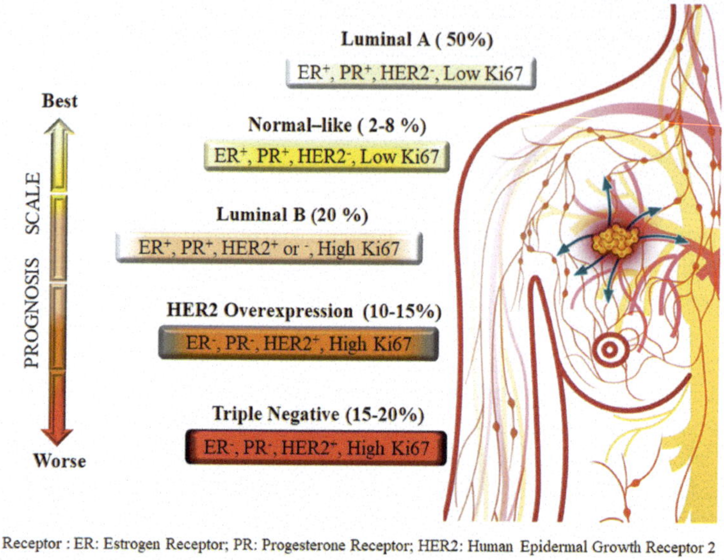

Fig. 1 Classification of the etiologic heterogeneity of BC and its progression

3 Conventional Treatment Strategy of BC Therapy

3.1 Surgical Process

Surgical procedures for BC have improved dramatically in recent years, with developments intended to reduce the long-term functional and esthetic effects of localized therapy. Standard treatments are mostly total mastectomy or radiotherapy, presuming clean margins are obtained. These two mechanisms have repeatedly been found to be equally effective in terms of relapse-free and improved overall survival [21, 29, 91]. Conservative surgery is mostly recommended to prevent the reappearance of micro-calcifications on breast tomography or positive pathological margins after lumpectomy [91]. The proportion of patients receiving neoadjuvant therapy has been increasing recently. A recent meta-analysis showed that neoadjuvant therapy enhances patients' selection for breast-conserving treatment and prolongs survival [8].

3.2 Radiation Therapy

In BC, radiation can be given to the affected area or whole breast (after a lumpectomy), chest walls (after a mastectomy), and lymph nodes. Whole-breast radiation following a lumpectomy is a very common part of breast-conserving treatment [91].

Radiation, like adjuvant systemic therapies, provides a relatively constant proportional benefit independent of total BC risk. The use of short-term radiation after a lumpectomy for selective patients reduces the high dose of chemotherapeutics. A hypo-fractionated regimen for whole-breast radiation is officially "recommended" by the prescribed guidelines. Post-lumpectomy radiation to the part of the breast, rather than the whole breast, has primarily been found in lower-risk adults under age 50 and the elderly population. Though past research linked part of breast radiation to an increased risk of local recurrence and slightly poorer symptom management, new research outcomes contradict this, and consensus recommendations suggest a non-intraoperative part of breast radiation in low-risk patients. Increasing radiation to the cancer cell improves localized hold-on cancer cells but does not improve overall viability. A low overall survival chances reduce application of the modal radiation routinely even in node-positive cases. But, it should be explored for individuals with a higher nodal lesion load or risky biological conditions soon.

3.3 Systemic Therapy for BC

Every year, almost 45,000 individuals die from BC, primarily as a result of the disease's uncontrolled and incurable metastasis to the visceral, brain, and several other metastatic sites. An early diagnosis at a young age may only enhance the chances of survival. Systemic therapy for BC includes standard early-phase treatment regimens as well as a long-term treatment cycle (i.e., later lines). Early treatment for HR^+/ERBB2 BC should be based on endocrine therapy with a cyclin-dependent kinase (CDK) 4/6 inhibitor, such as Abemaciclib, Palbociclib, or Ribociclib, commonly included in first- or second-line therapy. When patients develop resistance to standard HR therapies, they are treated with combination chemotherapy. Single-agent or combination chemotherapy that is less toxic and enhances the patient's quality of life is preferred as standard treatment for BC. For individuals with TNBC who do not have inherited (breast cancer gene) BRCA1/2 mutations, cytotoxic chemotherapy is the prime treatment choice. TNBC can be treated in a target-specific manner with poly (ADP-ribose) polymerase (PARP) enzyme inhibitors [40]. For ERBB2+ BC, a taxane combined with Trastuzumab and Pertuzumab is frequently prescribed as first-line therapy, while Trastuzumab and Emtansine, an antibody–drug combination [21], are used as second-line therapy. A new chemotherapeutic drug is usually combined with an ERBB2-targeted therapy as follow-up therapy (Fig. 2). TNBC desperately needs a fresh breakthrough [82]. Recent evidence suggests that new antibody–drug conjugates show promise of effectiveness in the pretreatment of metastatic TNBC, and last-phase clinical studies are now on the process. Women with ERBB2+ and TNBC are more likely to develop brain metastases and have a poor prognosis. As a result, treatment of BC and metastatic BC with specific targeted and novel biomarker-based strategies is now in great demand.

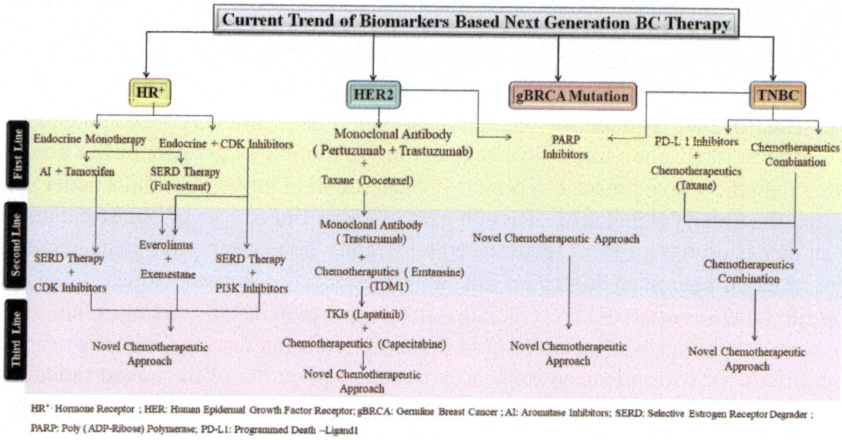

Fig. 2 Biomarker-based different management therapies of different types of breast cancer

4 Conventional Biomarker for BC Therapy

4.1 Luminal BC-Targeted Therapy

A significant proportion of BC cases (60–80%) is Luminal BC (known as HR$^+$), and this cancer case group is increasing, particularly among pre-menopausal females. The primary treatment for HR$^+$ BC is endocrine therapy, which blocks or reduces hormone levels. First-line therapy: Tamoxifen inhibits estrogen uptake by the ER; aromatase inhibitors (AIs) (Letrozole, Anastrozole, and Exemestane) prevent androgens from being converted to estrogens and cause estrogen depletion (Fig. 2). Luteinizing hormone-releasing analogs (Goserelin and Leuprolide) reduce hormone production from the ovary, and Fulvestrant (selective ER degrader) downregulated ER and was found to be biomarker-based therapy for BC [13, 21, 64]. Though researchers had great success in increasing the life span and treating successfully phases I–III HR$^+$ cancer, metastatic HR$^+$ BC is still the biggest limitation of the current therapy. So, a new approach for targeting is highly in demand. Several combination therapies are currently in trend (Table 1) with the existing therapies.

4.2 Hormone Receptor$^+$/ERBB2-Targeted Therapy

Endocrine therapy, which suppresses estrogen-induced cancer formation, is the standard treatment option for HR$^+$/ERBB2 BC. This standard therapy comprises a daily oral dose of anti-estrogen medicine for 5–6 years. But the therapy is dependent on the stage of menopause. Tamoxifen inhibits estrogen binding to the ER in both pre-menopausal and post-menopausal women. AIs (Anastrozole, Exemestane, and

Table 1 Biomarker-based combination drug therapy strategy for BC

Drug	Mode of action	Target	Therapy regimen
HR^+ BC			
Palbociclib	CDK4 and CDK6 inhibitors	Advance HER2−	Combination therapeutics with Letrozole and Fulvestrant
Ribociclib			
Everolimus			Combination therapeutics with Exemestane
Abemaciclib			Combination therapeutics with Letrozole and Fulvestrant
Pictilisib	Pan class I PI3K inhibitors	Early and Advance HER2−	Combination therapeutics with Letrozole and Fulvestrant
Voxtalisib	Pan class I PI3K and mTOR inhibitors	Advance HER2−	Combination therapeutics with Letrozole
Alpelisib	α-specific class I PI3K	Early and advance HER2−	Combination therapeutics with Letrozole and Fulvestrant
Entinostat	HDAC inhibitor	Advance ER−	Combination therapeutics with Exemestane
HER2+ BC			
Everolimus	mTOR inhibitors	Advance	Combination therapeutics with Trastuzumab and Vinorelbine
Neratinib	Irreversibly bind with HER1, HER2, and HER4	Early	Monotherapy
Patritumab	Anti-HER3 mAb	Advance	Combination therapeutics with Trastuzumab and Paclitaxel
Margetuximab	Anti-HER2 mAb	Advance	Monotherapy

(continued)

Table 1 (continued)

Drug	Mode of action	Target	Therapy regimen
Nelipepimut-S	Peptide vaccines	Early	Combination therapeutics with Trastuzumab
Trastuzumab and Emtansine	Antibody–drug conjugates	Early	Combination therapeutics with Trastuzumab and Paclitaxel
TNBC			
Olaparib	Oral PARP inhibitors	Advance stage HER2−, gBRCA+	Monotherapy
Talazoparib			
Niraparib			Combination therapeutics with Pembrolizumab

Abbreviations **CDK**: **Cyclin** dependent kinase, **mTOR**: Mammalian target of rapamycin, **PI3K**: hosphatidylinositol-3-kinase, **HDAC**: Istone deacetylases, **HER**: Human epidermal growth factor receptor, **PARP**: Oly (ADP-ribose) polymerase, **gBRCA**: Ermline reast cancer

Letrozole) are only effectual in post-menopausal females. The inhibitors decrease the level of circulating estrogen by suppressing androgen to estrogen conversion [44, 91]. Though hormone therapy has been demonstrated to be highly effective in the treatment of HR+/ERBB2 BC, the risk is still a concern. The doctor must decide whether to use gonadotropin-releasing hormone agonists like Leuprolide acetate and Goserelin to suppress the ovary or use oophorectomy to induce forced menopause. Furthermore, when menopause is induced, clinicians should choose between Tamoxifen and AIs to reduce the chances of metastasis [91]. Patients with HR+ BC are at risk of relapse of cancer even many years after their first diagnosis [85]. Gene-based adjuvant chemotherapy for HR^+/ERBB2 node-negative BC patients is under development based on targeted biomarker treatment [48, 91] and could be the next-generation therapy.

4.3 ERBB2+-Targeted Therapy

The discovery of ERBB2 treatment has been one of the most remarkable advancements in BC management. Trastuzumab, the monoclonal antibody (mAb) which targets the extracellular domain of ERBB2, was examined in humans for the first time in the 1990s. For stages II and III for $ERBB2^+$ BC, neoadjuvant chemotherapy regimens of Adriamycin/Cyclophosphamide-Paclitaxel and Docetaxel/Carboplatin were found to be very effective [75, 78]. Paclitaxel/Trastuzumab is currently considered to be the golden standard [97] of treatment for small, node-negative ERBB2+

cancer due to its improved long-term results and lower toxicity compared to single-agent chemotherapy. Pertuzumab combined with Neratinib [46] lowers the risk of recurrence in patients with high-risk ERBB2+ BC and is found to be more effective than standard therapy. Pertuzumab is a mAb that inhibits the dimer formation of the ERBB2 protein. Neratinib (Tyrosine Kinase Inhibitor (TKI)) given in combination with Capecitabine for the ERBB2 therapy. But till now, no chemotherapy has shown overall survival of the ERBB2+ BC, so advanced and more novel biomarkers are needed.

4.4 ERBB2-BC-Targeted Therapy

Chemotherapy remains an essential therapeutic option for the vast majority of stages I–III BC who seek to avoid recurrence. All three Docetaxel/Cyclophosphamide, Adriamycin/Cyclophosphamide, and Cyclophosphamide/Methotrexate/5-Fluorouracil are well-proven for lower-risk individuals' chemotherapy for ERBB2 BC and are considered to be the first-line biomarker-based therapy. Chemotherapy combining both anthracycline (e.g., Adriamycin) and taxane (e.g., Adriamycin/Cyclophosphamide followed by taxane) offers the most effective potential and continues to be the best option for high-risk patients. Anthracycline appears to be particularly significant in patients suffering from lymph node-positive cancer and TNBC [12, 91]. In search of alternative approaches, several combination therapies are in trend (Table 1).

4.5 Triple-Negative BC Targeted Therapy

Chemotherapy is routinely given to all patients with TNBC who have a tumor size bigger than 5 mm, even if it is axillary node-negative. Anticancer drugs are exclusive treatments recommended by the Food and Drug Administration (FDA) for treating non-metastatic TNBC conditions [41]. Because defective DNA damage repair is a natural feature of TNBC, DNA-crosslinking platinum chemotherapies have been investigated and are the first-line choice for the management of TNBC. To date, single-agent Capecitabine is the exclusive evidence-based adjuvant escalation technique available in this condition [47]. Multiple approaches are focused on the management of TNBC. Currently, monotherapy and combination therapy (Table 1) are in trend alongside conventional chemotherapies.

5 Pathway-Based Next-Generation Biomarker for BC Therapy

Several studies have been conducted in recent years to discover and verify molecular biomarkers specifically responsible for HR^+ BC and TNBC for use as prognostic and predictive biomarkers. These one-of-a-kind procedures are known as multi-parameter, multi-analyte, and multi-gene assays.

5.1 *ER Signaling Pathway for BC Therapy*

Tamoxifen, the first FDA-approved targeted cancer treatment, was a prescribed ER modulator (SERM) [74]. After the success of Tamoxifen, many more alternative endocrine therapies were discovered and are under clinical trials. Steroid-based (e.g., Exemestane) and non-steroid-based (e.g., Anastrozole, Letrozole) AIs and selective ER degraders (SERDs) such as Fulvestrant stand out as the first-line therapy to target ER BC [66]. Though ER targeting achieved great success after the second-line SERD and CDK4/6 combination therapy, the limitations of drug resistance and recurrence of BC needed a lot of research for a novel chemotherapeutic agent.

Many different target-specific novel biomarker-based therapy approaches have been developed and are in the development stage to address this constraint. SERD-targeted therapy; inhibition of ER transcriptional target genes (via CDK4 and CDK6 inhibitors) found an area of research for the next-generation novel biomarkers. The inhibition of upstream pathways (via mTOR/PI3K/protein kinase B [AKT] inhibitors) is one of the most investigated novels' approaches to target BC cells. Currently, research focused on the inhibition of ER phosphorylation (via CDK7 inhibitor) and activation and inhibition of ER transcription (via bromo-domain inhibitors and histone) biomarkers for the ER biomarker-based BC therapy.

Estrogen receptor 1 (ESR1) mutations remain the most problematic aspect of ER BC management [60]. Many ESR1 mutations cluster in the substance-binding region, resulting in ER transcriptional activation that is ER-independent. Advance SERMs and/or SERDs are being investigated as single or combined therapies for the next generation of BC management.

5.2 *CDK 4/6 Signaling Pathway for BC Therapy*

CDK 4/6 inhibitors used in composite with ER therapeutics produce superior outcomes and are utilized in standard clinical applications. CDK4/6 inhibitors block the CDK4/6-cyclin D complex from phosphorylating the retinoblastoma gene and causing G1 cell cycle arrest [32]. As of yet, there is no anticipating biomarker of counter-defiance. Crosstalk between the CDK4/6 and PI3K pathways opens up new therapeutic avenues. These specific CDK4/6 signaling biomarkers are the first- and second-line therapeutic approaches for HR^+ BC. Though these target-specific biomarkers showed excellent for combination therapy, a monotherapy still needs to prove its efficacy without severe side effects or high dose.

5.3 *PI3K/AKT/mTOR Signaling Pathway for BC Therapy*

The PI3K/AKT/mTOR biomarker-based signaling pathway is a significant and current trend in oncogenic pathways in BC research. These biomarkers are being studied specifically for drug resistance and advanced therapy. PI3K activation, downstream of TKIs activation and G protein-coupled receptor signaling, catalyzes the formation of phosphatidylinositol (3,4,5)-trisphosphate (PIP3), a lipidic second messenger [59]. In plasma, PIP3 binds and activates the proto-oncogene serine/threonine kinase AKT. Cancer suppressor genes such as phosphatase and tensin homolog (PTEN) and INPP4B [35] specifically control the termination of PI3K/AKT signaling and may be a significant biomarker for BC treatment. This pathway is one of the most significant biomarker-based pathways that controls cancer cell metabolism and cell stress, which directly regulates apoptosis and cell proliferation. These biomarkers are currently becoming the most predominant biomarkers for BC therapy.

mTOR is a well-known therapeutic downstream effector and one of the most investigated biomarkers. The catalytic component of the mTOR complex 1 (mTORC1) is a serine/threonine protein kinase [20, 96]. New possibilities for BC therapy arise from signaling interactions between the PI3K and the RAS/ERK/MAPK [98] signaling. The therapeutic effectiveness of the PI3K/AKT/mTOR pathway highly depends on the BC subtypes, the expression of proteins on the cell site, and the nature of anticancer drugs. PIK3CA-activating (PIK3 cancer-associated) [54] mutations have been found to stimulate the PI3KAKT-mTOR pathway in Luminal A, B, and HER2 subtypes. This mutation alters the p110α subunit, which is responsible for PI3K signaling without regulation and contributes to uncontrolled cancer cell proliferation and metastasis. To date, targeting these mutagenic biomarkers has not shown a successful outcome but has become the next generation of biomarkers for BC therapy.

Decreased PTEN expression has been documented in all subtypes of BC but is substantially more common in the basal-like subtype and acts as the prime biomarker for basal-type BC for next-generation therapy. After a decade of research, a list of promising preventive agents targeting segments of the PI3K pathway was developed as a promising therapeutic agent for BC. Among them, pan-PI3K inhibitors (e.g., CH5132799, BKM120,GDC-0941, XL-147, PX-866), p110-specific (BYL719, CAL-101, INK-1117, GDC-0032), specific-PI3K inhibitors, complex of PI3K/mTOR inhibitors (BGT226, BEZ235, PF-4691502, XL-765, GDC-0980, PKI-587, SF1126), mTORC1 inhibitors (Ridaforolimus, Everolimus, and Temsirolimus), mTORC1/2 inhibitors (INK-128, AZD-8055, OSI-027), and AKT inhibitors (Capivasertib, GSK690693, Ipatasertib) [80] play a very important role and are the most promising biomarkers for the next generation of biomarkers for BC therapy. Though research is in process to find a chemotherapeutic to target this pathway's specific biomarkers, finding a viable alternative therapy index for these downregulating agents is a major call for more research.

Research suggests that PI3K inhibitors enhance the susceptibility of BC cells to CDK4/6 inhibitors. PI3K inhibitors may also inhibit fast adaptability and trigger apoptosis in Palbociclib-arrested cells with PI3K-stimulating mutations. Though PI3K/AKT/mTOR with or without the combination of CDK4/6 has proven to be the most significant biomarker for the management of BC, the drug resistance and failure of the monotherapy still open up a lot of new therapeutic approaches for the targeting of BC.

5.4 HER2 Signaling Pathway for BC

Trastuzumab, a mAb targeting HER2, changed the therapeutic strategy for HER2+ BC [18]. The HER2 receptor is an ErbB receptor that is morphologically similar to the epidermal growth factor receptor (EGFR). HER1 (EGFR or ErbB1), HER3 (ErbB3), and HER4 (ErbB4) are also members of the ErbB family. HER2 lacks a biological ligand and self-dimerizes the ligand independently [55]. This enhances cell proliferation by activating downstream-indicating pathways such as the MAPK and PI3K/AKT pathways. Following Trastuzumab's success, other HER2 targeting therapeutics were developed, largely mAb (e.g., Trastuzumab and Pertuzumab, with or without chemotherapeutics. ADCs (e.g., ado-Trastuzumab-Emtansine [T-DM1] and TKIs (e.g., Lapatinib are novel biomarker-specific approaches for BC management. Trastuzumab and Pertuzumab are the two mAbs authorized by the FDA for HER2-overexpressed BC [38]. Trastuzumab's anticancer efficacy arises from antibody-dependent cellular cytotoxicity. Margetuximab (MGAH22) is a novel fragment crystallizable region (Fc)-optimized chimeric mAb [72]. Margetuximab is designed with a better Fc-domain capacity to attach to the immunoglobulin G Fc receptor on immune effector cells and generate antibody-dependent cellular toxicity. It comes under the next-generation therapy of BC management as this Fc targeting therapy is in a very nascent stage of research.

T-DM1 is the first ADC to be authorized by the FDA for the therapy of HER2+ BC. Targeting molecules like T-DM1 target-overexpressed HER2 receptors to deliver chemotherapeutics like Emtansine inside the tumor cells. The exact pathway for T-DM1 is not very clear, so a lot of research is needed for the development of more target-specific T-DM1 molecules for better therapeutic efficacy [49] in the near future. Furthermore, bispecific antibodies like PRS-343 and ZW-25 [30, 80] are also in the research pipeline for the HER2 BC treatment. These advanced bispecific antibodies are capable of producing therapeutic potential as well as acting as an immunologic memory builder for long-term anticancer impact.

TKIs are also used to combat HER2⁺ BC [18, 30, 38]. In the case of brain metastases, their ability to cross the blood–brain barrier was found to be promising. Lapatinib, the only TKI authorized by the FDA for BC, is an oral reversible TKI that targets both EGFR and HER2 [93]. Neratinib [93] is an advanced next-generation TKI that permanently inhibits EGFR and HER2. Neratinib is more powerful than Lapatinib but more hazardous. It comes from the next generation of therapeutics for BC treatment. The limitations of these TKIs always open a research area for more target-specific and safe therapeutics for BC cell targeting.

Another irreversible pan-HER inhibitor being investigated is Pyrotinib [93]. About half of the individuals with $HER2^+$ BC are also HR+. Research has shown that with effective HER2 inhibition, ER signaling can act as the principal onco-biomarker and produce an escape mechanism for acquired resistance. Endocrine therapy should be used with HER2-blocking treatments to improve BC management. In addition to ErbB2 overexpression, HER2 pathway upregulation includes PI3K dysregulation and HER2 receptor mutation. Multiple HER2 activating mutations have been identified in $HER2^+$, ER^+, and TNBC. According to the study, HER2-mutated BC is currently treated with mAbs, ADCs, or TKIs. The effectiveness and sensitivity of HER2 BC therapies highly depend on the nature of the mutation, HER2 overexpression, the BC microenvironment, and different BC molecular subtypes.

5.5 *Triple-Negative BC Signaling Pathways*

TNBC is one of the most aggressive and unmanageable BCs among all other subtypes. PARP inhibitors are the first-line choice for TNBC management and were developed to stimulate synthetic lethality in BRCA-deficient malignancies [31, 32]. PARP inhibitors are often used in TNBC based on three DNA-based homologous recombination deficit scores that have been linked to BRCA1/2 genetic abnormalities. Olaparib is one of the first FDA approved biomarker-based target agent for TNBC specifically for HER2- germline BRCA1/2-mutated cancer ($gBRACA^+$) [7, 83]. PARP inhibitors are now being researched in both mono and chemo-conjugated conditions and have become the most important next-generation biomarkers for TNBC management. Recently, apart from Olaparib, TNBC was similarly treated with FDA-approved Talazoparib with fewer side effects. Several clinical studies combining PARP therapies and immune checkpoint inhibitors are currently underway.

Glembatumumab vedotin, a well-known antibody–drug conjugate, was found to be beneficial in gpNMB-overexpressed BC [77]. This conjugated therapy was found to be target-specific for biomarker-based TNBC management. TKIs of VEGFR, EGFR, and SRC biomarkers have also been studied for the management of TNBC [10, 94] since these signaling receptors and the biomarkers regulate cancer cell proliferation and are overexpressed in TNBC. These novel markers are under study or in clinical trials and are effective next-generation pathway-based biomarkers to target TNBC.

Pembrolizumab, a mAb, may benefit patients regardless of PD-L1 expression [63] by triggering an immunogenic response that destroys tumor cells. Tumors with BRCA mutations are also more susceptible to platinum-based agents. Because PARP inhibitors and platinum salts both target areas of the same biochemical pathway, they prevent cell proliferation to induce cell apoptosis [17]. So, combination therapy of PARP and platinum-based chemotherapeutics may be a good approach to target TNBC. Other drugs that target genomic instability, such as CX5461, are being tested in clinical studies alongside PARP inhibitors and platinum salts.

The intra-tumoral overexpression of VEGF (angiogenic factor) is an important biomarker to distinguish between TNBC and non-TNBC and is the next generation for the targeting and management of TNBC. Bevacizumab (anti-VEGF mAbs) was found to be effective in preventing tumor neo-vasculature development and metastasis [39]. Currently, for metastatic TNBC, Bevacizumab and chemotherapeutics' (Docetaxel) combination therapy has been found to enhance the response rate.

TNBC has an overexpression of the EGFR and serves as a well-known biomarker [42]. Several phase II trials are underway or under consideration to study the effectiveness of Cetuximab (anti-EGFR mAbs) with adjuvant of Cisplatin for TNBC [67]. A current attempt is in the process of identifying a subset of TNBC patients who may react better to EGFR inhibitors. Lower expression of α-crystallin B chain, overexpression of PTEN, and absence of KRAS expression in cancer may be associated with a favorable response [89]. These specific biomarkers play a very significant role in the current as well as future chemotherapeutic development for the most complicated TNBC treatment.

SRC is a non-receptor signaling kinase overexpressed in TNBC and is downstream of EGFR, IGF-1R, PDGFR, and HGFR growth factor receptors [99]. When Dasatinib (inhibitor of SRC) was evaluated as a single treatment for TNBC, the outcome was disappointing. In cell line experiments, however, Dasatinib was found to have synergistic antitumor efficacy in a panel of TNBC cell lines in combination with Cetuximab and Cisplatin [88]. This drug-antibody complex induced apoptosis and decreased EGFR and MAPK phosphorylation more than monotherapy. After the success of this combination to combat TNBC, several clinical trials are under investigation to use Dasatinib-containing combination therapy for TNBC patients who overexpressed both EGFR and c-SRC for the next generation of TNBC therapeutics [89].

The lack of specific signaling receptors restricts the use of target-specific treatments in advanced TNBC. Currently, chemotherapeutics (typically taxanes, anthracyclines, and platinum-based therapeutics with or without Bevacizumab) are the only systemic treatment options. Due to the absence of any specific biomarkers and well-defined molecular targets, TNBC has the fewest options for treatment compared to any other BC subtype. It is crucial to find novel therapeutic targets and to successfully produce targeted therapies. To enhance this therapeutic method, clinically relevant molecular biomarkers of therapeutic responsiveness must be developed and characterized.

6 Next-Generation Non-invasive Biomarkers for Early Detection for BC

The most common confirmative test for BC is a needle or surgical biopsy. The procedure is both invasive and uncomfortable. As a result, research is currently focusing on non-invasive methods to identify biomarkers for early detection and treatment of BC. Body fluid-based tests are becoming increasingly common for detecting BC indicators (Fig. 3).

6.1 Circulating Carcinoma Proteins

Cell proliferation, metastasis, angiogenesis, cell signaling, and immunological control of BC are all well co-related with circulating carcinoma proteins [23]. Currently, these circulating carcinoma proteins are gaining the attention of scientists as potential biomarkers for early detection and therapeutics for BC. CA27-29, CA15-3, CA-125, carcino-embryonic antigen (CEA), circulating extracellular domain of HER2, tissue polypeptide-specific antigen (TPS), tissue polypeptide antigen (TPA) [24, 53] have all been identified as serum carcinoma protein indicators in BC. Rather than specific biomarker identification, a cluster of biomarkers may improve sensitivity and specificity for early detection as well as target-specific chemotherapeutics for the management of BC in the near future. Other secreted onco-proteins, such as TFF1, TFF3, ARTN, and SHON [2, 51], detected as circulating carcinoma proteins, are under investigation for therapeutic targets in the management of BC. Currently,

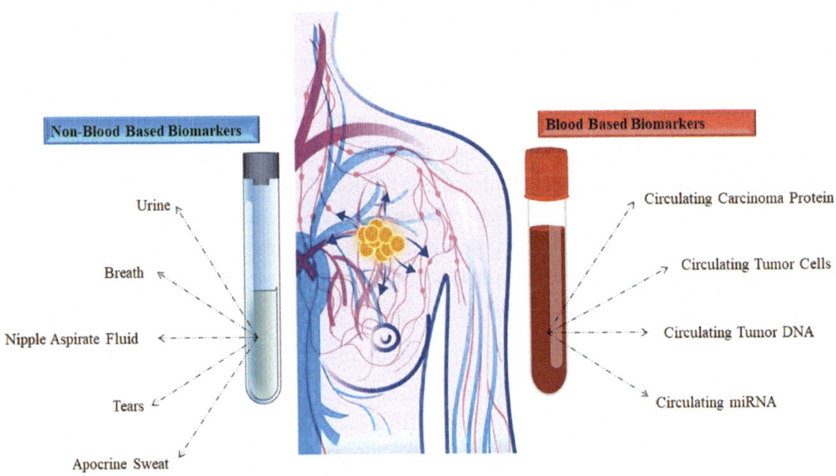

Fig. 3 Non-invasive body fluid-based biomarker for breast cancer management

these proteins are used for diagnosis, but they hold great potential for the treatment of BC shortly.

6.2 Circulating Tumor Cells

Circulating Tumor Cells (CTCs) increase the chance of recurrence and mortality in early-stage BC [61]. Many approaches have been under investigation for decades for counting and analyzing CTCs. Among them, physical features, immunomagnetic separation, immunosorbent assays, and RT-PCR tests show significant success in identifying and analyzing CTCs. CellSearch® and AdnaTest® [45, 68] are two CTC technologies that have been extensively explored in clinical studies. CTCs over a cut-off range of five cells per 7.5 mL of blood indicate a poor life span, making CTCs a potential prognosticative biomarker [71] for the early detection of BC. However, poor frequencies of CTCs and variability of antigens expressed on their surface make identification challenging and limit the application as a diagnostic tool. Although identification is challenging and the study is in its early stages, this blood-based biomarker might represent the future of BC care.

6.3 Circulating Tumor DNA

In BC, circulating tumor DNA (ctDNA) has come forth as a potential marker of BC progression [84]. Increased ctDNA levels have been correlated with advanced-stage BC and metastasis. Patients with BC exhibited considerably greater amounts of ctDNA than healthy controls. BC recurrence is more reliably diagnosed using chromosomal instability analysis of cfDNA using low-pass whole-genome sequencing than with traditional blood CA15-3 and CEA biomarkers [86, 92]. Changes in DNA methylation are frequently documented in cancer. Several genes have been identified to be hyper-methylated in BC, including p16, BRCA1, RASSF1A, APC, and GSTP1 [6, 34, 65]. cfDNA fragmentation patterns, in addition to methylation patterns, may be utilized to diagnose BC early. "DNA evaluation of fragments for early interception" (DELFI) [57] was developed using genome-wide fragmentation pattern analysis to uncover a huge number of faults in cfDNA. Early malignancies can be detected and the organ of tumor genesis is determined using ctDNA mutation and methylation tests.

6.4 Circulating miRNAs

When compared to healthy controls, a considerable number of upregulated circulated miRNAs [33] were discovered in circulation in the case of BC patients. Research has proven that a combination of a certain type and amount of circulating miRNAs differentiates BC from a healthy subject as well as BC from benign lesions. As a consequence, researchers believe that differently expressed circulating miRNAs might be utilized to detect BC and can act as next-generation biomarkers for the management of BC [33]. However, no clinically appropriate panels of circulating miRNAs for BC diagnosis are available to date [4].

Aside from blood, various bodily fluids such as urine, nipple aspirate fluid, tears, and perspiration, as well as patients' breath, have been studied for BC management.

6.5 Urine-Based Biomarkers

From metabolite profiling to proteome profiling to cellular and molecular analysis, research has demonstrated that urine may include valuable biomarkers for BC screening [19]. Phospholipid metabolism is increased in BC tissues, particularly of phosphatidylcholine (PC), phosphatidylethanolamine (PE), and sphingomyelin, or precursor chemicals. Urine samples were examined by nano-flow liquid chromatography/electrospray ionization tandem mass spectrometry before and after surgery to identify the change in the lipid profile of the BC patients [11]. Different urinary proteins are found in BC patients compared to healthy controls and act as potential biomarkers. A panel of 13 elevated proteins used for BC treatment includes AGRIN, DYH8, Filaggrin, FIBA, HBA, Keratin KIC10, Leucine, LRC36, MAST4, MMRN2, NEGR1, PEPA, and two uncharacteristically identified proteins, CI-131 and C4orf-14 [25, 62] which have gained popularity as urine-based biomarkers for the early-stage identification and analysis of BC. These stage-specific indicators have been linked to DCIS, early invading BC, and metastasis conditions of BC. Various analytical techniques such as nuclear magnetic resonance spectroscopy, gas/liquid chromatography, and capillary electrophoresis coupled with mass spectrometry have been used to study metabolomic and proteomic patterns [56] for different types and stages of BC. The research found a cancer-specific algorithm and applied computed models for early BC detection based on these profiles. Though urine-based biomarkers are not very clinically accepted, there might still be a future for the early detection and management of BC.

6.6 Breath-Based Biomarkers

Human breath comprises volatile and semi-volatile substances that can be the outcome of metabolic activity or pathological condition [14]. Research has currently proven that the circulating cancer cells or the tumor growth can alter the exhaled breath chemical composition. BC is a metabolic disorder and a change in the metabolic alteration in the human body can change the volatile and semi-volatile substances in human breath. Chemical tests of patients' exhaled breath have been used to diagnose a wide range of malignancies.

The application of breath analysis for BC revealed that BC-affected women contain a higher level of volatile pentane in their breath [52]. According to this study, a collection of volatile oxidative stress markers is useful to distinguish BC patients from healthy volunteers only through a simple breath analysis. This breath test properly detected women with BC, beating a screening mammography test in terms of negative predictive value. This makes breath testing a prime screening method for BC in the next generation of BC management. In a prospective investigation, breath samples from BC patients were studied using gas chromatography/mass spectrometry and identified five particular components such as 3-Methylhexane, Decene, Caryophyllene, Naphthalene, and Trichlorethylene that might be helpful for identify BC patients from healthy controls [51, 52]. Thanks to developments in breath collecting technology, clinical analysis, analytic approaches, and data modeling for BC are in a high progression stage. Recently, a clinical trial used commercially available economical electronic noses to distinguish different breath patterns among BC patients. The artificial neural network model was applied to distinguish the breath pattern of the healthy and BC subject. Scientist need to consider the various factors such as accuracy of breath test results, breaths' collection techniques, patient physiologic conditions, test settings, and methods of analysis, as well as standardized processes for better understand the breath biomarkers for therapeutic efficacy.

6.7 Nipple Aspirate Fluid Biomarkers

Nipple Aspirate Fluid (NAF) is a natural body secretion of breast epithelial duct cells. In healthy non-lactating mothers, NAF is collected through nipple aspiration or other means. NAF color and the presence of specific protein act as biomarkers for the early identification of BC. Women with bloody NAF had a more prominent risk of BC than those with whitish, creamy, greenish, or yellowish NAF. The NAFs obtained from the healthy subject were analyzed using NMR and gas chromatography/mass spectrometry. According to the analytical research, scientists discovered 38 metabolites, which included amino acids, organic acids, fatty acids, and carbohydrates. Among them, eight metabolites are unique to NAF, 19 to plasma, and 24 are shared metabolites [16], signaling that NAF has a different metabolic profile than comparable plasma samples. Exfoliated breast epithelial cells, which are the origin of BC, are also included in NAF. As a result, NAF may be a superior source of BC indicators to other body fluids.

6.8 *Tears' Biomarkers*

Tears are largely secreted from the lacrimal glands of the upper eyelids and circulate throughout our body's organs and tissues. As a result, tears are one of the most effective body fluids for external drug screening and pharmacokinetic research and are currently used as cancer biomarkers. Mammaglobin B (Lacryglobin) [43, 50] is overexpressed in BC and is employed as a biomarker in patients with axillary lymph node micrometastases. Mammaglobin B was initially detected in healthy human reflex tears. However, in different research studies, it was discovered that Lacryglobin is overexpressed in the tears of 88% of BC patients and acts as a biomarker [73]. A study employed surface-enhanced laser desorption/ionization time-of-flight mass spectroscopy to analyze the proteome profiles of tears (and blood) from BC patients and healthy controls, and differing protein expression levels were observed. The analysis revealed that a protein pattern in secreted tear fluid is distinguishable from non-BC patients and can be used as a high specificity and sensitivity tool for next-generation BC treatment. Using a semi-quantitative methodology, researchers observed that more than 20 different and unique proteins were expressed differentially in the tear fluid and were found positive for invasive BC. These findings show that eye fluids can be a possible source of non-invasive markers for BC screening in the next generation of BC management.

6.9 *Apocrine Sweat Biomarkers*

With the advancement of proteomic and metabolomics technologies, there has been an increase in interest in screening for sweat biomarkers. Sweat composition varies dramatically in a number of skin and other disorders. Metabolite analysis of sweat from lung cancer patients demonstrates cancer detection specificity and sensitivity. A similar BC sweat test showed that twenty measurable sweat markers [70] were sensitive and specific in diagnosing BC. Though sweat-based BC diagnosis has not been clinically validated, it holds high hope for future diagnostic and therapeutic biomarkers.

7 Formulation and Drug Repositioning

7.1 *Target-Specific Nano-formulation for Next-Generation BC Therapy*

Biomarker-based targeted drug delivery for BC overcomes the challenges of toxicity and enhances therapeutic efficacy. But the stability of the chemotherapeutics inside and outside of the body is challenging. Conventional and advanced nanotechnology-based drug delivery is one of the best approaches to delivering the target molecules to a selective location at the tumor mass level. Currently, most of the anticancer API-loaded nano-formulations fall under passive targeting. Passive drug delivery mainly depends on the enhanced permeability and retention effect (EPR) of the cancer microenvironment. The main limitations are poor bio-distribution and healthy tissue toxicity. Most of the FDA-approved formulations are passive targeted approaches. The best examples are PTX nano-formulations (Taxol®, (1) Abraxane®, (2) Genexol®, (3) Nanoxel®, and (4) Lipusu®) [58]. The route, ingredients, and other pre-formulation aspects are significant factors for passive nanotechnology-based drug delivery.

To overcome the limitations of the challenge currently, active drug delivery and stimuli-responsive drug delivery approaches advanced the nature of the formulation for BC management. For the next generation, BC nano-formulation needs to develop "more selective" response approaches. Trastuzumab in combination with paclitaxel for HER2+ BC is the first active targeted formulation for BC [27]. For enhancing the stability and efficacy of the formulation, different polymeric, lipidic, and hybrid nano-formulation approaches are under research and open up a future path for advanced research. Apart from the antibodies, a high number of ligands like proteins, peptides, aptamers, and carbohydrates are under investigation for the future active targeted nano-formulation for BC therapy.

Recently, the research on drug delivery for BC has focused on stimuli-responsive drug delivery [69]. The unique microenvironment of the cancer tissue opens up the approach to delivering the drug to the cancer site in a selective manner. This selective delivery of the APIs reduces the dose burden and healthy tissue toxicity. The stimuli which are currently being highly investigated are cancer cell pH, redox condition, enzyme expression, and temperature alteration. Despite a large number of successful pre-clinical nano-formulations, the clinical success is limited. The hope for the next generation's targeted specific drug delivery is high and promising.

7.2 *Drug Repositioning: Next Generation of BC Therapy*

Development of chemotherapeutics is a long and complex process. The cost and the time make the process more difficult in this cutting-edge competitive world. Moreover, the success rate of those new chemotherapeutics is not always very positive. So, research has focused on the repositioning of existing drugs for BC management [20]. A large number of the compounds are already approved or in clinical trials for reuse in BC therapy. The drug repositioning for the BC depends on the genomic nature, proteomics, transcription, and metabolomics of the cancer cells to be selected as anticancer drugs. A large number of drugs have already been repositioned for BC and shown excellent success. Alkylating agents (Cyclophosphamide, Thiotepa), Antimetabolite (Gemcitabine, Methotrexate), and mTOR inhibitors (Everolimus, Votubia, Evertor) [97] are better known as anticancer agents rather than for their actual purpose. The journey of drug repositioning for breast cancer has still got a long way to go and is next-generation therapeutics for BC management. The challenges the scientist should consider in the case of the drug repositioning are the dose concentration, dosing frequency, and monotherapy. To date, no single repositioning drugs have shown excellence in monotherapy and they are always given as neoadjuvant therapy. The next generation of a repositioned drug for BC needs to meet this challenge by maintaining its safety profile.

8 Conclusion

The death rate from BC has decreased in recent decades due to advances in treatment. Targeting ER has proven to be one of the most effective therapeutic strategies for HR+ BC. Furthermore, the success of biological therapies like anti-HER2 mAbs has demonstrated the possibility and importance of the molecular biomarker-directing method in BC treatment. However, TNBC is still a fatal illness with few therapy choices. The molecular mechanics behind the complex and genetically heterogeneous therapeutic reaction in BC have been better understood in recent years. This has fired up the research for innovative targeted drugs for exclusively targeting the specific molecular subtypes of BC, such as PARP inhibitors, PI3K/AKT/mTOR, CDK4/6 pathways, multiple kinases, or immune checkpoint inhibitor biomarkers. One of the most important difficulties in oncology is dealing with malignancy heterogeneity and genetic complications. A tiny clone can become predominant and play a significant role in the acquisition of resistance and metastasis under the selection force of a certain drug. The use of an effective combination of therapeutics to target multiple pathways and biomarkers of BC simultaneously might help to address heterogeneity and get rid of resistance. Furthermore, non-invasive methods of biomarker detection open up a new pathway for the early detection of new biomarker targets for the effective management of BC. More detailed research is required to confirm the treatment efficacy of these approaches in the hope that personalized therapy will become more approachable and incorporated at the therapeutic and clinical trial levels. Advancer stimuli specific formulation is the future for the BC management. Though the research for the nano-formulation has gained the most attention recently, the path for the most advanced and perfect nano-formulation for the BC manager is far away. Cost and value analyses are especially important when looking at how to transfer therapeutic advancements in BC to low-income and developing countries, which are already loaded down by the therapeutic burden and inadequate diagnostic and treatment capacity. The repositioning of existing drugs for the BC is gaining popularity. Still, high-dose toxicity, monotherapy, and the right choice of chemotherapy remain challenging for the next generation. All of these contextual challenges associated with BC therapy are expected to become more complicated as therapeutic customization becomes more frequent. Effective evidence collection, analysis, and application are necessary to help address these difficult challenges for better next-generation cancer management.

References

1. Abd-Elnaby M, Alfonse M, Roushdy M (2021) Classification of breast cancer using microarray gene expression data: a survey. J Biomed Inform 117:103764. https://doi.org/10.1016/j.jbi.2021.103764
2. Abdel-Fatah T, Broom RJ, Lu J, Moseley PM, Huang B, Li L, Liu S, Chen L, Ma RZ, Cao W (2019) SHON expression predicts response and relapse risk of breast cancer patients after anthracycline-based combination chemotherapy or tamoxifen treatment. Br J Cancer 120(7):728–745. https://doi.org/10.1038/s41416-019-0405-x
3. Abrahimi MS, Elwood M, Lawrenson R, Campbell I, Tin Tin S (2021) Associated factors and survival outcomes for breast conserving surgery versus mastectomy among New Zealand women with early-stage breast cancer. Int J Environ Health Res 18(5):2738. https://doi.org/10.3390/ijerph18052738
4. Aggarwal V, Priyanka K, Tuli HS (2020) Emergence of circulating MicroRNAs in breast cancer as diagnostic and therapeutic efficacy biomarkers. Mol Diagn Ther 24(2):153–173. https://doi.org/10.1007/s40291-020-00447-w
5. Ahmad A (2019) Breast cancer statistics: recent trends. Breast Cancer Metastasis Drug Resist:1–7. https://doi.org/10.1007/978-3-030-20301-6_1
6. Ambrosone CB, Young AC, Sucheston LE, Wang D, Li Y, Liu S, Tang L, Hu Q, Freudenheim JL, Shields PG (2014) Genome-wide methylation patterns provide insight into differences in breast tumor biology between American women of African and European ancestry. Oncotarget 5(1):237. https://doi.org/10.18632/oncotarget.1599
7. Andreopoulou E, Schweber SJ, Sparano JA, McDaid HM (2015) Therapies for triple negative breast cancer. Expert Opin Pharmacother 16(7):983–998. https://doi.org/10.1517/14656566.2015.1032246
8. Asselain B, Barlow W, Bartlett J, Bergh J, Bergsten-Nordström E, Bliss J, Boccardo F, Boddington C, Bogaerts J, Bonadonna G (2018) Long-term outcomes for neoadjuvant versus adjuvant chemotherapy in early breast cancer: meta-analysis of individual patient data from ten randomised trials. Lancet Oncol 19(1):27–39. https://doi.org/10.1016/S1470-2045(17)30777-5
9. Bal A, Joshi K (2022) Pathology of breast cancer. Breast Cancer, Springer:125–138. https://doi.org/10.1007/978-981-16-4546-4_8
10. Bayraktar S, Glück S (2013) Molecularly targeted therapies for metastatic triple-negative breast cancer. Breast Cancer Res Treat 138(1):21–35. https://doi.org/10.1007/s10549-013-2421-5
11. Beretov J, Wasinger VC, Graham PH, Millar EK, Kearsley JH, Li Y (2014) Proteomics for breast cancer urine biomarkers. Adv Clin Chem 63:123–167. https://doi.org/10.1016/B978-0-12-800094-6.00004-2
12. Blum JL, Flynn PJ, Yothers G, Asmar L, Geyer CE Jr, Jacobs SA, Robert NJ, Hopkins JO, O'Shaughnessy JA, Dang CT (2017) Anthracyclines in early breast cancer: The abc trials—usor 06–090, nsabp b-46-i/usor 07132, and nsabp b-49 (nrg oncology). J Clin Oncol 35(23):2647. https://doi.org/10.1200/JCO.2016.71.4147
13. Buqué A, Bloy N, Perez-Lanzón M, Iribarren K, Humeau J, Pol JG, Levesque S, Mondragon L, Yamazaki T, Sato A (2020) Immunoprophylactic and immunotherapeutic control of hormone receptor-positive breast cancer. Nat Commun 11(1):1–18. https://doi.org/10.1038/s41467-020-17644-0
14. Buszewski B, Kęsy M, Ligor T, Amann A (2007) Human exhaled air analytics: biomarkers of diseases. Biomed Chromatogr 21(6):553–566. https://doi.org/10.1002/bmc.835
15. Cejalvo JM, Pascual T, Fernández-Martínez A, Brasó-Maristany F, Gomis RR, Perou CM, Muñoz M, Prat A (2018) Clinical implications of the non-luminal intrinsic subtypes in hormone receptor-positive breast cancer. Cancer Treat Rev 67:63–70. https://doi.org/10.1016/j.ctrv.2018.04.015
16. Chan AA, Bashir M, Rivas MN, Duvall K, Sieling PA, Pieber TR, Vaishampayan PA, Love SM, Lee DJ (2016) Characterization of the microbiome of nipple aspirate fluid of breast cancer survivors. Sci Rep 6(1):1–11. https://doi.org/10.1038/srep28061

17. Chartron E, Theillet C, Guiu S, Jacot W (2019) Targeting homologous repair deficiency in breast and ovarian cancers: biological pathways, preclinical and clinical data. Crit Rev Oncol Hematol 133:58–73. https://doi.org/10.1016/j.critrevonc.2018.10.012
18. Costa RL, Czerniecki BJ (2020) Clinical development of immunotherapies for HER2+ breast cancer: a review of HER2-directed monoclonal antibodies and beyond. NPJ Breast Cancer 6(1):1–11. https://doi.org/10.1038/s41523-020-0153-3
19. Dayon L, Cominetti O, Affolter M (2022) Proteomics of human biological fluids for biomarker discoveries: technical advances and recent applications. Expert Rev Proteom 19(2):131–151. https://doi.org/10.1080/14789450.2022.2070477
20. De A, Kuppusamy G (2020) Metformin in breast cancer: preclinical and clinical evidence. Curr Probl Cancer 44(1):100488. https://doi.org/10.1016/j.currproblcancer.2019.06.003
21. De A, Kuppusamy G, Karri VVSR (2018) Affibody molecules for molecular imaging and targeted drug delivery in the management of breast cancer. Int J Biol Macromol 107:906–919. https://doi.org/10.1016/j.ijbiomac.2017.09.059
22. Ding Q, Huo L, Peng Y, Yoon EC, Li Z, Sahin AA (2022) Immunohistochemical markers for distinguishing metastatic breast carcinoma from other common malignancies: update and revisit. Semin Diagn Pathol Elsevier. https://doi.org/10.1053/j.semdp.2022.04.002
23. Dong MP, Enomoto M, Thuy LTT, Hai H, Hieu VN, Hoang DV, Iida-Ueno A, Odagiri N, Amano-Teranishi Y, Hagihara A (2020) Clinical significance of circulating soluble immune checkpoint proteins in sorafenib-treated patients with advanced hepatocellular carcinoma. Sci Rep 10(1):1–10. https://doi.org/10.1038/s41598-020-60440-5
24. Duffy MJ (2006) Serum tumor markers in breast cancer: are they of clinical value? Clin Chem 52(3):345–351. https://doi.org/10.1373/clinchem.2005.059832
25. Fakhar HBZ, Rezaei-Tavirani M, Akbari ME, Sajadi SN, HadiZadeh M. Proteins which have been found in breast cancer by proteomic's analyzer. https://doi.org/10.46718/JBGSR.2022.11.000271
26. Feng Y, Spezia M, Huang S, Yuan C, Zeng Z, Zhang L, Ji X, Liu W, Huang B, Luo W (2018) Breast cancer development and progression: Risk factors, cancer stem cells, signaling pathways, genomics, and molecular pathogenesis. Genes Dis 5(2):77–106. https://doi.org/10.1016/j.gendis.2018.05.001
27. Fraguas-Sánchez A, Martín-Sabroso C, Fernández-Carballido A, Torres-Suárez A (2019) Current status of nanomedicine in the chemotherapy of breast cancer. Cancer Chemother Pharmacol 84(4):689–706. https://doi.org/10.1007/s00280-019-03910-6
28. Garreffa E, Arora D (2021) Breast cancer in the elderly, in men and during pregnancy. Surg Infect (Larchmt):139–146. https://doi.org/10.1016/j.mpsur.2021.11.018
29. George JN, Nester CM (2014) Syndromes of thrombotic microangiopathy. N Engl J Med 371(7):654–666. https://doi.org/10.1056/NEJMra1312353
30. Gombos A, Franzoi MA, Awada A (2019) Investigational drugs in early stage clinical trials for the treatment of HER2+ breast cancer. Expert Opin Investig Drugs 28(7):617–627. https://doi.org/10.1080/13543784.2019.1633306
31. Guo G, Zhang F, Gao R, Delsite R, Feng Z, Powell SN (2011) DNA repair and synthetic lethality. Int J Oral Sci 3(4):176–179. https://doi.org/10.4248/IJOS11064
32. Guo Q, Lin X, Ye L, Xu R, Dai Y, Zhang Y, Chen Q (2019) Comparative efficacy of CDK4/6 inhibitors plus aromatase inhibitors versus fulvestrant for the first-line treatment of hormone receptor-positive advanced breast cancer: a network meta-analysis. Target Oncol 14(2):139–148. https://doi.org/10.1007/s11523-019-00633-9
33. Hamam R, Hamam D, Alsaleh KA, Kassem M, Zaher W, Alfayez M, Aldahmash A, Alajez NM (2017) Circulating microRNAs in breast cancer: novel diagnostic and prognostic biomarkers. Cell Death Dis 8(9):e3045–e3045. https://doi.org/10.1038/cddis.2017.440
34. Heichman KA, Warren JD (2012) DNA methylation biomarkers and their utility for solid cancer diagnostics. Clin Chem Lab Med 50(10):1707–1721. https://doi.org/10.1515/cclm-2011-0935
35. Hodgson MC, Shao L-J, Frolov A, Li R, Peterson LE, Ayala G, Ittmann MM, Weigel NL, Agoulnik IU (2011) Decreased expression and androgen regulation of the tumor suppressor gene INPP4B in prostate cancer androgen regulation of INPP4B. Cancer Res 71(2):572–582. https://doi.org/10.1158/0008-5472.CAN-10-2314

36. Horvath E (2021) Molecular subtypes of breast cancer-What breast imaging radiologists need to know. Rev Chil Radiol 27(1):17–26
37. Houghton SC, Hankinson SE (2021) Cancer progress and priorities: breast cancer. Cancer Epidemiol Biomarkers Prev 30(5):822–844. https://doi.org/10.1158/1055-9965.EPI-20-1193
38. Hurvitz SA, Hu Y, O'Brien N, Finn RS (2013) Current approaches and future directions in the treatment of HER2-positive breast cancer. Cancer Treat Rev 39(3):219–229. https://doi.org/10.1016/j.ctrv.2012.04.008
39. Huveldt D, Lewis-Tuffin LJ, Carlson BL, Schroeder MA, Rodriguez F, Giannini C, Galanis E, Sarkaria JN, Anastasiadis PZ (2013) Targeting Src family kinases inhibits bevacizumab-induced glioma cell invasion. PLoS ONE 8(2):e56505. https://doi.org/10.1371/journal.pone.0056505
40. Ilic L, Haidinger G, Simon J, Hackl M, Schernhammer E, Papantoniou K (2022) Trends in female breast cancer incidence, mortality, and survival in Austria, with focus on age, stage, and birth cohorts (1983–2017). Sci Rep 12(1):1–10. https://doi.org/10.1038/s41598-022-10560-x
41. Janeva S, Zhang C, Kovács A, Parris TZ, Crozier JA, Pezzi CM, Linderholm B, Audisio RA, Bagge RO (2020) Adjuvant chemotherapy and survival in women aged 70 years and older with triple-negative breast cancer: a Swedish population-based propensity score-matched analysis. Lancet Healthy Longevity 1(3):e117–e124. https://doi.org/10.1016/S2666-7568(20)30018-0
42. Jang MH, Kim HJ, Kim EJ, Chung YR, Park SY (2015) Expression of epithelial-mesenchymal transition–related markers in triple-negative breast cancer: ZEB1 as a potential biomarker for poor clinical outcome. Hum Pathol 46(9):1267–1274. https://doi.org/10.1016/j.humpath.2015.05.010
43. Jeyalatha MV, Qu Y, Liu Z, Ou S, He X, Bu J, Li S, Reinach PS, Liu Z, Li W (2017) Function of meibomian gland: contribution of proteins. Exp Eye Res 163:29–36. https://doi.org/10.1016/j.exer.2017.06.009
44. Joshi H, Press MF (2018) Molecular oncology of breast cancer. The Breast, Elsevier 282–307:e285. https://doi.org/10.1016/B978-0-323-35955-9.00022-2
45. Kasimir-Bauer S, Bittner A-K, König L, Reiter K, Keller T, Kimmig R, Hoffmann O (2016) Does primary neoadjuvant systemic therapy eradicate minimal residual disease? Analysis of disseminated and circulating tumor cells before and after therapy. Breast Cancer Res 18(1):1–15. https://doi.org/10.1186/s13058-016-0679-3
46. Kiepas A (2021) Defining the molecular mechanisms that regulate breast cancer cell migration, McGill University (Canada). https://www.proquest.com/dissertations-theses/defining-molecular-mechanisms-that-regulate/docview/2571092678/se-2
47. Korde LA, Somerfield MR, Carey LA, Crews JR, Denduluri N, Hwang ES, Khan SA, Loibl S, Morris EA, Perez A (2021) Neoadjuvant chemotherapy, endocrine therapy, and targeted therapy for breast cancer: ASCO guideline. J Clin Oncol 39(13):1485–1505. https://doi.org/10.1200/JCO.20.03399
48. Krop I, Ismaila N, Andre F, Bast RC, Barlow W, Collyar DE, Hammond ME, Kuderer NM, Liu MC, Mennel RG (2017) Use of biomarkers to guide decisions on adjuvant systemic therapy for women with early-stage invasive breast cancer: American Society of Clinical Oncology clinical practice guideline focused update. J Clin Oncol 35(24):2838. https://doi.org/10.1200/JCO.2017.74.0472
49. Lambert JM, Chari RV (2014) Ado-trastuzumab Emtansine (T-DM1): an antibody–drug conjugate (ADC) for HER2-positive breast cancer. ACS Publications 57(16):6949–6964. https://doi.org/10.1021/jm500766w
50. Lebrecht A, Boehm D, Schmidt M, Koelbl H, Schwirz RL, Grus FH (2009) Diagnosis of breast cancer by tear proteomic pattern. Cancer Genomics Proteomics 6(3):177–182
51. Li J, Guan X, Fan Z, Ching L-M, Li Y, Wang X, Cao W-M, Liu D-X (2020) Non-invasive biomarkers for early detection of breast cancer. Cancers 12(10):2767. https://doi.org/10.3390/cancers12102767
52. Li J, Peng Y, Duan Y (2013) Diagnosis of breast cancer based on breath analysis: an emerging method. Crit Rev Oncol Hematol 87(1):28–40. https://doi.org/10.1016/j.critrevonc.2012.11.007

53. Li X, Dai D, Chen B, Tang H, Xie X, Wei W (2018) Clinicopathological and prognostic significance of cancer antigen 15-3 and carcinoembryonic antigen in breast cancer: a meta-analysis including 12,993 patients. Disease Markers. https://doi.org/10.1155/2018/9863092
54. Liedtke C, Cardone L, Tordai A, Yan K, Gomez HL, Figureoa LJB, Hubbard RE, Valero V, Souchon EA, Symmans WF (2008) PIK3CA-activating mutations and chemotherapy sensitivity in stage II–III breast cancer. Breast Cancer Res 10(2):1–10. https://doi.org/10.1186/bcr1984
55. Linggi B, Carpenter G (2006) ErbB receptors: new insights on mechanisms and biology. Trends Cell Biol 16(12):649–656. https://doi.org/10.1016/j.tcb.2006.10.008
56. Lisec J, Schauer N, Kopka J, Willmitzer L, Fernie AR (2006) Gas chromatography mass spectrometry–based metabolite profiling in plants. Nat Protoc 1(1):387–396. https://doi.org/10.1038/nprot.2006.59
57. Lumbard K, Keefer LA, van't erve I, Carey J, Chesnick B, Butler D, Rongione M, Punt CJ, Dracopoli NC, Fijneman RJ (2022) DELFI as a real-time treatment response assessment for patients with cancer. Cancer Res 82(12_Supplement):2224–2224. https://doi.org/10.1158/1538-7445.AM2022-2224
58. Luo C, Wang Y, Chen Q, Han X, Liu X, Sun J, He Z (2012) Advances of paclitaxel formulations based on nanosystem delivery technology. Mini Rev Med Chem 12(5):434–444. https://doi.org/10.2174/138955712800493924
59. Luo J, Manning BD, Cantley LC (2003) Targeting the PI3K-Akt pathway in human cancer: rationale and promise. Cancer Cell 4(4):257–262. https://doi.org/10.1016/s1535-6108(03)00248-4
60. Luque-Bolivar A, Pérez-Mora E, Villegas VE, Rondón-Lagos M (2020) Resistance and overcoming resistance in breast cancer. Breast Cancer Targets Ther 12:211. https://doi.org/10.2147/BCTT.S270799
61. Matikas A, Kotsakis A, Apostolaki S, Politaki H, Perraki M, Kalbakis K, Nikolaou M, Economopoulou P, Hatzidaki D, Georgoulias V (2022) Detection of circulating tumour cells before and following adjuvant chemotherapy and long-term prognosis of early breast cancer. Br J Cancer 126(11):1563–1569. https://doi.org/10.1038/s41416-022-01699-5
62. Mazur M, Pyatchanina T (2016) The use of proteomic technologies in breast cancer research. Exp Oncol. http://dspace.nbuv.gov.ua/handle/123456789/137748
63. Mok TS, Wu Y-L, Kudaba I, Kowalski DM, Cho BC, Turna HZ, Castro G Jr, Srimuninnimit V, Laktionov KK, Bondarenko I (2019) Pembrolizumab versus chemotherapy for previously untreated, PD-L1-expressing, locally advanced or metastatic non-small-cell lung cancer (KEYNOTE-042): a randomised, open-label, controlled, phase 3 trial. The Lancet 393(10183):1819–1830. https://doi.org/10.1016/S0140-6736(18)32409-7
64. Morganti S, Curigliano G (2020) Moving beyond endocrine therapy for luminal metastatic breast cancer in the precision medicine era: looking for new targets. Expert Rev Precis Med Drug Dev 5(1):7–22. https://doi.org/10.1080/23808993.2020.1720508
65. Murria R, Palanca S, de Juan I, Egoavil C, Alenda C, García-Casado Z, Juan MJ, Sánchez AB, Santaballa A, Chirivella I (2015) Methylation of tumor suppressor genes is related with copy number aberrations in breast cancer. Am J Cancer Res 5(1):375. https://www.ncbi.nlm.nih.gov/pmc/articles/PMC4300703/
66. Nagaraj G, Ma CX (2021) Clinical challenges in the management of hormone receptor-positive, human epidermal growth factor receptor 2-negative metastatic breast cancer: a literature review. Adv Ther 38(1):109–136. https://doi.org/10.1007/s12325-020-01552-2
67. Nechushtan H, Vainer G, Stainberg H, Salmon AY, Hamburger T, Peretz T (2014) A phase 1/2 of a combination of cetuximab and taxane for "triple negative" breast cancer patients. The Breast 23(4):435–438. https://doi.org/10.1016/j.breast.2014.03.003
68. Onstenk W, Sieuwerts AM, Kraan J, Van M, Nieuweboer AJ, Mathijssen RH, Hamberg P, Meulenbeld HJ, De Laere B, Dirix LY (2015) Efficacy of cabazitaxel in castration-resistant prostate cancer is independent of the presence of AR-V7 in circulating tumor cells. Eur Urol Oncol 68(6):939–945. https://doi.org/10.1016/j.eururo.2015.07.007
69. Oshiro-Júnior JA, Rodero C, Hanck-Silva G, Sato MR, Alves RC, Eloy JO, Chorilli M (2020) Stimuli-responsive drug delivery nanocarriers in the treatment of breast cancer. Curr Med Chem 27(15):2494–2513. https://doi.org/10.2174/0929867325666181009120610

70. Patel A, Shah S (2022) Noninvasive biomarkers: emerging trends in early detection of breast cancer. Breast Cancer Bench Pers Med:125–143. https://doi.org/10.1007/978-981-19-0197-3_7
71. Paterlini-Brechot P, Benali NL (2007) Circulating tumor cells (CTC) detection: clinical impact and future directions. Cancer Lett 253(2):180–204. https://doi.org/10.1016/j.canlet.2006.12.014
72. Pegram MD, Tan-Chiu E, Miller K, Rugo HS, Yardley DA, Liv S, Stewart SJ, Erban JK (2014) A single-arm, open-label, phase 2 study of MGAH22 (margetuximab)[fc-optimized chimeric anti-HER2 monoclonal antibody (mAb)] in patients with relapsed or refractory advanced breast cancer whose tumors express HER2 at the 2+ level by immunohistochemistry and lack evidence of HER2 gene amplification by FISH. Am J Clin Oncol. https://doi.org/10.1200/jco.2014.32.15_suppl.tps671
73. Pieragostino D, D'Alessandro M, di Ioia M, Di Ilio C, Sacchetta P, Del Boccio P (2015) Unraveling the molecular repertoire of tears as a source of biomarkers: beyond ocular diseases. Proteom Clin Appl 9(1–2):169–186. https://doi.org/10.1002/prca.201400084
74. Pinkerton JV, Thomas S (2014) Use of SERMs for treatment in postmenopausal women. J Steroid Biochem Mol Biol 142:142–154. https://doi.org/10.1016/j.jsbmb.2013.12.011
75. Plevritis SK, Munoz D, Kurian AW, Stout NK, Alagoz O, Near AM, Lee SJ, Van Den Broek JJ, Huang X, Schechter CB (2018) Association of screening and treatment with breast cancer mortality by molecular subtype in US women, 2000–2012. JAMA 319(2):154–164. https://doi.org/10.1001/jama.2017.19130
76. Ravikumar M, Rachana P (2022) Study on different approaches for breast cancer detection: a review. SN Comput Sci 3(1):1–6. https://doi.org/10.1007/s42979-021-00898-w
77. Rose AA, Biondini M, Curiel R, Siegel PM (2017) Targeting GPNMB with glembatumumab vedotin: current developments and future opportunities for the treatment of cancer. Pharmacol Ther 179:127–141. https://doi.org/10.1016/j.pharmthera.2017.05.010
78. Rouzier R, Perou CM, Symmans WF, Ibrahim N, Cristofanilli M, Anderson K, Hess KR, Stec J, Ayers M, Wagner P (2005) Breast cancer molecular subtypes respond differently to preoperative chemotherapy. Clin Cancer Res 11(16):5678–5685. https://doi.org/10.1158/1078-0432.CCR-04-2421
79. Salama AM, Hanna MG, Giri D, Kezlarian B, Jean M-H, Lin O, Vallejo C, Brogi E, Edelweiss M (2022) Digital validation of breast biomarkers (ER, PR, AR, and HER2) in cytology specimens using three different scanners. Mod Pathol 35(1):52–59. https://doi.org/10.1038/s41379-021-00908-5
80. Savard M-F, Khan O, Hunt KK, Verma S (2019) Redrawing the lines: the next generation of treatment in metastatic breast cancer. Am Soc Clin Oncol Educ Book 39:e8–e21. https://doi.org/10.1200/EDBK_237419
81. Scheltens P, Vijverberg E (2021) Aducanumab: appropriate use recommendations. Springer. https://doi.org/10.14283/jpad.2021.45
82. Schettini F, Brasó-Maristany F, Kuderer NM, Prat A (2022) A perspective on the development and lack of interchangeability of the breast cancer intrinsic subtypes. NPJ breast cancer 8(1):1–4. https://doi.org/10.1038/s41523-022-00451-9
83. Sharma P (2016) Biology and management of patients with triple-negative breast cancer. Oncologist 21(9):1050–1062. https://doi.org/10.1634/theoncologist.2016-0067
84. Shoukry M, Broccard S, Kaplan J, Gabriel E (2021) The emerging role of circulating tumor DNA in the management of breast cancer. Cancers 13(15):3813. https://doi.org/10.3390/cancers13153813
85. Stade A, Stade B (2017) Sélection et décryptage. J Med 377:1836–1846. https://doi.org/10.1684/ito.2017.0081
86. Suppan C, Brcic I, Tiran V, Mueller HD, Posch F, Auer M, Ercan E, Ulz P, Cote RJ, Datar RH (2019) Untargeted assessment of tumor fractions in plasma for monitoring and prognostication from metastatic breast cancer patients undergoing systemic treatment. Cancers 11(8):1171. https://doi.org/10.3390/cancers11081171

87. Tabár L, Dean PB, Tucker FL, Yen AM-F, Chen SL-S, Jen GHH, Wang JW-C, Smith RA, Duffy SW, Chen TH-H (2022) A new approach to breast cancer terminology based on the anatomic site of tumour origin: the importance of radiologic imaging biomarkers. Eur J Radiol 149:110189. https://doi.org/10.1016/j.ejrad.2022.110189
88. Tian J, Raffa FA, Dai M, Moamer A, Khadang B, Hachim IY, Bakdounes K, Ali S, Jean-Claude B, Lebrun J-J (2018) Dasatinib sensitises triple negative breast cancer cells to chemotherapy by targeting breast cancer stem cells. Br J Cancer 119(12):1495–1507. https://doi.org/10.1038/s41416-018-0287-3
89. Tong CW, Wu M, Cho WC, To KK (2018) Recent advances in the treatment of breast cancer. Front Oncol 8:227. https://doi.org/10.3389/fonc.2018.00227
90. Veeraraghavan J, De Angelis C, Mao R, Wang T, Herrera S, Pavlick A, Contreras A, Nuciforo P, Mayer I, Forero A (2019) A combinatorial biomarker predicts pathologic complete response to neoadjuvant lapatinib and trastuzumab without chemotherapy in patients with HER2+ breast cancer. Ann Oncol 30(6):927–933. https://doi.org/10.1093/annonc/mdz076
91. Waks AG, Winer EP (2019) Breast cancer treatment: a review. JAMA 321(3):288–300. https://doi.org/10.1001/jama.2018.19323
92. Watanabe K, Nakamura Y, Low S-K (2021) Clinical implementation and current advancement of blood liquid biopsy in cancer. J Hum Genet 66(9):909–926. https://doi.org/10.1038/s10038-021-00939-5
93. Xuhong JC, Qi XW, Zhang Y, Jiang J (2019) Mechanism, safety and efficacy of three tyrosine kinase inhibitors lapatinib, neratinib and pyrotinib in HER2-positive breast cancer. Am J Cancer Res 9(10):2103. https://www.ncbi.nlm.nih.gov/pmc/articles/PMC6834479/
94. Yadav BS, Chanana P, Jhamb S (2015) Biomarkers in triple negative breast cancer: a review. World J Clin Oncol 6(6):252. https://doi.org/10.5306/wjco.v6.i6.252
95. Yang SX, Polley EC (2019) Systemic treatment and radiotherapy, breast cancer subtypes, and survival after long-term clinical follow-up. Breast Cancer Res Treat 175(2): 287–295. https://doi.org/10.1007/s10549-019-05142-x
96. Yoon M-S (2017) The role of mammalian target of rapamycin (mTOR) in insulin signaling. Nutrients 9(11):1176. https://doi.org/10.3390/nu9111176
97. Yu KD, Liu GY, Chen CM, Li JW, Wu J, Lu JS, Shen ZZ, Shao ZM (2013) Weekly paclitaxel/carboplatin/trastuzumab therapy improves pathologic complete remission in aggressive HER2-positive breast cancers, especially in luminal-B subtype, compared with a once-every-3-weeks schedule. Oncologist 18(5):511–517. https://doi.org/10.1634/theoncologist.2012-0057
98. Zhang F, Ni Z-J, Ye L, Zhang Y-Y, Thakur K, Cespedes-Acuña CL, Han J, Zhang J-G, Wei Z-J (2021) Asparanin A inhibits cell migration and invasion in human endometrial cancer via Ras/ERK/MAPK pathway. Food Chem Toxicol 150:112036. https://doi.org/10.1016/j.fct.2021.112036
99. Zhang S, Yu D (2012) Targeting Src family kinases in anti-cancer therapies: turning promise into triumph. Trends Pharmacol Sci 33(3):122–128. https://doi.org/10.1016/j.jbi.2021.103764

Nanostructures-Based Polymeric Composite for Theranostic Applications

Poonam Jain, K. Gireesh Babu, Alle Madhusudhan, and Mitchell Lee Taylor

Abstract Nanomaterials are synthesized by fusing together a variety of fantastic physical and chemical properties that enable the delivery of responsive, intelligent, and smart nanomedicine. Till now, various polymeric and metal-based nanomaterials have so far been developed for the biomedical applications such as drug/gene delivery, sensing, immune cells delivery, imaging, and wound healing etc. As polymeric nanoparticles exhibit extraordinary of high drug loading and sustained release property, on the other hand, metal nanoparticles have lacking this property, although these nanoparticles are still of great use because of the unique property such as contrast imaging property under electromagnetic field, photothermal property toward near-infrared radiation, and hyperthermia effect in the region of tumor site upon injected the magnetic active nanoparticles under magnetic field. However, using metal-based nanoparticles alone limits their utility in biomedical applications due to their instability in biological fluids. Unquestionably, the scientist modified the surface of metal nanoparticles utilizing several biocompatible polymers and proteins to create multifunctional theranostic metal–organic nanoparticles. Thus, this approach not only enhances colloidal stability for longer periods of time, but also enhances bioavailability of synthesized nanoparticles, improves the payload of many therapeutic agents via electrostatic and covalent conjugation, enhances blood plasma half-life, releases the drug in a sustained manner, and simultaneously aids in disease diagnosis. Theranostic formulations may be delivered specifically using functionalized nanoparticles, which also lower the possibility of any potential toxicity problems brought on by metal ions. Thus, in this book chapter, we have covered various theranostic polymeric nanostructures, their synthesis process, general chemical, physical, and biological characteristics, and their biomedical applications.

P. Jain · K. G. Babu (✉)
Department of Life Sciences, Parul Institute of Applied Sciences, Parul University, Vadodara, Limda, India
e-mail: allemadhusudhan@gmail.com

A. Madhusudhan (✉) · M. L. Taylor
Department of Chemistry, The University of Memphis, Memphis, TN 38152, USA
e-mail: allemadhusudhan@gmail.com

A. Madhusudhan et al. (eds.), *Functional Smart Nanomaterials and Their Theranostics Approaches*, Smart Nanomaterials Technology,
https://doi.org/10.1007/978-981-99-6597-7_6

Keywords Nanomaterials · Theranostic · Photothermal effect · Hyperthermia · Immune cells

1 Introduction

Cancer is the second leading cause of death in which patients at their third and fourth stages are metastasized and very difficult to treat and cure. Therefore, patients often undergo many painful surgeries and chemotherapy, which can significantly impact their health. There are several ways of treatment. Chemotherapy is the primary approach for regulating the proliferation of cancer cells before surgery and radiotherapy. However, after clinical treatment, chemotherapy often leads to drug resistance and causes brutal side effects such as anemia, nausea, hair loss, dizziness. Furthermore, the recurrence of tumor growth is a significant concern both during and after treatment. According to GLOBOCAN report 2022, 1,918,030 new cancer cases and 609,360 cancer deaths have been expected to take place in the USA [100], while in 2021, estimates from World Health Organization in 2019 stated that the person below the age of ~65 years in >100 out of 180 countries is more pronounced to incidence and mortality rate [105].

In 2020, estimated data showed 19.3 million new cases, 10 million cancer deaths, for both sexes male and female. Only in Asian country, out of total global population (59.5%), about 58% of death cases were recorded for both sexes [100]. If we talk about particularly about cancer disease, in 112 countries, prostate cancer is most occurrence cancer in men, lung cancer in 36 countries, and liver cancer in 11 countries. Although lung cancer has a high mortality rate in men, lung cancer is the most common cancer form in women, followed by breast and cervical cancer in terms of death rates. Beside this, colorectal, non-melanoma, stomach, thyroid, bladder, pancreatic, leukemia, ovary cancer types are other top-most cancers that being clinically diagnosed and reported for death worldwide in case of women and men both and separately [9, 25, 100, 105]. Therefore, cancer is one of the diseases that causes morbidity and mortality globally and its cumulative risk of death that is non-communicably spreading in the human community due to the irregular lifestyle risk factors such as smoking, drinking alcohol, physical inactivity, obesity, and adaptation of urban lifestyle and also due to sudden gene mutation [54, 98, 108].

Nevertheless, early diagnosis of cancer and drug/gene therapy can prevent the tumor from spreading. However, numerous studies have found that chemotherapy alone exhibits a wide range of side effects because of its low bioavailability, undesired biodistribution, and requirement of high doses to achieve highest therapeutic index [96]. The better understanding of nanotechnology and nanotherapeutics appeared as most beneficial field of sciences over the traditional cancer therapy [80]. Basically, nanotherapeutics is a combination of nanocarrier and therapeutic ingredients that have been accepted and approved by Food and Drug Administration (FDA) for several cancer treatments. As the number of nanomedicine-based treatments under clinical trials increases, some of them are being approved for biomedical applications [120].

The main benefit of using nanostructures is that when bioactive molecules or drugs, are transformed into nanosize range or incorporated into nanocarrier, it actually improves the blood circulation time of that therapeutic agents, their localization and efficacy of delivered drug at the targeted cite [57, 83]. In addition, nanotheranostic allows the detection or diagnosis, delivery of drug, and treatment of cancer disease through various nanomaterials.

Typically, the term theranostic, was first time introduced in 2002, which means any substances for a combine purpose i.e., treatment and diagnostics or image-guided drug delivery (DD) [42]. Numerous nanomaterials have been synthesized and designed for biomedical applications, for example, sensing, imaging, and therapy [70, 87, 102]. These imaging materials are utilized in the various forms for biomedical purposes, such as fluorescent groups, zero-dimension nanostructure like quantum dots for tracking [1, 109], magnetic [62, 112] and paramagnetic nanomaterial [66, 74], also called contrast agents (CAs) for magnetic resonance imaging (MRI), computed tomography image (CT), and positron emission tomography (PET).

Despite this, treatment approaches involve conjugating and loading drugs for active and passive delivery, respectively, in biological systems via various administration route (e.g., oral intravenous or intramuscular). Another therapeutic strategy for treating cancer is gene therapy, which works by silencing, editing, and/or replacing the primary diseased gene [47, 77, 106]. The effect of photothermal [20, 81] and hyperthermia are another therapeutic approach that utilizes especial metallic nanostructures capable of absorbing particular wavelength of light. This temperature increase induces apoptosis or cancer cell death [44]. Thus, wisely developed theranostic nanomaterials are being used for treatment of many diseases including cancer and some of them already in market. Therefore, the main focus of this book chapter is on newly established polymeric nanostructures that show combined therapeutic and imaging capabilities simultaneously for efficient cancer treatment.

2 Classification of Nanoparticles

Nanoscience is a multidisciplinary branch of science that employs materials having at least one dimension, a nanoscale size range between 1 and 100 nm, and a spectrum of physical, chemical, and biological properties. In other words, the manipulation of matter at the atomic level involves synthesis, characterization, and the potential application of nanostructured materials in various industries such as electrical, data storage, sensing, medicine and textile, painting, and so on. Furthermore, the potential knowledge of developed nanotechnology-based materials can provide benefits to human-society, animals and environment. In one of this fields, nanomedicine specifically utilizes biocompatible nanomaterials and therapeutic ingredient (drug/gene) for the health benefits. The European Sciences' Foundation stated that nanomedicine is the science and technology of diagnosing, treating, and preventing disease and injury, relieving pain, improving the human health through utilizing molecular knowledge and tools [103].

Thus, nanotechnology is essentially divided into two categories: (i) nanomaterial and (ii) nanodevices. Henceforth, nanomaterials are those nanostructures that can be nanoparticles, dendrimer, micelles, and nanodrug conjugates or prodrug [43].

The classification of nanoparticles is then categorized on the basis of dimensions. Hence, nanoparticles are classified into four dimensions.

(a) One-dimensional nanostructured particles (e.g., optical fibers [55]).
(b) Two-dimensional nanostructured particles (e.g., carbon nanotubes [8, 22], graphene sheets [17, 33, 65]).
(c) Three-dimensional nanostructured particles (e.g., metal nanocuboids [47], sphere [95], and other x, y, z-axis directional nanoparticles [123]).
(d) Zero-dimensional nanostructured particles are those materials which do not exhibit any dimension but have great importance in nanomedicine (e.g., quantum dots [53]).

Despite dimension, nanomaterials are divided into inorganic and organic NPs. By composition, organic NPs involve polymeric, solid lipid NPs, liposomes, niosomes, emulsions, micelles, and dendrimers, while inorganic-based NPs are metal and metal oxide NPs that are usually formed from transition metal or rare earth metals. For example, gold (Au), silver (Ag) iron (Fe), manganese (Mn), gadolinium (Gd), and cerium oxide (CeO_2) nanoparticles (NPs) are most accepted and famous NPs for various biomedical applications (Fig. 1). With the advancement of nanotechnology research, both organic and metallic or metal oxide framework have also been developed and are referred to as hybrid NPs. These hybrid NPs exhibit extraordinary properties and are used for various applications.

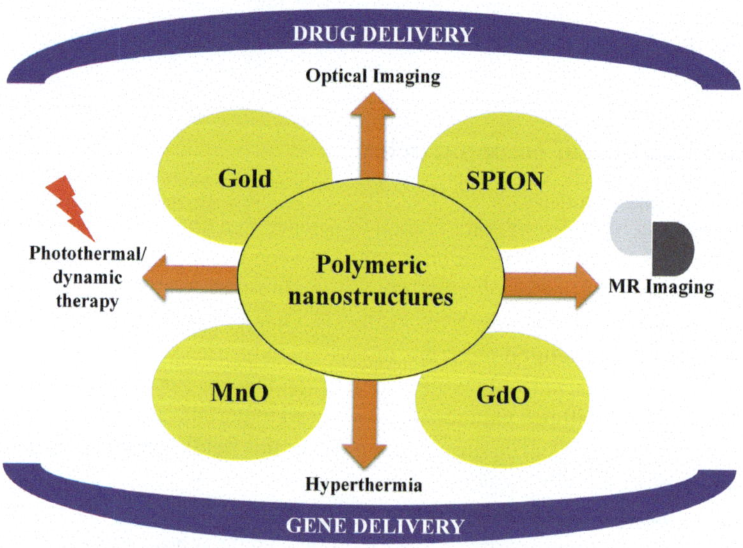

Fig. 1 Polymer-coated metal and metal oxide nanostructures for various applications

2.1 *Polymeric Nanoparticles*

Polymeric NPs are made up of short or long chain of synthetic and natural polymers such as poly(lactic-co-glycolic acid) (PLGA), poly(lactic)acid (PLA), polyalkyl cyanoacrylates, poly-Ɛ-caprolactone [27, 45, 89]. These NPs are recognized as safe and biocompatible, and the US FDA and EMA have granted them approval [61] for use in clinical trials and other applications. Polymeric NPs show high colloidal stability in the biological fluid, and bioavailability with improved drug efficacy than alone drug. The peculiar property of polymeric NPs is the offering of functionalization the surface using different targeting ligands (biotin, folic acid, cRGD peptide, bombesin) and aptamers [33, 36, 46, 48, 49, 92, 91]. Various chemotherapeutic drugs are encapsulated with high payload and into the polymeric NPs and controlled release for cancer treatment [61]. Thus, polymeric NPs are synthesized by various methods which are described below.

2.1.1 Solvent Evaporation Method

Solvent evaporation is widely employed technique used to prepare NPs of polymers. This method of polymeric NPs is mainly depended on the volatile solvents and the product form as emulsion. Previously, the major solvents were chloroform and other such as dichloromethane; however, due to their toxicity effect, the scientists started using ethyl acetate. Nanoemulsions are typically formed during high-speed stirring, and by that time, the polymerand the organic solvent have evaporated, resulting in the formation of NPs. These NPs then diffuse during the phase conversion. The well-known method for the formation of polymeric NPs is preparation of single emulsions and double emulsion [60]. The former method is based on oil-in-water also known as (o/w) or water-in-oil (w/o), and later method follows water-in-oi-in water (w/o/w) [16, 122]. Therefore, nanosize particles are formed at RT via high-speed magnetic stirring, homogenization, and ultrasonication. Among three, first two forces are mostly utilized than ultrasonication due to breakdown of therapeutic molecules or genes [28]. Hence, to remove the waste and unreacted substance, washing process is performed by ultracentrifuge via several times with water, and finally, the product is solidified by lyophilization.

Beside the single and double emulsion methods of synthesis, polymeric NPs are synthesized by more than two polymers in which one as main polymer and another as a copolymer that gives the strength to NPs and to hold the different natures of drug compounds. For example, the well-known drug paclitaxel (PTX) which belongs to taxol family, is hydrophobic in nature. It is loaded into the poly(ethylene glycol)-block-poly(D,L-lactide acid) (PEG-PDLLA) micelles and studied in clinical trials, demonstrating negligible side effect of the drug [75]. It is very important to maintain the integrity and structure of NPs before reaching to diseased site because most of the drug shows premature release in GIT, or at low pH. Numerous stabilizers, like Poloxamer, are thus added to the preparation phase in order to get the desired results.

PLGA-based NPs were synthesized which were wrapped with protein shell, namely bovine serum albumin (BSA), and further functionalized through acidity-triggered rational membrane peptide. This system has shown the reduction of serious adverse effect of doxorubicin (DOX) without losing the integrity of particle's nanostructure in the low pH and the resultant formulation showed excellent therapeutic efficacy against breast cancer cells [79].

2.1.2 Emulsification or Solvent Diffusion Method

In this method, polymeric NPs are synthesized by o/w emulsion between a partially aqueous-miscible solvent (polymer + therapeutic drug) along with the surfactant for stabilization. Henceforth, the formed emulsion has two phases. The internal phase consists of a half hydro-miscible organic solvent. Further adding of water in the initial phase causes solvent diffusion into the external phase of dispersed droplets or formed colloidal particles [16]. Also, adding a few amounts of oil, e.g., C6, 8, 10, and 12 in the organic phase attributes the formation of nanocapsules. Hereafter, the organic solvent is evaporated or filtered to produce 50–900 nm NPs or capsules.

2.1.3 Emulsification or Reverse Salting-Out Method

This method is again similar to described two methods with a slight change. In this method, nanospheres are formed via separating the aqueous-miscible solvent (in above, partially aqueous miscible) from the aqueous solution with the help of salting-out technique [16]. The most common salting-out agents are magnesium and calcium chloride which are basic electrolytes. The non-electrolyte salting-out agent such as sucrose is used in the synthesis of nanosphere. Thus, several water-soluble organic solvents such as acetone, methanol, and ethanol are extremely utilized for the composition of o/w emulsion, and finally, a gel-type particle, salting-out agent, and a colloidal stabilizer are formed in the aqueous phase. Hence, size, chain length, concentration of polymer, volume of solvent play an important role in the formation of different ranges of nanosphere. At high-speed magnetic stirring and room temperature (RT) condition, the o/w emulsions are prepared and diluted by using deionized water to diffuse into the organic solvent to the exterior phase, which results in polymer precipitation and so-called nanosphere [38]. In this process, cross-filtration is utilized to remove the unwanted solvent and other impurities. Therefore, reverse salting-out method provides nanoscale sphere with mean particle size >130 nm and <600 nm.

2.1.4 Nanoprecipitation

In this process, nanospheres and nanocapsules are synthesized using solvent displacement method where in initial step two miscible solvents are utilized for the synthesis of nanoparticles. The two phases, i.e., internal phase is composed of polymer are dissolved in organic solvents such as acetone and ethanol that evaporated at RT under high-speed stirring or homogenization and external immiscible water phase. Hence, the polymer is added and dissolved in a water-miscible phase with intermediate polarity that further added dropwise or with controlled rate into another aqueous solution. Owing to rapid diffusion property of polymer, the round-shape NPs are formed instantly without any phase change, and when the solvent diffuses out from the preformed nanodrops, then finally the precipitates of polymeric NPs are formed which are generally lyophilize or freeze dry. However, the formed nanoprecipitates do not show stability in colloidal solution. Therefore, surfactant is added during the NPs' formation which improves the colloidal stability for longer time of period. Since many years, nanoprecipitates are used in the formation of nanospheres and nanocapsules which also have high drug payload than emulsification and evaporation method. Despite solvent displacement property, both the NPs are formed by distinct method. For instance, capsular structure is produced via already dissolved therapeutic agent in an oil and then emulsified in the polymeric (organic) solution via avoiding the dispersion of internal phase and external phase of formed emulsion, while spherical structures are formed via addition of active molecules and dispersed in the polymeric solution.

Besides the above-described method, polymeric nanoparticles are also synthesized via dialysis method, supercritical fluid technology, polymerization of monomers (emulsion and mini-and microemulsion polymerization and interfacial polymerization method).

2.1.5 Advantages of Polymeric Nanoparticles

The polymeric NPs are highly recommended as nanocarrier due to their biocompatible nature, and however, use of unappropriated amount of surfactant may cause adverse effect. Furthermore, these NPs show high drug payload and efficacy against cancer cells as proved by various in vitro and in vivo studies. As NPs also improve the efficacy of drug through passive delivery and modulate the pharmacokinetics and dynamic properties. Although some studies have reached clinical trials, they have not showen a greater anticancer effect than pristine drug. The main reason of failure of drug-loaded nanocarrier includes stability of synthesized nanospheres and nanocapsules in the stomach pH 1.2, interaction of blood serum protein due to which the nanoformulations are washed out by reticuloendothelial system (RES) or Kupffer cells. Hence, based on the interaction property of drug-loaded nanoformulation toward the physical barrier such as blood circulation, the proteins, cell membranes, and stomach pH in case of oral administration, alternative method like surface modification for active delivery is promoted.

Conjugation of Targeting Ligand

Some examples of targeted ligand such as biotin, folic acid, and cRGD peptide are covalently conjugated or functionalized on the surface of polymeric NPs which specially interact with receptors that overexpressed on cancer cells than normal cells; for instance, MCF-7 cancer cells overexpress receptors of biotin, folate, and integrin proteins and improve the cellular uptake[119]

Polysaccharide Conjugation or Shell Structure

Furthermore, several natural polysaccharides are used for formation of shell structure and introduced surrounding the formed polymeric NPs. As these polysaccharides show higher stability in GIT pH and one of the successful studies reported by our group in which the inulin was utilized as shell structure to stabilize the nanoformulation in the GIT tract [32, 37].

For many years, scientists and researchers have discovered the extraordinary properties of polyethylene glycol (PEG); therefore, it has received Generally Regarded as Safe (GRAS) from the FDA [104]. In biological system, it gives steric barrier to formed NPs, therefore also known as "stealth polymer". The main function of PEGylation is to avoid unnecessary interaction of enzymes and proteins, e.g., opsonin and further enhance half-life of nanoformulation in the blood circulation. In 1977, Davis and Abuchowsky described about the conjugation of PEG to BSA and liver catalases in order to increase the half-life of the proteins with no modulation of their activity.

Hence, polymeric nanoparticles show several advantages such as:

- Biocompatibility and high blood circulation half-life.
- High payload.
- Reduces the doses' frequency of therapeutic agent.
- Provide surface functionality.
- Prodrug formation.
- Biodegradation.

2.2 *Metallic Nanoparticles and non-metallic Moieties*

Due to the potential utility in every branch of science and technology, d-block elements have recently attracted a great deal of attention worldwide. Metal NPs, which include gold, silver, copper, zinc, titanium, selenium, manganese, and iron oxide NPs, among many other elements, are mostly used in biological applications [31, 73]. The main role of these synthesized NPs is as DD vehicle, for imaging, sensing, and as self-therapeutic agent. Scientists are also exploring other potential uses because metal NPs possess several unique physical properties like surface plasmon resonance, photoluminescence, chemical such as catalytic, sensing, produce reactive oxygen species (ROS), as well as biological properties like DD, contrast imaging, hyperthermia, photothermal effect. These unique properties arise from factors such as (i) high surface-to-volume ratio, (ii) the nature of the metal used, (iii) the size and shape of the nanoparticle, and (iv) their surface properties. These qualities make metal NPs suitable candidate for the diagnosis, tracking, and treatment of various diseases that can be life-threatening. As a result, in this book chapter, the most popular and FDA-approved metal and metal oxide NPs used in biomedical applications are discussed.

Depending on the native nature of metal, various physical and chemical methods have been used to synthesize these nanomaterials. The most common methods are chemical precipitation/reduction, hydrothermal, and sol–gel methods which are majorly employed to synthesize and stabilize of the metal-based NPs. Furthermore, selection of synthesis method of metal-based NPs has high importance which also depends on various factors. These factors are:

- The kinetics of interaction of precursor metal ions (Au chloride) with reducing agent (sodium borohydrate) to produce Au NPs,
- Adsorption or interaction of stabilizing agent (e.g., PVA) with NPs during synthesis,
- Temperature,
- Reaction time,
- Agitation speed in case of stirring methods, etc.

Hence, various experimental techniques and optimizations show a significant impact on the stability, physicochemical properties, and morphology (structure and size) of the NPs.

2.2.1 Synthesis Method for Metal-Based Nanoparticles for Potential Use in Biomedical Applications

Broadly, metallic NPs are synthesized by technique which is divided into two approaches—bottom-up approach and top-down approach (Fig. 2). In former approach, the raw material is used to develop the nanomaterials and particle size is reduced to nanoscale particles, while atoms or molecules are the starting material and kept for nucleation and growth in case of later approach. Thus, in this chapter, we shall explain about the bottom approach. These methods are:

Chemical Precipitation Method

In this method, metal NPs such as gold, silver, platinum, and other NPs are synthesized in the presence of appropriate solvent, and ionic salt such as Au^{3+}, Ag^{+}, and Pt^{4+} is reduced by adding reducing agents (sodium borohydrate and hydrazine hydrate) with or without temperature [2, 97]. For stabilizing, sodium citrate and other stabilizing agents are added which further monitored via visual color change or by analysis

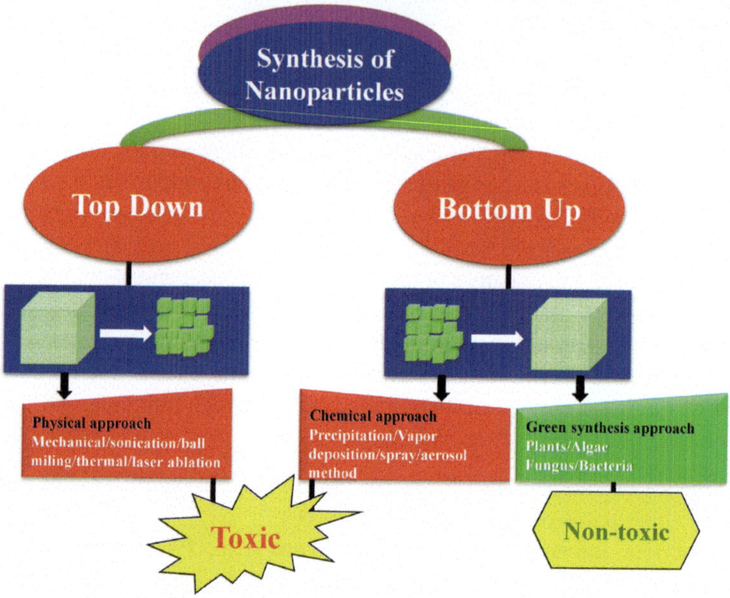

Fig. 2 Top-down and bottom-up approaches for the synthesis of various nanoparticles via physical, chemical, and green methods

through absorption at particular wavelength. As metal NPs possess surface plasmon resonance (SPR) property, change in the particle size shows district color.

Despite this, metal oxide NPs are synthesized by adding oxidizing agent such as sodium hydroxide (NaOH) and potassium oxide (KOH) [90]. For instance, our team have synthesized manganese NPs by taking manganese chloride as ionic salt and NaOH. In both cases, the precipitate is washed several times with double-distilled water and ethanol to remove impurities and dry through lyophilization and vacuum dry. In another example, magnetite and maghemite NPs were synthesized by dissolving the iron precursor which was precipitated by adding NaOH (1.5 M) or 25% ammonia under magnetic stirring. This study revealed that addition of base significantly affects the particle size of both the iron-based NPs, while very small-sized maghemite NPs are obtained because of Ostwald ripening effect and further oxidation produces magnetite NPs [85]. Hence, the main advantage of chemical precipitation is simplest method for synthesis of metal NPs.

Hydrothermal Method

In this method, at high pressure along with temperature, reaction of aqueous solution vapors with solid material occurs which ultimately produces very small scale of NPs. Ideally, the ionic product has maximum value of <300 or 250 °C, the NPs are synthesized in the range of 180–300 °C, and the critical temperature along with pressure of distilled water is reached up to ~370 °C and ~22 MPa, respectively, and therefore under supercritical circumstances, the solvent properties for many substances, such as the dielectric constant and solubility, alter significantly. There is a relationship between water density and the dielectric constant in relation to temperature and pressure. Polar inorganic salts can be dissolved in water (dielectric constant of 78) at ambient temperature. Thus, with rising temperature and falling pressure, the dielectric constant of water falls.

Similar, in hydrothermal synthesis, first metal precursor or polymer, e.g., zinc chloride or starch, respectively, is hydrated up to certain temperature like 50–70 °C, then hydrolyzed via excess of heating (180–300 °C) in hydrothermal autoclave device. Thereafter via dehydration, metal hydroxide converts into metal oxide (ZnO) product or graphene-based QDs [11, 69]. Depending on the basis of rate of hydrolysis and solubility of salt precursor, NPs or QDs of different particle size ranges are produced. Since many years, hydrothermal method is utilized to develop various metal and metal oxide NPs due to rapid synthesis, uniform particle size with less nanoscale and purity.

Sol–Gel Method

This method is used to prepare metal oxide NPs and their nanocomposite. There are five major steps through which NPs are synthesized: i. hydrolysis, ii. polycondensation, iii. aging, iv. drying, and v. thermal decomposition. For synthesis, metal precursor is first dissolved and hydrolyzed in water or in other organic solvents (alcohol) which oxidize and after that, the condensation of inorganic aerosol polymers and a metal oxide gel with a three-dimensional network is produced or metal oxide NPs produce in the form of gel, which are washed several times to remove impurities [13]. As iron oxide NPs are versatile nanomaterials used for theranostic application, the morphology is influenced by change in various factors. For example, successful synthesis of iron oxide NPs is occurred by sol–gel method and this study showed the salt precursor played an important role in change in morphology and dispersity [19]. In one experiment, manganese dioxide NPs were synthesized by reducing the permanganate using fumaric acid. Similarly, different shapes of ultrafine MnO_2 nanorods and nanowires were produced and stabilized by adding surfactants such as cetyltrimethylammonium bromide (CTAB), polyvinylpyrrolidone (PVP). In this process, first surfactant was dissolved in ethanol and then manganese acetate was added followed by nitric acid at 80 °C [107]. Therefore, several studies revealed the advantage of sol–gel technique to synthesize NPs of different shapes or morphologies.

3 Strategy for Production of Theranostic Polymeric Nanocomposite

3.1 Gold Nanoparticles (AuNPs)

Gold NPs have attracted significant attention in the biomedical industry for many years due to its advanced qualities and therapeutic capabilities, [3, 6, 88]. These characteristics include high colloidal stability, hydrophilicity, and the capacity to interact with polymers, biomolecules, and dyes on their surface that are chemically engage as secondary design and functioning.

It has been reported that AuNPs with spherical shape perform inadequately toward strong near-infrared (NIR) absorption due to deep tissue penetration in photoacoustic imaging (PAI) and photothermal therapy (PTT). Furthermore, the nanosized range means <100 nm which is more preferably show strong NIR absorptions and then typically used for tumor site imaging and further treatment. Along with that, nanoscale AuNPs also show prolong blood circulation time and strong enhanced permeability effect (EPR) at the target sites. Different structures of AuNPs such as rod, stars, and cages shapes with relatively small particles have been shown strong NIR absorptions (Fig. 3). Despite these extraordinary properties, Au-based NPs of different sizes are also possessing some other attractive characteristics including high drug payload loading, high therapeutic impact against cancer cells without leaching, and bioaccumulation in the body tissues. However, some recent reports also suggested the potential toxic effects like inflammatory response via immune cells which are supposed to depending on the particle size, morphology, concentration, etc. For instance, treatment with rod shape AuNPs produces ROS species due to disrupting mitochondria followed by apoptosis against various cancer and normal cells [29].

Thus, to avoid the unwanted cytotoxicity, AuNPs are surface modified using natural and biocompatible biomolecules such as PEG, amino acids, and small polysaccharide chains. For example, AuNPs were coated with PEG chain, which provided high colloidal stability and declared their non-toxic behavior under clinical trial [4, 72]. Likewise, some other Au-based nanocarrier, e.g., Auroshell was found as biocompatible in clinical trials, and even it has been proven that AuNPs are not accumulate in the liver tissue and resultant in vivo and in vitro studies have shown that administered formulation not causes any liver dysfunction. Due to biocompatible nature of Au-based nanostructures, researchers are still designing and developing surface-modified AuNPs to enhance the stability and therapeutic efficacy in the biological system and for treatment of malignant tumors [26, 51]. To improve the colloidal stability and payload, polysaccharide nanosphere-coated Au nanorods were developed with a Janus shape.

The synthesized Janus NPs exhibited PA-guided potential PTT and gene therapy against breast cancer [18]. The prepared NPs showed excellent localized surface plasma resonance (LSPR) peaks in the NIR area which is also responsible to enhance the quality of photo-based temperature for PTT (approximately 52.8 °C). The fluorescence property was also measured after loading red fluorescent CdTe

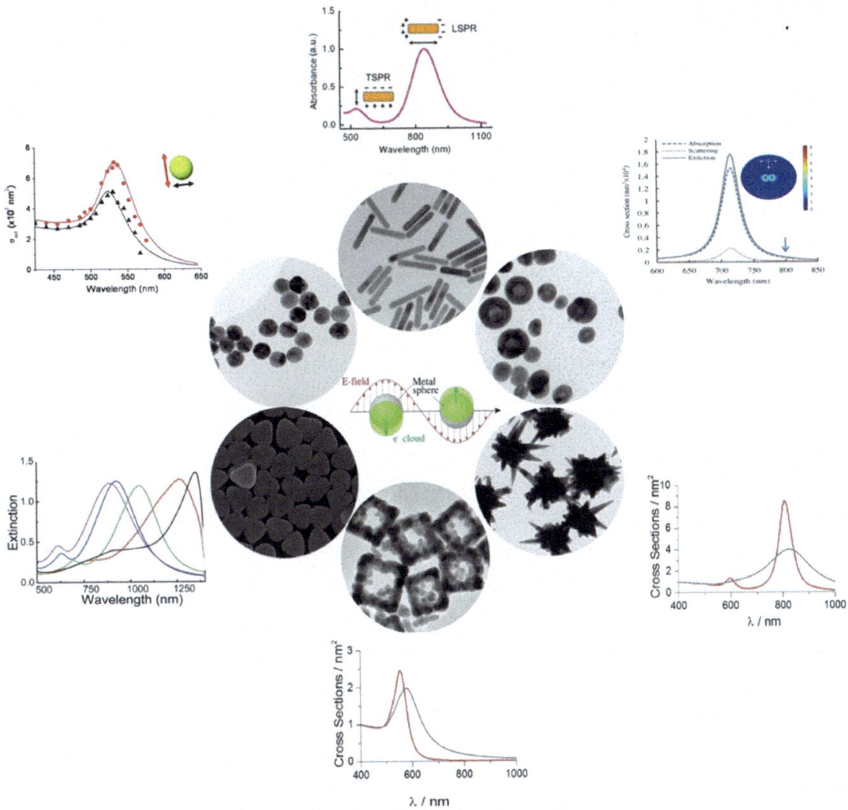

Fig. 3 Different sizes of gold nanoparticles as observed by electron microscopy images with the absorption and scattering graph adopted from [94]

QDs onto chitosan bounded Au nanorods for an animal imaging experiment with deep tissue penetration in a time dependent manner. Hence, AuNPs showed excellent contrast properties and the contrast imaging was confirmed in the dark field microscopy due to the coupling effect of chitosan AuNPs[18]. Hence, various strategy has been reported on modification of AuNPs surface, which enhanced the cytotoxic effect via delivering of gene at cancer site and cell imaging.

3.2 *Superparamagnetic Nanoparticles*

The superparamagnetic NPs, known as magnetic, or iron oxide-based NPs are available in many phases such as Fe_3O_4, Fe_2O_3, and Fe_2O_4. Among them, Fe_3O_4 NPs are well-known iron-based NPs due to the high magnetic property without external magnetic field. Magnetite and magnemite NPs are extremely used for the biomedical applications such as CAs in MRI applications and magnet-based DD. To improve the stability in the biological media, these NPs are also coated with polymers and other biocompatible biomolecules and finally guided for DD at the target tissue site. Hence, these iron-based particles possess superparamagnetism, which under an external magnetic field become magnetized up to the saturation level of magnetization, and after the remove of the magnetic field, loss their residual magnetic properties. Previous studies report that the saturation of magnetization and contrast property depend on the particle size and morphology.

For example, synthesis of size-dependent (from 16 to 20 nm) properties of magnetic iron oxide NPs for MRI relaxation, and cellular uptake [12]. In another example, 22 nm and 30 nm Fe_2O_3 NPs were fabricated in which less size Fe_2O_3 showed shorter dephasing than higher size for MRI relaxation [63]. Beside the magnetic property, magnetic NPs also possess hyperthermia property; therefore, one team have synthesized drug-loaded magnetic solid lipid NPs for magnetic-tracking-based DD. In this study, the release of paclitaxel was observed by hyperthermia action of magnetic NPs up to 4 h [78]. Another reported study fabricated lipid/silica coatings of 100–200 nm size on SPIONs for theranostic applications, such as contrast properties and hyperthermia applications. As shown in Fig. 4, the bare SPIONs were synthesized in the presence of ferrous and ferric chloride and ammonium hydroxide so called as co-precipitation method under an inert atmosphere, whereas lipid coating was done by thin-film hydration method. Finally, an anticancer drug, DOX was successfully loaded up to 35%, while further coating with silica increases encapsulation efficiency about ~50% [82].

The co-precipitation was also followed by one another team by reaction between ferrous chloride, NaOH, and ammonium oxalate to produce porous hematite α-Fe_2O_3 NPs. The surface coating was done using surfactant sodium dodecyl sulfate and L-cysteine, an amino acid which showed largest surface area of 614 m^2/gm and suggested the possible applications like pigments, sensors, electrode materials, and catalysts [5, 86]. With the aim of the gene delivery via oral route, one research group synthesized SPIONs using ferrous, ferric chloride, and NaOH as an alkaline precipitator and in situ surface modified with ascorbic acid and chitosan conjugation. The obtained particles were spherical in shape, 100 nm in size with positive zeta potential value of +25 mV utilized for plasmid DNA delivery toward endothelial pro-genitor cells (EPC) and poikilothermic (Vero) cells [40].

3.3 Manganese Oxide (MnO_x) Nanoparticles

Manganese oxides (MnO_x) have received much research as potential multifunctional materials. Several works have performed into developing new nanostructures with novel morphologies in order to increase specific property and related applications. Different structural characteristics and morphologies might be produced by modifying the synthesis conditions, for Mn_3O_4, MnO_2, and MnO. MnO-based nanostructure possesses various biomedical applications such as DD agent, CAs and is highly

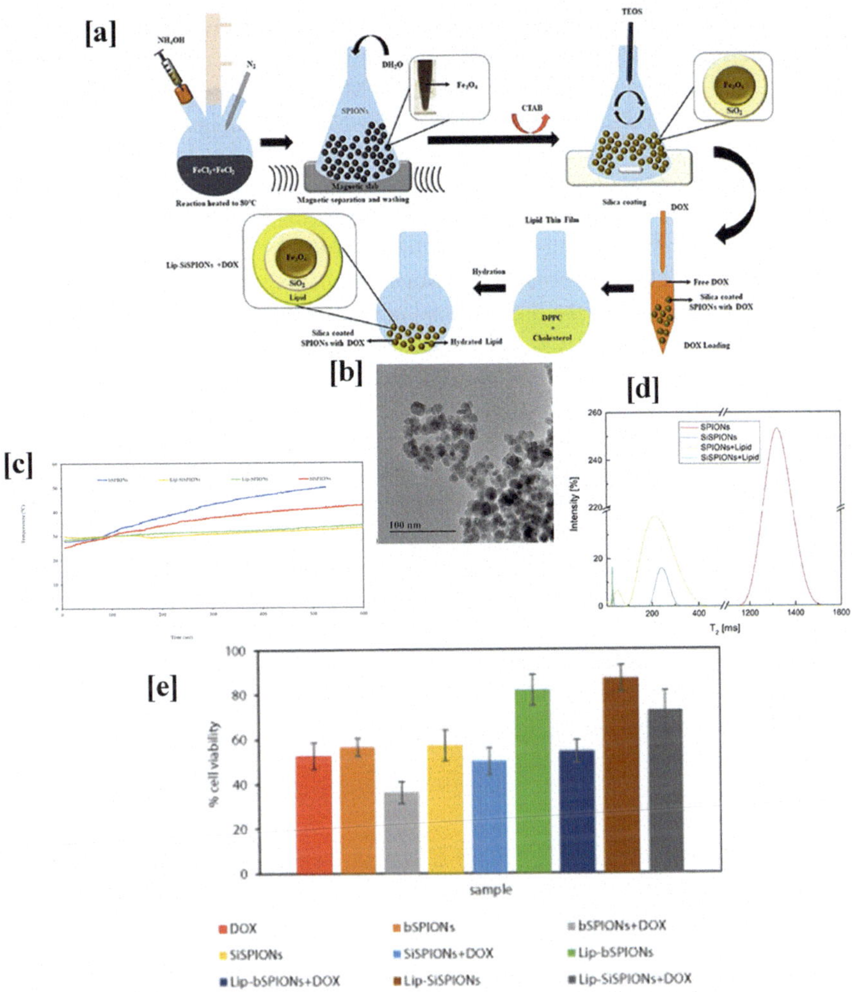

Fig. 4 **a** Schematic diagram of synthesis method for the SPION core and core–shell SPIONs, **b** TEM images for bare SPIONs, **c** Time-dependent temperature rise curve for bSPIONs, **d** T_2 characteristic relaxation time distributions for the prepared particles, **e** effect of bare and DOX-loaded coated NPs for various SPIONs by in vitro cytotoxicity assay. (Adopted from [82]

accepted in the biomedical field [14, 21, 35]. Furthermore, Mn-based NPs show good biocompatibility and bright images' property. Hence, MnO, MnO_2, Mn_3O_4, and MnOx have shown excellent contrast property mainly due to short circulation time of Mn ions which are eliminated from the body without accumulating in the tissue.

Besides this, Mn-based NPs have also shown PT properties under the NIR region, which, with and without anticancer drug, demonstrate excellent inhibition of cell growth. For example, sheet-shaped MnO_2 NPs were produced by a single-step hydrothermal reaction using $KMnO_4$ and distilled water first by simple stirring and then transferred to Teflon lined autoclave, washing with DW, and drying at 60 °C [110]. In addition to tumor microenvironment (TME)-specific imaging and stimulate responsive release of drug, biodegradable hollow MnO_2 nanoplatform is created for modulating hypoxic TME for boosting of cancer therapy, leading to complete effects that favor anti-tumor immune effect. Along with this, hollow structures allow for the co-loading of chemotherapeutic DOX and the photodynamic agent chlorine e6 (Ce6) in H-MnO_2 nanoshells after PEG modification [116].

In case of the high-quality magnetic resonance imaging (MRI) applications, CAs are utilized to differentiate the diseased and normal sites. Some generally used CAs are gadolinium (Gd)-based CAs which give bright contrast images and SPIONs which provide dark images. Likewise, Mn-based oxide NPs are also shown positive bright images with longer relaxation. Perhaps, our team have synthesized Mn_3O_4 NPs (20 nm) by precipitation method by mixing manganese chloride and NaOH at 80 °C up to 1 h. After synthesis, the NPs are washed using ethanol and dried at 60 °C in the hot dry air oven. The obtained particles were surface-modified with L-arginine, which showed excellent MRI imaging capabilities and suggested in vivo contrast properties(Fig. 5) [35].

In another study, to observe the in vivo contrast property of Mn_3O_4, NPs were synthesized by solvothermal decomposition. Hence, in the occurrence of tris (2, 4-pentanedionato) and Mn trisodium citrate, surface modification with PEG and L-cysteine was done improve the colloidal stability [111].

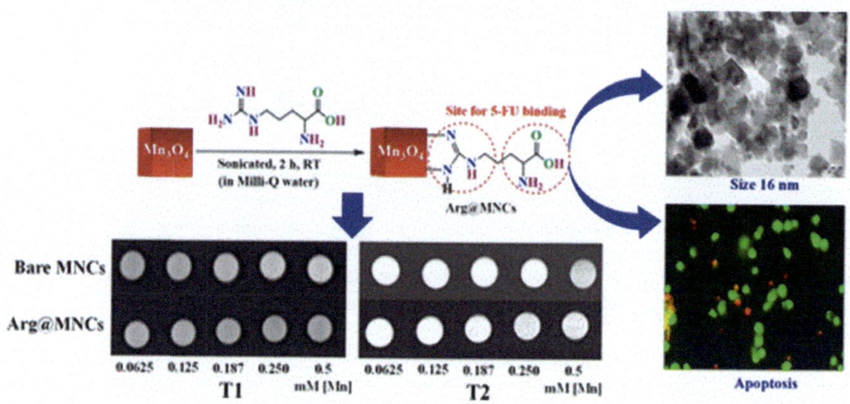

Fig. 5 Image showed surface modification *using* L-arginine of manganese oxide nanoparticles and suggested their application for colon cancer death and MRI application (Adopted from [35])

3.4 Gadolinium-Based Nanoparticles

One of the frequently used spin–lattice (T_1) CAs is Gd^{3+} ion for MRI applications. By speeding up the longitudinal relaxation rate of water protons, contrast is produced. As compared to how tissues of fats seem bright and fluid-based tissues dark in T_1-weighted images? The H_2O-enriched cellular tissues appear mid-gray in these pictures. Most Gd-chelates are used as CAs in MRI. The Gd-based CAs are generally composed of chelated molecules such as Gd–tetraazacyclododecane tetraacetic acid (Gd–DOTA) and Gd–diethylenetriamine pentaacetic acid (Gd–DTPA). The complexes are generally synthesized and modified using polymers to reduce the toxic effect due to leaching and accumulation of Gd ions. At present, various FDA-approved Gd-based CAs are utilized for imaging purposes such as gadobenate (MultiHance®), gadodiamide (Omniscan™), gadopentetate (Magnevist®), gadoterate (Dotarem®), gadoversetamide (OptiMARK™).

Hence, Gd has a strong paramagnetic property, which allows Gd^{3+} ions to exhibit strong T_1 relaxation under a magnetic field, resulting in bright contrast images, unlike SPIONs. Structurally, Gd^{3+} ions are similar to Ca^{2+} ions and frequent use of Gd^{3+} ions causes biological competition and interference to Ca^{2+} ions. Therefore, in order to reduce the toxic effects, Gd-based NPs are surface modified by various biological molecules that not improve the bioavailability but also improve the stability in biological system. For example, extremely small Gd oxide NPs were synthesized by a wet-chemical method by adding stabilizing agent, i.e., poly (acrylic acid) into gadolinium nitrate ($Gd(NO_3)_3$) and ammonium ($NH_3{\cdot}H_2O$) solution. Furthermore, the obtained NPs are stabilized by reduced BSA and functionalized by lactoferrin and RGD2 dimers for targeted DD [99]. Likewise, a targeted ligand hyaluronic acid (HA) conjugated HA-Gd_2O_3 NPs, of 105 nm particle size through one-pot hydrothermal method for the application of effective MRI and radio-sensitization of tumors [113]. Thus, gadolinium oxide-based nanostructures have synthesized to target the solid tumor and rare disease.

4 Application of Polymeric Nanocomposite and Its Application in Biomedical Field

Metal-based NPs possess unique theranostic applications which mean that the same formulation or system is designed to deliver the drug and simultaneously help in diagnosis of diseases. Early diagnosis of any disease is very important for the better treatment of particular disease properly. Some diseases such as uncontrolled tumor growth at their early stages are important to diagnose and manage by diagnostic approaches. For instance, FDA-approved Gd-based, SPION-based CAs are frequently used in the diagnostic centers. In addition to this, development of plasmonic NPs, such as Au-based NPs, has also been taken more seriously in the clinical field for applications like optical imaging, DD, and anticancer effects in the treatment of various solid tumors. Hence, in section we are discussing the various applications of polymeric NPs for theranostic applications.

4.1 *Polymeric Nanoparticles: As Drug Delivery and Imaging Agent*

Diagnostics with extraordinarily high precision, non-invasiveness, and real-time imaging is one of the biggest challenges that is developing as the plasmonic capabilities of Au-based nanostructures are developed. Using fluorescence centered signaling and imaging is a frequent and very sensitive medical procedure. Some novel imaging strategies, such as the exploitation of the plasmonic characteristics of AuNPs, are necessary because of very less knowledge about molecular information, quick photobleaching of the organics structure-based fluorophores, extremely high toxicity, and lack of depth tissues imaging in vivo model [10]. More light intensity is furthermore generated from a fluorescent color due to the dispersion of Au nanostructures. For example, AuNPs were synthesized and delivered as MMP-responsive candidate for malignant-targeted PA along with imaging-guided PTT and DD.

In this study, complementary DNA strands were first grafted onto the surface of AuNPs, and further DOX was conjugated to PEG through thermal labile linkers and finally functionalized via MMP-cleavable peptide to target the murine squamous cell carcinoma cells (SSC-7, overexpress high MMP-2 receptors). The finalized system showed higher uptake of PEGy-DOX-AuNPs up to 2.72 folds than PEGy-pep-DOX-AuNPs due to occurrence of extracellular MMPs and MTT assay which showed IC_{50} values of 28.6 and 12.9 $\mu g\ mL^{-1}$, for PEGy-pep-AuNPs + NIR and DOX-PEGy-pep AuNPs + NIR groups, respectively, further suggested the dose-dependent toxicity. Hence, this work contends that effective photon conversion and drug release need the assembly of AuNPs caused by MMP. In the presence of MMP-2, the intense NIR absorption of aggregation of NPs resulted in dramatically increased heat generation to destroy tumor cells. Furthermore, in vivo studies on biodistribution and tri-modality imaging were performed using PEG-pep-DOX AuNPs as CAs via PA and PT imaging. Images of SCC-7 tumors were captured before PA imaging and at different time intervals after the systemic injection of both formulations. Hence, after 24, the intensity of PA was maximum (Fig. 6) [117].

In another study, doxorubicin (DOX) and docetaxel (DTX) therapeutics were loaded on to the polydopamine-coated SPION which showed an excellent cytotoxicity effect against PC-3 cancer cells, IC_{50} value < 836.1 nM than for pure DOX 1200 nM and apoptosis cells, while DTX showed great cell growth inhibition with less concentration, IC_{50} value 100 nM [101]. (3-aminopropyl) triethoxysilane (APTES) was grafted on the external surface of SPION for delivery of Nintedanib which was released in controlled manner at pH 5.5 up to 85% than in pH 7.4 up to 48 h.

Further, in this study, the dose-dependent toxicity was observed against human lung cancer cells (L-132) with about 70% reduction in cell growth at concentration

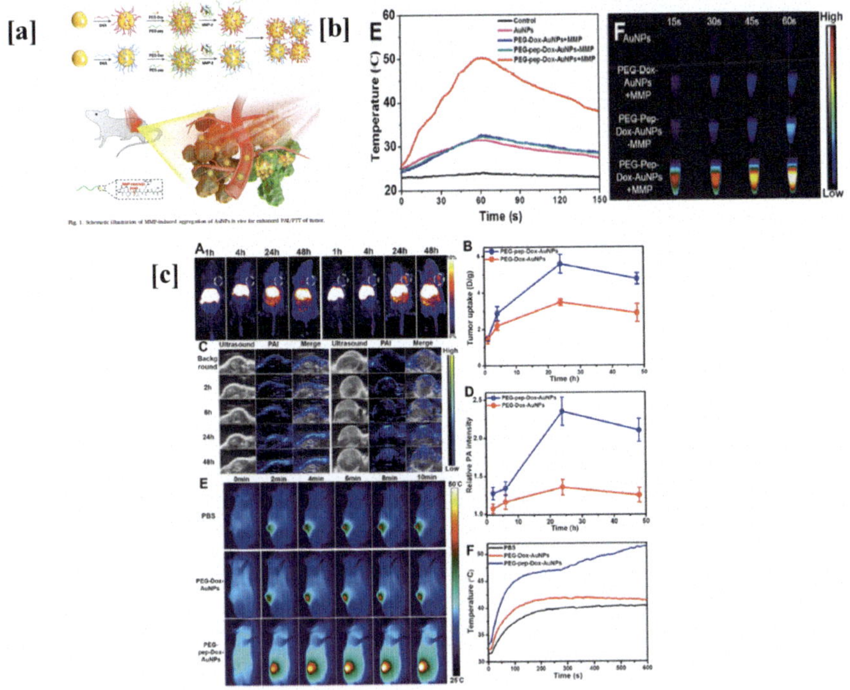

Fig. 6 Synthesis of **a** MMP-induced aggregation of AuNPs in vivo for enhanced PAI/PTT of tumor, **b** the characterization of MMP-responsive nanoprobes and **c** in vitro imaging distribution and the tri-modality imaging of PEG-pep-DOX-AuNPs. (Adopted from [117])

of 100 μg/mL [39]. Similarly, PTX was delivered for the treatment of A549 cancer cells through linoleic acid modified SPION. The high encapsulation efficiency was observed up to 98.5%, along with site specific drug delivery which showed excellent anti-proliferative up to <18% than pristine drug [76]. Besides drug delivery, several studies have also reported theranostic applications of NPs (Table 1).

Magnetized SPIONs interact with the proton and generate the T_2 relaxation time, which thereby create darker grayscale image with the increasing particle concentration. For example, CT/MR detectable magnetic microsphere was prepared by emulsified method using Gel, SA, and SPIONs and encapsulated the DOX. This study also suggests that the hyperthermia effect induces the Warburg effect in the tumor site due to the slow desorption of DOX at low pH. The resulting to delivery of DOX in the targeted blood vessels without causing any hemolytic toxicity. This study further demonstrated the concentration-dependent brightening effect in in vivo experiment which has also provided the T_2-weighted MR imaging capacity. Besides the SPIONs, MnO-based NPs also show MR imaging effect along with therapeutic property. For example, albumin-bound PTX NPs were synthesized to improve chemotherapeutic effect via release O_2 from MnO_2. Enhancing the abundance of oxygen in the tumor

Table 1 Various polymeric-based metallic nanomaterials, their surface modification, functionalization, chemotherapeutic drug loading for treatment and diagnosis of cancer, and other diseases

Nanosystems	Surface modifying moieties/functionalizing agent	Disease	Applications	References
Paramagnetic ferric iron(III) ion	Poly(lactic-co-glycolic)acid	Breast cancer	MRI Cell viability study In vivo vascular imaging	Marasini et al. [64]
Iron oxide nanoparticles	Porphyrin-doped poly(9,9-dioctylfluorene-alt-benzothiadiazole)	Brain tumors	Magnetic resonance and fluorescent imaging	Arias-Ramos et al. [7]
Superparamagnetic iron oxide nanoparticles	PLGA	Cervical cancer	Ultrasound-induced reactive oxygen species-mediated therapy and imaging	Li et al. [52]
AuNPs	PSMA	Prostate tumor	MRI	Luo et al. [59]
AuNCs	Alpha-lactalbumin	Breast tumor	Fluorescence	Yang et al. [115]
Au-Mesoporous Silica Nanocomposites	Thiol groups, mercaptopropyl-triethoxysilane, CTAB	Breast cancer	Two-photon-excited (TPE) photodynamic therapy (PDT)	Croissant et al. [15]
Fluorescent silica nanoparticles	Doping with gadolinium and Ce6, templates of epitope of CD59 protein YNCPNPTADCK, and doxorubicin	Breast tumor	Fluorescence imaging, MRI, drug delivery, NIR irradiation	Peng et al. [84]
Mn_3O_4 NPs	L-arginine, inulin and shRNA	Colon tumor	Gene delivery, cytotoxicity study, MRI, and biodistribution study	Kourani et al. [47]
MnO_2 NPs	PEG	Renal carcinoma	Glutathione-based drug delivery, MRI	Liu et al. [56]

(continued)

Table 1 (continued)

Nanosystems	Surface modifying moieties/functionalizing agent	Disease	Applications	References
AgNPs	PEG and aptamer As1411	Glioma cells	Cell viability, glioma irradiation-based treatment	Zhao et al. [121]
AgNPs	Monoclonal antibodies (mAb)	Viral infection	Viral diagnosis	Yen et al. [118]

microenvironment (TME) can help improve tumor oxygenation and potentially overcome the hypoxia seen in solid tumors. Hence, this study showed remarkable O_2 generation in the presence of H_2O_2 in every 10 min than albumin NPs. In this study, the effect on bright contrast MRI in the tumor regions was observed at different time interval up to 2 days, with a relaxivity rate of 2.34 mM^{-1} s^{-1} [68].

The combination of two transition metal oxide NPs has been discovered for the MRI activity and drug delivery. For example, magnetic MnO_2 sweet gum ball nanospheres were developed for cancer theranostic. For this, initially MnO_2 nanosheets are assembled into the iron oxide which is having gumball-like structure, and finally, DOX was loaded onto it which was released intracellularly in the U87MG cancer cells [24]. In addition to the many benefits of transition metal-based drug delivery, lanthanide nanostructure has also attracted a significant amount of attention for the treatment of various diseases. For example, DOX as a fluorescent drug showed anticancer effect and loaded on to the w/o emulsion-mediated synthesized PEG-coated Gd_2O_3 NPs. The developed system has been used not only for drug delivery but also used for two different types of imaging purpose that optical and MRI. In A549 cancer cells, the higher cellular uptake was recorded which also showed excellent cytotoxicity effect against A549, PANC-1, and U-87 cancer cells [50]. In a different work, chitosan-modified Gd_2O_3 NPs (138 nm) for the delivery of the anticancer medication 5-aminolevulinic acid were developed.

The prepared nanoformulation demonstrated concentration-dependent cell growth suppression, pH-responsive dual cellular and metabolic CAs, and outstanding T_1 and T_2 contrast images for glioblastoma C6 cells with r^2 value of 4.89 than only Gd_2O_3 NPs (2.34 mM^{-1} s^{-1}). Hence, this study suggested the possible interaction behavior between ions of Gd^{3+} and H_2O molecules conceivably initiate water exchange property and resulted into longitudinal water relaxivity than the clinically used CAs [30, 71]. Similarly, tumor-targeted drug delivery and MRI property were observed by Gd_2O_3 NPs which were developed using cyclic arginylglycylaspartic acid (cRGD) and/or folic acid (FA) as targeting ligand and hydrophilic polyacrylic acid for surface modification. As the folate receptor overexpresses on to the many cancer cells, hence the conjugate was delivered successfully by active targeting into the human malignant glioma (U87MG) cell lines and normal mouse hepatocytes (NCTC1469), which showed excellent cell growth inhibition due to successfully uptake by the cancer cells than the normal cell. Further, in vivo MRI property was observed with a bright contrast image and confirmed the T_1 contrast property in Gd-based nanostructures that could be utilized for imaging of tumor area in U87MG tumor-bearing mice [34].

4.2 *Polymeric Nanoparticles for Photothermal and Photodynamic and Imaging-Guided Drug Delivery*

The theranostic polymeric nanoparticles have also shown self-treatment efficacy *via* thermally degradation of cancer cells. As metallic NPs have absorption property for

NIR, hence several studies have reported the anticancer effect without any therapeutic drug delivery. A possible generation of inorganic and biodegradable theranostic nanoplatform for successful cancer treatment has been proposed by the Mn-LDH NPs, considering their outstanding T_1-weighted MRI contrast effect, enhanced PT performance, biosafety, and degradation. In particular, Mn(ox)-LDH NPs (oxidized Mn-LDH NPs) demonstrated extremely effective cancer cells killing capabilities by exposing them with image-guided exterior IR 808-nm laser effect. More significantly, the Mn-LDH NPs showed a high level of biosafety and were able to disintegrate under physiological settings, giving to both the MRI and the in vitro/vivo tests. Hence, T_1-weighted MRI of Mn-LDH NPs was decreased in slow rate with the increase of oxidation time and the relaxation rate from 1.48 $mM^{-1}s^{-1}$ to 1.03 $mM^{-1}s^{-1}$, especially in 24 h.

Furthermore, excellent in vivo MR imaging with a bright signal was observed at the tumor site in MCF-7 tumor-bearing mice [114]. The release of Ce6 from MnO_2 sweet gum ball nanospheres was tested for its photosensitizing effects. The 9,10-anthracenediyl- bis(methylene) dimalonic acid (ABDA) was expected to release with singlet O_2, leading to a reduction in absorbance which acted as a probe. This reduction was determined by monitoring the alteration in absorbance at 378 nm. Hence, this study showed release of O_2 as the absorbance of ABDA was decreased which suggested that along with chemotherapy, MnO_2-based nanocarrier can induce the apoptosis in the cancer cells under PDT process [24]. In another study, the autophagy effect by albumin-coated gold nanorods (GNRs) was developed for PTT effects on human neuroblastoma CD133+ cancer stem cells (CSCs). As to alter the toxic effect and to progress the colloidal stability, the albumin was coated on the surface of GNRs which triggered temperature-dependent cell growth inhibition due to exposure of NIR at the intensity of 0.5, 1, and 1.5 W/cm^2 for 480 sec [93].

Beside this, magnetite-Au nanocrystals were developed and conjugated with two fluorescent dyes. This study also suggested that the elevation of magnetization and octahedral shape of the developed nanosystem contributed to its biocompatibility, as confirmed by cell viability assay on 4T1 cells up to concentration of <350 μg/mL. In vitro relaxivity measurements demonstrated a superior r^2 value of 498.1 $mM^{-1}s^{-1}$ for NP-PEG in an aqueous medium, which was three times higher than that of commercial CAs agent. Additionally, in vivo T_2 relaxivity in 4T1 cells showed an r^2 value of 276.9 $mM^{-1}s^{-1}$. Hence, this study suggests the potential for future application in simultaneous MRI and treatment techniques. These techniques could be built on active drug administration, magnetic hyperthermia, or magneto-mechanical actuation, harnessing the cutting-edge multitasking capabilities of hybrid NPs for therapy and visualization [23].

Currently, bioavailability of drugs and the duration of their targeted release both can be increased *via* the use of chemotherapy and nanotechnology. Polymeric nanocomposite is without loading of therapeutic agents, and plasmonic nanomaterials such as Au-based particles have been showed excellent anticancer effect against many cancer lines. For instance, aptamer-functionalized and PEG-coated AuNPs were developed for the PTX delivery against breast cancer cells along with PTT. Several formulations of core AuNPs, such as PEG-AuNPs, PTX/PEG-AuNPs,

PEG-AuNPs@antiMUC1, and PTX/PEG-AuNPs@antiMUC1, were used to evaluate their anticancer effect an under continuous NIR laser irradiation at 810 nm with a power density of 160 mW/cm^2 for 60 s. In this study, PTX was released in sustained way from PEG-AuNPs@antiMUC1 nanoformulation which also showed synergistic effect as the same formulation showed selective uptake in the MUC1-positive cells than MUC1-negative breast cancer cells. The developed synergistic combinational system was evidenced by the enhanced appearance of apoptotic Caspase3 and the reduced expression of the anti-apoptotic Bcl-xL. Additionally, NIR irradiation also provided significant cellular absorption and had a strong anticancer impact.

In recent few years, Gd-based NPs are considered as multifunctional nanosystem as these NPs not only show contrast imaging property but also possess PT property under NIR range. In order to create a novel theranostic material with increased relaxation rate in MRI-guided synergetic chemodynamic/PTT, the most common MRI CAs, Gd_2O_3, play an important role. Due to the abundance of effective paramagnetic metal ions and the simultaneous improvement brought on by the presence of Mn and Gd cluster, the produced Mn-doped Gd_2O_3 nanoplate display increased T_1 contrast ability. One side, Mn element provides this formulation with extraordinary Fenton-like capability for producing the $^{\bullet}OH$ from H_2O_2 in the tumor area, and on the other hand, effect of PT conversion ability was raised due to coating of polydopamine on to the Mn-doped Gd_2O_3 nanoplate. Hence, the assessment of T_1 CAs shows 6.7 mM^{-1} s^{-1} with a high-quality contrast image with brighter one than only Gd oxide, which further showed the T_1 signal enhancement in hepatocellular area of mice. Without any doubt, the same formulation showed significant cell growth inhibition due to polydopamine coating and conversion of NIR laser. Furthermore, this study also suggested the diagnosis of tumor using treatment via PTT which helps to inhibit the tumor growth without causing any side effects. Hence, the results from this study proven that Mn-doped GdOP@PDA-PEG nanosystem could be a suitable candidate in a synergistic CDT/PTT.

4.3 *Polymeric Nanoparticles for Immune Cell-Based Therapy and Imaging*

Currently, application of immune cell therapy is gaining great attention due to the excellent treatment mechanism toward cancer. This type of therapy is highly challenging in daily clinical trials, and in most situations, it demands for better understanding in order to achieve accurate outcomes. Although immune-based cancer therapy is challenging, scientists are working to develop immune cell-based nanomedicine that has therapeutic properties. For instance, AuNPs were labeled by T cells that were injected intravenously to treat melanoma tumor. From CT scan, the in vivo xenografted mice possess enhanced T cells' accumulation at cancer site, which was not even affected by AuNPs and their function. Hence, this study successfully suggested cell tracking and inhibition of cancer cells by released cytokines and suggested the possible fate of other immune cells for cancer treatment [67].

In another study, for siRNA delivery, the AuNPs-attached hPDL-1 was synthesized to downregulate the program death ligand (PDL-1). Hence, the prepared PDL-1-targeted metal nanostructure acted as theranostic agents for gene immunotherapy showed gene silencing and PA-guided PTT toward HCC827 cells [58]. SPIONs have also been utilized for immunotherapy via surface coating by chitosan lactate, TAT peptide functionalization, and siRNA loading against PDL-1 and A2aR for suppressing the T cell responses toward cancer cells. This study has demonstrated the gene silencing effect in CT26 and 4T1 tumor-bearing mice, along with enhanced anti-cancer immune responses and increased survival rate [41]. Hence, several studies on cancer therapy as well as imaging purpose have been reported, in which some of them are under clinical trials.

5 Conclusion and Future Aspects

Polymeric nanoparticles have been found as outstanding nanocarrier for many hydrophilic or hydrophobic drugs and have shown high biocompatibility and biodegradability; therefore, polymeric NPs are excellent choice to boost the payload of pharmaceuticals. The fundamental disadvantage of polymeric NPs is their lack of additional properties, such as the ability to image the site of a disease at an early or late stage. In contrast, inorganic NPs have good CA imaging capabilities, but due to hydrophobic nature, the synthesize nanostructure is found to be unstable in colloidal solutions. In order to create a hybrid nanostructure that not only exhibits CAs properties, but also lessens the toxic effects of metals, polymeric nanostructure and biocompatible metal NPs are considered to be combined as hybrid theranostic nanomaterials. From this point forward, in this chapter, we have outlined the general synthesis protocol and development of organic-based NPs of metal/metal oxide nanoparticles and variety of methods for surface engineering. Additionally, we have discussed how polymeric NPs might enhance drug loading, PTT and PDT action, as well as the identification of uncontrolled malignant conditions, among other things. Although the metal and organic-based hybrid NPs have demonstrated excellent biological properties, their instability within the body still calls for more in-depth understanding at the cellular level via in vivo study for not just one or two cancers but for numerous cancers with a single nanosystem. Finally, numerous studies that demonstrated the favorable advantages and disadvantages of this theranostic nanomedicine have been conducted in order to better understand polymeric nanoparticles. These studies are particularly helpful in the delivery of the therapeutic agent, as they have demonstrated different stimuli-responsive drug release at the disease site. They function as excellent active mode delivery systems rather than passively. Additionally, the research has shown potential diagnostic data that aids in the treatment of certain diseases, offers the opportunity to treat without horrifying side effects, and aids in practical applications.

References

1. Abdellatif AAH, Younis MA, Alsharidah M, Al Rugaie O, Tawfeek HM (2022) Biomedical applications of quantum dots: overview, challenges, and clinical potential. Int J Nanomed 17:1951–1970. https://doi.org/10.2147/IJN.S357980
2. Alle M, Sharma G, Lee S-H, Kim J-C (2022b) Next-generation engineered nanogold for multimodal cancer therapy and imaging: a clinical perspectives. J Nanobiotechnol 20(1):222. https://doi.org/10.1186/s12951-022-01402-z
3. Alle M, G B reddy, Kim TH, Park SH, Lee S-H, Kim J-C (2020) Doxorubicin-carboxymethyl xanthan gum capped gold nanoparticles: microwave synthesis, characterization, and anti-cancer activity. Carbohydr Polym 229:115511. https://doi.org/10.1016/j.carbpol.2019.115511
4. Alle M, Bandi R, Sharma G, Dadigala R, Husen A, Kim JC (2021a) Current trends in engineered gold nanoparticles for cancer therapy BT. In: Alle M, Husen A, Kim JC (eds) Smart nanomaterials in biomedical applications. Springer International Publishing, Cham, pp 1–40
5. Alle M, Park SC, Bandi R, Lee S-H, Kim JC (2021b) Rapid in-situ growth of gold nanoparticles on cationic cellulose nanofibrils: Recyclable nanozyme for the colorimetric glucose detection. Carbohydr Polym 253:117239. https://doi.org/10.1016/j.carbpol.2020.117239
6. Alle M, Bandi R, Sharma G, Dadigala R, Lee S-H, Kim JC (2022a) Gold nanoparticles spontaneously grown on cellulose nanofibrils as a reusable nanozyme for colorimetric detection of cholesterol in human serum. Int J Biol Macromol 201:686–697. https://doi.org/10.1016/j.ijbiomac.2022.01.158
7. Arias-Ramos N, Ibarra LE, Serrano-Torres M, Yagüe B, Caverzán MD, Chesta CA, Palacios RE, López-Larrubia P (2021) Iron oxide incorporated conjugated polymer nanoparticles for simultaneous use in magnetic resonance and fluorescent imaging of brain tumors. Pharmaceutics 13
8. Bandi R, Tummala S, Dadigala R, Alle M, Lee S-H (2021) Role of Metal-doped carbon dots in bioimaging and cancer therapy BT. In: Alle M, Husen A, Kim JC (eds) Smart nanomaterials in biomedical applications. Springer International Publishing, Cham, pp 101–123
9. Bray F, Ferlay J, Soerjomataram I, Siegel RL, Torre LA, Jemal A (2018) Global cancer statistics 2018: GLOBOCAN estimates of incidence and mortality worldwide for 36 cancers in 185 countries. CA Cancer J Clin 68(6):394–424. https://doi.org/10.3322/caac.21492
10. Chang D, Ma Y, Xu X, Xie J, Ju S (2021) Stimuli-Responsive polymeric nanoplatforms for cancer therapy. Front Bioeng Biotechnol 9. https://doi.org/10.3389/fbioe.2021.707319
11. Chen W, Li D, Tian L, Xiang W, Wang T, Hu W, Hu Y, Chen S, Chen J, Dai Z (2018b) Synthesis of graphene quantum dots from natural polymer starch for cell imaging. Green Chem 20(19):4438–4442. https://doi.org/10.1039/C8GC02106F
12. Chen L, Xie J, Wu H, Li J, Wang Z, Song L, Zang F, Ma M, Gu N, Zhang Y (2018a) Precise study on size-dependent properties of magnetic iron oxide nanoparticles for in vivo magnetic resonance imaging. J Nanomater 2018:3743164. https://doi.org/10.1155/2018/3743164
13. Chen Z, Wu C, Zhang Z, Wu W, Wang X, Yu Z (2018c) Synthesis, functionalization, and nanomedical applications of functional magnetic nanoparticles. Chin Chem Lett 29(11):1601–1608. https://doi.org/10.1016/j.cclet.2018.08.007
14. Choi JY, Lee SH, Bin NH, An K, Hyeon T, Seo TS (2010) In vitro cytotoxicity screening of water-dispersible metal oxide nanoparticles in human cell lines. Bioprocess Biosyst Eng 33(1):21–30. https://doi.org/10.1007/s00449-009-0354-5

15. Croissant JG, Qi C, Maynadier M, Cattoën X, Wong Chi Man M, Raehm L, Mongin O, Blanchard-Desce M, Garcia M, Gary-Bobo M, Durand JO (2016) Multifunctional gold-mesoporous silica nanocomposites for enhanced two-photon imaging and therapy of cancer cells. Front Mol Biosci 3. https://doi.org/10.3389/fmolb.2016.00001
16. Crucho CIC, Barros MT (2017) Polymeric nanoparticles: a study on the preparation variables and characterization methods. Mater Sci Eng C 80:771–784. https://doi.org/10.1016/j.msec.2017.06.004
17. Dadigala R, Bandi R, Alle M, Gangapuram BR, Lee SH (2021) Graphene-based smart nanomaterials for photothermal therapy BT. In: Kim JC, Alle M, Husen A (eds) Smart Nanomaterials in Biomedical Applications. Springer International Publishing, Cham, pp 125–153
18. Dai X, Zhao X, Liu Y, Chen B, Ding X, Zhao N, Xu F-J (2021) Controlled synthesis and surface engineering of janus chitosan-gold nanoparticles for photoacoustic imaging-guided synergistic gene/photothermal therapy. Small 17(11):2006004. https://doi.org/10.1002/smll.202006004
19. Dhanasekaran V, Anandhavelu S, Polychroniadis EK, Mahalingam T (2014) Microstructural properties evaluation of Fe_2O_3 nanostructures. Mater Lett 126:288–290. https://doi.org/10.1016/j.matlet.2014.04.065
20. Dheyab MA, Aziz AA, Khaniabadi PM, Jameel MS, Oladzadabbasabadi N, Rahman AA, Braim FS, Mehrdel B (2023) Gold nanoparticles-based photothermal therapy for breast cancer. Photodiagnosis Photodyn Ther :103312. https://doi.org/10.1016/j.pdpdt.2023.103312
21. Ding B, Zheng P, Ma P, Lin J (2020) Manganese oxide nanomaterials: synthesis, properties, and theranostic applications. Adv Mater 32(10):1–35. https://doi.org/10.1002/adma.201905823
22. Dutta SD, Ganguly K, Bandi R, Alle M (2021) A new era of cancer treatment: carbon nanotubes as drug delivery tools BT. In: Kim JC, Alle M, Husen A (eds) Smart nanomaterials in biomedical applications. Springer International Publishing, Cham, pp 155–171
23. Efremova MV, Naumenko VA, Spasova M, Garanina AS, Abakumov MA, Blokhina AD, Melnikov PA, Prelovskaya AO, Heidelmann M, Li Z-A, Ma Z, Shchetinin IV, Golovin YI, Kireev II, Savchenko AG, Chekhonin VP, Klyachko NL, Farle M, Majouga AG, Wiedwald U (2018) Magnetite-Gold nanohybrids as ideal all-in-one platforms for theranostics. Sci Rep 8(1):11295. https://doi.org/10.1038/s41598-018-29618-w
24. Feng Y, Ding D, Sun W, Qiu Y, Luo L, Shi T, Meng S, Chen X, Chen H (2019) Magnetic manganese oxide sweetgum-ball nanospheres with large mesopores regulate tumor microenvironments for enhanced tumor nanotheranostics. ACS Appl Mater Interfaces 11(41):37461–37470. https://doi.org/10.1021/acsami.9b11843
25. Ferlay J, Soerjomataram I, Dikshit R, Eser S, Mathers C, Rebelo M, Parkin DM, Forman D, Bray F (2015) Cancer incidence and mortality worldwide: sources, methods and major patterns in GLOBOCAN 2012. Int J Cancer 136(5):E359–E386. https://doi.org/10.1002/ijc.29210
26. Gad SC, Sharp KL, Montgomery C, Payne JD, Goodrich GP (2012) Evaluation of the toxicity of intravenous delivery of auroshell particles (gold–silica nanoshells). Int J Toxicol 31(6):584–594. https://doi.org/10.1177/1091581812465969
27. Gagliardi A, Giuliano E, Venkateswararao E, Fresta M, Bulotta S, Awasthi V, Cosco D (2021) Biodegradable polymeric nanoparticles for drug delivery to solid tumors. Front Pharmacol 12:17. https://doi.org/10.3389/fphar.2021.601626
28. Ganapuram BR, Alle M, Dadigala R, Kotu GM, Guttena V (2013) Development, evaluation and characterization of surface solid dispersion for solubility and dispersion enhancement of irbesartan. J Pharm Res. https://doi.org/10.1016/j.jopr.2013.06.012
29. Gerber A, Bundschuh M, Klingelhofer D, Groneberg DA (2013) Gold nanoparticles: recent aspects for human toxicology. J Occup Med Toxicol 8(1):32. https://doi.org/10.1186/1745-6673-8-32
30. Gholibegloo E, Ebrahimpour A, Mortezazadeh T, Sorouri F, Foroumadi A, Firoozpour L, Shafiee Ardestani M, Khoobi M (2022) pH-Responsive chitosan-modified gadolinium oxide

nanoparticles delivering 5-aminolevulinic acid: a dual cellular and metabolic T1-T2* contrast agent for glioblastoma brain tumors detection. J Mol Liq 368:120628. https://doi.org/10.1016/j.molliq.2022.120628
31. Guadie Assefa A, Adugna Mesfin A, Legesse Akele M, Kokeb Alemu A, Gangapuram BR, Guttena V, Alle M (2017) Microwave-assisted green synthesis of gold nanoparticles using olibanum gum (Boswellia serrate) and its catalytic reduction of 4-nitrophenol and hexacyanoferrate (III) by sodium borohydride. J Clust Sci. https://doi.org/10.1007/s10876-016-1078-8
32. Gupta N, Jangid AK, Pooja D, Kulhari H (2019) Inulin: a novel and stretchy polysaccharide tool for biomedical and nutritional applications. Int J Biol Macromol 132:852–863. https://doi.org/10.1016/J.IJBIOMAC.2019.03.188
33. Gupta N, Bhagat S, Singh M, Jangid AK, Bansal V, Singh S, Pooja D, Kulhari H (2020) Site-specific delivery of a natural chemotherapeutic agent to human lung cancer cells using biotinylated 2D rGO nanocarriers. Mater Sci Eng C 112:110884. https://doi.org/10.1016/j.msec.2020.110884
34. Ho SL, Yue H, Lee S, Tegafaw T, Ahmad MY, Liu S, Saidi AK, Zhao D, Liu Y, Nam S-W, Chae KS, Chang Y, Lee GH (2022) Mono and multiple tumor-targeting ligand-coated ultrasmall gadolinium oxide nanoparticles: enhanced tumor imaging and blood circulation. Pharmaceutics 14
35. Jain P, Patel K, Jangid AK, Guleria A, Patel S, Pooja D, Kulhari H (2020) Modulating the delivery of 5-fluorouracil to human colon cancer cells using multifunctional arginine-coated manganese oxide nanocuboids with MRI properties. ACS Appl Bio Mater. https://doi.org/10.1021/acsabm.0c00780
36. Jain P, Patel K, Jangid AK, Guleria A, Patel S, Pooja D, Kulhari H (2021) Biotinylated Mn_3O_4 nanocuboids for targeted delivery of gemcitabine hydrochloride to breast cancer and MRI applications. Int J Pharm 606:120895. https://doi.org/10.1016/j.ijpharm.2021.120895
37. Jangid AK, Patel K, Jain P, Patel S, Gupta N, Pooja D, Kulhari H (2020) Inulin-pluronic-stearic acid based double folded nanomicelles for pH-responsive delivery of resveratrol. Carbohydr Polym 247:116730. https://doi.org/10.1016/J.CARBPOL.2020.116730
38. Jenjob R, Phakkeeree T, Seidi F, Theerasilp M, Crespy D (2019) Emulsion techniques for the production of pharmacological nanoparticles. Macromol Biosci 19(6):1900063. https://doi.org/10.1002/mabi.201900063
39. Karade VC, Sharma A, Dhavale RP, Dhavale RP, Shingte SR, Patil PS, Kim JH, Zahn DRT, Chougale AD, Salvan G, Patil PB (2021) APTES monolayer coverage on self-assembled magnetic nanospheres for controlled release of anticancer drug Nintedanib. Sci Rep 11(1):5674. https://doi.org/10.1038/s41598-021-84770-0
40. Karimi Jabali M, Allafchian AR, Jalali SAH, Shakeripour H, Mohammadinezhad R, Rahmani F (2022) Design of a pDNA nanocarrier with ascorbic acid modified chitosan coated on superparamagnetic iron oxide nanoparticles for gene delivery. Colloids Surf A Physicochem Eng Asp 632:127743. https://doi.org/10.1016/j.colsurfa.2021.127743
41. Karoon Kiani F, Izadi S, Ansari Dezfouli E, Ebrahimi F, Mohammadi M, Chalajour H, Mortazavi Bulus M, Nasr Esfahani M, Karpisheh V, Mahmoud Salehi Khesht A, Abbaszadeh-Goudarzi K, Soleimani A, Gholizadeh Navashenaq J, Ahmadi M, Hassannia H, Hojjat-Farsangi M, Shahmohammadi Farid S, Hashemi V, Jadidi-Niaragh F (2022) Simultaneous silencing of the A2aR and PD-1 immune checkpoints by siRNA-loaded nanoparticles enhances the immunotherapeutic potential of dendritic cell vaccine in tumor experimental models. Life Sci 288:120166. https://doi.org/10.1016/j.lfs.2021.120166
42. Kevadiya BD, Ottemann BM, Thomas M Ben, Mukadam I, Nigam S, McMillan J, Gorantla S, Bronich TK, Edagwa B, Gendelman HE (2019) Neurotheranostics as personalized medicines. Adv Drug Deliv Rev 148:252–289. https://doi.org/10.1016/j.addr.2018.10.011
43. Kim TH, Alle M, Kim JC (2019) Oxidation and temperature-responsive poly(hydroxyethyl acrylate-co-phenyl vinyl sulfide) micelle as a potential anticancer drug carrier. Pharmaceutics 11

44. Kim TH, Alle M, Park SC, Zhao F, Long W, Samala S, Kim JC (2021) Self-assembly prepared using an ion pair of poly(ethylene imine) and (phenylthio) acetic acid as a drug carrier for oxidation, temperature, and NIR-responsive release. Chem Eng J 415:128954. https://doi.org/10.1016/j.cej.2021.128954
45. Kopeček J, Yang J (2020) Polymer nanomedicines. Adv Drug Deliv Rev. https://doi.org/10.1016/j.addr.2020.07.020
46. Koudrina A, DeRosa MC (2020) Advances in medical imaging: aptamer and peptide-targeted MRI and CT contrast agents. ACS Omega 5(36):22691–22701. https://doi.org/10.1021/acsomega.0c02650
47. Kourani K, Jain P, Kumar A, Jangid AK, Swaminathan G, Durgempudi VR, Jose J, Reddy R, Pooja D, Kulhari H, Kumar LD (2022) Inulin coated Mn_3O_4 nanocuboids coupled with RNA interference reverse intestinal tumorigenesis in Apc knockout murine colon cancer models. Nanomedicine 40:102504. https://doi.org/10.1016/j.nano.2021.102504
48. Kulhari H, Pooja D, Singh MK, Kuncha M, Adams DJ, Sistla R (2015) Bombesin-conjugated nanoparticles improve the cytotoxic efficacy of docetaxel against gastrin-releasing but androgen-independent prostate cancer. Nanomedicine 10(18):2847–2859. https://doi.org/10.2217/nnm.15.107
49. Kulhari H, Telukutla SR, Pooja D, Shukla R, Sistla R, Bansal V, Adams DJ (2017) Peptide grafted and self-assembled poly(γ-glutamic acid)-phenylalanine nanoparticles targeting camptothecin to glioma. Nanomedicine 12(14):1661–1674. https://doi.org/10.2217/nnm-2017-0067
50. Kumar S, Meena VK, Hazari PP, Sharma RK (2017) PEG coated and doxorubicin loaded multimodal Gadolinium oxide nanoparticles for simultaneous drug delivery and imaging applications. Int J Pharm 527(1):142–150. https://doi.org/10.1016/j.ijpharm.2017.05.027
51. Li SD, Huang L (2008) Pharmacokinetics and biodistribution of nanoparticles. Mol Pharm 5(4):496–504. https://doi.org/10.1021/mp800049w
52. Li WP, Su CH, Chang YC, Lin YJ, Yeh CS (2016) Ultrasound-induced reactive oxygen species-mediated therapy and imaging using a fenton reaction activable polymersome. ACS Nano 10:2017
53. Li J, Zhu J-J (2013) Quantum dots for fluorescent biosensing and bio-imaging applications. Analyst 138(9):2506–2515. https://doi.org/10.1039/C3AN36705C
54. Li X, Deng Y, Tang W, Sun Q, Chen Y, Yang C, Yan B, Wang Y, Wang J, Wang S, Yang F, Ding Y, Zhao G, Cao G (2018) Urban-rural disparity in cancer incidence, mortality, and survivals in Shanghai, China, during 2002 and 2015. Front Oncol 8. https://doi.org/10.3389/fonc.2018.00579
55. Li M, Singh R, Wang Y, Marques C, Zhang B, Kumar S (2022) Advances in novel nanomaterial-based optical fiber biosensors—a review. Biosensors (Basel) 12
56. Liu J, Chen Q, Feng L, Liu Z (2018) Nanomedicine for tumor microenvironment modulation and cancer treatment enhancement. Nano Today 21:55–73. https://doi.org/10.1016/j.nantod.2018.06.008
57. Liu Y, Wang J, Xiong Q, Hornburg D, Tao W, Farokhzad OC (2021) Nano-bio interactions in cancer: from therapeutics delivery to early detection. Acc Chem Res 54(2):291–301. https://doi.org/10.1021/acs.accounts.0c00413
58. Liu B, Cao W, Qiao G, Yao S, Pan S, Wang L, Yue C, Ma L, Liu Y, Cui D (2019) Effects of gold nanoprism-assisted human PD-L1 siRNA on both gene down-regulation and photothermal therapy on lung cancer. Acta Biomater 99:307–319. https://doi.org/10.1016/j.actbio.2019.08.046
59. Luo D, Wang X, Zeng S, Ramamurthy G, Burda C, Basilion JP (2019) Prostate-specific membrane antigen targeted gold nanoparticles for prostate cancer radiotherapy: does size matter for targeted particles? Chem Sci 10(35):8119–8128. https://doi.org/10.1039/C9SC02290B
60. Madhusudhan A, Reddy GB, Venkatesham M, Veerabhadram G (2012) Design and evaluation of efavirenz loaded solid lipid nanoparticles to improve the oral bioavailability. Int J Pharm Pharm Sci Res 2(4):84–89

61. Madhusudhan A, Reddy G, Venkatesham M, Veerabhadram G, Kumar D, Natarajan S, Yang M-Y, Hu A, Singh S (2014) Efficient pH dependent drug delivery to target cancer cells by gold nanoparticles capped with carboxymethyl chitosan. Int J Mol Sci 15(5):8216–8234. https://doi.org/10.3390/ijms15058216
62. Mahmoudi M, Sant S, Wang B, Laurent S, Sen T (2011) Superparamagnetic iron oxide nanoparticles (SPIONs): development, surface modification and applications in chemotherapy. Adv Drug Deliv Rev 63(1–2):24–46. https://doi.org/10.1016/j.addr.2010.05.006
63. Marashdeh MW, Ababneh B, Lemine OM, Alsadig A, Omri K, El Mir L, Sulieman A, Mattar E (2019) The significant effect of size and concentrations of iron oxide nanoparticles on magnetic resonance imaging contrast enhancement. Results Phys 15:102651. https://doi.org/10.1016/j.rinp.2019.102651
64. Marasini R, Rayamajhi S, Moreno-Sanchez A, Aryal S (2021) Iron(iii) chelated paramagnetic polymeric nanoparticle formulation as a next-generation T1-weighted MRI contrast agent. RSC Adv 11(51):32216–32226. https://doi.org/10.1039/D1RA05544E
65. Masteri-Farahani M, Ghorbani F, Mosleh N (2021) Boric acid modified S and N co-doped graphene quantum dots as simple and inexpensive turn-on fluorescent nanosensor for quantification of glucose. Spectrochim Acta A Mol Biomol Spectrosc 245:118892. https://doi.org/10.1016/j.saa.2020.118892
66. McDonagh BH, Singh G, Hak S, Bandyopadhyay S, Augestad IL, Peddis D, Sandvig I, Sandvig A, Glomm WR (2016) L-DOPA-coated manganese oxide nanoparticles as dual mri contrast agents and drug-delivery vehicles. Small 12(3):301–306. https://doi.org/10.1002/smll.201502545
67. Meir R, Shamalov K, Betzer O, Motiei M, Horovitz-Fried M, Yehuda R, Popovtzer A, Popovtzer R, Cohen CJ (2015) Nanomedicine for cancer immunotherapy: tracking cancer-specific T-cells in vivo with gold nanoparticles and CT imaging. ACS Nano 9(6):6363–6372. https://doi.org/10.1021/acsnano.5b01939
68. Meng L, Cheng Y, Gan S, Zhang Z, Tong X, Xu L, Jiang X, Zhu Y, Wu J, Yuan A, Hu Y (2018) Facile deposition of manganese dioxide to albumin-bound paclitaxel nanoparticles for modulation of hypoxic tumor microenvironment to improve chemoradiation therapy. Mol Pharm 15(2):447–457. https://doi.org/10.1021/acs.molpharmaceut.7b00808
69. Mohan S, Vellakkat M, Aravind A, Reka U (2020) Hydrothermal synthesis and characterization of Zinc Oxide nanoparticles of various shapes under different reaction conditions. Nano Express 1(3):30028. https://doi.org/10.1088/2632-959X/abc813
70. Mollarasouli F, Zor E, Ozcelikay G, Ozkan SA (2021) Magnetic nanoparticles in developing electrochemical sensors for pharmaceutical and biomedical applications. Talanta 226:122108. https://doi.org/10.1016/j.talanta.2021.122108
71. Mortezazadeh T, Gholibegloo E, Alam NR, Dehghani S, Haghgoo S, Ghanaati H, Khoobi M (2019) Gadolinium (III) oxide nanoparticles coated with folic acid-functionalized poly(β-cyclodextrin-co-pentetic acid) as a biocompatible targeted nano-contrast agent for cancer diagnostic: in vitro and in vivo studies. Magn Reson Mater Phys Biol Med 32(4):487–500. https://doi.org/10.1007/s10334-019-00738-2
72. Muddineti OS, Ghosh B, Biswas S (2015) Current trends in using polymer coated gold nanoparticles for cancer therapy. Int J Pharm 484(1):252–267. https://doi.org/10.1016/j.ijpharm.2015.02.038
73. Murali Krishna I, Bhagavanth Reddy G, Veerabhadram G, Madhusudhan A (2016) Eco-friendly green synthesis of silver nanoparticles using salmalia malabarica: synthesis, characterization, antimicrobial, and catalytic activity studies. Appl Nanosci. https://doi.org/10.1007/s13204-015-0479-6
74. Narmani A, Farhood B, Haghi-Aminjan H, Mortezazadeh T, Aliasgharzadeh A, Mohseni M, Najafi M, Abbasi H (2018) Gadolinium nanoparticles as diagnostic and therapeutic agents: their delivery systems in magnetic resonance imaging and neutron capture therapy. J Drug Deliv Sci Technol 44:457–466. https://doi.org/10.1016/j.jddst.2018.01.011

75. Nel AE, Mädler L, Velegol D, Xia T, Hoek EMV, Somasundaran P, Klaessig F, Castranova V, Thompson M (2009) Understanding biophysicochemical interactions at the nano–bio interface. Nat Mater 8(7):543–557. https://doi.org/10.1038/nmat2442
76. Ngema LM, Adeyemi SA, Marimuthu T, Ubanako P, Wamwangi D, Choonara YE (2022) Synthesis of novel conjugated linoleic acid (CLA)-coated superparamagnetic iron oxide nanoparticles (SPIONs) for the delivery of paclitaxel with enhanced in vitro anti-proliferative activity on A549 lung cancer cells. Pharmaceutics 14
77. Ogris M, Walker G, Blessing T, Kircheis R, Wolschek M, Wagner E (2003) Tumor-targeted gene therapy: strategies for the preparation of ligand–polyethylene glycol–polyethylenimine/DNA complexes. J Control Release 91(1):173–181. https://doi.org/10.1016/S0168-3659(03)00230-X
78. Oliveira RR, Carrião MS, Pacheco MT, Branquinho LC, de Souza ALR, Bakuzis AF, Lima EM (2018) Triggered release of paclitaxel from magnetic solid lipid nanoparticles by magnetic hyperthermia. Mater Sci Eng C 92:547–553. https://doi.org/10.1016/j.msec.2018.07.011
79. Palanikumar L, Al-Hosani S, Kalmouni M, Nguyen VP, Ali L, Pasricha R, Barrera FN, Magzoub M (2020) pH-responsive high stability polymeric nanoparticles for targeted delivery of anticancer therapeutics. Commun Biol 3(1):95. https://doi.org/10.1038/s42003-020-0817-4
80. Pandit P, Bhagat S, Rananaware P, Mohanta Z, Kumar M, Tiwari V, Singh S, Brahmkhatri VP (2022) Iron oxide nanoparticle encapsulated; folic acid tethered dual metal organic framework-based nanocomposite for MRI and selective targeting of folate receptor expressing breast cancer cells. Microporous Mesoporous Mater 340:112008. https://doi.org/10.1016/j.micromeso.2022.112008
81. Patel V, Rajani C, Tambe V, Kalyane D, Anup N, Deb PK, Kalia K, Tekade RK (2022) Nanomaterials assisted chemo-photothermal therapy for combating cancer drug resistance. J Drug Deliv Sci Technol 70:103164. https://doi.org/10.1016/j.jddst.2022.103164
82. Patil-Sen Y, Torino E, De Sarno F, Ponsiglione AM, Chhabria V, Ahmed W, Mercer T (2020) Biocompatible superparamagnetic core-shell nanoparticles for potential use in hyperthermia-enabled drug release and as an enhanced contrast agent. Nanotechnology 31(37):375102. https://doi.org/10.1088/1361-6528/ab91f6
83. Peng X, Chen K, Liu W, Cao X, Wang M, Tao J, Tian Y, Bao L, Lu G, Teng Z (2020b) Soft Mesoporous organosilica nanoplatforms improve blood circulation, tumor accumulation/penetration, and photodynamic efficacy. Nanomicro Lett 12(1):137. https://doi.org/10.1007/s40820-020-00465-7
84. Peng H, Qin Y-T, He X-W, Li W-Y, Zhang Y-K (2020a) Epitope molecularly imprinted polymer nanoparticles for chemo-/photodynamic synergistic cancer therapy guided by targeted fluorescence imaging. ACS Appl Mater Interfaces 12(11):13360–13370. https://doi.org/10.1021/acsami.0c00468
85. Peternele WS, Monge Fuentes V, Fascineli ML, Rodrigues da Silva J, Silva RC, Lucci CM, Bentes de Azevedo R (2014) Experimental investigation of the coprecipitation method: an approach to obtain magnetite and maghemite nanoparticles with improved properties. J Nanomater 2014:682985. https://doi.org/10.1155/2014/682985
86. Phul R, Shrivastava V, Farooq U, Sardar M, Kalam A, Al-Sehemi AG, Ahmad T (2019) One pot synthesis and surface modification of mesoporous iron oxide nanoparticles. NanoStruct NanoObjects 19:100343. https://doi.org/10.1016/j.nanoso.2019.100343
87. Pirzada M, Altintas Z (2019) Nanomaterials for healthcare biosensing applications. Sensors 19
88. Pooja D, Panyaram S, Kulhari H, Rachamalla SS, Sistla R (2014a) Xanthan gum stabilized gold nanoparticles: characterization, biocompatibility, stability and cytotoxicity. Carbohydr Polym 110:1–9. https://doi.org/10.1016/j.carbpol.2014.03.041
89. Prabhu RH, Patravale VB, Joshi MD (2015) Polymeric nanoparticles for targeted treatment in oncology: current insights. Int J Nanomed 10:1001–1018. https://doi.org/10.2147/IJN.S56932
90. Selvan ST, Tan TTY, Yi DK, Jana NR (2010) Functional and multifunctional nanoparticles for bioimaging and biosensing. Langmuir 26(14):11631–11641. https://doi.org/10.1021/la903512m

91. Shah SA, Majeed A, Rashid K, Awan SU (2013b) PEG-coated folic acid-modified superparamagnetic $MnFe_2O_4$ nanoparticles for hyperthermia therapy and drug delivery. Mater Chem Phys 138(2–3):703–708. https://doi.org/10.1016/J.MATCHEMPHYS.2012.12.044
92. Shah SA, Majeed A, Rashid K, Awan SU (2013a) PEG-coated folic acid-modified superparamagnetic $MnFe_2O_4$ nanoparticles for hyperthermia therapy and drug delivery. Mater Chem Phys 138(2–3):703–708. https://doi.org/10.1016/j.matchemphys.2012.12.044
93. Shahabad ZA, Avci CB, Bani F, Zarebkohan A, Sadeghizadeh M, Salehi R, Ghafarkhani M, Rahbarghazi R, Bagca BG, Ozates NP (2022) Photothermal effect of albumin-modified gold nanorods diminished neuroblastoma cancer stem cells dynamic growth by modulating autophagy. Sci Rep 12(1):11774. https://doi.org/10.1038/s41598-022-15660-2
94. Sharifi M, Hosseinali SH, Hossein Alizadeh R, Hasan A, Attar F, Salihi A, Shekha MS, Amen KM, Aziz FM, Saboury AA, Akhtari K, Taghizadeh A, Hooshmand N, El-Sayed MA, Falahati M (2020) Plasmonic and chiroplasmonic nanobiosensors based on gold nanoparticles. Talanta 212:120782. https://doi.org/10.1016/J.TALANTA.2020.120782
95. Sharma PK, Dorlikar S, Rawat P, Malik V, Vats N, Sharma M, Rhyee JS, Kaushik AK (2021c) Nanotechnology and its application: a review. In: Khondakar KR, Kaushik AK (eds) Nanotechnology in cancer management. Elsevier, pp 1–33
96. Sharma G, Alle M, Chakraborty C, Kim J-C (2021a) Strategies for transdermal drug delivery against bone disorders: a preclinical and clinical update. J Control Release 336:375–395. https://doi.org/10.1016/j.jconrel.2021.06.035
97. Sharma G, Park SC, Bandi R, Ahn J, Alle M, Kim J (2021b) platinum nanoparticles and their antibacterial activity Polyquaternium enhances the colloidal stability of chitosan-capped platinum nanoparticles and their antibacterial activity
98. Sharp L, Donnelly D, Hegarty A, Carsin A-E, Deady S, McCluskey N, Gavin A, Comber H (2014) Risk of several cancers is higher in urban areas after adjusting for socioeconomic status. resúlts from a two-country population-based study of 18 common cancers. J Urban Health 91(3):510–525. https://doi.org/10.1007/s11524-013-9846-3
99. Shen Z, Liu T, Yang Z, Zhou Z, Tang W, Fan W, Liu Y, Mu J, Li L, Bregadze VI, Mandal SK, Druzina AA, Wei Z, Qiu X, Wu A, Chen X (2020) Small-sized gadolinium oxide based nanoparticles for high-efficiency theranostics of orthotopic glioblastoma. Biomaterials 235:119783. https://doi.org/10.1016/j.biomaterials.2020.119783
100. Siegel RL, Miller KD, Fuchs HE, Jemal A (2022) Cancer statistics. CA Cancer J Clin 72(1):7–33. https://doi.org/10.3322/caac.21708
101. Singh N, Sallem F, Mirjolet C, Nury T, Sahoo SK, Millot N, Kumar R (2019) Polydopamine modified superparamagnetic iron oxide nanoparticles as multifunctional nanocarrier for targeted prostate cancer treatment. Nanomaterials 9
102. Singh G, Kushwaha A, Sharma M (2022) Ultra-trace detection of caffeine and theophylline with high sensitivity and selectivity using $Gd2(MoO_4)_3$ nanosheets. Mater Today Commun 31:103390. https://doi.org/10.1016/j.mtcomm.2022.103390
103. Sorg C (2005) Scientific forward look on nanomedicine. Eur Sci Found Policy Briefing 23(February):1–6
104. Suk JS, Xu Q, Kim N, Hanes J, Ensign LM (2016) PEGylation as a strategy for improving nanoparticle-based drug and gene delivery. Adv Drug Deliv Rev 99:28–51. https://doi.org/10.1016/j.addr.2015.09.012
105. Sung H, Ferlay J, Siegel RL, Laversanne M, Soerjomataram I, Jemal A, Bray F (2021) Global cancer statistics 2020: GLOBOCAN estimates of incidence and mortality worldwide for 36 cancers in 185 countries. CA Cancer J Clin 71(3):209–249. https://doi.org/10.3322/caac.21660
106. Taghavi S, Abnous K, Taghdisi SM, Ramezani M, Alibolandi M (2020) Hybrid carbon-based materials for gene delivery in cancer therapy. J Control Release 318:158–175. https://doi.org/10.1016/j.jconrel.2019.12.030
107. Tang W, Shan X, Li S, Liu H, Wu X, Chen Y (2014) Sol–gel process for the synthesis of ultrafine MnO_2 nanowires and nanorods. Mater Lett 132:317–321. https://doi.org/10.1016/j.matlet.2014.05.211

108. Tybjerg AJ, Friis S, Brown K, Nilbert MC, Morch L, Køster B (2022) Updated fraction of cancer attributable to lifestyle and environmental factors in Denmark in 2018. Sci Rep 12(1):549. https://doi.org/10.1038/s41598-021-04564-2
109. Wagner AM, Knipe JM, Orive G, Peppas NA (2019) Quantum dots in biomedical applications. Acta Biomater 94:44–63. https://doi.org/10.1016/j.actbio.2019.05.022
110. Wang L, Guan S, Weng Y, Xu S-M, Lu H, Meng X, Zhou S (2019) Highly efficient vacancy-driven photothermal therapy mediated by ultrathin MnO_2 nanosheets. ACS Appl Mater Interfaces 11(6):6267–6275. https://doi.org/10.1021/acsami.8b20639
111. Wang P, Yang J, Zhou B, Hu Y, Xing L, Xu F, Shen M, Zhang G, Shi X (2017a) Antifouling manganese oxide nanoparticles: synthesis, characterization, and applications for enhanced MR imaging of tumors. ACS Appl Mater Interfaces 9(1):47–53. https://doi.org/10.1021/acsami.6b13844
112. Weerathunge P, Pooja D, Singh M, Kulhari H, Mayes ELH, Bansal V, Ramanathan R (2019) Transferrin-conjugated quasi-cubic SPIONs for cellular receptor profiling and detection of brain cancer. Sens Actuators B Chem 297(May):126737. https://doi.org/10.1016/j.snb.2019.126737
113. Wu C, Cai R, Zhao T, Wu L, Zhang L, Jin J, Xu L, Li P, Li T, Zhang M, Du F (2020a) Hyaluronic acid-functionalized gadolinium oxide nanoparticles for magnetic resonance imaging-guided radiotherapy of tumors. Nanoscale Res Lett 15(1):94. https://doi.org/10.1186/s11671-020-03318-9
114. Xie W, Guo Z, Gao Q, Wang D, Liang K, Gu Z, Zhao LY (2020) Manganese-doped layered double hydroxide: a biodegradable theranostic nanoplatform with tumor microenvironment response for magnetic resonance imaging-guided photothermal therapy. ACS Appl Bio Mater 3(9):5845–5855. https://doi.org/10.1021/acsabm.0c00564
115. Yang J, Wang T, Zhao L, Rajasekhar VK, Joshi S, Andreou C, Pal S, Hsu H, Zhang H, Cohen IJ, Huang R, Hendrickson RC, Miele MM, Pei W, Brendel MB, Healey JH, Chiosis G, Kircher MF (2020) Gold/alpha-lactalbumin nanoprobes for the imaging and treatment of breast cancer. Nat Biomed Eng 4(7):686–703. https://doi.org/10.1038/s41551-020-0584-z
116. Yang G, Xu L, Chao Y, Xu J, Sun X, Wu Y, Peng R, Liu Z (2017) Hollow MnO_2 as a tumor-microenvironment-responsive biodegradable nano-platform for combination therapy favoring antitumor immune responses. Nat Commun 8:1
117. Yang K, Liu Y, Wang Y, Ren Q, Guo H, Matson JB, Chen X, Nie Z (2019) Enzyme-induced in vivo assembly of gold nanoparticles for imaging-guided synergistic chemo-photothermal therapy of tumor. Biomaterials 223:119460. https://doi.org/10.1016/j.biomaterials.2019.119460
118. Yen C-W, de Puig H, Tam JO, Gómez-Márquez J, Bosch I, Hamad-Schifferli K, Gehrke L (2015) Multicolored silver nanoparticles for multiplexed disease diagnostics: distinguishing dengue, yellow fever, and Ebola viruses. Lab Chip 15(7):1638–1641. https://doi.org/10.1039/C5LC00055F
119. Yoo J, Park C, Yi G, Lee D, Koo H (2019) Active targeting strategies using biological ligands for nanoparticle drug delivery systems. Cancers (Basel) 11(5). https://doi.org/10.3390/cancers11050640
120. Zhang P, Xiao Y, Sun X, Lin X, Koo S, Yaremenko AV, Qin D, Kong N, Farokhzad OC, Tao W (2023) Cancer nanomedicine toward clinical translation: obstacles, opportunities, and future prospects. Med. https://doi.org/10.1016/j.medj.2022.12.001
121. Zhao J, Liu P, Ma J, Li D, Yang H, Chen W, Jiang Y (2019) Enhancement of radiosensitization by silver nanoparticles functionalized with polyethylene glycol and aptamer As1411 for glioma irradiation therapy. Int J Nanomed 14:9483–9496. https://doi.org/10.2147/IJN.S224160
122. Zielińska A, Carreiró F, Oliveira AM, Neves A, Pires B, Venkatesh DN, Durazzo A, Lucarini M, Eder P, Silva AM, Santini A, Souto EB (2020) Polymeric Nanoparticles: production, characterization, toxicology and ecotoxicology. Molecules 25
123. Zuddas E, Lentijo-Mozo S, Casu A, Deiana D, Falqui A (2017) Building composite iron-manganese oxide flowerlike nanostructures: a detailed magnetic study. J Phys Chem C 121(31):17005–17015. https://doi.org/10.1021/acs.jpcc.7b04915

Functional Biomaterials for Image-Guided Therapeutics of Solid Tumor

Sauraj

Abstract Image-guided therapeutic agents show the great potential to facilitate the precise delivery of drugs to tumor tissue/cancer cells along with real-time imaging of therapeutic agents at targeted site. Theranostics biomaterials are becoming popular in field of diagnosis as well as cancer therapy due to their efficient tumor-targeting ability, strong magnetization, and fluorescent properties. In past few years, a variety of image-guided biomaterials have been investigated in biomedical applications specially for cancer therapies that can be classified into two categories organic and inorganic nanomaterials. Inorganic nanomaterials are usually created with inorganic materials such as quantum dot and metal oxides' NPs, whereas the organic biomaterials' nanoparticles are created with organic materials including micelles, liposomes, and polymeric NPs via loaded the specific fluorescent dyes or therapeutic agents. This chapter covers fundamental aspects, classification, properties, and biomedical applications of organic and inorganic image-guided biomaterials along with the recent advancements to overcome the limitations of the conventional cancer therapy, clinical status, and future perspective.

Keywords Biomaterials · Theranostics agents · Bioimaging · Cancer therapy

1 Introduction

Nanosystems for multipurpose diagnostics as well as treatment response monitoring are being widely investigated by researchers in the field of drug delivery. In addition to enhancing the efficacy of therapeutic drugs, theranostics nanoparticles may be employed for pre- and post-treatment tumor evaluation using imaging methods

---

This chapter dedicated to Prof. Yuvraj Singh Negi on the occasion of his retirement.

Sauraj (✉)
Department of Pharmacy, Gachon University, Inchone 21912, South Korea
e-mail: saurajpolymeriitd@gmail.com

A. Madhusudhan et al. (eds.), *Functional Smart Nanomaterials and Their Theranostics Approaches*, Smart Nanomaterials Technology,
https://doi.org/10.1007/978-981-99-6597-7_7

such as computed tomography (CT), magnetic resonance imaging (MRI), and ultrasonography [25]. In recent years, theranostics nanoparticles showed the great potential biomedical applications; thereby, various kinds of theranostics nanoparticles have been developed for the simultaneous cancer imaging and therapy [24, 33]. Theranostics, which integrates diagnosis and therapy, can decrease treatment delays and enhance therapeutic outcomes, thus become a crucial component of customized cancer treatment. Early identification and decisive treatment of cancer using theranostic techniques can enhance patient survival. Tumor imaging is an important aspect of tumor diagnosis, staging, radiation planning, and treatment efficacy evaluation. Lately, molecular imaging, which combines molecular biology and in vivo imaging, has allowed for the analysis of biological processes and the detection of illnesses [3]. In contrast to ex vivo biopsy or cell culture procedures, theranostics NPs can enable the investigation of treatment response or disease development through in vivo imaging. A variety of nanomaterials, including organic nanomaterials (liposomes, dendrimers, polymers) and inorganic nanomaterials, are already being used as imaging agents in cancer therapy [8]. Currently, these nanoparticles are being used to developing theranostics nanomaterials. These nanoparticles can modulate the pharmacokinetics of therapeutic medicines, thus improved the permeability and retention impact in tumors, resulting avoid the unfavorable side effects associated with systematic delivery of therapies to the target location [30, 36]. In this chapter, covers advancements in development of image-guided organic and inorganic nanoparticles.

2 Functional Nanomaterials Based Image-Guided Therapeutics

Nanotheranostics has the potential not only to enable the delivery of therapeutic agent to cancer cells, but also to use imaging to disclose the effectiveness of drug delivery, off-target effects, potential toxicity. The nanomaterials used to build nanotheranostics fall into two categories: organic NPs and inorganic NPs. Liposomes, dendrimers, and polymers are examples of organic nanoparticles. Organic NPs are the most often utilized kind because they are biocompatible, biodegradable, easy to make, and cost-effective. Inorganic NPs are made up of physiologically inert inorganic components such as silicon, gold, silver, carbon, and iron. The most common organic and inorganic nanomaterials that have been extensively used to construct nanotheranostics are shown in Fig. 1.

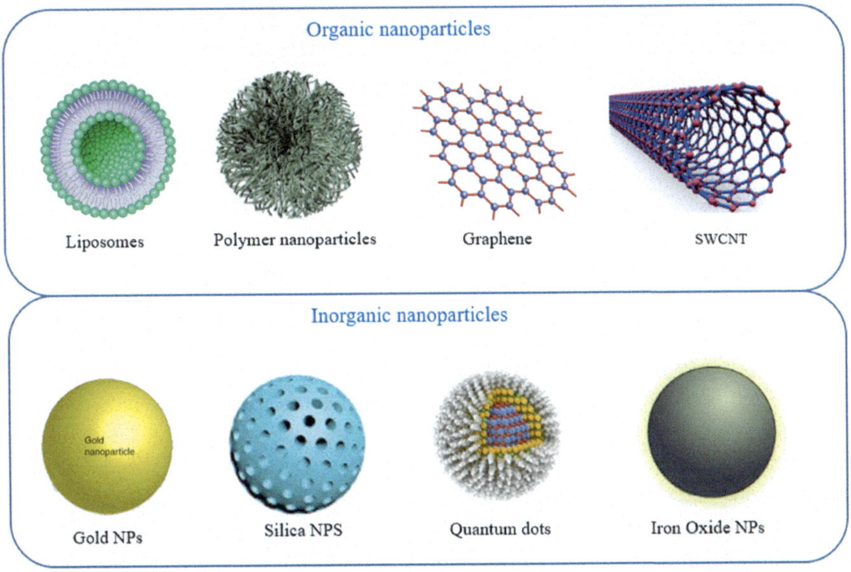

Fig. 1 Commonly used organic and inorganic nanomaterials for the development of image-guided nanoparticles

2.1 *Organic Nanomaterials Based Image-Guided Therapeutics*

Over the past few decades, the liposomes and polymers (natural as well as synthetic) based nanocarries have been extensively used in biomedical application to improve the water solubility, encapsulation efficiency, accumulation, and site target specificity of the payloads. In recent years, a variety of image-guided therapeutic nanomaterials based on organic nanomaterials have been investigated for efficacies of image-guided drug delivery [22].

2.1.1 Liposomes Based Therapeutic

Liposomes have been regarded a potent medication carrier in cancer therapy due to similar resemble with the normal cell membrane. Liposomes have a lipid bilayer structure that can solubilize the hydrophilic as well as hydrophobic substances. Therefore, the superior biocompatibility makes them powerful sources in drug delivery. Generally, liposomal drug delivery vehicles were created by using the combination of natural lipids or synthetic lipids and the surfactants. By modifying the lipid components and the associated ratio, the physiochemical characteristics may be tailored to particular uses. Over the past few years, a variety of liposomal such as pH-sensitive liposomal, cation-responsive and immune-responsive liposomal have been devolved and utilized in drug delivery applications. Urtti et al. created indocyanine green-loaded liposomes to improve the solubility and the thermal stability of the ICG. The offered liposomes release the ICG in a light triggered manner, thus significantly improve the therapeutic outcomes of ICG [17].

Similarly, Vasilis et al. designed polyethylene glycol (PEG) grafting liposomes for the dynamic imaging of ICG in the tumor microenvironment. The PEG component enhances the circulation time of liposomes and avoids capture by the endothelial system. The in vivo results revealed that the liposomes-ICG system, a very effective optoacoustic agent witch produced greater optoacoustic signals than the gold nanoparticles (AuNPs). Moreover, enhance the tissue accumulation of ICG in tumor bearing mice with remarkable sensitivity [4]. In another study, lipid nanoparticles were prepared using cholesteryl-1-14C-oleate [14C]CHO, a constituent of the nanoparticle for the loading of ICG. The ICG exhibited lipids nanoparticles showed a different biodistribution pattern as compared to free ICG, which indicates the improve the encapsulation and bio distribution efficacy of efficiency ICG compared to free ICG [21]. Hu et al. created liposomes by encapsulating Ce6 in liposomes with ICG in order to turn "OFF" Ce6. The ICG molecules substantially inhibited $^1O_2$ production by Ce6 after irradiation at 660 nm. ICG was degraded and the phototoxicity of Ce6 was activated when exposed to 808 nm light. Furthermore, when exposed to NIR light, ICG may generate heat, which can be employed as a photothermal treatment to destroy cancer cells (PTT). As a result, our developed liposomes containing ICG and Ce6 might be employed in cancer treatment as an activatable PDT paired with PTT [38].

2.1.2 Polymer-Based Nanomaterials

Polymer nanoparticles (NPs) are ideal candidates for bio-applications due to their excellent biocompatibility, low cost, and ease of functionalization [5]. Polymeric nanoparticles have the potential to improve the stability and solubility of encapsulated cargos, facilitate transport across membranes, and extend circulation periods, all of which have the potential to improve safety and efficacy. Therefore, a variety of therapeutic agents have been developed by utilizing the polymer nanoparticles (NPs) [22]. In this contribution, Liu et al. constructed theranostic nanoparticle by encapsulating a contrast agent (ICG) into the folic acid-decorated FA-PEG-PLGA–paclitaxel conjugate for safe chemo/photothermal therapy. The presence of folate on the surface of the NPs would contribute to the NPs specific binding to folate-receptor-positive cancer cells, resulting in improved intracellular absorption and anticancer effectiveness using a laser-triggered drug delivery system paired with PTT [19].

However, a large number of theranostic nanoparticles have been developed in past years by using the traditional photosensitizers (PSs) such as porphyrins, chlorins, and phthalocyanines derivatives. These PSs demonstrate Aggregation-caused Quenching (ACQ) which occurs during the aggregation stage, resulting in a significant drop in fluorescence intensity and ROS production efficiency. Tang et al. developed aggregation-induced emission (AIE)-based photosensitizers with strong fluorescence and high photosensitization efficiency in aggregate due to restriction of intra molecule mobility to overcome the inherent quenching problem experienced by classic photosensitizers [31].

Two kinds of stimuli-responsive and AIE-featured nanoparticles were prepared by Tang et. al. via encapsulation of a AIE-photosensitizer (MeTTMN) into the two PEG-based stimuli-responsive polymers. The newly developed AIE nanoparticles have great biocompatibility, outstanding stability, remarkable image quality, and significantly improved PDT performance [34]. An another study, kai et al. developed AIE-based multifunctional nanoparticles for targeted photodynamic therapy. For this purpose, an AIE-PS (namely, TTD) was encapsulated into poly (ethylene glycol)-lipid as the encapsulation matrix. The platform has a strong fluorescence and can rapidly produce ROS when exposed to light.

Cell cytotoxicity experiments revealed that produced NPs may specifically and effectively destroy cancer cells [14]. Moreover, an AIE-based GSH activable photosensitizer was developed by Tang et al., for imaging-guided photodynamic therapy, where Ferrocene was attached to TPEPY-SH through S–S bond. Ferrocene suppresses the fluorescence and singlet oxygen production of TPEPY-S-F via a PeT mechanism. When GSH is added, the S–S bond in TPEPY-S-Fc is broken, resulting in TPEPY-SH, which causes intense red fluorescence and significant singlet oxygen production. Under light irradiation, TPEPY-S-Fc showed excellent biocompatibility, strong cytomembrane penetration, and the ability to create singlet oxygen to trigger CT-26 cell death [39]. To broaden the use of photodynamic therapy in clinical uses, two-photon PDT (TP-PDT) is proposed, which uses two NIR photons for PS activation instead of one single visible photon. This phenomenon can overcome the limitation of conventional PDT, because majority of the AIE-PSs generally absorbed the light of UV–visible range that limits their application in deep imaging. On the other AIE-PSs with greater two-photon absorption cross-section provide better and controlled spatial selectivity with significant and penetration depth of tissue.

Bin Liu et al. developed a novel AIE-PS with large 2PA two-photon. Further to realize the water solubility as well as biocompatibility, the AIE-dots were developed by loaded it in to DSPE-PEG-polymer. Due the large 2PA, the developed offered large TPA cross-section AIE-dots produce high quantity of ROS with excellent photostability. Importantly, the offered AIE-dots showed better deep-image-guided potential, thus overcome the limitation of PDT-related deep penetration [12].

To further explore the area of two-photon dynamic therapy, PSs with AIE features were generated by polymerizing a short molecule PS, resulting in high $^1O_2$ production efficiency and large 2PA. The two AIE-PSs, namelyPTPEDC1 and PTPEDC2, were synthesized by connecting the monomer unit of the AIE-PS to obtain the high 2PA cross-section area. Further, the PTPEDC1 and PTPEDC2 dots were produced via encapsulated in DSPE-PEG-Mal matrix to achieve water dispersibility and biocompatibility. PTPEDC2 dots had a 548% higher $^1O_2$ generation efficiency and a 651% bigger 2PA cross-section than TPEDC dots, resulting in outstanding 2PE-PDT efficacy cancer tumor therapy. The discovery of conjugated polymer PSs with increased two-photon photosensitization sheds light on the development of extremely effective photosensitizers for two-photon stimulated photodynamic treatment [32]. In this contribution, tetraphenylethene (TPE-red), a kind of red emissive AIE-PSs, was synthesized and subsequently encapsulated into PSMA polymer to offer nanoparticles. The 2PA of the synthesized photosensitizer was measured by using a femtosecond (fs) laser at 1040 nm. The in vivo results are demonstrating that the developed AIE-PS NPs with larger 2PA showed great potential in deep tissue imaging PDT [2].

2.1.3 Carbon-Based Nanomaterials

The nanomaterials based on carbon have been widely exploited in biomedical applications including drug delivery and tissues' engineering because of their distinctive structure. Carbon-based nanomaterials have considerable interest in PDT because a variety of the nanomaterials have been created either encapsulated or chemical conjugation of carbon nanomaterials.

Moreover, carbon-based materials might be conjugated with diverse functional polymers to provide biocompatibility and water dispersibility. Fullerene, carbon nanotubes (single wall as well as multi-walled), and graphene are extensively used carbon-based nanomaterials in PDT. In this contribution, Giyoong Tae et al. created a novel catalytic nanosystem (nGO-hemin-Ce6) via co-loaded hemin and chlorin e6. To provide the biocompatibility, the GO was further coated with Pluronic. Since the photosentitizer (Ce6) is highly oxygen dependent which utilize the intracellular oxygen to produce the ROS, resulting impart the tumor hypoxia. Thus, to enhance the intracellular concentration of oxygen (O_2), hemin was used as a catalase-mimetic nanozyme which generates intracellular O_2 in the presence of H_2O_2, thus enhance the overall PDT effects of the photosensitizer (Ce6). By the catalase-mimetic activity of hemin, the nGO-hemin-Ce6 demonstrated excellent singlet oxygen generation efficiency in the presence of H_2O_2. Moreover, in vitro cell tests show that the nGO-hemin-Ce6 is capable of producing intracellular O_2, which increases the PDT. Because of their tiny size, the nGO-hemin-Ce6 nanosystem demonstrated significant tumor accumulation via passive targeting. The catalytic nanosystem of nGO-hemin-Ce6 suppressed tumors much better than a control nanosystem lacking hemin (nGOeCe6) [23].

Because of its enormous surface area, cheap cost, broad NIR absorption, and photothermal heating capabilities, graphene is a very attractive nanomaterial for the development of PDT-PTT nanomedicine platforms. Taratula et al. created a Phthalocyanine-loaded graphene nanoplatform for imaging-guided combinatorial phototherapy by exploiting the virtue graphene. Additionally, the created nanoplatform was coupled with PEG, and luteinizing hormone-releasing hormone (LHRH) peptide, to increase the biocompatibility as well as tumor-targeted ability. Combinatorial phototherapy resulted in increased ovarian cancer cell demise, with a lethal effectiveness of 90–95% at low Pc and low-oxygen graphene doses, likely because tothe synergistic effects of produced ROS and moderate hyperthermia [29].

Since, most of the photosensitizers used in PDT therapy exhibited poor depth of penetration due their absorption in UV–visible range, which limits the clinic applications of PDT. To achieve the better depth of penetration, the photosensitizes should absorbed the light of NIR range. However, it is very difficult to synthesize the photosensitizer that exhibited the absorption in NIR range is difficult.

To address this problem, upconverting nanoparticles were used, which use near-infrared excitation rather than UV excitation and have unique qualities such as high enhanced raised penetration depth in biological samples, decreased background auto-fluorescence, photobleaching, and photo damage of biological specimens. In this consequence, Yang et al. fabricated nanographene oxide-UCNPs-Ce6 nanocomposite, where NGO was finalized with PEG-polymer and then covalently conjugated with UCNPs. To realize the PDT effect, the well-known photosensitizer (Ce6) was loaded. The UCNPs can amplify the emission of Ce6 in the NIR region. Thus, utilization of UCNPs to enhance the emission of convention photosensitizer in the NIR region is a unique strategy to improve the bioimaging capability of the photosensitizer. The developed NGO-UCNPs-Ce6 nanocomposites significantly generates the ROS upon irradiated with 808 nm, thus showed great potential in image-guided PDT [13] (Table 1).

Table 1 Classification of organic nanomaterials used in image-guided cancer therapy

Organic nanoparticles	Nanomaterials compositions	Contrast agent	Refs.
Liposomes	Indocyanine green (ICG)-loaded liposomes' nanoparticles	Indocyanine green (ICG)	Lajunen et al. [17]
Liposomes	ICG-loaded polyethylene glycol (PEG) grafting liposomes	Indocyanine green (ICG)	Beziere et al. [4]
Liposomes	Indocyanine green-loaded cholesteryl-1-14C-oleate nanoparticles	Indocyanine green	Mérian et al. [21]
Liposomes	Ce6 and ICG co-loaded liposomes nanoparticles	Ce6 and ICG	Yuan et al. [38]
PEG-PLGA-co-polymer	ICG-encapsulated FA-(PEG-PLGA)–paclitaxel conjugate nanoparticles	ICG	Liu et al. [19]
Polyethylene glycol-*b*-polycaprolactum	AIE-photosensitizer (MeTTMN)-loaded mPEG-*Hyd*-PCL-CIN and mPEG-*SS*-PCL-CIN nanoparticles	AIE-photosensitizer (MeTTMN)	Xu et al. [34]
Poly(ethylene glycol)-lipid	AIE-PS TTD-loaded Poly(ethylene glycol)-lipid nanoparticles		Jin et al. [14]
DEPE-PEG-MAL matrix	AIEgens (TPEDC)-loaded DEPE-PEG-MAL dots	Two photon AIE-PS (TPEDC)	Gu et al. [12]
DSPE-PEG-Mal matrix	PTPEDC-loaded DEPE-PEG-MAL dots	Two photon AIE-PS (TPEDC)	Wang et al. [32]
Poly(styrene-co-maleic anhydride) (PSMA)	TPE-red-encapsulated PSMA nanoparticles	AIEgens (TPE-red)	Alifu et al. [2]
Pluronic and graphene oxide	Ce6 loaded Pluronic-coated nanographene oxide	Ce6	Sahu et al. [23]
Graphene	Phthalocyanine-loaded graphene nanoplatform	Phthalocyanine	Taratula et al. [29]
Ferrocene	GSH-active Ferrocene-AIE-PS	AIE-photosensitizer (TPEPY-SH)	Zhang et al. [39]
Graphene oxide	NGO-UCNPs-Ce6 nanocomposites	Ce6 and UCNPs	Gulzar et al. [13]

2.2 *Inorganic Nanomaterials Based Image-Guided Therapeutics*

Beside the organic nanomaterials (liposomes, micelles, and polymer nanomaterials), inorganic nanomaterials have been extensively employed in medical application due to the simplicity of synthesis, size, shape, and surface characteristics which can be facilely manipulated. Furthermore, by using the intrinsic optical, electrical, and magnetic capabilities of diverse inorganic nanomaterials, they have been extensively used in various biomedical applications including PDT, PTT, and the combinational of PDT/PTT. Thus, by the virtue of the nanomaterials, some of them have been approved by FDA and being used in clinical. On the basis of this significant advantage, a large number of innovative and effective nanosystem have been developed over the last few years in cancer treatment [18]. Some important theranostic inorganic nanomaterial fabricated by using inorganic nanoparticles has been discussed in below.

2.2.1 Gold Nanoparticles Based Therapeutics

Gold nanoparticles have received considerable interest in biomedicine and drug delivery applications as carrier of gene, drugs, and contrast agents. Furthermore, because of their ease of synthesis, high surface area, and adaptable surface chemistry, smart polymers may be used to develop a range of smart drug delivery systems that release the payload in presence to external stimuli. Additionally, because of its high molar absorption generally appeared in broad visible-NIR region, GNPs can be employed as a photothermal agent in cancer treatment [27].

Aioub et al. generated nuclear-targeted gold NPs (AuNPs) by the surface modification of AuNPs with various peptides to explore the cellular uptake as well as monitoring their localization within live cells. Nuclear-targeted AuNPs, significantly directed to nucleus of the cells due to their nucleus targeting capability, were observed to accumulate more quickly in cancer cells than those that were not directed to the nucleus [1]. Like gold NPs, the nanostructures are considered a good choice for photothermal treatment due to their high molar absorption coefficient and great selectivity via both passive and active targetings. The photothermal responses are widely used in cancer treatment. Chen et al. produced Au nanocages through surface PEGylating to sustain a long circulation period in the bloodstream which thereby increased permeability and retention (EPR) effect following irradiation, and the GNPs in the labeled cells/tissues create local heating, which leads in cell death via damage of biomolecules and the cell membrane [6]. Combining photothermal treatment with photodynamic therapy (PDT) and/or chemotherapy can boost the therapeutic efficiency of GNPs. Liu et al. also created similar kind of system by utilizing the surface of GNPs with the hydrophilic PEG-polymer to amplify the biocompatibility as well delivery of the hydrophobic drug at the desire site. The vivo studies were performed in cancer-bearing animals which reveal their efficiency of cancer-related disease [20].

2.2.2 Silica Nanoparticles Based Therapeutics

Silica nanoparticles (SNPs), like gold nanoparticles, are an important family of inorganic nanomaterials. Due to the viability of silanol groups, the silica is considered a very excellent carrier in drug delivery. The silanol groups can be easy modified with other polymer, therapeutic agents, or targeting moiety. Furthermore, the MSNPs use the enormous pore capacity to increase the distribution of numerous hydrophobic anticancer medicines inside the circulation. This is critical because the efficacy of such medications may be limited by their low water solubility. MSNPs' mesoporous cavities may hold a wide range of organic compounds (for example, medicines, SiRNA, proteins, or photosensitizers), making them intriguing agents in bioimaging applications. Sun et al. created triple-modal imaging-guided nanoplatform based on DOX @ Gd–doped mesoporous silica nanoparticles loaded with ICG. The developed thermosensitive liposomes revealed outstanding anticancer efficiency and adequate imaging effect of NIRF, PA, and MRI in vitro and in vivo experiments [28]. An another study, Shi et al., designed a radiolabeled antibody-conjugated quantum dot and Cai et al. created a radiolabeled antibody-conjugated quantum dot yolk/shell nanosystem (QD@HMSN) for fluorescence imaging in a separate investigation. The developed nanosystem (QD@HMSN) dramatically increased tumor uptake (1.5-fold), and high imaging contrast was produced with selective tumor vascular targeting [26].

2.2.3 Quantum Dots Based Therapeutic Biomaterials

Carbon dots (CDs), a novel family of carbon nanomaterials, have piqued the interest of biomedical researchers, particularly in tumor bioimaging. CDs have been authorized as innovative biomaterials due to their remarkable performance in precise targeted bioimaging and tumor treatment. Because of their superior photophysical characteristics, CDs have been used as contrast agents in vivo optical imaging. Light-activated nanotheranostics, such as photothermaltherapy (PTT) and photodynamic therapy, have recently emerged as the most promising cancer treatment technique in clinical trials (PDT). PTT, on the other hand, is a potential treatment technique that is used in a variety of cancer therapies. PTT's therapeutic effectiveness is greatly dependent on the photothermal agent for successful photothermal communication. Nonetheless, rational photothermal agent design remains a crucial difficulty. The PTT characteristics of photothermal agents are typically determined by their absorption cross-section and light absorption selectivity. Under the stimulation of 808 nm near-infrared light, the CDs had a photothermal conversion efficiency (PCT) of 78%. Moreover, the CDs may absorb light in the NIR-II range. Because of these fundamental properties, CDs are a good choice for photothermal treatment (PTT). In recent years, a variety of CDs-based medicinal chemicals for image-guided PTT therapy in cancer diagnostics and treatment have been discovered. To assess the inherent photothermal conversation efficacy of CDs, Ge et al. created a series of CDs by selecting precursor molecules of polythiophene derivatives (PT2) and then carbonizing the polymer of polythiophene phenylpropionic acid (PPA). The produced CDs demonstrated excellent photothermal conversion efficiency with a wide absorption band ranging from visible to NIR and a red emission [11].

To obtain high photothermal conversion efficiency, it might be used with other photothermal materials. For instant, Yu et al. fabricated a hollow-structured CDs based on CuS by loaded a tumor suppressor protein. The offered CuSCDs showed the excellent photothermal conversion efficiency and tumor targeting capability due to decorated the T7 peptide on the surface of the CuSCDs which enhance the selectivity to the cancer cells. The resultant CuSCDB@MMT7 NPs release the polyubiquitinated tumor suppressor protein via triggered PTT mechanism which revealed the accumulation of polyubiquitinated in tumor tissue allowing for increased tumor cell death and decreased metastasis [37].

Generally, fluorescent dyes were used to evaluate the intracellular microenvironment like (pH GSH, cell death process, etc.). However, these probes have significant disadvantages, such as needing lengthy synthetic route, having an inadequate response range, and having an excessively long emission wavelength. Ye et al. create a red emissive two-photon excited fluorescence probe based on carbon dots (pH-CDs) for measuring intracellular pH. The discovered pH-CDs have a number of remarkable features, including good selectivity, great photostability, and minimal cytotoxicity. They also have a pH-sensitive behavior in the pH range between (3.5 and 6.5), which is ideal for detecting the pH value in live cells. Moreover, fluorescence signal may be controlled by switching the aggregation and disaggregation between CDs [35].

2.2.4 Magnetic Nanoparticles Based Therapeutic Biomaterials

Another major family of inorganic nanomaterials is magnetic nanoparticles (MNPs). MNPs have found broad uses in biological and medical science due to their unique magnetic characteristics and conjugation with numerous biological and pharmacological molecules, such as multimodal bioimaging, hyperthermia, and tissue healing. Yet, magnetic resonance imaging (MRI) has been extensively employed in medical research as a diagnostic which suffers from inadequate sensitivity as well as poor spatial resolution. To address this problem, two or more imaging techniques have been used in tandem for instant MRI, or MRI/CT imaging. The MRI and CT imaging are generally utilized in combination to accomplish high resolution and sensitivity of soft-tissue with high contrast. Furthermore, the creation of specialized fluorescent probes for both optical and MRI imagings is extremely desirable to amplifying the resolution and selectivity. In this contribution, Foy et al. created MNPs with long-term imaging characteristics using NIR dyes, which showed excellent tumor colocalization in mice with xenograft breast cancers with and without an external magnetic field [10].

Magnetic thermotherapy that utilized MNPS is a novel approach for interstitial hyperthermia and thermoablation. When a local tumor site is subjected to an external magnetic field, temperature rises to 42–43 °C; thus, cancer cells die via necrotic pathway, and however, there is no damage of normal tissue. So far, a variety of MNPs have been developed for tumor thermotherapy though their surface modifications. Jordan et al. developed SPIONs for thermotherapy by coating with aminosilane and dextran. Treatment efficiency was measured by animal survival time and histological exams of the brain as well as tumor. Thermotherapy of aminosilane-coated MNPs resulted in a 4.5-fold increase in lifespan, as compared to dextran-coated particles

Table 2 Classification of inorganic nanomaterials used in image-guided cancer therapy

Inorganic nanoparticles	Nanomaterials' compositions	Contrast agent	Refs.
Gold nanoparticles (AuNPs)	Surface modified of AuNPs with different peptides	AuNPs	Aioub et al. [1]
Gold nanoparticles (AuNPs)	PEG-modified AuNPs	AuNPs	Chen et al. [6]
Gold nanoparticles (AuNPs)	Dox-loaded PEG-modified gold nanoparticles	AuNPs	Cheng et al. [7]
Gold nanoparticles (AuNPs)	Hydrophilic PEG-polymer-coated gold nanoparticles	ICG	Liu et al. [20]
Silica nanoparticles	DOX @ Gd-doped silica nanoparticles coated	ICG	Sun et al. [28]
Silica nanoparticles/CDs	Radiolabeled antibody-conjugated silica nanosystem Cu-NOTAQD@HMSN-PEG-TRC105	CDs	Shi et al. [26]
Quantum dots (CDs)	CDs developed by polythiophene derivatives (PT2) and polythiophene phenylpropionic acid (PPA) polymer	CDs	Ge et al. [11]
Quantum dots (CDs) and CuS nanoparticles	T7 peptide-coated CDs and CuS nanoparticles composited (CuSCDs)	CDs	Yu et al. [37]
Quantum dots (CDs)	CDs developed by using the precursors (*p*-phenylenediamine, dopamine, and *o*-phenylenediamine)	CDs	Ye et al. [35]
Magnetic nanoparticles (MNPs)	MNPs with excellent optical imaging	MNPs	Jones and Zweier [15]
Superparamagnetic iron oxide nanoparticles (SPIONs)	SPIONs were developed by using aminosilane and dextran	SPIONs	Jordan et al. [16]
Anionic SPIONs	Iron oxide colloidal nanomagnets	SPIONs	Fortin et al. [9]

which had no benefit [16]. Fortin et al. explored anionic SPIONs with homogeneous size and magnetic anisotropy. The findings indicate that SPIONs can act as promising mediators of magnetic hyperthermia in a variety of environments and consider as an excellent platform for the attachment of diverse targeted molecules. As a result, when paired with other treatment procedures such as radiation therapy or chemotherapy, magnetic hyperthermia will produce a synergistic therapeutic benefit [9] (Table 2).

3 Conclusion and Future Perspective

Over the last few years, a range of nanomaterials (organic/inorganic) coupled with fluorescent probes, medicines, or other agents have been created for image-guided targeted drug administration. These nanoparticles dramatically increase the payload's water solubility, biocompatibility, and target specificity. Currently, the majority of nanoparticles have been made by utilizing polymers or lipids. However, they demonstrate great advantages in disease therapy, there still facing the difficulty in scaling up where organic solvents mostly usage of in their fabrication process which causes toxicity and reduce their biocompatibility. Moreover, most of the therapeutic nanoparticles are mainly developed for the treatment or prevention of only one disease. On the other hand, theranostic nanomaterials based on inorganic nanomaterials showed significant biotoxicity. Therefore, the researchers focus to develop biocompatible therapeutic nanoparticles for multi-therapeutic or more than one disease. Fortunately, various PSs with aggregation-induced emission (AIE) attributes have been flourishing in recent years, which showed weakly emissive in solution phase but exhibited strong luminescence in aggregates state, due to the mechanism of restriction of intramolecular motion (RIM). In contrast to ACQ photosensitizers, the AIE-based photosensitizers exhibited high brightness in aggregates, high photostability, and excellent ROS generation efficiency. Therefore, the AIE-based photosensitizers (PSs) with high ROS generation efficiency and strong emission in red/near-infrared region (NIR) are considered as promising agents for image-guided photodynamic therapy (PDT). In addition to high ROS generation efficacy, the therapeutic efficacy of PDT is also depending on the subcellular localization of PSs, due to high reactivity and short diffusion distance (<20 nm) of the ROS. For this reason, great efforts have been made to develop the cellular organelle-specific AIE-PSs to improve the efficacy of image-guided photodynamic therapy (PDT). As a result, AIE-based nanomedicine will be the medicine of the future, with nanoparticle-based therapeutics playing a key role.

References

1. Aioub M, Kang B, Mackey MA, El-Sayed MA (2014) Biological targeting of plasmonic nanoparticles improves cellular imaging via the enhanced scattering in the aggregates formed. J Phys Chem Lett 5:2555–2561. https://doi.org/10.1021/jz501091x
2. Alifu N, Dong X, Li D, Sun X, Zebibula A, Zhang D, Zhang G, Qian J (2017) Aggregation-induced emission nanoparticles as photosensitizer for two-photon photodynamic therapy. Mater Chem Front 1:1746–1753. https://doi.org/10.1039/c7qm00092h
3. Bai JW, Qiu SQ, Zhang GJ (2023) Molecular and functional imaging in cancer-targeted therapy: current applications and future directions. Springer, US
4. Beziere N, Lozano N, Nunes A, Salichs J, Queiros D, Kostarelos K, Ntziachristos V (2015) Dynamic imaging of PEGylated indocyanine green (ICG) liposomes within the tumor microenvironment using multi-spectral optoacoustic tomography (MSOT). Biomaterials 37:415–424. https://doi.org/10.1016/j.biomaterials.2014.10.014

5. Calzoni E, Cesaretti A, Polchi A, Di Michele A, Tancini B, Emiliani C (2019) Biocompatible polymer nanoparticles for drug delivery applications in cancer and neurodegenerative disorder therapies. J Funct Biomater 10:1–15. https://doi.org/10.3390/jfb10010004
6. Chen J, Glaus C, Laforest R, Zhang Q, Yang M, Gidding M, Welch MJ, Xia Y (2010) Gold nanocages as photothermal transducers for cancer treatment. Small 6:811–817. https://doi.org/10.1002/smll.200902216
7. Cheng Y, Samia AC, Meyers JD, Panagopoulos I, Fei B, Burda C (2008) Highly efficient drug delivery with gold nanoparticle vectors for in vivo photodynamic therapy of cancer. J Am Chem Soc 130:10643–10647. https://doi.org/10.1021/ja801631c
8. Chow JCL (2022) Special issue: application of nanomaterials in biomedical imaging and cancer therapy. Nanomaterials 12:1–40. https://doi.org/10.3390/nano12050726
9. Fortin JP, Wilhelm C, Servais J, Ménager C, Bacri JC, Gazeau F (2007) Size-sorted anionic iron oxide nanomagnets as colloidal mediators for magnetic hyperthermia. J Am Chem Soc 129:2628–2635. https://doi.org/10.1021/ja067457e
10. Foy SP, Manthe RL, Foy ST, Dimitrijevic S, Krishnamurthy N, Labhasetwar V (2010) Optical Imaging and Magnetic Field Tumors. 4:5217–5224
11. Ge J, Jia Q, Liu W, Guo L, Liu Q, Lan M, Zhang H, Meng X, Wang P (2015) Red-emissive carbon dots for fluorescent, photoacoustic, and thermal theranostics in living Mice. Adv Mater 27:4169–4177. https://doi.org/10.1002/adma.201500323
12. Gu B, Wu W, Xu G, Feng G, Yin F, Chong PHJ, Qu J, Yong KT, Liu B (2017) Precise two-photon photodynamic therapy using an efficient photosensitizer with aggregation-induced emission characteristics. Adv Mater 29:1–7. https://doi.org/10.1002/adma.201701076
13. Gulzar A, Xu J, Yang D, Xu L, He F, Gai S, Yang P (2018) Nano-graphene oxide-UCNP-Ce6 covalently constructed nanocomposites for NIR-mediated bioimaging and PTT/PDT combinatorial therapy. Dalt Trans 47:3931–3939. https://doi.org/10.1039/c7dt04141a
14. Jin G, He R, Liu Q, Lin M, Dong Y, Li K, Tang BZ, Liu B, Xu F (2019) Near-infrared light-regulated cancer theranostic nanoplatform based on aggregation-induced emission luminogen encapsulated upconversion nanoparticles. Theranostics 9:246–264. https://doi.org/10.7150/thno.30174
15. Jones R, Zweier (2014) 基因的改变 NIH Public Access. Bone 23:1–7. https://doi.org/10.1021/nn101427t.Optical
16. Jordan A, Scholz R, Maier-Hauff K, van Landeghem FKH, Waldoefner N, Teichgraeber U, Pinkernelle J, Bruhn H, Neumann F, Thiesen B, von Deimling A, Felix R (2006) The effect of thermotherapy using magnetic nanoparticles on rat malignant glioma. J Neurooncol 78:7–14. https://doi.org/10.1007/s11060-005-9059-z
17. Lajunen T, Kontturi LS, Viitala L, Manna M, Cramariuc O, Róg T, Bunker A, Laaksonen T, Viitala T, Murtomäki L, Urtti A (2016) Indocyanine green-loaded liposomes for light-triggered drug release. Mol Pharm 13:2095–2107. https://doi.org/10.1021/acs.molpharmaceut.6b00207
18. Liang R, Wei M, Evans DG, Duan X (2014) Inorganic nanomaterials for bioimaging, targeted drug delivery and therapeutics. Chem Commun 50:14071–14081. https://doi.org/10.1039/c4cc03118k
19. Liu F, Chen Y, Li Y, Guo Y, Cao Y, Li P, Wang Z, Gong Y, Ran H (2018a) Folate-receptor-targeted laser-activable poly(Lactide-co-glycolic acid) nanoparticles loaded with paclitaxel/indocyanine green for photoacoustic/ultrasound imaging and chemo/photothermal therapy. Int J Nanomed 13:5139–5158. https://doi.org/10.2147/IJN.S167043
20. Liu F, Li Y, Li P (2018b) Poly (Lactide-Co-Glycolic Acid) nanoparticles loaded with paclitaxel/Indocyanine green for photoacoustic/ultrasound imaging and chemo/photothermal therapy:5139–5158
21. Mérian J, Boisgard R, Bayle PA, Bardet M, Tavitian B, Texier I (2015) Comparative biodistribution in mice of cyanine dyes loaded in lipid nanoparticles. Eur J Pharm Biopharm 93:1–10. https://doi.org/10.1016/j.ejpb.2015.03.019
22. Mitchell MJ, Billingsley MM, Haley RM, Wechsler ME, Peppas NA, Langer R (2021) Engineering precision nanoparticles for drug delivery. Nat Rev Drug Discov 20:101–124. https://doi.org/10.1038/s41573-020-0090-8

23. Sahu A, Min K, Jeon J, Yang HS, Tae G (2020) Catalytic nanographene oxide with hemin for enhanced photodynamic therapy. J Control Release 326:442–454. https://doi.org/10.1016/j.jconrel.2020.07.023
24. Shahbazi R, Ozpolat B, Ulubayram K (2016) Oligonucleotide-based theranostic nanoparticles in cancer therapy. Nanomedicine 11:1287–1308. https://doi.org/10.2217/nnm-2016-0035
25. Shi D (2009) Integrated multifunctional nanosystems for medical diagnosis and treatment. Adv Funct Mater 19:3356–3373. https://doi.org/10.1002/adfm.200901539
26. Shi S, Chen F, Goel S, Graves SA, Luo H, Theuer CP, Engle JW, Cai W (2018) In vivo tumor-targeted dual-modality PET/optical imaging with a yolk/shell-structured silica nanosystem. Nano-Micro Lett 10:1–11. https://doi.org/10.1007/s40820-018-0216-2
27. Si P, Razmi N, Nur O, Solanki S, Pandey CM, Gupta RK, Malhotra BD, Willander M, De La Zerda A (2021) Gold nanomaterials for optical biosensing and bioimaging. Nanoscale Adv 3:2679–2698. https://doi.org/10.1039/d0na00961j
28. Sun Q, You Q, Wang J, Liu L, Wang Y, Song Y, Cheng Y, Wang S, Tan F, Li N (2018) Theranostic nanoplatform: triple-modal imaging-guided synergistic cancer therapy based on liposome-conjugated mesoporous silica nanoparticles. ACS Appl Mater Interfaces 10:1963–1975. https://doi.org/10.1021/acsami.7b13651
29. Taratula O, Patel M, Schumann C, Naleway MA, Pang AJ, He H, Taratula O (2015) Phthalocyanine-loaded graphene nanoplatform for imaging-guided combinatorial phototherapy. Int J Nanomed 10:2347–2362. https://doi.org/10.2147/IJN.S81097
30. Terreno E, Uggeri F, Aime S (2012) Image guided therapy: the advent of theranostic agents. J Control Release 161:328–337. https://doi.org/10.1016/j.jconrel.2012.05.028
31. Wang S, Wang X, Yu L, Sun M (2021) Progress and trends of photodynamic therapy: from traditional photosensitizers to AIE-based photosensitizers. Photodiagnosis Photodyn Ther 34:102254. https://doi.org/10.1016/j.pdpdt.2021.102254
32. Wang S, Wu W, Manghnani P, Xu S, Wang Y, Goh CC, Ng LG, Liu B (2019) Polymerization-enhanced two-photon photosensitization for precise photodynamic therapy. ACS Nano 13:3095–3105. https://doi.org/10.1021/acsnano.8b08398
33. Xie J, Lee S, Chen X (2010) Nanoparticle-based theranostic agents. Adv Drug Deliv Rev 62:1064–1079. https://doi.org/10.1016/j.addr.2010.07.009
34. Xu S, Wu W, Cai X, Zhang CJ, Yuan Y, Liang J, Feng G, Manghnani P, Liu B (2017) Highly efficient photosensitizers with aggregation-induced emission characteristics obtained through precise molecular design. Chem Commun 53:8727–8730. https://doi.org/10.1039/c7cc04864e
35. Ye X, Xiang Y, Wang Q, Li Z, Liu Z (2019) A red emissive two-photon fluorescence probe based on carbon dots for intracellular pH detection. Small 15:1–9. https://doi.org/10.1002/smll.201901673
36. Yoon HY, Jeon S, You DG, Park JH, Kwon IC, Koo H, Kim K (2017) Inorganic nanoparticles for image-guided therapy. Bioconjug Chem 28:124–134. https://doi.org/10.1021/acs.bioconjchem.6b00512
37. Yu Y, Song M, Chen C, Du Y, Li C, Han Y, Yan F, Shi Z, Feng S (2020) Bortezomib-encapsulated CuS/carbon dot nanocomposites for enhanced photothermal therapy via stabilization of polyubiquitinated substrates in the proteasomal degradation pathway. ACS Nano 14:10688–10703. https://doi.org/10.1021/acsnano.0c05332
38. Yuan A, Tang X, Qiu X, Jiang K, Wu J, Hu Y (2015) Activatable photodynamic destruction of cancer cells by NIR dye/photosensitizer loaded liposomes. Chem Commun 51:3340–3342. https://doi.org/10.1039/c4cc09689d
39. Zhang YH, Li X, Huang L, Kim HS, An J, Lan M, Cao QY, Kim JS (2020) AIE based GSH activatable photosensitizer for imaging-guided photodynamic therapy. Chem Commun 56:10317–10320. https://doi.org/10.1039/d0cc02045a

Development in Biosensor-Based Diagnostics for Bacterial Diseases: Opportunities and Challenges

Arka Sanyal, Priya Mitra, Tanima Dey, Debatri Dutta, Koustav Saha, Arunima Pandey, and Ritesh Pattnaik

Abstract The recurring upsurge in mortality and morbidity intimidates the detection of bacterial pathogens crucial in the disciplines of food safety, health care, and medicine. There is an essential requirement to improve existing antimicrobial therapeutic strategies due to exponential growth in antimicrobial resistance and the frozen pool for antibiotic medication, and therefore accurate and rapid bacterial identification is critical. Several approaches like polymerase chain reactions, immunoassays, and culture procedures—the three current gold standards for detecting bacteria—have the conspicuous drawbacks of prolonged turnaround times and poor accuracy. Hence, their potential usage as point-of-care devices is substantially constrained. Current-edge detection methods have been recommended and constructed over the period to address these shortcomings. Biosensors are one of those pioneering detection tools that are capable of identifying a variety of biomolecules, which are extensively utilized for the effective detection of pathogens. This chapter provides a comprehensive overview of several biosensing systems with a greater emphasis on technological advancements in improving detection sensitivity and specificity. The mechanism of detection along with the advantages and disadvantages of each kind of biosensor is carefully discussed, with the variables impacting their perspective as potential point-of-care devices.

Keywords Infectious diseases · Biosensors · Point-of-care · Antibiotic therapy · Rapid diagnostics

Abbreviations

MDR	Multidrug-resistant
ELISA	Enzyme-linked immunosorbent assay

A. Sanyal · P. Mitra · T. Dey · D. Dutta · K. Saha · A. Pandey · R. Pattnaik (✉)
KIIT School of Biotechnology, Kalinga Institute of Industrial Technology (KIIT-DU), Bhubaneswar, Odisha, India
e-mail: rpattnaik@kiitbiotech.ac.in

A. Madhusudhan et al. (eds.), *Functional Smart Nanomaterials and Their Theranostics Approaches*, Smart Nanomaterials Technology,
https://doi.org/10.1007/978-981-99-6597-7_8

PCR	Polymerase chain reaction
qPCR	Quantitative polymerase chain reaction
LOD	Limit of detection
CFU	Colony forming unit
DNA	Deoxyribonucleic acid
SPR	Surface plasmon resonance
FRET	Forster resonance energy transfer
IRIS	Interferometric reflectance imaging sensor
pSi	Porous silicon layer
mAbs	Monoclonal antibodies
ZnNPs	Zinc nanoparticles
AuNPs	Gold nanoparticles
HIV	Human immunodeficiency virus
TB	Tuberculosis
OLED	Organic light-emitting diode
Rct	Charge transfer resistance
ESAT-6	Early Secretary Antigen Target-6
FET	Field-effect transistor
LAPS	Light-addressable potentiometric sensor
ZnO	Zinc oxide
IGZO	Indium gallium zinc oxide
SiNWs	Silicon nanowires
CNTs	Carbon nanotubes
OBPs	Odorant binding proteins
QCM	Quartz crystal microbalance
SAW	Surface acoustic wave
STW	Shear transverse wave
LSPR	Localized surface plasmon resonance
GQDs	Graphene quantum dots
POCT	Point-of-care testing
WHO	World Health Organization
TPP	Target product profiles

1 Introduction

In the past decade, although significant progress has occurred in disease diagnosis, prevention, and treatment, bacterial infections pose a major global health issue [1, 2]. The crucial part of the clinical intervention of such infectious diseases includes the timely diagnosis and initiation of targeted antimicrobial treatment [3]. However, there exist several critical challenges for the effective management of bacterial infections such as the proliferation of multidrug-resistant (MDR) pathogens, the application of effective antimicrobials, and the emergence of novel infectious agents that contribute toward the diagnosis of bacterial infections in humans and other animals. The current diagnostic methods for clinically relevant infectious diseases brought on by a wide range of agents, including bacteria, viruses, fungi, and parasite pathogens, rely on a number of laboratory-based procedures, such as culturing, microscopy, nucleic-acid amplification, and immunoassays. Though in use, all such techniques showcase shortcomings including the lack of sensitivity and predictability in many clinical scenarios studied while using microscopy. Likewise, immunoassay approaches such as enzyme-linked immunosorbent assay (ELISA) even albeit extremely sensitive, multiplex detection methods are frequently labor-intensive and difficult to execute.

Advanced techniques based on nucleic-acid amplification like polymerase chain reaction (PCR) despite offering molecular specificity have intensive sample preparation and the risk of obtaining unreliable findings. As a promising diagnostic approach, the biosensing technique in recent times has gained popularity due to its potential benefits and efficacy in clinical care, with particular traits including responsiveness, specificity, and quickness as essential components in starting early diagnosis and treatment [4, 5]. The application of biomarkers along with developments in cutting-edge technologies like nanotechnology and microfluidics increases the efficiency of the healthcare industry. Importantly, the incorporation of transducers into biomaterials has sparked the creation of interfaces that can generate signals, sometimes known as "biosensors" [6]. Recently, there has been a paradigm shift in disease detection through the use of biosensors.

This book chapter gives an introduction to biosensors, their operating principle, characteristics, and types. We have also highlighted the advantages and shortcomings of each type with respect to bacterial disease diagnosis. Additionally, we introduce our readers to the concept of point-of-care testing (POCT) and how biosensors have found applications as POCT devices, highlighting the commercially available POCT devices available in the market for bacterial detection. As a concluding remark, we also elucidate the challenges and future perspectives surrounding this rapidly evolving field.

2 Biosensors and it's Operating Principle

Biosensors are analyte detectors, comprising a biological base along with a physicochemical detection system, where the analytes are determined by the design and functioning of the sensors. Smartphones, smart-watches for example, are nowadays used as a biosensor by installing simple attachments, enabling for trouble-free initial screening. Biosensors generally monitor biomolecules specific to disease conditions, such as nucleic acids and proteins, which are detected using antibodies, enzymes, microorganisms, and so on [7–9]. These sensors primarily detect biochemical signs of illness in body fluids such as urine, blood, and saliva [10]. It can also be used for a variety of clinical diagnoses, including cholesterol, cardiovascular disease indicators, cancer or tumor biomarkers, and allergic reactions. Moreover, biosensors are also used to detect food-borne and water-borne bacterial and viral pathogens

[11–13]. Hence, in clinical applications, these disposable devices provide a potential alternative. The biological sensor, the detector, and the reader are the three fundamental components of a biosensor. Interdisciplinary investigations are essential before designing these biosensors to identify the appropriate and cost-effective material, biological element, and transducing devices.

As mentioned earlier biosensors are made up of three parts (Fig. 1). First, the sensing components are also known as bio-receptors, which mimic in vivo molecular recognition behavior. These sensing elements include cells, cell receptors, enzymes, antibodies, nucleic acids, and microorganisms. Depending on the type of biosensor, these biological sensing components identify the analyte and interact with it. The use of bacterial or viral sequences of nucleic acids is one of the most used biorecognition techniques in this milieu [14]. The second component is the transductor, also known as the detector, which detects the signal associated with the physicochemical alteration due to the interaction of the analyte with the bioreceptor. It converts the signal into an analyzable and measurable form [15]. The reader is the last component of a biosensor, which includes a display that is powered by the software along with hardware to provide the outputs [16].

The functionality and effectiveness of a biosensor are determined by several essential aspects. To begin, selectivity refers to a bioreceptor's ability to recognize a particular bio-entity when examining a multi-component sample. This is most certainly the crucial trait to define the required bioreceptor. The second trait is repeatability, which defines the capacity to achieve the very same response for several instances

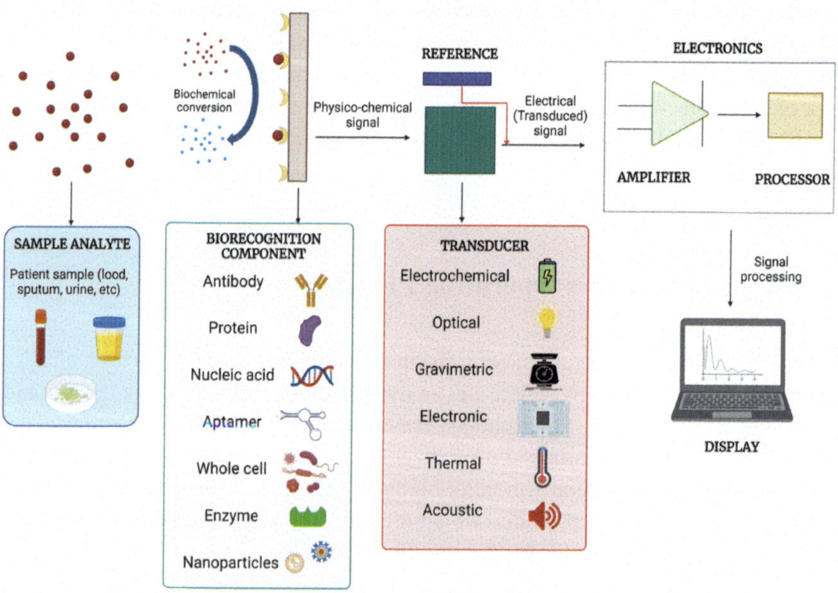

Fig. 1 A schematic depiction of the essential components (and their types) for a typical biosensor. [Created with BioRender.com]

Table 1 Essential features of an ideal bacterial biosensor

Sl. no	Features	Characteristics
1	Sensitivity	The limit of detection should be less that 10^3 CFU per ml
2	Specificity	Able to differentiate different species of bacteria, should function well in complicated matrices containing different components of samples together such as sputum and blood or food and beverages
3	Rapidness	Not more than 5–10 min for each test
4	Size	Small in size and easily portable in nature
5	Sample processing	Simple process and not much time or skills required
6	Stability	To allow for longer shelf life, the bio-recognition element must be stable at high temperatures (till 45 °C) in some countries
7	Skills required	Professional training is not necessary to perform the test, and patients can perform it themselves

in a specific experimental setup. Reliability and robustness are achieved by reproducible signals. The third one is stability which refers to the ability to withstand interference's that might impact the precision as well as the accuracy of the equipment. Fourth is sensitivity, defined as the limit of detection (LOD), which represents the smallest measure of the detectable analyte. Lastly, linearity assesses the accuracy of the measurements which are within the analyte concentration range (linear range), and it should lead to a shift in output (resolution) in response to the slightest fluctuation in analyte concentration [17]. All these essential criteria's desirable in a biosensor have been summarized (Table 1).

3 Biosensors for Bacterial Detection

Infectious diseases continue to threaten human society dreadfully; for example, the recent outbreak of COVID-19 took the lives of more than 5 million people in the past two years. Bacterial pathogens are important and easy targets for detection and identification in medicine, food safety, public health, and security. Around 1962, Clark and Lyons first described the concept of a biosensor when they developed an oxidase enzyme electrode for glucose detection [18]. These user-friendly, highly specific and highly sensitive, rapidly used, portable devices are also used in the detection of microorganisms like bacteria and viruses, in food processing industries for the detection of toxins, environmental screening, clinical diagnosis, and bioterrorism. Application of these state-of-the-art technologies in day-to-day screening and detection helps to substitute the old-age traditional time-consuming methods like sample culturing, quantitative polymerase chain reactions, and immunoassays (Table 2).

Table 2 Traditional bacterial detection methods

Sl. no	Traditional bacterial detection method	Examples of pathogens detected	Advantages	Disadvantages	References
1.	Culture of bacterial samples	*E. coli*, *F. parahaemolyticus*, *Salmonella sp.*, *L. monocytogenes*	• Culture methods are generally very sensitive • LOD ranges from 1–10 CFU mL^{-1} to 2000 CFU mL^{-1} • Universal method which can be performed easily • Cheap • Can detect specific bacterial or different bacterial species at the same time using non-selective media	• It is a time-consuming process and yields results in a few days • Initially, higher concentration of bacteria strain is required for the culture • If bacterial have the ability to switch to dormancy under favorable conditions, it is difficult to culture and identifies the bacteria • Antibiotic-resistant bacteria are difficult to identify by culturing	[19, 20]

(continued)

Table 2 (continued)

Sl. no	Traditional bacterial detection method	Examples of pathogens detected	Advantages	Disadvantages	References
2.	Polymerase chain reactions	*Escherichia coli* O157:H7, *Listeria monocytogenes Staphylococcus aureus*, *Salmonella*, and *Vibrio parahaemolyticus*	• PCR amplifies the targeted bacteria's DNA sequences into numerous copies for identification • qPCRs typically detect bacteria ranging between 10^3 and 10^4 CFU mL^{-1} • Different bacterial species can be simultaneously detected by multiplexed qPCR using multiple DNA sequences • Results are available in <24 h; less time taking	• qPCR is highly specific, detecting one bacterial species at a time • Several qPCR assays or multiplexed qPCR are used to detect different bacterial species, which is time-consuming and labor-extensive • Trained personnel required to operate the equipment • Can sometimes lead to false positive results	[21]
3.	Immunology-based assays	*Escherichia coli*, *Salmonella enterica*	• It works on the principle of highly specific antigen–antibody interactions and enzyme-linked immunosorbent assay (ELISA) is the most common • Its limit of detection ranges from10^4 to10^5 CFU mL^{-1} • A small sample volume is required for the assay • High reproducibility • Reagents are long-lasting when kept radiation-free and produce non-toxic wastes • Uses simple equipment • Low operating costs	• Multistep processes, such as standard solution preparation, antibody incubation, washing, and blocking, prior to actual experiment • Time-consuming • Low sensitivity compared to culture and qPCR methods	[22]

In the current chapter, we have discussed biosensors that are classified into optical, electrochemical, mechanical, microbe-based biosensors, nano-biosensors, and more over used for the detection of bacterial-mediated diseases.

3.1 Optical Biosensors

Optical biosensors sometimes referred to as "optodes" use optic fibers as transduction elements, thereby performing various optic mechanisms for analyte detection such as absorption, colorimetry, fluorescence, or luminescence in the ultraviolet, visible, and near-infrared regions. They provide lesser unnecessary fluctuations in signals which boosts up electromagnetic interference, hence outbalancing electrochemical and piezoelectric biosensors [23]. They have emerged recently as a potential solution for disease diagnostics and as the point-of-care devices for therapeutics, thereby overcoming some of the challenges faced in conventional disease diagnostics by delivering immediate results and employing small and user-friendly devices. These biosensors basically rely on optical phenomena such as surface plasmon resonance (SPR) and Forster resonance energy transfer (FRET) [24]. The various subtypes of optical biosensor and their uses in the detection of infectious illnesses are covered in detail in the sections that follow.

3.1.1 Interferometric Reflectance-Based Biosensors

The interferometric reflectance imaging sensor (IRIS) signal is characterized by the interference of the fields reflected off a layered substrate that is typically comprised of a porous silicon layer (pSi) with the presence of air-filled pores having diameters less than 150 nm [25, 26]. Using the basic principle of reflection of light, these types of biosensors are specifically used for the detection of bacteria. Since 2007, IRIS has undergone many commercial prototype development phases for biosensors employing an optical bench and a continuously adjustable laser source. Within a year, Zoiray Technologies marketed a prototype device for protein microarrays [27]. The material has been widely utilized as an optical biosensor since the early 2000s due to its huge surface area, cost, complexity, simplicity of manufacturing, and functionalization. The multifunctional IRIS platform has been extensively used in various applications.

The interferometric reflectance phenomenon and subsequent pSi biosensor design largely depend on the analyte binding to the pSi surface, which causes a change in the medium's refractive index and the detection of a shift in the interference pattern as well as a decrease in the intensity of the reflected light. Urmann et al. have described the above-mentioned phenomenon by devising a simple Hemag1P aptamer-functionalized pSi optical biosensor for capture and detection of the probiotic, *Lactobacillus acidophilus* by reflective interferometric FTIR with a LOD of 106

CFU/ml [28]. On the surface of pSi, *E. coli* monoclonal antibodies were immobilized in order to directly capture the associated bacteria. At various bacterial concentrations, the intensity change was also observed [29]. To detect and target *E. coli* with the lowest LOD, another system based on optical interferometry, a method frequently employed with porous silicon-based optical biosensors, and nanomaterials was conceived [30].

3.1.2 Fluorescence-Based Biosensors

Fluorescence intensity has a direct impact on the measure of bacterial concentration; hence, fluorescence-based biosensing platforms have shown widespread use in the detection of infectious diseases. The fluorescence method has become the most dominant analytical tool in medical testing, biotechnology, and fields of drug discovery [31]. A large number of fluorescent-based biosensors have already been developed in the last 30 years for a variety of cations, anions, or neutral molecules although sensing large biomolecules such as nucleic acids and proteins remained a major challenge. Based on the structure, fluorescent biosensors can be segregated into two classes: (i) sensors in which fluorophore directly interacts with the analyte where the fluorescent biosensors operate mainly by enhancing or quenching emission and (ii) sensors in which fluorophore indirectly interacts with the analyte where the fluorophore is directly to the molecule that specifically binds to the target analyte without the direct involvement of the fluorophore in the binding process [32].

A typical fluorescence measurement is done with the excitation of the fluorophore radiating at many wavelengths and being illuminated by a particular wavelength. For sensitive detection of the targeted analytes present in the sample in trace amounts, reporter molecules tagged with fluorescent dyes are required. It was established that functionalized *Staphylococcus aureus* cells with fluorescence could be transformed into nanobioprobes that could target and detect pathogens, which is an intriguing use of fluorescence-based detection. *S. aureus's* surface protein A interacts with particular monoclonal antibodies' Fc regions, allowing these mAbs to target a variety of viruses and bacteria [33]. Perception of fluorescence was accomplished by synthesizing CdS0.5Se0.5 within the cell, and these nanobioprobes offer a method for developing biosensors employing a functionalized organism while lowering toxicity hazards.

Another fluorescence-based optical system was developed by combining both the advantages of stimuli-responsive biosensors and antibodies for enhanced selectivity. The goal was to create a hybrid nanoparticle system made of mesoporous silica-curcumin NPs that were 500 nm in size and were functionalized with 6 nm ZnNPs that were modified with polyclonal antibodies against *S. typhi* [34]. A "turn on fluorescence" biosensing platform was also invented that used a FRET system with dopamine-stabilized AuNPs and copper clusters with citrate functionalization to detect microorganisms [35], while a smartphone-based fluorescence device was designed for the detection of E.coli 0157:H7 via sandwich assay which involved a field portable fluorescent imager consisting of an excitation light source, sample loading chamber, a long-pass filter, and a focusing lens for collection of signal attached with a cell phone [36]. Magnetic beads labeled with mAbs against a specific bacteria are utilized to create a sandwich test that uses an external magnet to detect and isolate the pathogen from the sample combination.

3.1.3 Plasmonic Biosensors

Plasmonic biosensors have shown to be quite effective for highly sensitive real-time and label-free biosensing applications and are divided into two categories—one which uses a thin metal-based film and the other which uses nanostructured-based inorganic plasmon resonance. Surface plasmon resonance is a phenomenon where resonant coupling occurs between electromagnetic waves and collective oscillations of free electrons in metals. The surface plasmonic biosensor has been the most commonly used biosensor made up of thin metal film mostly made of gold and mainly used to characterize biomolecular interactions. The main principle of surface plasmon resonance (SPR) biosensing lies in the propagation of surface plasmons along a metal–dielectric interface. The efficiency of a plasmonic biosensor can be increased by combining it with surface-enhanced Raman scattering (SERS) which is used as a non-invasive label-free diagnostic biosensor, thereby attaining information from one measurement. The popularity of nanoplasmonic sensors with a focus on noble nanoparticles such as gold and silver and their inherent localized surface plasmon resonance properties have already been established in the realm of biosensors [37].

There are currently many available diagnostics for tuberculosis such as tuberculin skin test, interferon-gamma release assays, and sputum smear microscopy, which are considered the gold standard methods. However, due to the low sensitivity and specificity, higher cost, turnover time and requirement of skilled workers associated with such platforms, SPR biosensors have taken an edge over the commonly used diagnostic methods by distinguishing active TB infections through various biomarkers eliminating the need for expensive labels, cutting down skill dependency in obtaining accurate results and also decreasing turnover time. A portal SPR device with a LOD of 10ng/ml uses antibody responses against Ag85 secretory protein, while an organic light-emitting diode (OLED) prism-coupled SPR system targets IS6110 DNA which showed LOD of 63pg/ml is able to diagnose TB infections in a highly sensitive and specific system [38].

3.1.4 Colorimetric Biosensors

Optical biosensors also include colorimetric biosensors which are recently flourishing due to its clarity in result visualization, feasibility, and cost-effectiveness [39]. The detection events are based on the redox and the enzyme-catalyzed reactions which are happening in the colorimetric sensors, causing a color change which is detectable by the naked eye. A redox-responsive biosensor that combines the benefits of colorimetric analysis and electrochemical sensing mainly through a reduction reaction of p-benzoquinone to hydroquinone conversion in response to the enzymatic reaction of bacterial respiration is used for producing a distinct red color and thereby detecting bacterial concentration up to 10^4 CFU/ml [40]. Recently, peroxidase is impersonated by gold nanoparticles, platinum NP (nanoparticles), metal nanomaterials (MNMs), nanohybrids based on carbon nanotubes (CNTs), and graphene oxide for colorimetric-based detection which degrade the O–O bond of H_2O_2 into a high oxidizing $OH^{\bullet}$ radical [41, 42]. The radical produced can oxidize a peroxidase substrate such as 3,3′,5,5′-tetramethyl-benzidine (TMB), and 2,2′-azino-bis (3-ethylbenzothiazoline-6-sulfonic acid) ($ABTS^{2-}$) with H_2O_2, generating blue and green colors when in aqueous solution [5, 8, 43, 44].

3.2 Electrochemical Biosensors

Electrochemical biosensors as the name suggests are devices that are capable of quantifying biochemical and biological reactions via conversion into an electrical signal. As simple as the definition seems, the concept was difficult to bring into reality, since connecting a biological system to an electrical circuit is complex. However, a simplified version has been developed with a bio-analyte receptor, providing a surface for a biological or biochemical reaction to take place, following a transducer to convert the signal, and a detector to finally detect and quantify the result [45, 46]. The development of an oxygen monitor of tissue/blood was the first significant milestone followed by the development of a glucose sensor in the broad map of electrochemical biosensors [18, 47]. Recent times have shown the petrifying side of infectious diseases and their rise seems to be unstoppable, which concerns the finding of different solutions for the identification and detection of the disease and causative agent. The principle following which the electrochemical biosensors work for infectious disease is based on the alterations happenings at the interacting surface of the sensor sample matrix [45]. The molecules considered here as samples are DNA, RNA, enzymes, tissues, etc. The sample interacts with the analyte resulting in an electrical signal, which is then amplified and noise is removed to get an accurate result [4, 48].

Typically, a working electrode serves as the primary component of an electrochemical cell. An electrode is an electrical conductor through which electrons and holes flow to carry the charge. Thus, conducting and semiconducting materials, including metals like gold (Au) and nonmetals like carbon, are used to produce electrodes. Multiple materials and manufacturing techniques can be used to create electrodes. A potentiostat system often uses a three-electrode format (working, auxiliary, and reference), whereas conductometry and electrochemical impedance spectroscopy frequently utilize a two-electrode format (working and auxiliary) (EIS).

The two types of bio-recognition components for electrochemical biosensors are (i) bio-catalytic and (ii) biocomplexing. The biosensor response in the case of bio-catalytic bio-recognition elements is based on a reaction catalyzed by macromolecules. The three most frequently employed bio-catalytic bio-recognition molecules are tissues, entire cells, and enzymes. The biosensor response for biocomplexing-based bio-recognition elements is dependent on how analytes interact with structured molecular assemblies or macromolecules. Antibodies, peptides, and phages are the most often employed during biocomplexing-based bio-recognition components for pathogen detection. The immobilization of bio-recognition components on electrodes is crucial to the design, production, and effectiveness of electrochemical biosensors for pathogen detection since biosensors are self-contained devices. Immobilization aims to create a solid, irreversible link between the bio-recognition and the target [49]. Electrochemical biosensors are categorized based on parameters measured such as current (amperometric/voltammetric biosensors), impedance (impedance biosensors), or potential (potentiometric biosensors). These biosensors differ in electrode configuration, applied signals, measured signals, mass transport regimes, binding information provided, and target size selectivity.

3.2.1 Impedance-Based Biosensors

The EIS approach uses variations in interfacial capacitance or charge transfer resistance (Rct) to identify biological changes occurring at the sensor surface. The advantages of using these biosensors include measurement over a broad range of periods and frequencies and highly accurate time-averaged responses. Impedance can be thought of as resistance to the flow of current in an electrical circuit. Impedance measurement data primarily comprises of resistive and capacitive components, on which faradaic and non-faradaic sensors are built. For the measurement of the electrode's electron transfer resistance in faradaic biosensors, the electrode holding the bioreceptor is submerged in a solution of an electrochemical redox probe, as opposed to non-faradaic sensors, which measure the interfacial capacitance instead. A change in the Rct value of the redox probe is related to the concentration of the target analyte and the change happens when the bioreceptor immobilized on the electrode surface collects the target analyte [50].

For example, Early Secretary Antigen Target-6 (ESAT-6) produced by *Mycobacterium tuberculosis* is treated as a target analyte and has been detected with independent electro-immunosensors that are designed, built, and tested by bio-microsystems based on printed circuit board platforms. The structure and arrangement of an electrode affect a sensor's effectiveness rather than only serving as a platform for sensing, while interdigitated microelectrodes are proven to be more effective than standard electrodes in downsizing [51]. Impedance biosensors for infectious diseases have made significant strides, but there are still challenges to be overcome and more research needs to be done before these sensors can be used in real time in the future.

3.2.2 Capacitance-Based Biosensors

Capacitive affinity biosensors can be constructed by immobilizing recognition components in thin layers on an electrode or between two electrodes, which monitors the changes in the dielectric characteristic upon the attachment of an analyte. A big receptor molecule being displaced by a smaller analyte can also cause these changes. To detect minute changes, a capacitive sensor is incorporated using flexible printed circuit board technology. The suggested platform includes a capacitive sensor, function generator, readout circuit, and personal computer [52]. In order to indirectly quantify the number of antibodies and glucose for a particular pathogen-associated reaction, oxygen bubbles are monitored at an interdigitated electrode, where antibody-containing samples first specifically attach to the antigens and are immobilized on the sensor surface; then, in a subsequent stage, catalase is coupled to an antibody binding protein associated with the antibodies.

In the next step, the exposure of the electrode to H_2O_2 produces oxygen, where the quantity of oxygen bubbles serves as a proxy for antibody levels for detection [53]. Concanavalin A was used in a similar manner to measure glucose. Mahadhy et al. have come up with an ultrasensitive capacitive biosensor for selective detection and quantification of *Enterobacteriaceae* DNA [54]. Based on the same technology, another biosensor has been designed and brought to the market to differentiate between infected and uninfected cells [55].

3.2.3 Potentiometric and Amperometric Biosensors

Potentiometric sensors operate on the concept that, with zero current flow, the potential difference between the working and reference electrodes varies directly with the concentration of the analyte being studied. The Nernst equation governs the connection between concentration and potential for potentiometric measurements, where E_{cell} is the observed cell potential at zero current [56]. Currently, ion-selective electrodes are frequently used to obtain the lowest detection limits by potentiometric devices (ISE). The detection limit is analyte-specific by definition, and modern devices have LOD ranging from 10^8 to 10^{11} M. Potentiometric sensors are useful for measuring low concentrations in small sample quantities since they ideally have the advantage of not affecting the sample chemically [57]. To detect pH variations, selective ion concentrations, and the kinetics of bio-catalytic events involving enzymes, numerous potentiometric devices are also based on different types of field-effect transistor (FET) devices [58].

A light-addressable potentiometric sensor (LAPS) is another illustration of a unique optical/electrochemical hybrid method. LAPS is a silicon-based detector that uses the photovoltaic effect to pinpoint the measurement location with precision. The surface potential distribution along the interface of the sample and substrate surfaces can be found by scanning with a focused light source. Aptamers have already been demonstrated to be efficient bio-recognition components for the detection of a variety of infectious diseases, and therefore, their combination with a possible FET transducer may one day result in the creation of accurate, quick, and portable analysis instruments for infectious diseases [59]. In a biological reaction, it constantly detects the current that results from the oxidation or reduction of an electroactive molecule. Recently, researchers have developed a novel potentiometric biosensor based on graphene and aptamers. Graphene (graphene oxide and reduced graphene oxide) is an excellent biosensor transducer.

The potentiometric aptasensor is based on chemically modified graphene as the aptasensor's transducer layer, and aptamers act as a sensing layer. Two variants of the aptasensor for the detection of *Staphylococcus aureus* were built using graphene oxide (GO) and reduced graphene oxide (RGO). In both cases, the biosensor was able to detect a single CFU/mL of *S. aureus* in a matter of minutes. In terms of simplicity and performance, the results exceeded existing detection approaches [60]. The simplest amperometric biosensors, in which a current is generated in proportion to the oxygen concentration, may be based on Clark oxygen electrodes. The decrease of oxygen at a platinum working electrode in comparison with an Ag/AgCl reference electrode at a specific voltage is used to measure this. Specific bioreceptor-target binding results in an amperometric signal either directly or indirectly in amperometric biosensors. Direct signals may result from charge transfer from an electron-rich label affixed to the target molecule. On the other hand, redox reactions that are catalyzed by enzyme labels on the target molecule allow for indirect signaling [47]. The glucose biosensor, which relies on the amperometric detection of hydrogen peroxide, is an illustration of an amperometric device. In advanced detection techniques, amperometry is used in conjunction with immunosensing methods to assess levels of the targeted molecules from the sample [61].

3.2.4 Field Effect Transistor Biosensors

A field effect transistor (FET) is made up of a metal oxide, a semiconductor channel, and three electrodes: the source, drain, and gate. Traditionally, bio-FET sensors have been based on newly developed functional materials, such as metal-oxide semiconductors like zinc oxide (ZnO) and indium gallium zinc oxide (IGZO); organic semiconductors like poly(3,4-ethylene dioxythiophene) (PEDOT); one-dimensional (1D) nanostructured materials like silicon nanowires (SiNWs) and carbon nanotubes (CNTs), and two-dimensional (2D) nanostructured materials like graphene [62]. Immobilization of antibodies, nucleic acids, aptamers, enzymes, microbes, or synthetic biomaterials on the sensing channels could enable label-free detection of infectious illnesses. When the chemical reaction occurs or the target molecule hits the surface, immobilization of these bio-species will aid in the label-free detection of a specific disease biomarker in the solution.

A capacitive double layer (DL), made up of layers such as an inner Helmholtz (IHL), an outer Helmholtz, and an electrolyte solution, is produced when an electrolyte solution interacts with a solid surface (such as silicon dioxide or any other oxides) (OHL) [63, 64]. The principle underlying nearly all Chem/BioFET operations is the capacity of these sensors to detect charge effects on the surface through the induced electric field caused by the presence of target molecules in the solution. FET-based sensors have drawn a lot of interest because of their distinctive qualities, including their label-free nature, ease of surface functionalization, precision, and low cost due to established chip manufacturing processes.

3.3 *Gravimetric Biosensors*

Since mass is a fundamental characteristic of any analyte, acoustic or mass-sensitive devices provide label-free detection of analytes. The acoustic wave devices are regarded as all-purpose gravimetric or mass-sensitive transducers. A change in mass causes a signal to be produced by gravimetric transducers. These transducers typically consist of a bulk acoustic wave instrument such as a quartz crystal microbalance (QCM), surface acoustic wave (SAW) resonators, and shear transverse wave (STW) resonators, which are frequently investigated in combination with various receptor materials for molecular recognition and biomedical diagnosis [65]. In the occasion of receptor-analyte interaction and analyte recognition, gravimetric biosensors produce a direct shift in frequency. QCM-based biosensors have been intensively investigated for the identification and quantification of a diverse range of analytes ranging from small molecules and ions to biomolecules and pathogens. In theory, QCM is highly sensitive to changes in mass loading, and depending on the nature of the analyte, these minor changes in mass caused by selective pathogen binding can be easily detected [65]. Gravimetric diagnostic systems have a selective layer or receptor that captures pathogens and a QCM transducer that translates pathogen binding events into legible sensor signals [66]. There are two other categories known as piezoelectric and magnetoelastic biosensors derived from gravimetric biosensors.

3.3.1 Piezoelectric Biosensors

A material's capacity to generate voltage when mechanically stressed is referred to as piezoelectricity or the piezoelectric effect. Two electrodes apply alternating voltage to the surface of the biosensor or sensor to excite it. When a crystal is placed in an oscillation circuit, alternating voltage causes it to mechanically oscillate, and the frequency of the oscillations is measured. A change in oscillation frequency is caused by the analyte or any other mass bonded to the surface of the crystal, or more specifically to the surface of the electrodes positioned on the crystal. Piezoelectric immunosensors are biosensors that include an antibody as a bio-recognition component, and the specificity of the antibody has a major impact on the whole immunosensor's specificity. The piezoelectric biosensor for the detection of *Staphylococcus aureus* includes an aptamer crosslinked by 4-mercaptobenzene-diazonium tetrafluoroborate on graphene interdigitated gold electrode on a crystal. Due to the precise capture of the bacteria on the aptamer, which allowed the DNA bases to engage with *S. aureus* instead of the graphene surface, the piezoelectric biosensors detected the presence of *S. aureus* [67].

3.3.2 Magnetoelastic Biosensors

In a typical magnetoelastic (ME) sensor, there is a superposition of alternating and static magnetic fields that cause the vibration of ferromagnetic metallic glass ribbons made ME sensor in a longitudinal manner at its resonance frequency. The sensor platform is crafted using a magnetoelastic material which undergoes a change of shape in the presence of a magnetic field. A layer of biomolecular recognition element is immobilized onto the surface of the ME biosensor platform to detect the target pathogen. In the presence of the target pathogens, these recognition elements on the biosensor with the pathogen interact and capture it, causing it to bind to the biosensor's surface. This binding causes an increase in the mass of the biosensor, causing its resonant frequency to decrease [68].

The magnetic flux is detected by a coil without any wire connection, and a network analyzer detects the resonance frequency. The sensor is designed in a way that there is a biologically sensitive element to detect the bacteria or protein. Thick-film ME sensors respond to a changing magnetic excitation field by mechanical vibration. Recently, platelet activities in blood and platelet-rich plasma have been successfully analyzed and quantified using magnetoelastic sensors. Guntupalli et al. designed an ME biosensor to selectively detect *Salmonella typhimurium* in a mixed microbial population [69]. For the detection of food-borne contamination, Park et al. developed a phage-based ME biosensor having an LOD in the range of 1.94 to 2.17 with a repeatability of 6.38% for the detection of *S. typhimurium* on spinach leaves [70]. Chai et al. developed a real-time in situ bacterial detection system from food surfaces using a combination of ME biosensor and surface-scanning coil detector [71].

3.4 Microbe-Based Biosensors

There are specific forms of biosensors that utilize microbes as the sensing element, which can be directly immobilized on a surface and have direct contact with the transducer. The term "microbial biosensor" refers to an analytical tool with a biologically integrated transducer that produces a quantifiable signal representing the concentration of the analyte. This approach is best suited for environmental and extracellular chemical analysis along with metabolic sensory modulation [72]. Various standard optical devices (such as fluorescence or bioluminescence) and sensory-regulated devices are used as the sensing and recognition mechanisms of microbial biosensors. Microorganisms have been developed as biosensors, and they offer numerous advantages over other platforms that use enzymes and mammalian cells, such as the ability to recognize a broad range of substrates, cost-effectiveness, rapid manufacturing, and easier genetic modification [73–75].

The use of microbes like yeast and bacteria to create microbial biosensors with a variety of benefits is becoming a viable approach in the present scenario. Compared to other approaches involving traditional instruments, such as gas chromatography, liquid chromatography, and mass spectrophotometry, and also taking into consideration of other types of biosensors, microorganism-based sensors offer significantly lower analytical costs for sensing elements [72]. The cost of analysis can be significantly decreased because microorganisms can be generated in large quantities by employing a straightforward culture procedure in an inexpensive liquid nutrient media. They have the capacity to detect a variety of target elements and can be easily engineered for particular substrates using current molecular biology techniques [76].

Engineering microbes through genetic modification appears to be more precise and targeted than creating mammalian, plant, or other types of biosensors. Other prominent biosensor types, like those based on enzymes or antibodies, are likewise unstable and extremely sensitive to pH and temperature. Nevertheless, more durable microbial biosensors exhibit excellent resistance to a variety of environmental stresses. In spite of the numerous benefits of microbial biosensors over traditional sensing devices, their widespread usage is constrained by a few inherent characteristics of microbes. These inherent characteristics include comparatively low sensitivity, which is closely correlated with both cell population size and an optical signal [77, 78], poor selectivity in multiplex detection [79], and intrinsic cellular heterogeneity both genotypically.

To get over these restrictions, a variety of recently created micro/nanotechnologies have been combined with microbial biosensors, which are used for a variety of detecting purposes [80]. For example, microfluidic devices demonstrated numerous benefits by reducing the volume of sample and reagent needed, speeding up analysis with excellent resolution and repeatability, and showcasing multiple experiments on a chip in a high-throughput way [78]. However, there are still a lot of intriguing possibilities and exciting potential in the field of microbial biosensors, even though the sophisticated micro/nanotechnologies will usher in a new age for this technology.

3.5 *Nano-biosensors*

It has been claimed that biosensors are capable of analyzing even the smallest biological analyte with extreme precision, adaptability, and sensitivity [81]. The creation of a nano-biosensor through the fusion of biosensors and nanotechnology is a common example of this application, which is currently being looked at as a potential possibility to accelerate the development of quick, particularly sensitive, and specialized tools for pathogen (bacterial and viral) detection. In order to detect biomolecules and pathogens, nano-biosensors leverage the chemical, optical, electrical, and magnetic properties of materials (Chamorro-Garcia and Merkoçi). A nano-biosensor is made up of a nanomaterial with a size between 1 and 100 nm, multiple atomic components arrayed on or at the surface, and notable physical, chemical, and physicochemical characteristics that are unique from those of the same materials at the bulk level.

Irrespective of the homogeneous or heterogeneous properties of nano-biosensors, they both contribute to biosensing technology [82]. The latter focuses on a solid substance, such as a selected receptor containing an antigen, whereas the former occurs in solution without phase separation and aids in the identification and capture of the analyte to be detected. Furthermore, the high surface area of nano-biosensor materials is thought to be a significant advantage over conventional devices, which significantly contributes to the sensitivity and quick reaction of the devices. These are introduced as great tools for the recognition, and interaction of proteins and nucleic acids, which enhances the accuracy and efficacy of the diagnosis of infectious pathogens [14]. A summary of some prospective nanotechnological components in biosensors is provided in the following sections.

3.5.1 Nanoparticles

Nanoparticles are atom clusters with dimensions comparable to the de Broglie wavelength associated with valence electrons, which is at least 100 [83] and with the variation of size and shapes, these NPs display distinctive colors, which can be used in bio-imaging applications [84]. Because of their enhanced properties, these NPs have possible future applications in a variety of fields, such as the medical, pharmaceutical, environmental, electronics, mechanical, and materials industries. The function of metal NPs (gold, silver, and nickel) sensors is based on the localized surface plasmon resonance (LSPR) spectrum's high sensitivity to change the local refractive index of nanoparticles when a foreign element attaches to them. These NPs have minimal toxicity and a range of exciting biological, physical, and chemical characteristics, including biocompatibility, optical, photo-thermal, and electrochemical properties, besides their inert nature in biological fluid. Furthermore, because of the existence of functional groups for building ligand-binding activities with a range of molecules, such as antibodies or genetic material, these NPs are easy to produce and functionalize [14].

In order to use magnetic NPs for biomedical applications, the analytes must be enhanced, either in the form of superparamagnetic magnetite (Fe_3O_4), greigite (Fe_3S_4), or maghemite (γ- Fe_2O_3), or as ferrites like metal oxides Fe_2O_3 (M = Ni, Zn, Mg, Mn, and Co). Noble metal NPs are now becoming promising candidates for optical nanosensors due to their particle size-dependent characteristics [85, 86]. Water-borne and food-borne diseases have a significant application for metal NP-based nano-biosensors [80]. In such circumstances, the infection is frequently connected to microbial contamination brought on by a variety of diseases, including bacteria. These pathogens have been detected using optically based nanotechnological detection methods. The biosensing process of a particular analyte starts with recognition of the analyte by the receptor, followed by transduction of the change resulting from a binding event, and its conversion to measurable signal, signal processing, and ultimately signal display.

In addition, nanomaterials including quantum dots, carbon nanotubes, and gold nanoparticles, for biosensor applications, have potential materials and are employed successfully in research and development [87]. Expanding the number of diverse biomolecules that may be sensed or detected by improving the sensitivity and offering more quick and flexible detection methods is typically one of the next challenges in NP-based biosensors. However, creating additional chances for these nano-biosensors to be translated from laboratory use to clinical use by utilizing samples from patients is the most challenging component of this field's advancement [88].

3.5.2 Electrospun Nanofibers

Nanofibers have emerged as a standout nanostructure among the many others that have been created over the past few decades as a result of their unique properties, such as having a large surface area and fibrous holes [8, 89]. The electrospinning technique, which dates back to the latter half of the twentieth century, involves applying an electric field to a polymer solution to induce a charged liquid jet to flow in the direction of an oppositely charged collector, where thin fibers are deposited. It has been used to successfully create nanofibers from a wide range of materials, including ceramics, polymers, tiny molecules, and composites of these, within 100 nm diameter. Electrospinning has been used to produce solid nanofibers with smooth surfaces in addition to a variety of secondary features, such as porous, hollow, or core-sheath nanofibers. During or after an electrospinning process, the inner layer and surface of such nanofibers can be further functionalized with other nanoparticles or molecules [90]. These nanostructures can be used to create nano-biosensors using two different strategies.

One method involves electrospinning functional polymers to create nanofibers that are then immediately incorporated into the relevant biosensor's inducing element. These biosensors will have high sensitivity, quick latency, and good biocompatibility. In contrast, electrospun nanofibers are utilized as templates for sensitive materials to be deposited on their surfaces, and then the system is chemically modified to create a composite film on an electrode with nanostructures that have the desired sensing properties [91, 92]. The electrospun nanofibers' morphology can be altered by adjusting the voltage that is applied, the needle tip to collector distance, the flow rate, morphology of the collector, the solvent used, conductivity, viscosity, molecular weight, temperature, and humidity. Furthermore, developed electrospun nanofibers with a diverse array of material interfaces, such as those that contain metals, metal oxides, carbon-based materials, advanced materials, polymers, and biodegradable biopolymers, can avoid agglomeration, which improves the electron-transport and electron transfer kinetics and enhances the efficiency of the fibers in sensing and biosensing. In the realm of biosensing of glucose, dopamine, ascorbic acid, H_2O_2, and other molecules, electrospinning is one of the most recently developed methods for creating electrospun nanofibers.

As a result, nanofiber-based biosensors offer benefits over traditional ones, including a variety of polymers for manufacturing, a high specific surface area with high responsiveness, and exceptional sensitivity. Nevertheless, advancements address the difficulties of large-scale or high-volume production, functionality and variety enhancement of the electrospun nanofibers, environmental concerns, and improvement in cell infiltration ability to make them 3D scaffolds. Future, developments in the manufacture of needleless or multi-needle nanofiber electrospinning sets will pose significant manufacturing challenges in terms of addressing safety and environmental concerns relating to solvent evaporation during the spinning process [93].

3.5.3 Graphene Quantum Dots

Due to their excellent physical, chemical, and electronic transport capabilities, recently discovered graphene quantum dots, which are two-dimensional (2D) graphene blocks of <100 nm, are useful in the creation of next-generation biosensors [94]. These materials exhibit exceptional qualities such as variable size, signal amplification capabilities, biocompatibility, electro-catalytic activity, and the ability to detect numerous biomolecules. In addition, they are extremely useful for biomedical applications due to their tiny size, non-toxicity, inertness, long-term chemical and water stability. Top-down and bottom-up procedures can be used to categorize the currently used techniques for creating GQDs. The synthesis of GQDs using bottom-up techniques makes it challenging to adjust the conditions because it involves intricate reaction pathways and certain organic components.

Thus, it is recommended to employ the top-down strategy, which entails fragmenting big blocks of carbon material. This approach requires abundant carbon sources as raw materials that are affordable and easy to obtain [95]. In biosensing applications, GQDs synthesized using various strategies are able to directly detect DNA, proteins, enzymes, antigens, antibodies, and other biomolecules due to the oxide components that are created on their surface during the manufacturing process. The recent creation of PEC and ECL biosensors utilizing these nanodots is equally noteworthy. In this field of study, the electrical and luminescent characteristics of GQDs/CQDs are combined, exploiting their unique properties to distinguish them from other nanomaterials [94].

3.6 Biosensors for the Detection of Bacterial Biomarkers

Bacterial biosensors, also known as bactosensors, offer incredible potential for environmental and medical diagnosis. Chronic diseases can be monitored and early disease discovery can assist lower mortality and enhancing patient quality of life [96]. At several points throughout the healthcare supply chain, in vitro diagnostic technologies are crucial in that situation [97]. Various techniques and procedures, such as colorimetric approaches, fluorescence polarization, and electrochemical analysis, have been created over time for the detection of pathogens including viruses and bacteria. However, they are quite expensive and have drawbacks such as high time consumption, poor stability, and short life duration.

Besides identifying entire bacterial species and using bacteria to detect antibiotics, the identification of specific biomarkers such as enzymes secreted by the bacteria, cellular constituents of bacteria, or bacteriophages can be helpful in further validating the bacteria's presence and directing toward therapeutic options. The majority of these biosensors are based on electrochemical and optical biosensing methods, thanks to their exceptional specificity and sensitivity, which impose a lot of emphasis on surface functionalization and target molecule capture. The incorporation of substrates like AAO and pSi aids in the versatility in the detection of bacterial biomarkers (Table 3).

4 Point of Care for Bacterial Detection System

Point-of-care testing (POCT) is a type of diagnostic technique in which the analysis is carried out near the patient. In simple terms, a POC diagnostic test is conducted near the patient or healthcare facility, with a quick turnaround time to allow significant clinical interventions which need to be implemented at the point of care and may result in a change in patient care management [102]. Such tests usually do not require access to traditional centralized laboratory facilities and are carried out in many locations such as our homes, clinics, pharmacies, or an ambulance. The concept of POCT, by an inexperienced person having no prior laboratory training, is not new and traces its origin way back to the fifteenth century when urine was tested at the patient's home [103]. However, it was during the last few decades that there has been an upsurge of research work focused on such POCT devices.

The most common justification for using POCT devices over traditional methods is the fact that it is streamlined and economically efficient since sample transportation to the laboratory is no longer required, and the laborious culturing analysis can be avoided. The purpose of POCT varies from triage and referral to diagnosis, treatment, and health monitoring, depending on the end-user and the ambient settings where it is carried out (Fig. 2). Typically, these diagnostic systems provide rapid results, enabling the prompt initiation of appropriate therapy and the facilitation of care along with referral to other medical practitioners for their opinion on the management of the disease. Nowadays, technological advancements have broadened the concept of POCT devices, which are no longer limited to healthcare institutions [104]. As a result, they are finding growing acceptance not only in clinical diagnostics but also in biosensing platforms, food quality and safety, environmental sensing, and other fields. From this chapter's perspective, we are going to focus exclusively on the current landscape of POCT devices in clinical diagnostic settings for infectious diseases from the viewpoint of bacterial detection.

The growing potential for POCT devices aiming to improve healthcare in developing countries, particularly through early detection of infectious diseases, has driven researchers and engineers to use a more specific design criterion. These are to

Table 3 Advantages and disadvantages of different categories of biosensors and their subtypes for bacterial detection

Types of biosensors	Characteristics			Advantages	Disadvantages
	Subtypes of biosensors	*Principle of biosensors*	*Pathogen detected by biosensors*		
Optical biosensors	Interferometric Reflectance-based biosensors	Based on the principle of reflection of light	– *Lactobacillus acidophilus* – *E.coli*	– Low noise – Lable-free – Rapid detection – User-friendly device – Large surface area – Easy functionalization of surface – Cost-effective – Simple – High sensitivity – Repeatability – High specificity	– Fabrication of biomaterials is difficult – Due to leakage in design, contamination can occur – Maintaining uniformity in the design is challenging – Biomolecules that are to be detected may be destroyed due to the use of unsterilized probes – Require precise alignment of light coupling to the sensing area
	Fluoroscence-based Biosensors	Fluoroscence-based biosensing device measure bacterial concentration	– *Staphylococcus aureus* – *S.typhi* – *E.coli* 0157:H7		
	Plasmonic Biosensors	Based on principle of surface plasmon resonance	– *Mycobacterium tuberculosis*		
	Colorimetric Biosensors	Color changes due to redox and the enzyme-catalyzed reactions happening in these colorimetric sensors are observed by the naked eyes	Bacterial concentration is measured		

(continued)

Table 3 (continued)

Types of biosensors	Characteristics			Advantages	Disadvantages
	Subtypes of biosensors	*Principle of biosensors*	*Pathogen detected by biosensors*		
Electrochemical biosensors (electrochemical interface)	Impedance-Based Biosensors	Electrochemical impedance spectroscopy (EIS) approach uses variations in interfacial capacitance or charge transfer resistance (Rct) to identify biological changes occurring at the sensor surface	– Early Secretary Antigen Target-6 (ESAT-6) produced by *Mycobacterium tuberculosis*	– Measurement over a broad range of periods and frequencies – Highly accurate – Time-averaged responses – Label-free – Simple sensor design – Inexpensive – Stable – Sensitive – Used in POC devices	– No real-time detection – Sensitive to sample matrix – Immobilization of bio-recognition components on electrodes is crucial to the design
	Capacitance-Based Biosensors	Dielectric characteristics changes based on capacity affinity between the analyte and immobilized recognition components in thin layers on an electrode or between two electrodes	– Enterobacteriacea DNA)		

(continued)

Table 3 (continued)

Types of biosensors	Characteristics			Advantages	Disadvantages
	Subtypes of Biosensors	*Principle of biosensors*	*Pathogen detected by biosensors*		
	Potentiometric Biosensors	Measures potential across electrodes which directly proportionate with the concentration of analyte	– *Staphylococcus aureus*		
	Field Effect Transistor Biosensors	These sensors detect charge on the surface through the induced electric field caused by the presence of target molecules in the solution	independent DNA detection of pathogen		

(continued)

Table 3 (continued)

Types of biosensors	Characteristics			Advantages	Disadvantages
	Subtypes of biosensors	*Principle of biosensors*	*Pathogen detected by biosensors*		
Gravimetric biosensors	Piezoelectric Biosensors	Generate voltage when mechanically stressed	*Staphylococcus aureus*	– Real-time detection multiplex – High throughput – Used in POC devices	– Extremely sensitive to sample matrix effects – Careful control of temperature is essential – Bulky equipment
	Magnetoelastic Biosensors	Respond to a changing magnetic excitation field by mechanically vibrating	*Salmonella typhimurium*		
Microbe-based biosensors	*Type of biosensor*	*Detection method*	*Application*	– Detects a wide range of substrates – Easier genetic modification for detecting certain analytes – Significantly lower cost – Capacity to detect a variety of target elements – More durable – Resistant to a variety of environmental stresses such as pH and temperature	– Selectivity in multiplex detection – Intrinsic cellular heterogeneity
	S. cerevisiae integrated automated device	Fluorescence	6 times faster than Ames Test		
	E.coli integrated portable device	Luminescence	Portable toxicity biosensor		

(continued)

Table 3 (continued)

Types of biosensors	Characteristics			Advantages	Disadvantages
	Subtypes of biosensors	*Principle of biosensors*	*Pathogen detected by biosensors*		
Nano-biosensors	*Subtypes of biosensors*	*Principle of biosensors*	*Pathogen detected by biosensors*	– Extreme precision – Adaptability – High sensitivity – Real-time detection – The high surface area of nano-biosensor materials is thought to be a significant advantage over conventional devices and contributes significantly to functionalization – Accuracy and efficacy of diagnosis for bacterial and viral disorders – Quick response time – Biocompatibility – Non-toxic	– Resolution of the nanosensors due to current cutting and masking technologies is not very good – Developing reproducible calibration methods is difficult

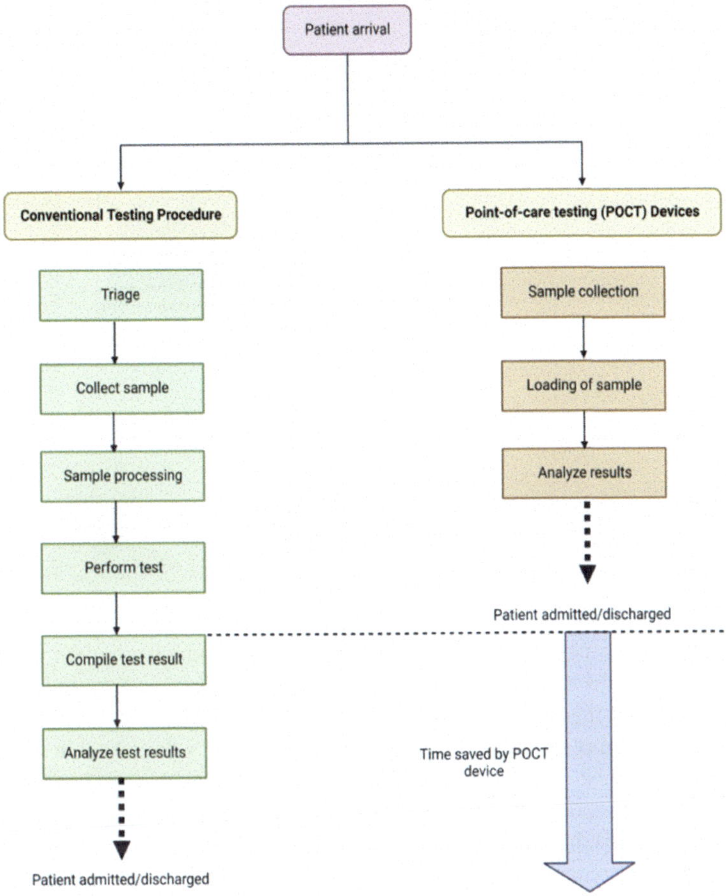

Fig. 2 A comparison between conventional testing procedures and point-of-care testing (POCT) devices showing POCT devices achieving rapid results hence saving time. [Created with BioRender.com]

guarantee that the technology can address the user's needs in a clinical and cost-effective manner, avoiding the potential expensive devices that fail to deliver the desired results. In view of this scenario, the World Health Organization (WHO) has issued guidelines addressing those developing POCT devices. These guidelines are referred to as ASSURED (Fig. 3) which stands for "A–affordable, S–sensitive, S–specific, U–user-friendly, R–rapid and robust, E–equipment-free, and D–deliverable to end-users". In order to assess a safe and dependable POCT, consideration of its sensitivity, specificity, and positive and negative predictive values in relation to disease occurrence in the target population is a mandate [105].

The POCT devices are available either in the form of small handheld devices, which include quantitative and qualitative strips that can be used even at home, or

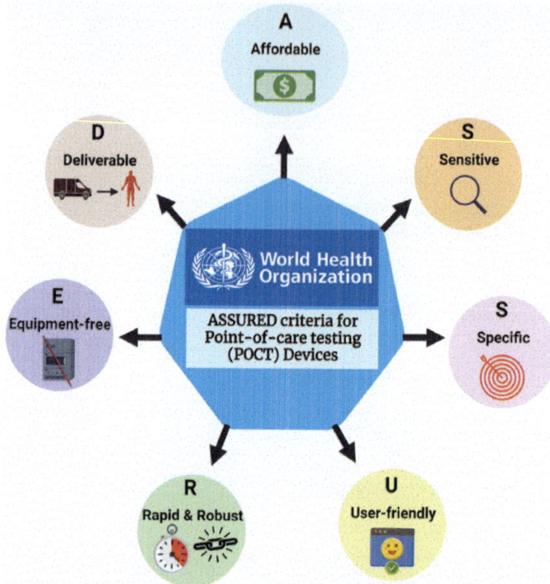

Fig. 3 ASSURED guidelines presented by the World Health Organization (WHO) highlighting the criteria's that need to be followed by point-of-care testing (POCT) devices. [Created with BioRender.com]

larger bench-top devices with more sophisticated designs, which are often variations of those used in conventional laboratory facilities [106]. There are a number of key design components that are incorporated into all POCT devices, the collective goal of which is to accomplish all of the desirable characteristics discussed in the above section to the greatest extent possible. These components include (1) an operating interface, (2) a bar-code system for identification, (3) the sample collector, (4) the reagent reservoir, (5) the reaction chamber, (6) sensors to detect the measurement of reaction, (7) signal reader to view the result [106]. It is important to note that not all of these components are present in simpler devices such as dipsticks. However, with the trend of increasing device miniaturization, it is becoming increasingly feasible to make smaller-sized devices that integrate all of these essential components.

4.1 *Biosensors as POCTs for Bacterial Detection*

In terms of human health, bacterial infections caused primarily by Gram-negative microorganisms pose a particular challenge. Bacterial species belonging to the ESKAPE group such as *Enterococcus faecium, Klebsiella pneumoniae, Acinetobacter baumannii, Staphylococcus aureus, Pseudomonas aeruginosa, and Enterobacter spp.* are of particular concern in the growing trend of antimicrobial resistance. These multidrug-resistant pathogens harbor plenty of drug-resistance mechanisms and pose a serious threat to healthcare [5, 102, 107] Apart from these strains, *Escherichia coli, Salmonella typhi, Clostridium perfringens, Mycobacterium tuberculosis, Legionella pneumophila, Chlamydia trachomatis, Treponema pallidum, and Shigella spp.* are examples of common pathogenic bacteria that can result in a number of life-threatening conditions in humans, such as typhoid, pneumonia, tuberculosis, syphilis, and sepsis [108].

Over the past few years, there has been a paradigm shift toward the development and application of novel detection techniques in order to overcome some of the limitations of associated techniques. To meet this demand, biosensors have been extensively researched as detection devices for pathogenic bacteria to provide a rapid and accurate assessment of infectious diseases. Biosensors are quite versatile, and robust devices that can be used to detect the presence of specific microorganisms or biochemical markers [109]. Biochemical signals are produced when bacteria interact with the bio-recognition element. These signals are then converted into physically detectable signals and measured through a transducer, to infer whether a patient has been infected with a particular bacterial pathogen. The POCT devices for bacterial detection have the flexibility of using antibodies, antigens, nucleic acids, or proteins, as the bio-recognition element [102]. Each type of biomolecule used as the bio-recognition element comes with its own set of advantages and limitations.

Recent developments in nanotechnology, microfluidics, and synthetic biology, have enabled biosensors to emerge as novel and promising diagnostic devices that can be effectively used for point-of-care testing [111]. These biosensors have received considerable attention from clinicians over the past decade, in an effort to develop point-of-care biosensors for various bacterial diseases (Fig. 4). Typically, such biosensors offer several advantages when compared to traditional detection systems. Apart from being inexpensive, portable, compact, and simple to use, they are reliable and safe, require a minimal amount of the analyte, and give quicker results.

The combination of different technologies may result in the development of novel POC biosensors with improved functional and detection properties, allowing better control of special and temporal factors [112]. These proposed and improved designs

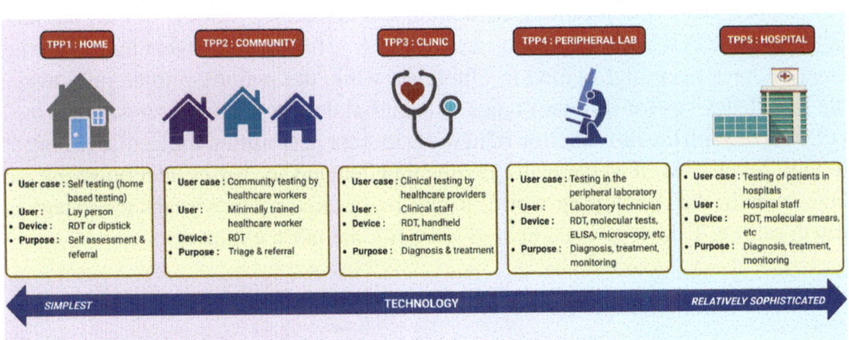

Fig. 4 A schematic of the different target product profiles (TPP) under the spectrum of point-of-care testing (POCT) devices. Adapted from [110], PLoS Med. 9(9)e1001306, copyright held by the authors under Creative Commons License.)

mostly integrate disciplines like nanotechnology and microfluidics in pursuit of fulfilling a critical property, that is to achieve comparable or even better sensitivity and selectivity than conventional laboratory diagnostic techniques so that these biosensors can be at the forefront of clinical and healthcare industry to detect pathogens of bacterial origin [96]. The immunochromatographic detection of particular bacterial antigens or, more rarely, an antibody in the patient sample (urine, swab, whole blood) utilizing the principle of enzyme-linked immunosorbent assay (ELISA), is the general operating principle of most of these devices. Attempts have also been made to use molecular biological procedures, most notably the polymerase chain reaction (PCR), for such POCT devices. However, given that they are technically challenging assessments, they are not considered rapid testing in the truest sense as yet [113].

4.2 Commercially Available POCT Devices

The necessity to make the healthcare sector more patient-centered has become a global trend, and it is predicated on the idea that healthcare should be structured around the patient rather than the practitioner. For many patients, centralized testing is inconvenient since the testing procedure is typically cumbersome and isolated from the consultation process, necessitating more than one visit to the doctor to complete the evaluation process. With the growing popularity of POCT devices in the market, there has been observable growth in the revenue of such devices from about 24 million in 2016 to over 40 billion as of 2021. This growing market for POCT devices reflects the need for devices that are more convenient to use and, in some cases, more effective in their work.

The POCT market is divided into two parts: "over-the-counter" products like glucose monitoring and pregnancy kits and the "professional market," which takes into account all other testings like critical care devices, disease detection, cardiac markers, diabetes, lipid profile, coagulation, and hematology. Some of the commercially available POCT devices which are widely used for purposes other than bacterial detection have been highlighted in (Table 4), while the major commercially available POCT devices for different types of bacterial detection have been summarized in (Table 5). With the introduction of new devices for performing molecular testing at the POC level, as well as the well-documented demand for better infectious disease diagnostics in the developing world, it will be fascinating to witness the extent of growth of POCT devices as compared to traditional laboratories.

Table 4 Commercially available POCT devices for various applications

Sl. no	Application	Name of POCT device	Measuring parameters	References
1	Biochemistry	Clini5	TC, HDL cholesterol, LDL Cholesterol, TG, Hb, HCT, RBC, UA, lac hct, TCO_2, pO_2, pCO_2	[114–117]
		ISTAT creatinine	Na^+, K^+, Cl^-, lac, cardiac markers,	
		test®	Coagulation factors	
2	Diabetes	StatStrip®	Glucose	[118]
3	Cardiology	I-Stat® Troponin I	CTnI	[119–121]
		Cardiac Reader® troponin T	Myoglobin and troponin T	
4	Pregnancy and infertility	SpermCheck®	Sperm count	[122, 123]
		Clinitest® hCG pregnancy test	HCG	
5	Viral infection detection	Alere q 1/2 Detect	HIV nucleic acids	[124–126]
		INSTI™	HIV-1/2 antibodies	
6	Drugs detection	Cozart®RapiScan	Cocaine, benzoylecgonine, ecgonine methyl ester	[127]
7	Cancer detection	FastPack® IP System	PSA, free PSA	[127]
8	Genetic disorder detection	GeneXpert®	*BCR-ABL* gene fusion in leukemia cells	[128]
9	Ultrasonology	FibroScan	Liver disease	[129]
10	Veterinary	Lactate Pro	Cattle blood L-lactate concentration	[130]

4.3 Challenges in Bacterial Detection at the POCT Level

Despite all the attractive advantages associated with POCT devices and a promising number of commercial products available in the market for bacterial disease diagnostics, it has its own set of challenges that need to be addressed. Errors can happen at any stage in the testing cycle, just as they can in any laboratory test. Recent reports have compared the error rates for pre-analytical, analytical, and post-analytical factors between POCT and central laboratory tests. POCT devices were found to have a higher percentage of pre-analytical errors than central laboratory testing [152]. The extent of accuracy for POCT is also related to the fact that they are typically performed by engaged employees of the clinical team rather than laboratory-trained personnel. Individuals who are not trained do not understand the significance of quality assurance (QA) and quality control (QC).

Furthermore, these individuals' primary focus is on patient care, and the procedures associated with POCT frequently add an additional burden to an already busy and stressful role. POCT implementation in a clinical unit affects the workflow of

Table 5 Commercially available POCT devices for bacterial detection

Sl. No	Name of POCT device	Measured parameters	Methods used	Accuracy	Advantages	References
1	Circulating cathodic antigen urine cassette test	Schistosoma mansoni	ELISA	r = 0.99	More accurate in low infection level Faster results Easily obtained samples such as urine can be used	[131–133]
2	Determine® Syphilis TP	*Treponema pallidum* Ab	– Rapid plasma reagin (RPR) test – Hemagglutination assay – Fluorescent treponemal antibody adsorption test – Venereal disease research laboratory (VDRL) test	r = 0.98	Much faster results than existing methods (15min) Easy to use and flexibile in sample that can be used	[134]
3	ACCEAVA® Strep A	Streptococcal A antigen	Rapid chromatographic immunoassay	r = 0.98	High accuracy Gives result in only 5 min No refrigeration required	[135]
4	BinaxNOW®	Pneumococcal and L. pneumophila serogroup 1 antigens	Urinary antigen tests (UATs)	r = 0.97	Almost 85% accurate Digitally connected test kit	[136, 137]
5	TrueNat MTBTM	*Mycobacterium tuberculosis* nucleic Acid	Chip-based nucleic-acid amplification test	r = 0.96	Highly sensitive Highly specific	[138]
6	HELIPROBE®	14C-Urea	Urea breath testing (UBT)	r = 0.95	Can be used in clinical setting Non-invasive	[139, 140]
7	Filariasis test strip	Wuchereria Bancrofti circulating filarial antigen	Immunochromatography card test [ICT]	r = 0.94	Increased sensitivity Longer shelf life	[141]

(continued)

Table 5 (continued)

Sl. No	Name of POCT device	Measured parameters	Methods used	Accuracy	Advantages	References
8	Immunoquick®	Pan malaria-specific pLDH and Pf HRP-2	Lateral-flow immunochromatographic test	r = 0.93	No special facility or equipment needed	[142, 143]
9	Signify Strep A	Group A Streptococcus carbohydrate antigen	Lateral immuno assay	r = 0.93	Rapid accurate	[144]
10	SD BioLine® Syphilis Duo	IgG, IgM, and IgA *Treponema pallidum*	ELISA	r = 0.85	Affordable, easy to use, and ability to rapidly detect HIV/syphillis	[145–147]
11	RAPIRUN® S. pneumonia	Pneumococcal C-ps	Immunochromatographic test	r = 0.77	Rapidly detects the cell wall, cell membrane, and capsular antigens of S. pneumoniae	[148, 149]
12	Determine™ TB LAM	LAM of *Mycobacterium tuberculosis*	Urine lipoarabinomannan (LAM) test based on immunochromatographic Test	r = 0.68	Can diagnose both pulmonary and extra-pulmonary TB Can use unprocessed sample Optimal performance with HIV patients	[150, 151]

staff who must integrate POCT-related processes such as daily quality control testing, instrument maintenance, and troubleshooting issues with POCT devices [153]. These are some basic challenges that need to be addressed with regard to the use of POCT for bacterial disease diagnostics. Looking at it from a different perspective, the most effective POCT will have no effect unless it is broadly used and backed up with appropriate therapeutic approaches [110]. Though some of the POCT devices for bacterial diagnostics meet all of the ASSURED criteria, these devices are rarely used to make clinical decisions in community or clinic settings. POCT appears to occur infrequently in homes, communities, or clinics. The majority of testing is done in laboratory facilities and hospitals. Based on several reports on this topic, it can be understood that there are a plethora of obstacles toward successful POCT use for bacterial diagnostics, ranging from economic, regulatory, and policy barriers, and infrastructure to user/provider preconceptions and social norms. These associated barriers have been summarized in (Table 6).

Table 6 Challenges of POCT devices

Challenges	Example
Expensive	POCT devices can be expensive and thus unable to afford them
Quality check	Quality control and quality assurance should be done prior to every single test, which is difficult for the untrained person to perform
Infrastructure	Lack of infrastructure, likely lack of consistent power supply, less storage space, absence of proper waste disposal units, lack of phlebotomy equipment, and difficulty in controlling temperature, make some forms of POCTs difficult to administer
Space in the laboratory	Poor laboratory capacity does not allow for the installation of POCT, which requires additional lab
Awareness	Lack of awareness of POCT devices among medical professionals and common people
Training	POCT device performance might suffer from a lack of regular, continuing competence training
Uncertainty of medical professionals	In some contexts, doctors and frontline caregivers may prioritize clinical diagnosis and empiric therapy above diagnosis by POCT devices
Policy regulation	Poor policy upgrades and poor regulation of POCT devices make it challenging to scale up the use of these devices
Social stigma	The relative lack of privacy and stigma may limit community adoption of POC testing for infectious diseases like HIV

5 Conclusion and Future Perspectives

Biosensors are described as a unique and promising tool for detecting biological protein markers linked with a variety of bacterial infections. It is important to note that by combining nanotechnology, their precision and accuracy may be improved, paving the way for point-of-care testing biosensors. The wide variety of communicable illnesses of bacterial origin provides a unique perspective for understanding the disease pathogenesis and potential signals for positive outcomes. Because these intermediate molecules are distinct to the route in which they are engaged, building a tailored biosensor for detecting them may be advantageous. Furthermore, advances in nanotechnology have allowed innovative nanostructured materials such as nanofibers, quantum dots, photonic crystals, and others to emerge toward the cause and improve the accuracy and persistence of such biosensors. One of the most important aspects to look into is the commercial perspective of such POCT biosensors, which need enough money funding, and skilled labor to facilitate novel research and growing demand for already commercialized products. Future projected conceptualizations and/or implementations of innovative computing approaches such as data analytics, the accessibility of AI with deep learning, and microchip-based techniques connected with biosensors might essentially be cutting-edge illness detection instruments. As a result, research into these smart systems coupled with sophisticated biosensors must be fostered in order to produce next-generation POCT diagnostics.

Author Contributions Priya Mitra and Arka Sanyal contributed equally to the conception of the study, manuscript writing, editing, and illustration preparation. Ritesh Pattnaik conceptualized, reviewed, and edited the entire manuscript. Tanima Dey, Debatri Dutta, Koustav Saha, and Arunima Pandey contributed significantly to manuscript writing with constructive discussions.

Conflict of Interest The authors declare that the work was conducted in the absence of any commercial or financial relationships that could be construed as a potential conflict of interest.

References

1. França RFO, da Silva CC, de Paula SO (2013) Recent advances in molecular medicine techniques for the diagnosis, prevention, and control of infectious diseases. Eur J Clin Microbiol Infect Dis 32:723–728. https://doi.org/10.1007/S10096-013-1813-0/TABLES/1
2. Yager P, Domingo GJ, Gerdes J (2008) Point-of-care diagnostics for global health. 10:107–144. https://doi.org/10.1146/ANNUREV.BIOENG.10.061807.160524
3. Mabey D, Peeling RW, Ustianowski A, Perkins MD (2004) Diagnostics for the developing world. Nat Rev Microbiol 2:231–240. https://doi.org/10.1038/nrmicro841
4. de Rezende Rodovalho V, Maria Alves L, Graci Madurro A, et al (2015) Applied to diagnosis of infectious diseases-an update. Austin J Biosens Bioelectron 1
5. Alle M, Park SC, Bandi R et al (2021) Rapid in-situ growth of gold nanoparticles on cationic cellulose nanofibrils: recyclable nanozyme for the colorimetric glucose detection. Carbohydr Polym 253:117239. https://doi.org/10.1016/J.CARBPOL.2020.117239
6. Datta M, Desai D, Kumar A (2017) Gene specific DNA sensors for diagnosis of pathogenic infections. Indian J Microbiol 57:139–147. https://doi.org/10.1007/S12088-017-0650-8/METRICS

7. Mehrotra P (2016) Biosensors and their applications—a review. J Oral Biol Craniofac Res 6:153–159. https://doi.org/10.1016/J.JOBCR.2015.12.002
8. Alle M, Bandi R, Lee S-H, Kim J-C (2020a) Recent trends in isolation of cellulose nanocrystals and nanofibrils from various forest wood and nonwood products and their application. In: Nanomaterials for agriculture and forestry applications. Elsevier, pp 41–80
9. Alle M, Bandi R, Sharma G et al (2022b) Gold nanoparticles spontaneously grown on cellulose nanofibrils as a reusable nanozyme for colorimetric detection of cholesterol in human serum. Int J Biol Macromol 201:686–697. https://doi.org/10.1016/J.IJBIOMAC.2022.01.158
10. Malon RSP, Sadir S, Balakrishnan M, Córcoles EP (2014) Saliva-based biosensors: noninvasive monitoring tool for clinical diagnostics. Biomed Res Int 2014 https://doi.org/10.1155/2014/962903
11. Bandi R, Alle M, Dadigala R et al (2022) Integrating the high peroxidase activity of carbon dots with easy recyclability: immobilization on dialdehyde cellulose nanofibrils and cholesterol detection. Appl Mater Today 26:101286. https://doi.org/10.1016/J.APMT.2021.101286
12. Krejcova L, Michalek P, Merlos Rodrigo M et al (2015) Nanoscale virus biosensors: state of the art. Nanobiosens Dis Diagn 2015:4–47. https://doi.org/10.2147/NDD.S56771
13. Alle M, Reddy GB, Kim TH, et al (2020c) Doxorubicin-carboxymethyl xanthan gum capped gold nanoparticles: Microwave synthesis, characterization, and anti-cancer activity. Carbohydr Polym 229:115511. https://doi.org/10.1016/j.carbpol.2019.115511
14. Castillo-Henríquez L, Brenes-Acuña M, Castro-Rojas A, et al (2020) Biosensors for the detection of bacterial and viral clinical pathogens. Sensors 20:6926. https://doi.org/10.3390/S20236926
15. Skládal P (2016) Piezoelectric biosensors. TrAC Trends Anal Chem 79:127–133. https://doi.org/10.1016/J.TRAC.2015.12.009
16. Bhalla N, Jolly P, Formisano N, Estrela P (2016) Introduction to biosensors. Essays Biochem 60:1. https://doi.org/10.1042/EBC20150001
17. Fu Z, Lu Y-C, Lai JJ (2019) Recent advances in biosensors for nucleic acid and exosome detection. Chonnam Med J 55:86–98. https://doi.org/10.4068/CMJ.2019.55.2.86
18. Clark LC, Lyons C (1962) Electrode systems for continuous monitoring in cardiovascular surgery. Ann N Y Acad Sci 102:29–45. https://doi.org/10.1111/J.1749-6632.1962.TB13623.X
19. de Boer E, Beumer RR (1999) Methodology for detection and typing of foodborne microorganisms. Int J Food Microbiol 50:119–130. https://doi.org/10.1016/S0168-1605(99)00081-1
20. Velusamy V, Arshak K, Korostynska O et al (2010) An overview of foodborne pathogen detection: In the perspective of biosensors. Biotechnol Adv 28:232–254. https://doi.org/10.1016/J.BIOTECHADV.2009.12.004
21. Jeong SK, Gang GL, Jong SP et al (2007) A novel multiplex PCR assay for rapid and simultaneous detection of five pathogenic bacteria: Escherichia coli O157:H7, Salmonella, Staphylococcus aureus, Listeria monocytogenes, and Vibrio parahaemolyticus. J Food Prot 70:1656–1662. https://doi.org/10.4315/0362-028X-70.7.1656
22. Galikowska E, Kunikowska D, Tokarska-Pietrzak E et al (2011) Specific detection of Salmonella enterica and Escherichia coli strains by using ELISA with bacteriophages as recognition agents. Eur J Clin Microbiol Infect Dis 30:1067–1073. https://doi.org/10.1007/S10096-011-1193-2/FIGURES/5
23. Borisov SM, Wolfbeis OS (2008) Optical biosensors. Chem Rev 108:423–461. https://doi.org/10.1021/CR068105T/ASSET/CR068105T.FP.PNG_V03
24. Yoo SM, Lee SY (2016) Optical biosensors for the detection of pathogenic microorganisms. Trends Biotechnol 34:7–25. https://doi.org/10.1016/j.tibtech.2015.09.012
25. Arshavsky-Graham S, Massad-Ivanir N, Segal E, Weiss S (2019) Porous silicon-based photonic biosensors: current status and emerging applications. Anal Chem 91:441–467. https://doi.org/10.1021/ACS.ANALCHEM.8B05028/ASSET/IMAGES/ACS.ANALCHEM.8B05028.SOCIAL.JPEG_V03

26. Jane A, Dronov R, Hodges A, Voelcker NH (2009) Porous silicon biosensors on the advance. Trends Biotechnol 27:230–239. https://doi.org/10.1016/j.tibtech.2008.12.004
27. Avci O, Ünlü NL, Özkumur AY, Ünlü MS (2015) Interferometric reflectance imaging sensor (IRIS)—a platform technology for multiplexed diagnostics and digital detection. Sensors 15:17649–17665. https://doi.org/10.3390/S150717649
28. Urmann K, Arshavsky-Graham S, Walter JG et al (2016) Whole-cell detection of live lactobacillus acidophilus on aptamer-decorated porous silicon biosensors. Analyst 141:5432–5440. https://doi.org/10.1039/C6AN00810K
29. Massad-Ivanir N, Shtenberg G, Tzur A et al (2011) Engineering nanostructured porous SiO2 surfaces for bacteria detection via "direct cell capture." Anal Chem 83:3282–3289. https://doi.org/10.1021/AC200407W/SUPPL_FILE/AC200407W_SI_001.PDF
30. Dey P, Fabri-Faja N, Calvo-Lozano O et al (2019) Label-free Bacteria quantification in blood plasma by a bioprinted microarray based interferometric point-of-care device. ACS Sens 4:52–60. https://doi.org/10.1021/ACSSENSORS.8B00789/SUPPL_FILE/SE8B00789_SI_001.PDF
31. Li B, Yu Q, Duan Y (2015) Fluorescent labels in biosensors for pathogen detection. Crit Rev Biotechnol 35:82–93. https://doi.org/10.3109/07388551.2013.804487
32. Pazos E, Vázquez O, Mascareñas JL, Vázquez ME (2009) Peptide-based fluorescent biosensors. Chem Soc Rev 38:3348–3359. https://doi.org/10.1039/B908546G
33. Xu L, Lu Z, Cao L et al (2017) In-field detection of multiple pathogenic bacteria in food products using a portable fluorescent biosensing system. Food Control 75:21–28. https://doi.org/10.1016/J.FOODCONT.2016.12.018
34. Huang F, Guo R, Xue L et al (2020) An acid-responsive microfluidic salmonella biosensor using curcumin as signal reporter and ZnO-capped mesoporous silica nanoparticles for signal amplification. Sens Actuat B Chem 312:127958. https://doi.org/10.1016/J.SNB.2020.127958
35. Guo Y, Li J, Song X et al (2021) Label-free detection of staphylococcus aureus based on bacteria-imprinted polymer and turn-on fluorescence probes. ACS Appl Bio Mater 4:420–427. https://doi.org/10.1021/ACSABM.0C00897/SUPPL_FILE/MT0C00897_SI_001.PDF
36. Zeinhom MMA, Wang Y, Song Y et al (2018) A portable smart-phone device for rapid and sensitive detection of E. coli O157:H7 in Yoghurt and Egg. Biosens Bioelectron 99:479–485. https://doi.org/10.1016/J.BIOS.2017.08.002
37. Anker JN, Hall WP, Lyandres O, et al (2008) Biosensing with plasmonic nanosensors. Nat Mater 7(6):442–453. https://doi.org/10.1038/nmat2162
38. Prabowo BA, Chang YF, Lai HC et al (2018) Rapid screening of Mycobacterium tuberculosis complex (MTBC) in clinical samples by a modular portable biosensor. Sens Actuat B Chem 254:742–748. https://doi.org/10.1016/J.SNB.2017.07.102
39. Song Y, Wei W, Qu X (2011) Colorimetric biosensing using smart materials. Adv Mater 23:4215–4236. https://doi.org/10.1002/ADMA.201101853
40. Sun J, Warden AR, Huang J et al (2019) Colorimetric and electrochemical detection of escherichia coli and antibiotic resistance based on a p-benzoquinone-mediated bioassay. Anal Chem 91:7524–7530. https://doi.org/10.1021/ACS.ANALCHEM.8B04997/SUPPL_FILE/AC8B04997_SI_001.PDF
41. Adnan M, Azad MOK, Ju HS et al (2019) Development of biopolymer-mediated nanocomposites using hot-melt extrusion to enhance the bio-accessibility and antioxidant capacity of kenaf seed flour. Appl Nanosci. https://doi.org/10.1007/s13204-019-01205-z
42. Sharma G, Park SC, Bandi R, et al (2021) platinum nanoparticles and their antibacterial activity Polyquaternium enhances the colloidal stability of chitosan-capped platinum nanoparticles and their antibacterial activity
43. Choi Y, Hwang JH, Lee SY (2018) Recent trends in nanomaterials-based colorimetric detection of pathogenic bacteria and viruses. Small Methods 2:1700351. https://doi.org/10.1002/SMTD.201700351
44. Bandi R, Alle M, Park CW et al (2021) Cellulose nanofibrils/carbon dots composite nanopapers for the smartphone-based colorimetric detection of hydrogen peroxide and glucose. Sens Actuat B Chem 330:129330. https://doi.org/10.1016/J.SNB.2020.129330

45. Chaubey A, Malhotra BD (2002) Mediated biosensors. Biosens Bioelectron 17:441–456. https://doi.org/10.1016/S0956-5663(01)00313-X
46. Kasemo B (2002) Biological surface science. Surf Sci 500:656–677. https://doi.org/10.1016/S0039-6028(01)01809-X
47. Luppa PB, Sokoll LJ, Chan DW (2001) Immunosensors—principles and applications to clinical chemistry. Clin Chim Acta 314:1–26. https://doi.org/10.1016/S0009-8981(01)00629-5
48. Pejcic B, de Marco R, Parkinson G (2006) The role of biosensors in the detection of emerging infectious diseases. Analyst 131:1079–1090. https://doi.org/10.1039/B603402K
49. Thvenot DR, Toth K, Durst RA, Wilson GS (1999) Electrochemical biosensors: recommended definitions and classification (Technical Report). Pure Appl Chem 71:2333–2348. https://doi.org/10.1351/PAC199971122333/MACHINEREADABLECITATION/RIS
50. Fernández-Sánchez C, McNeil CJ, Rawson K (2005) Electrochemical impedance spectroscopy studies of polymer degradation: application to biosensor development. TrAC Trends Anal Chem 24:37–48. https://doi.org/10.1016/J.TRAC.2004.08.010
51. Boggaram V, Gottipati KR, Wang X, Samten B (2013) Early secreted antigenic target of 6 kDa (ESAT-6) protein of mycobacterium tuberculosis induces interleukin-8 (IL-8) expression in lung epithelial cells via protein kinase signaling and reactive oxygen species. J Biol Chem 288:25500–25511. https://doi.org/10.1074/jbc.M112.448217
52. Couniot N, Afzalian A, van Overstraeten-Schlögel N et al (2015) Capacitive biosensing of bacterial cells: analytical model and numerical simulations. Sens Actuat B Chem 211:428–438. https://doi.org/10.1016/J.SNB.2015.01.108
53. Bresler HS, Lenkevich MJ, Murdock JF, et al (1992) Application of capacitive affinity. Biosensors 89–104. https://doi.org/10.1021/BK-1992-0511.CH009
54. Mahadhy A, Mamo G, Wernersson ES-, et al (2014) PCR-free ultrasensitive capacitive biosensor for selective detection and quantification of enterobacteriacea DNA. J Anal Bioanal Techn 5:1–10. https://doi.org/10.4172/2155-9872.1000210
55. Abdelhamid H, Salem TZ, Wahba M A, et al (2022) A capacitive sensor for differentiation between virus-infected and uninfected cells. Sens Biosensing Res 36 :100497. https://doi.org/10.1016/J.SBSR.2022.100497
56. Bi H, Han X (2019) Chemical sensors for environmental pollutant determination. In: Chemical, gas, and biosensors for internet of things and related applications, pp 147–160. https://doi.org/10.1016/B978-0-12-815409-0.00010-3
57. Pisoschi AM (2016) Potentiometric biosensors: concept and analytical applications-an editorial. Biochem Anal Biochem 5(3). https://doi.org/10.4172/2161-1009.1000e164
58. Caras S, Janata J (1980) Field effect transistor sensitive to penicillin. Anal Chem 52:1935–1937
59. Yoshinobu T, Schöning MJ (2021) Light-addressable potentiometric sensors for cell monitoring and biosensing. Curr Opin Electrochem 28:100727. https://doi.org/10.1016/J.COELEC.2021.100727
60. Hernández R, Vallés C, Benito AM et al (2014) Graphene-based potentiometric biosensor for the immediate detection of living bacteria. Biosens Bioelectron 54:553–557. https://doi.org/10.1016/J.BIOS.2013.11.053
61. Santandreu M, Alegret S, Fàbregas E (1999) Determination of β-HCG using amperometric immunosensors based on a conducting immunocomposite. Anal Chim Acta 396:181–188. https://doi.org/10.1016/S0003-2670(99)00436-5
62. Manimekala T, Sivasubramanian R, Dharmalingam G (2022) Nanomaterial-based biosensors using field-effect transistors: a review. J Electronic Mater 51:1950–1973. https://doi.org/10.1007/S11664-022-09492-Z
63. Li W, Gao Y, Zhang J et al (2020) Universal DNA detection realized by peptide based carbon nanotube biosensors. Nanoscale Adv 2:717–723. https://doi.org/10.1039/C9NA00625G
64. Vu CA, Chen WY (2019) Field-effect transistor biosensors for biomedical applications: recent advances and future prospects. Sensors 19:4214. https://doi.org/10.3390/S19194214

65. Afzal A, Mujahid A, Schirhagl R, et al (2017) Gravimetric viral diagnostics: QCM based biosensors for early detection of viruses. Chemosensors 5:7. https://doi.org/10.3390/CHEMOSENSORS5010007
66. Walton PW, O'Flaherty MR, Butler ME, Compton P (1993) Gravimetric biosensors based on acoustic waves in thin polymer films. Biosens Bioelectron 8:401–407. https://doi.org/10.1016/0956-5663(93)80024-J
67. Pohanka M (2017) The piezoelectric biosensors: principles and applications, a review. Int J Electrochem Sci 12:496–506. https://doi.org/10.20964/2017.01.44
68. Shen W, Li S, Park M-K et al (2012) Blocking agent optimization for nonspecific binding on phage based magnetoelastic biosensors. J Electrochem Soc 159:B818–B823. https://doi.org/10.1149/2.057210JES
69. Guntupalli R, Lakshmanan RS, Hu J et al (2007) Rapid and sensitive magnetoelastic biosensors for the detection of Salmonella typhimurium in a mixed microbial population. J Microbiol Methods 70:112–118. https://doi.org/10.1016/J.MIMET.2007.04.001
70. Park MK, Park JW, Wikle HC, Chin BA (2013) Evaluation of phage-based magnetoelastic biosensors for direct detection of Salmonella Typhimurium on spinach leaves. Sens Actuat B Chem 176:1134–1140. https://doi.org/10.1016/J.SNB.2012.10.084
71. Chai Y, Wikle HC, Wang Z et al (2013) Design of a surface-scanning coil detector for direct bacteria detection on food surfaces using a magnetoelastic biosensor. J Appl Phys 114:104504. https://doi.org/10.1063/1.4821025
72. Lim JW, Ha D, Lee J et al (2015) Review of micro/nanotechnologies for microbial biosensors. Front Bioeng Biotechnol 3:61. https://doi.org/10.3389/FBIOE.2015.00061/BIBTEX
73. D'Souza SF (2001) Microbial biosensors. Biosens Bioelectron 16:337–353. https://doi.org/10.1016/S0956-5663(01)00125-7
74. Lei Y, Chen W, Mulchandani A (2006) Microbial biosensors. Anal Chim Acta 568:200–210. https://doi.org/10.1016/J.ACA.2005.11.065
75. Su L, Jia W, Hou C, Lei Y (2011) Microbial biosensors: a review. Biosens Bioelectron 26:1788–1799. https://doi.org/10.1016/J.BIOS.2010.09.005
76. Zhang F, Keasling J (2011) Biosensors and their applications in microbial metabolic engineering. Trends Microbiol 19:323–329. https://doi.org/10.1016/J.TIM.2011.05.003
77. Casavant NC, Thompson D, Beattie GA et al (2003) Use of a site-specific recombination-based biosensor for detecting bioavailable toluene and related compounds on roots. Environ Microbiol 5:238–249. https://doi.org/10.1046/J.1462-2920.2003.00420.X
78. Kim BC, Gu MB (2003) A bioluminescent sensor for high throughput toxicity classification. Biosens Bioelectron 18:1015–1021. https://doi.org/10.1016/S0956-5663(02)00220-8
79. Jouanneau S, Durand MJ, Courcoux P et al (2011) Improvement of the identification of four heavy metals in environmental samples by using predictive decision tree models coupled with a set of five bioluminescent bacteria. Environ Sci Technol 45:2925–2931. https://doi.org/10.1021/ES1031757/SUPPL_FILE/ES1031757_SI_001.PDF
80. Bhattacharjee R, Nandi A, Mitra P et al (2022) Theragnostic application of nanoparticle and CRISPR against food-borne multi-drug resistant pathogens. Mater Today Bio 15:100291. https://doi.org/10.1016/J.MTBIO.2022.100291
81. Malik P, Katyal V, Malik V et al (2013) Nanobiosensors: concepts and variations. ISRN Nanomater 2013:1–9. https://doi.org/10.1155/2013/327435
82. Wang G (2018) Nanotechnology: the new features. https://doi.org/10.48550/arxiv.1812.04939
83. Malhotra BD, Ali MDA (2018) Nanomaterials in biosensors: fundamentals and applications. Nanomater Biosens 1. https://doi.org/10.1016/B978-0-323-44923-6.00001-7
84. Dreaden EC, Alkilany AM, Huang X et al (2012) The golden age: gold nanoparticles for biomedicine. Chem Soc Rev 41:2740–2779. https://doi.org/10.1039/C1CS15237H
85. Madhusudhan A, Reddy GB, Krishana IM (2019) Green synthesis of gold nanoparticles by using natural gums. Nanomaterials and plant potential. Springer International Publishing, Cham, pp 111–134
86. Murty BS, Shankar P, Raj B, et al (2013) Unique properties of nanomaterials. In: Textbook of nanoscience and nanotechnology, pp 29–65. https://doi.org/10.1007/978-3-642-28030-6_2

87. Shrestha B (2022) Nanotechnology for biosensor applications. In: Sustainable nanotechnology for environmental remediation, pp 513–531. https://doi.org/10.1016/B978-0-12-824547-7.00013-8
88. Malekzad H, Sahandi Zangabad P, Mirshekari H et al (2017) Noble metal nanoparticles in biosensors: recent studies and applications. Nanotechnol Rev 6:301–329. https://doi.org/10.1515/NTREV-2016-0014/ASSET/GRAPHIC/J_NTREV-2016-0014_FIG_002.JPG
89. Murthe SS, Mohamed Saheed MS, Perumal V, et al (2019) Electrospun nanofibers for biosensing applications Nanobiosensors for Biomolecular Targeting 253–267. https://doi.org/10.1016/B978-0-12-813900-4.00011-7
90. Xue J, Xie J, Liu W, Xia Y (2017) Electrospun nanofibers: new concepts, materials, and applications. Acc Chem Res 50:1976–1987. https://doi.org/10.1021/ACS.ACCOUNTS.7B00218/ASSET/IMAGES/LARGE/AR-2017-002185_0011.JPEG
91. Kafi AKM, Wali Q, Jose R et al (2017) A glassy carbon electrode modified with SnO2 nanofibers, polyaniline and hemoglobin for improved amperometric sensing of hydrogen peroxide. Microchim Acta 184:4443–4450. https://doi.org/10.1007/S00604-017-2479-6/TABLES/1
92. Tang L, Xie X, Zhou Y et al (2017) A reusable electrochemical biosensor for highly sensitive detection of mercury ions with an anionic intercalator supported on ordered mesoporous carbon/self-doped polyaniline nanofibers platform. Biochem Eng J 117:7–14. https://doi.org/10.1016/J.BEJ.2016.09.011
93. Kailasa S, Sai Bhargava Reddy M, Raj Maurya M et al (2021) Electrospun nanofibers: materials, synthesis parameters, and their role in sensing applications. Macromol Mater Eng 306:2100410. https://doi.org/10.1002/MAME.202100410
94. Walther BK, Dinu CZ, Guldi DM et al (2020) Nanobiosensing with graphene and carbon quantum dots: recent advances. Mater Today 39:23–46. https://doi.org/10.1016/J.MATTOD.2020.04.008
95. Zhao C, Song X, Liu Y, et al (2020) Synthesis of graphene quantum dots and their applications in drug delivery. J Nanobiotechnol 18:1–32. https://doi.org/10.1186/S12951-020-00698-Z
96. Gopal A, Yan L, Kashif S et al (2022) Biosensors and point-of-care devices for bacterial detection: rapid diagnostics informing antibiotic therapy. Adv Healthc Mater 11:2101546. https://doi.org/10.1002/ADHM.202101546
97. Chang HJ, Zúñiga A, Conejero I, et al (2021) Programmable receptors enable bacterial biosensors to detect pathological biomarkers in clinical samples. Nat Commun 12:1–12. https://doi.org/10.1038/s41467-021-25538-y
98. Shosholcheva M, Kartalov A, Kuzmanovska B, et al Editor-in-Chief Production Editor
99. Berggren C, Bjarnason B, Johansson G (2001) Capacitive biosensors. https://doi.org/10.1002/1521-4109
100. Tamayo AIB, Rizo LSL, de Armas MB et al (2020) Biotin self-assembled monolayer for impedimetric genosensor for direct detection of HIV-1. Microchem J 153:104462. https://doi.org/10.1016/J.MICROC.2019.104462
101. Guo X, Sang S, Guo J, et al (2017) A magnetoelastic biosensor based on E2 glycoprotein for wireless detection of classical swine fever virus E2 antibody. Sci Rep 7:1–8. https://doi.org/10.1038/s41598-017-15908-2
102. Reali S, Najib EY, Treuerné Balázs KE et al (2019) Novel diagnostics for point-of-care bacterial detection and identification. RSC Adv 9:21486–21497. https://doi.org/10.1039/C9RA03118A
103. Venner AA, Beach LA, Shea JL et al (2021) Quality assurance practices for point of care testing programs: recommendations by the Canadian society of clinical chemists point of care testing interest group. Clin Biochem 88:11–17. https://doi.org/10.1016/J.CLINBIOCHEM.2020.11.008
104. Gutierres SL, Welty TE (2004) Point-of-care testing: an introduction. 38:119–125. https://doi.org/10.1345/APH.1D212
105. McNerney R, Daley P (2011) Towards a point-of-care test for active tuberculosis: obstacles and opportunities. Nat Rev Microbiol 9:204–213. https://doi.org/10.1038/nrmicro2521

106. John AS, Price CP (2014) Existing and emerging technologies for point-of-care testing. Clin Biochem Rev 35:155
107. Madhusudhan A, Reddy G, Venkatesham M et al (2014) Efficient pH dependent drug delivery to target cancer cells by gold nanoparticles capped with carboxymethyl chitosan. Int J Mol Sci 15:8216–8234. https://doi.org/10.3390/ijms15058216
108. Dahl CA, Burgess DCH, Wasserman J (2007) Estimating the global health impact of improved diagnostic tools for the developing world
109. Zhang C, Paria D, Semancik S et al (2019) Composite-scattering plasmonic nanoprobes for label-free, quantitative biomolecular sensing. Small 15:1901165. https://doi.org/10.1002/SMLL.201901165
110. Pai NP, Vadnais C, Denkinger C et al (2012) Point-of-care testing for infectious diseases: diversity, complexity, and barriers in low- and middle-income countries. PLoS Med 9:e1001306. https://doi.org/10.1371/JOURNAL.PMED.1001306
111. Mayer M, Baeumner AJ (2019) A megatrend challenging analytical chemistry: biosensor and chemosensor concepts ready for the internet of things. Chem Rev 119:7996–8027. https://doi.org/10.1021/ACS.CHEMREV.8B00719/ASSET/IMAGES/LARGE/CR-2018-00719G_0020.JPEG
112. Tay A, Pavesi A, Yazdi SR et al (2016) Advances in microfluidics in combating infectious diseases. Biotechnol Adv 34:404–421. https://doi.org/10.1016/J.BIOTECHADV.2016.02.002
113. Stürenburg E, Junker R (2009) Point-of-care testing in microbiology: the advantages and disadvantages of immunochromatographic test strips. Dtsch Arztebl Int 106:48. https://doi.org/10.3238/ARZTEBL.2009.0048
114. Lindquist KA, Chow K, West A, et al (2014) The StatStrip glucose monitor is suitable for use during hyperinsulinemic euglycemic clamps in a pediatric population. 16:298–302. https://doi.org/10.1089/DIA.2013.0288. https://home.liebertpub.com/dia
115. Amundson BE, Apple FS (2015) Cardiac troponin assays: a review of quantitative point-of-care devices and their efficacy in the diagnosis of myocardial infarction. Clin Chem Lab Med 53:665–676. https://doi.org/10.1515/CCLM-2014-0837/MACHINEREADABLECITATION/RIS
116. Apple FS, Murakami MAM, Christenson RH et al (2004) Analytical performance of the i-STAT cardiac troponin I assay. Clin Chim Acta 345:123–127. https://doi.org/10.1016/J.CCCN.2004.03.010
117. Goldmann BU, Langenbrink L, Matschuck G et al (2004) Quantitative bedside testing of troponin T: is it equal to laboratory testing? The cardiac reader troponin T (CARE T) study. Clin Lab 50:1–10
118. Coppola MA, Klotz KL, Kim KA et al (2010) SpermCheck® Fertility, an immunodiagnostic home test that detects normozoospermia and severe oligozoospermia. Hum Reprod 25:853–861. https://doi.org/10.1093/HUMREP/DEP413
119. Ledden DJ, Novamo AK, Schulman LS (2014) Evaluation of the CLINITEST® Human Chorionic Gonadotropin (hCG) pregnancy test for susceptibility to the hook effect by the hCG β core fragment. Clin Chem 60:1578–1580. https://doi.org/10.1373/CLINCHEM.2014.231498
120. Adams S, Luo W, Wesolowski L et al (2017) Performance evaluation of the point-of-care INSTI™ HIV-1/2 antibody test in early and established HIV infections. J Clin Virol 91:90–94. https://doi.org/10.1016/J.JCV.2017.03.012
121. Chang M, Steinmetzer K, Raugi DN et al (2017) Detection and differentiation of HIV-2 using the point-of-care Alere q HIV-1/2 Detect nucleic acid test. J Clin Virol 97:22–25. https://doi.org/10.1016/J.JCV.2017.10.013
122. Hsiao NY, Dunning L, Kroon M, Myer L (2016) Laboratory evaluation of the Alere q point-of-care system for early infant HIV diagnosis. PLoS ONE 11:e0152672. https://doi.org/10.1371/JOURNAL.PONE.0152672
123. Kolbrich EA, Kim I, Barnes AJ et al (2003) Cozart® RapiScan oral fluid drug testing system: an evaluation of sensitivity, specificity, and efficiency for cocaine detection compared with

ELISA and GC-MS following controlled cocaine administration. J Anal Toxicol 27:407–411. https://doi.org/10.1093/JAT/27.7.407
124. Jobbagy Z, van Atta R, Murphy KM et al (2007) Evaluation of the cepheid GeneXpert BCR-ABL assay. J Mol Diagn 9:220–227. https://doi.org/10.2353/JMOLDX.2007.060112
125. Tapper EB, Castera L, Afdhal NH (2015) FibroScan (Vibration-Controlled Transient Elastography): where does it stand in the United States practice. Clin Gastroenterol Hepatol 13:27–36. https://doi.org/10.1016/J.CGH.2014.04.039
126. Karapinar T, Kaynar O, Hayirli A, Kom M (2013) Evaluation of 4 point-of-care units for the determination of blood l-lactate concentration in cattle. J Vet Intern Med 27:1596–1603. https://doi.org/10.1111/JVIM.12204
127. Bezerra FSM, Leal JKF, Sousa MS et al (2018) Evaluating a point-of-care circulating cathodic antigen test (POC-CCA) to detect Schistosoma mansoni infections in a low endemic area in north-eastern Brazil. Acta Trop 182:264–270. https://doi.org/10.1016/J.ACTATROPICA.2018.03.002
128. Diaz T, Almeida MDGB, Georg I et al (2004) Evaluation of the determine rapid syphilis TP assay using sera. Clin Vaccine Immunol 11:98–101. https://doi.org/10.1128/CDLI.11.1.98-101.2004
129. El-Ghareeb AS, Abd El Motaleb GS, Waked NM et al (2016) Circulating cathodic antigen cassette test versus haematuria strip test in diagnosis of urinary schistosomiasis. J Parasit Dis 40:1193–1198. https://doi.org/10.1007/S12639-015-0648-2/TABLES/5
130. Rogo T, Schwartz RH, Ascher DP (2010) Comparison of the inverness medical acceava strep a test with the genzyme OSOM and quidel QuickVue strep a tests. 50:294–296. https://doi.org/10.1177/0009922810385675
131. Athlin S, Iversen A, Özenci V (2017) Comparison of the ImmuView and the BinaxNOW antigen tests in detection of Streptococcus pneumoniae and Legionella pneumophila in urine. Eur J Clin Microbiol Infect Dis 36:1933–1938. https://doi.org/10.1007/S10096-017-3016-6/TABLES/2
132. Eletu SD, Sheppard CL, Thomas E et al (2017) Development of an extended-specificity multiplex immunoassay for detection of streptococcus pneumoniae serotype-specific antigen in urine by use of human monoclonal antibodies. Clin Vaccine Immunol 24:262–279. https://doi.org/10.1128/CVI.00262-17/SUPPL_FILE/ZCD012175543SD1.XLSX
133. Nikam C, Jagannath M, Narayanan MM et al (2013) Rapid diagnosis of Mycobacterium tuberculosis with Truenat MTB: a near-care approach. PLoS ONE 8:e51121. https://doi.org/10.1371/JOURNAL.PONE.0051121
134. Alarcón-Rivera G, Vázquez-Jiménez G, de La Abarca M et al (2011) Comparative analysis between breath test, serological immunoassay and rapid-urease test for detection of Helicobacter pylori infection in Mexican patients with non-investigated dyspepsia. Rev Gastroenterol Mex 76:322–329
135. Mansour-Ghanaei F, Joukar F, Sheykhian MR et al (2013) Effect of gastric acidification on the 14C-UBT HELIPROBE® accuracy during Pantoprazole treatment in Helicobacter pylori positive patients. Int J Clin Exp Med 6:185
136. Chesnais CB, Vlaminck J, Kunyu-Shako B et al (2016) Measurement of circulating filarial antigen levels in human blood with a point-of-care test strip and a portable spectrodensitometer. Am J Trop Med Hyg 94:1324. https://doi.org/10.4269/AJTMH.15-0916
137. Martín-Díaz A, Rubio JM, Herrero-Martínez JM et al (2018) Study of the diagnostic accuracy of microbiological techniques in the diagnosis of malaria in the immigrant population in Madrid. Malar J 17:1–8. https://doi.org/10.1186/S12936-018-2459-2/TABLES/2
138. van Dijk DP, Gillet P, Vlieghe E et al (2010) Evaluation of the Immunoquick+4 malaria rapid diagnostic test in a non-endemic setting. Eur J Clin Microbiol Infect Dis 29:577–583. https://doi.org/10.1007/S10096-010-0898-Y/TABLES/3
139. Clegg HW, Dallas SD, Roddey OF et al (2003) Extrapharyngeal group A streptococcus infection: diagnostic accuracy and utility of rapid antigen testing. Pediatr Infect Dis J 22:726–731. https://doi.org/10.1097/01.INF.0000078835.72497.AB

140. Holden J, Goheen J, Jett-Goheen M, et al (2017) An evaluation of the SD Bioline HIV/syphilis duo test. Int J STD AIDS 29:57–62. https://doi.org/10.1177/0956462417717649
141. Shakya G, Singh DR, Ojha HC et al (2016) Evaluation of SD bioline HIV/syphilis duo rapid test kits in Nepal. BMC Infect Dis 16:1–5. https://doi.org/10.1186/S12879-016-1694-9/TABLES/3
142. Shimelis T, Tadesse E (2015) The diagnostic performance evaluation of the SD BIOLINE HIV/syphilis Duo rapid test in southern Ethiopia: a cross-sectional study. BMJ Open 5:e007371. https://doi.org/10.1136/BMJOPEN-2014-007371
143. Mukae H, Yatera K, Noguchi S et al (2015) Evaluation of a rapid immunochromatographic ODK0501 assay for detecting Streptococcus pneumoniae antigens in the sputum of pneumonia patients with positive S. pneumoniae urinary antigens. J Infect Chemother 21:176–181. https://doi.org/10.1016/J.JIAC.2014.11.003
144. Suzuki S, Nishimura N, Jinta T et al (2016) Evaluation of the rapid immunochromatographic ODK0501 assay for Streptococcus pneumoniae antigen detection with nasopharyngeal swabs: preliminary report. Multidiscip Respir Med 11:1–5. https://doi.org/10.1186/S40248-016-0060-5/TABLES/3
145. Lawn SD, Kerkhoff AD, Burton R et al (2017) Diagnostic accuracy, incremental yield and prognostic value of Determine TB-LAM for routine diagnostic testing for tuberculosis in HIV-infected patients requiring acute hospital admission in South Africa: a prospective cohort. BMC Med 15:1–16. https://doi.org/10.1186/S12916-017-0822-8/FIGURES/4
146. Salou M, Badjassem DG, Dossim S et al (2019) Onsite performance verification of DETERMINE^{TM} TB LAM Ag: a rapid diagnostic test for tuberculosis screening in urine. Open J Med Microbiol 09:105–115. https://doi.org/10.4236/OJMM.2019.93011
147. Cantero M, Redondo M, Martín E et al (2015) Use of quality indicators to compare point-of-care testing errors in a neonatal unit and errors in a STAT central laboratory. Clin Chem Lab Med 53:239–247. https://doi.org/10.1515/CCLM-2013-1053/MACHINEREADABLECITATION/RIS
148. Zydron CT, Woodworth A, Storrow AB (2011) The future of point-of-care testing in emergency departments. 5:175–181. https://doi.org/10.1517/17530059.2011.567263
149. Alle M, Sharma G, Lee SH, Kim JC (2022b) Next-generation engineered nanogold for multimodal cancer therapy and imaging: a clinical perspectives. J Nanobiotechnol 20:1–34. https://doi.org/10.1186/S12951-022-01402-Z
150. Chamorro-Garcia A, Merkoçi A (2016) Nanobiosensors in diagnostics. https://doi.org/10.1177/1849543516663574
151. Gault MH, Seymour ME, Howell WE (2001) Evaluation of i-STAT creatinine assay. Nephron 88:178–182. https://doi.org/10.1159/000045982
152. Moreno M, Schwartz A, Dvorkin R (2015) The accuracy of point-of-care creatinine testing in the emergency department. Adv Emerg Med https://doi.org/10.1155/2015/965368
153. Shephard MDS (2011) Point-of-care testing and creatinine measurement. Clin Biochem Rev 32:109

Nanostructured Electrodes as Electrochemical Biosensors for Biomedical Applications

Rajlakshmi Chetty, Varun Pratap Singh, Alle Madhusudhan, Raymond Wilson, and Alberto Rodriguez-Nieves

Abstract In recent decades with breakthrough achievements in nanotechnology, nanostructured-based electrode biosensors have emerged as widely used techniques considering its sensitivity and selectivity. Recently, electrochemical biosensors are representing exceptional attributes such as easy to operate, economical, simple to construct, and integration of nanostructured material of interest. Electrochemical biosensors hold its specificity in detection of biological targets by means of electrochemical change of interface in nanostructured electrodes. By carefully regulating the electrode interface design at the nanoscale and utilizing the inherent qualities of nanomaterials, new biosensing platforms with improved functionality are produced. This includes porous nanostructured materials and nanoscale components like nanotubes, nanowires, nanorods, nanoparticles, and nanosheets. This chapter's objective is to provide a thorough and insightful overview of the numerous strategies for creating electrochemical nanostructured electrodes for sensor improvement; hence, these techniques are utilized for detection of biochemical analysis in human physiology clinics and hospitals.

Keywords Biosensors · Nanomaterials · Electrochemical · Biomedical diagnosis · Nanostructured electrode

R. Chetty
School of Forensic Sciences, National Forensic Sciences University, Gandhinagar 38007, Gujarat, India

V. P. Singh
School of NanoSciences, Central University of Gujarat, Gandhinagar 382030, Gujarat, India

A. Madhusudhan (✉) · R. Wilson · A. Rodriguez-Nieves
Department of Chemistry, The University of Memphis, Memphis, TN 38152, USA
e-mail: allemadhusudhan@gmail.com

A. Madhusudhan et al. (eds.), *Functional Smart Nanomaterials and Their Theranostics Approaches*, Smart Nanomaterials Technology,
https://doi.org/10.1007/978-981-99-6597-7_9

1 Introduction

Biosensor uses in the hospitals and pharmaceutics fields have produced a great agreement in the area of study. The sensors and biosensors are successfully used for illness discovery, prevention, restoration, diagnosis and monitoring, and management of human being health and disease suffering patient. Microbiota such as bacteria, viruses, and pathogens can also be found using biosensors [35, 65, 82]. People can develop and enhance healthier behaviors with the help of activity and step trackers. Recently, research sciences and medical organizations have switched to adopting affordable biosensors to test for toxins in food, drinking water, juice, milk, etc., manage human biologic processes, evaluate accurate health diagnosis, and more [51, 94]. Biosensors utilized for various applications include screening for infections, treating chronic infections, health management, and controlling progress [32]. It is now promising to identify ailments and monitor the biological response to cure [99]. Because these techniques are easy to handle, adaptable, and efficient in industrial processes, biosensors have good promise. The considerable advantages of biosensors in the medical industry are covered in this research [24, 109].

Furthermore, biosensors have unique capabilities that are offered and graphically displayed for cardiovascular disease and healthcare services [36, 49]. The paper discusses the innovative use of "Nanostructured Electrodes" as electrochemical biosensors in biomedical applications. Thus, paper contributes significantly to the development of biosensors in the medical industry. Older individuals may now manage their health with less intervention and thanks to biosensors' sophisticated wearable features, which also immediately communicate medical information with healthcare practitioners to cut down on hospital visits [42, 90]. There are endless opportunities for consumer and commercial usage of biosensors in wellness, physical fitness, athletics, etc. Thus, a biosensor is a device that uses one or more biorecognition components and a signaling pathways' compartment to provide quantitative and selective monitoring of one or more analyte(s), such as cancer biomarkers, DNAs, poisons, heavy metals, medicines, poisonous gases.

1.1 Fundamentals of Biosensors

The high accuracy and tolerable sensitivity of conventional chemical analysis techniques are just two of their numerous benefits. However, those analytical procedures frequently need complicated equipment, pricey reagents, a lot of sample consumption, and are not portable. They also cannot allow on-site monitoring, necessitating the use of a lab professional with high skills [11, 75]. Consequently, chemical and biological sensors must be developed. Thus, first glucose oxidase-based sensor has been developed in 1962, and Clark and Lyons began the rapid processing of biosensors and its applications. A biosensor can be defined as "an analytical device which includes a biologically active element or component in close contact with an appropriate physicochemical transducer to generate measurable signal (optical, electrical, or electrochemical) directly proportional to the concentration of target substance(s)" in the simplest terms [11].

Thus, biosensors are massively important flexible detection techniques that can address a variety of potential issues and challenges in a variety of fields, including homeland security, drug and pharmaceutical detection, safety of goods and foods, and defense system [14]. Regarding sampling circumstances, sensors can be employed nonstop to detect the target substance or analyte in a range of complicated models without requiring any further pre-treatments of sample [34]. In addition, simultaneous multi-target recognition with high selective credentials and cost and sample size reduction may be achieved [54, 72]. The glucometer (blood glucose monitoring) is used as the most prominent illustration of this in daily life.

1.2 Basic Structural Unit of Biosensor

A biosensor is consisted of three main parts: a transducer, a processor of signaling, and a bio-recognition component (also called a bio-receptor or sensing components) (a figure of the design and the key structure consists of typical sensors which is established in Fig. 1). A biological receptor is synthesized to specifically detect a designated analyte in the first compartment. The bio-direct receptor's connection to the transducer allows it to translate both the biological and biochemical reactions into a quantitative indicator. The transducer's unseen/very small indication inputs are eventually enlarged to produce superior signals that include the key characteristics of the input signals.

As a result, a digital indicator or processor gathers the enlarged motions so they may be seen, examined, and kept in the device's internal memory [72]. These electronic components allow the detection, recording, and transmission of the collected data. The choice of a biological receptor and a workable way to added it onto a surface of producing component, i.e., transducer are crucially regulated and affect the primary biosensor benefits (i.e., the high selectivity and the sensitivity characteristics) [73]. In terms of performance variables, interactions among the biological detection points and the objective being sensed must be resistant to changes in temperature, stirring circumstances, pH of the measurement, and alien species addition (i.e., non-targeting analyte). Therefore, choosing an efficient and proper method to immobilize the sensing portion is highly important, and controlling the working of biosensors [79]. Otherwise, the transducer's surface would have undergone deactivation, leaching out, or faulty orientation. The two most popular immobilization strategies in this context are conjugation chemistry and physical chemistry like adsorption on the surface. The optimum immobilization method is then selected built on the properties of the analyte and the type of bio-recognition element [95, 98]. Moreover, advancements in biosensor technology are anticipated in a number of areas, including point-of-care (POC), implantable biosensor devices including wearable

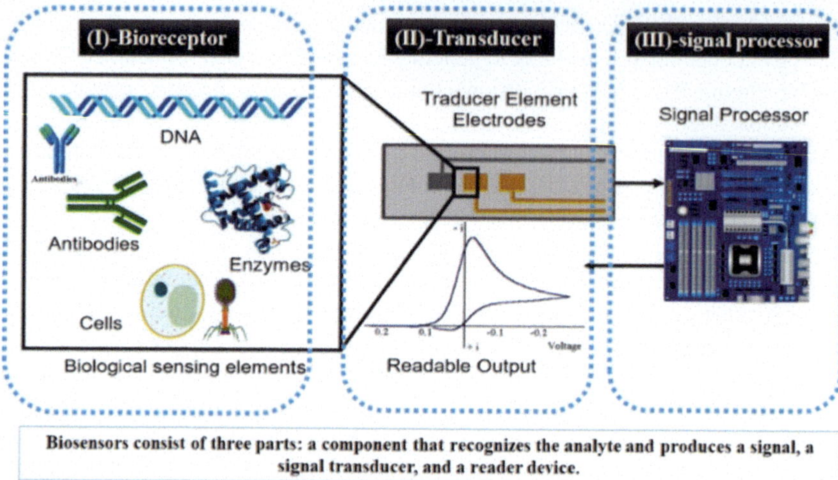

Fig. 1 Simple diagram showing the important building blocks such as: (I) a biological receptor or bioreceptors, (II) a transducer, and (III) signal processor. This image is adopted from Hassan [36]

devices. Hence, the basics' effects of nanosize on the structure and conception of sensors and biosensors were thus illustrated in this chapter.

2 Electrochemical Biosensors

Electrochemical sensing and biological sensing schemes provide the best ability for numerous analyte(s) diagnosis in compound or mixtures (for example, blood serum and several clinical samples). At the crossing point of the electrochemical transducer of this detecting platform, biological/electrochemical circumstances (facilitated or non-facilitated processes) occur (working electrode surface). Additionally, because the resulting electrochemical signals could be recorded, the chosen binding effect as well as the catalytic process among an analyte and immobilized bio-receptor may possibly screened, detected, and identified. The main electrochemical approaches for building, correcting, and improving excellent biosensors of electrochemical sensors and are potentiometric, amperometric, conductometric, impedimetric, and voltammetric procedures [54]. In subsequent subsections, a summary of each of these electrochemical methods will be mentioned.

2.1 Potentiometric Biosensors

This type of biosensor is an electrochemical mode of device that combines a biotic sensing component with transducer or functional probe to produce a alteration in electrical potential [77]. Ion-sensitive field transistors or some ion-based electrodes, also known as ion-selective electrodes (ISEs), may use to translate a specific biological effect into an electric signal. Adaptable potentiometric sensing techniques can be created for different analytes via chosen expressed bio-receptor (for example, small peptides unit, aptamers, antibodies, or enzymes). Beside this, potentiometric nanobiosensors have recently been synthesized employing a three-dimension-based molecular-imprinted polymer (MIP) for the quick diagnosis of corona virus or COVID-19 [27]. Middle-East respiratory disease (MERS) and the targeted coronavirus (H1N1 and H3N2) were distinguished, demonstrating the great discernment of our COVID-19-MIP biosensor. Additionally, the virus biosensors have very low detection limits of 100 picogram/milliliter and 200 PFU/mL for the spike made up of proteins and suspension of virus-related particles, respectively.

2.2 Amperometric Biosensors

For using an amperometric approach, a three- or two-electrode setup could be used. The sensor generates currents of faradaic caused by potential redox reactions at the exterior of the functional electrode at a static voltage while bringing up to an Ag/AgCl or calomel, common reference electrodes. The concentration or amount of the sample containing analyte that is present in a secondary electrolyte determines the value of the generated redox current [23]. The working probe is consisted of nanomaterials such as metallic oxides (such as zinc oxide [7]), carbon-based structure (for example, graphite [76], grapheme [78], carbon nanotubes [21], or noble metals (such as gold [18, 19]), platinum [83], and copper [107] (ITO). Techniques for amperometric sensing have a broad active-linear response, high accuracy, quick values for reading, and high sensitivity. However, the substantial interaction with other electroactive chemicals indicates a weak selectivity, which is viewed as a major disadvantage and downside [84]. One another scientist recently created an amperometric biosensor for the assessment in the clinics of creatinine in the testing of kidney function. The sensors displayed promising application for in-person and at-home testing, as demonstrated by Yaping's group [88].

2.3 Conductometric Biosensors

Any electrochemical reactive change that is taking place in a solution can be detected by conductometric biosensors. This made it possible to identify any changes in the ionic composition of the tested sample brought on by chemistry and biochemistry [59]. The conductometric sensors demonstrated a number of benefits: Because (i). no requirement of a reference electrode, (ii). Thin-film electrodes are suitable for downsizing and mass manufacture utilizing affordable materials, and (iii). Conductometric transducers do not require much power. A thin coating of gold, chromium, copper, and nickel produced on the electrode surface was used to create conductometric biosensors.

2.4 *Impedimetric Biosensors*

The impedimetric biosensors capture the electrical impedance produced on the edge of a firm electrode surface during supply of a tiny AC voltage [55]. Next, the variations in resistance are calculated as a function of frequency. Figure 2 provides a straightforward illustration of what is happening in the systems of the EIS throughout the performance of experiments and prior to producing the EIS signals [36].

The plot of Nyquist, an illustration shown in Fig. 3, is made by plotting a pretended impedance (Zimag) on the Y-axis and a real impedance (Z real) on the X-axis to represent the EIS representation. While the Zimag (the imaginary impedance) is stated in negative numbers, every point on the plot of Nyquist represents the impedance at a single frequency point. Low-frequency data are located on the right side of X-axis, while higher frequencies are assigned to the left.

EIS is one of the most sensitive methods for examining the interfacial properties that can be carried out at functionalized surfaces among electrochemical biosensors. Therefore, bio-recognition events like the capture of an antigen by an antibody, a protein–protein interaction, a drug's engagement with its target, or the activity of the

Fig. 2 Schematic illustration of EIS systems, course of experiments and prior to the generation of EIS signals. (The figure is adopted from Hassan [36])

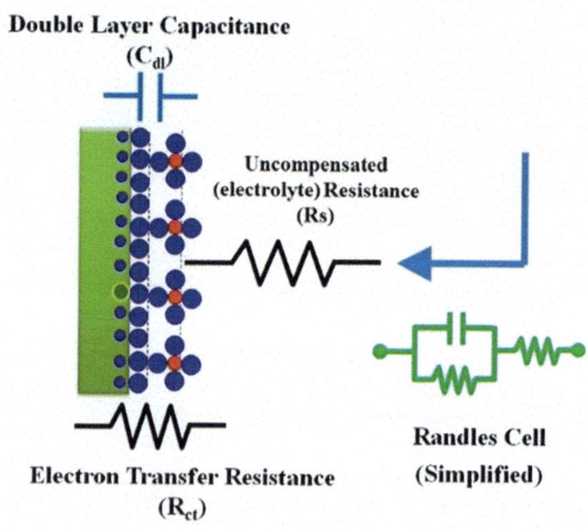

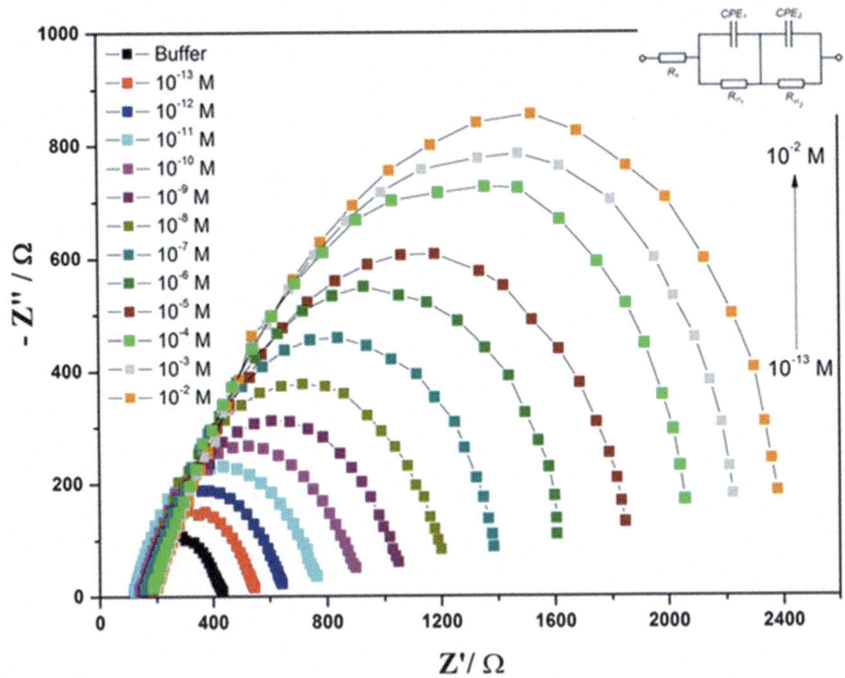

Fig. 3 Electrochemical cell is coupled with a one-use screen-printed electrode that has been modified along with a ligand/nano-cellulose nanostructure to display the change in charge transfer resistances (Rct) caused by a varied cadmium (Cd) ions' concentration. (The equivalent circuit, shown in the inserted figure, was employed in the quantitative study [60]

entire microbial cell could be kinetically observed [97, 107]. To create impedimetric immunosensors or impedimetric aptasensors, respectively, several impedimetric biosensors were constructed employing antibodies or aptamers [10, 68]. The impedimetric immunosensor forms an immunological compound at the surface of sensor, which is the location of the EIS extents, as a result of antigen–antibody-specific immune interactions. The impedimetric aptasensor, on the other hand, displays the attachment of short single-stranded oligonucleotides, also known as aptamers, which have excellent stability and potent binding affinity (for example, ribonucleic acid or deoxyribonucleic acid) [50]. The impedance to electron and charge transport rises as a result. Thus, label-free diagnosis of biomolecular identification actions is made possible by EIS biosensors [67]. Due to their inclination for lab-on-a-chip synthesis with simple operation, EIS systems are increasingly being used in biomedical and environmental analysis [63]. However, EIS may be used to characterize new substances by giving electrochemical knowledge on ongoing processes like fuel cell, development or corrosion inhibition, the charging and discharging of various electrochemical devices such as batteries, or processes [60, 101].

2.5 Voltammetric Biosensors

Voltammetric biosensors measure the changes in electric potential to determine the generated faradaic current, which is used to examine the target concentration. The capability of monitoring numerous analytes simultaneously (simultaneous analysis) is one of these sensors' advantages [37]. Electric current and changeable electrical potential (X-axis) are both measured using voltammetry (Y-axis). The peak point at a specific value is utilized for documents due to the redox reaction of electroactive analyte at the functional surface of the electrode (analyte character cathodic/anodic peak potentials), and also intensity of the peak current reflects the amount of the corresponding species (Faradic current). Cyclic voltammetry (CV) is the famous voltammetric method for gathering both quantitative and qualitative data. The CV data might be used to get a wealth of knowledge on the kinetics of electron transfer, the redox reactions thermodynamics, the diffusion rate, and the adsorption rate. The identification of heavy metals, such as lead (Pb(II)), chromium (Cr(VI)), and cadmium (Cd(II)), in environmental samples [1, 28, 40], the monitoring of sugar molecules and urea in the circulatory fluid through sensor chips, the presence of pharmaceutical chemicals in plasma and dosage forms, and the detection of pathogenic biomarkers in microorganism cultures employing crown-ethers [92] have all been reported and reviewed.

3 Role of Nanomaterials on Biosensor Performance

Metal electrodes (platinum, silver, gold, and zinc oxide), glassy carbon electrode (GCE), carbon paste electrode (CPE) [57], and graphite electrode (GE) are the most often employed working electrode materials. Polarography has traditionally employed Hg electrodes, such as a dropping Hg electrode (DME) as well as a hanged Hg drop electrode (HMDE) [70]. Starting with the DME and completing with the disposable screen-printed electrodes, Fig. 3 is compiled the most popular working electrodes. The other electrodes had certain undesirable characteristics, such as their huge sizes, which necessitate big electrochemical cells and an excessive amount of the sample to be analyze, also to the problem with transportation and ability to dispose. Mercury's toxicity restricted and prohibited its employment. Therefore, these traditional working electrodes are employed in laboratory research and substance characterization.

Nanomaterials play a significant role in a wide range of applications such as biomaterials, catalysis, nanobiosensors, and nano-bioelectronics [39, 40]. Nanotubes, wires, pores, or any other nanostructure in used in sensors and the synthesis of diagnostic tools has received a lot of attention [38, 104]. Nanostructure is utilized in the fabrication of transducers and the decoration of solid surfaces. Because of their nanoscale size, engineered materials with required biocompatibility have an large surface-to-volume ratio, as well as high mechanical strength, electrical conductivity, and catalytic activity [41]. As a result, they are frequently used to improve biosensor working by augmenting electrochemical information, increment in sensitivity, and given that a low LOD that is various orders of magnitudes lower. Nanoparticle's sensor stages are good materials to effectively retain the arrangements and alignment while supporting the biological activity of the biomolecules which are immobilized in terms of the molecular orientation and structure stability. Furthermore, the use of certain nanomaterials facilitates direct charge/electron transfer [106]. As an illustration, DNA, antibodies, enzymes, and entire cells have all been conjugated with electrochemical transducers using carbon nanomaterials.

Nanomaterials are divided into four varieties based on their dimensions and sizes: zero dimensions, one dimensions, two dimensions, and three dimensions. Metallic nanoparticles (such as silver (Ag), gold (Au), platinum (Pt), and quantum dots (QDs)) possess all three dimensions of their materials present at the nanorange in zero dimensions (0D). Nanoparticles (NPs) can be spherical in shape and have a 1–50 nm average diameter. One dimension (1D) has three dimensions, two of which can be macroscale and one of which is only in the range of 1–100 nm. Nanowires, fibers, rods, and tubes are a few types of 1D nanomaterials. Two dimensions (2D) consist of one macroscale dimension and two nanoscale dimensions. Nanomaterials can be used to coat solid surfaces to produce 2D-like structures, such as nanosheets, nanowalls, thin films, or nanothin multilayers [100]. The entire field of the 2D-like structures in nanomaterials can be within a few square micrometers, assuming that thickness in the nanoscale region remains constant.

In the 3D nanostructures of nanoscale materials, there are no dimensions, while all dimensions are at the macroscale. Consequently, 3D nanostructures are known for bulk substances that incorporate discrete nanoscale blocks (<100 nm). Due to their mobility and ease of use, optical and electrochemical nanosensors are the utmost popular detection unit [53]. Widely used in the sensing field, various nanostructured materials offer expanded surface areas to enhance the receptor-analyte interface and improve electrochemical signal outcomes. Owing to benefits of great electrical conductivity, facile decoration, and negligible toxicity, organic or carbon-based nanostructures are ideal materials for production of sensor, especially due to vast surface area. Because of the strong electrocatalytic activity and propensity to directly promote the transfer of electrons or charges in mediator-less biosensing systems, metal and metal oxide nanomaterials are the most common materials employed for sensor alteration. In order to boost rapid transfer of electron mechanism between the electrode and the bioreceptors' active sites, ZnO, CuO, NiO, TiO_2, and Fe_3O_4 have therefore been widely utilized in electrochemical biosensing [48]. This has connected to a synergistic improvement in sensing behavior.

Due to their resistance to oxidation reactions and advantageous biocompatibility characteristics, great metallic nanomaterials, including Au, Pt, Ag, and Pd, are used as electrode [5, 17, 58, 81]. These nanostructures might be combined with other elements (such as polymeric nanomaterials or hydrogels) in the electrode bed or could be placed on the electrode surface [5, 6]. Nanosensors' devices ultimately enable the production of sensing chips and portable nanosensing devices for quick and precise multi-target in complicated living and surrounding ecological matrices based on the benefits of those special features of nanomaterials.

3.1 Metal-Based Nanostructured Biosensor

Metal-based nanostructures find utility due to their functional capabilities, including enzyme-like activity, electrical conductivity, surface plasmon resonance states, charge transfer abilities, and sensing properties. These qualities offer distinct advantages, particularly in biosensor applications (Fig. 4). Core–shell nanostructures are produced using several processes, including hydrothermal [6], sonochemical [20], microwave-assisted extraction [25], laser ablation [8], and others. Mallick et al. emphasized the need for core–shell metal oxide nanomaterials (NPs) with chemical, physical, and distinctive nanostructures at the nanosize in increased applications biosensor in 2022 [62] (Table 1). It has been demonstrated that the detecting efficacy of immunosensors and improvement in biosensors with the dimensions and morphology of nanomaterials such as TiO_2, ZnS, CuO, ZnO, WO_3, and hybrid nanocomposite with varied morphologies [9, 69, 71, 89].

A novel platform for both diagnostic and therapeutic applications was made possible by the improved optical characteristics of NSMs when combined with metal oxide [16, 43-45, 52, 56]. $NiWO_4$, TiO_2, Fe_2O_3, and other transition metal oxides, among others, can be utilized to create analyte-targeted nanocomposite catalyst in order to merging ZnO-CuO hierarchical nanocomposites, doping of fluorine onto tin oxide for the purpose of detecting sugar, i.e., glucose. [108] suggested a non-enzymatic sugar biosensor. The electrode displayed increased sensitivity with a 3066 AmM/1 cm^2 detection limit [108]. FTE-based insulin sensor was created by [46]. Sputtering was used to add NiO QDs to the ZnO nanorods that were formed between the sources and the drains. Along with great specificity of 13.14 A cm^2 mM^{-1} and 7.31 A cm^2 mM^{-1}, respectively, the developed sensor demonstrated a broad linear amplitude of 0.001^{-10} mM but also 10–50 mM [46]. The technique was employed to analyze human whole blood and serum samples. A FET-based, vertically situated, glucose sensor was described by Ahmad et al. in contrast to NPs, glucose sensors, nanorods of ZnO were treated with Fe_2O_3, that improved the functionality of the biological sensor [2].

This biosensor was improved, and Fe_2O_3's strong electrocatalytic properties were credited with this. The manufactured sensor demonstrated outstanding enzymatic performance with a LOD of 12 μM to 18 mM. The sensors were utilized to identify collected fresh glucose from mouse total blood and biological fluids, indicating the practical usability of FET-based sensor in actual sample analysis. Numerous additional metal oxide ZnO nanoparticles have also been employed as sensors in addition to the results mentioned above. The appealing characteristics of ZnO NSM-based biosensors that include metal oxide modifications make the design of multifunctional NSM-based sensors easier. The use of these nanohybrids in biosensing might alter the way environmental and diagnostic samples are used.

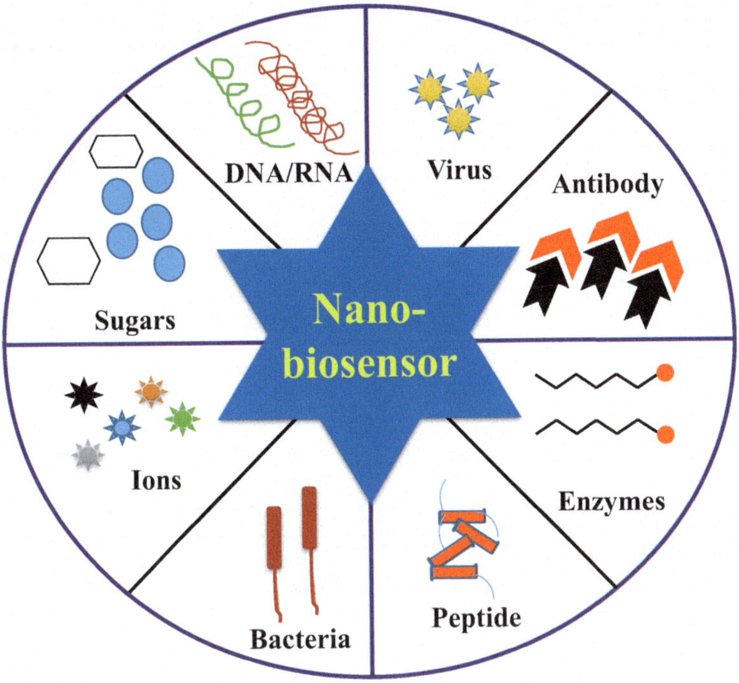

Fig. 4 Applications of nanosized biosensor for sensing application for various biological molecules' detection including enzymes, peptide or protein, bacteria, ions, non-enzymatic sugars, nucleic acids, viruses, and antibody for betterment via electrochemical biosensing

Metal oxide NPs are another metal-based nanostructure that is commonly employed in biosensor applications. Various types of manufactured nanosized metallic NPs have shown numerous applications as transducers in nanosensors, and further numerous investigations on them are being done [26]. Tasaltin et al. used a low-cost, simple, green, and rapid sonochemical approach to synthesis unique green polyphenol matrix NPs with matcha tea powder extract and then researched the construction of a biosensor for mobile phone digital spectrophotometric determination of ammonia [93]. CuO NPs were drop-cast onto the gold galvanic transducer in the investigation. The produced biosensor conducted color sensing by performing doing red–green–blue (RGB) assessment in the tests using a smartphone, provides a simple method.

3.2 *Carbon-Based Nanostructured Biosensors*

The unique elements in the field of material science are carbon. Due to its many atomic array forms, including graphene, diamond, fullerenes, and nanotubes, it

Table 1 List of various metal-based nanostructures as an excellent biosensor for detecting many pharmaceutical drugs, biochemicals, and viruses

Nanostructured electrodes	Technique	Analyte for testing	Detection of limits	Linear range (nanometer)	References
ZnO NPs nano-silica composite	Differential pulse voltammetry	Atorvastatin	0.70 nm	2.5–6.5 × 10^5	[12]
ZnO NPs/1-butyl-3-methylimidazolium tetrafluoroborate/CPE	Square-wave voltammetry	Doxorubicin	9.0 nm	70–5.0 × 10^5	[4]
ZnO NPs/1-butyl-3-methylimidazolium tetrafluoroborate/CPE	Square-wave voltammetry	Dasatinib	0.5 nm	1000–12.0 × 10^5	[4]
5% Barium-doped zinc oxide nanoparticle/GCE	Differential pulse voltammetry	Nimesulide	1.794 nm	1.0 × 10^4–100	[13]
Cu-doped ZnO NPs	Impedimetric	Glucose	10^{-9} M	10^{-9}–10^{-5} M	[61]
ZnO/Pt–Pd nanocomposite	Voltammetric	Dengue virus	4.3 × 10^{-5}	1 × 10^{-6} M to 100 × 10^{-6} M	[87]
Silica-coated ZnO QDs	Differential pulse voltammetry	Acephate	10^{-9} M	10^{-11} M to 10^{-3} M	[74]
CuO-modified ZnO nanorods	Amperometric	Glucose	0.40 μM	8.45 mM	[3]
Fe_2O_3-functionalized ZnO nanorods	Amperometric	Glucose	0.95 mmol L^{-1}	100–400 mg dL^{-1}	[64]
Zinc oxide nanoflowers	–	Uric acid	0.13 mM	–	[25]

could be applied in a broad range of area. It was employed in many sectors because of its capacity to enhance electromechanical, physical, and chemical qualities. As shown in Table 2, numerous investigations on carbon-based nanobiosensors for use in biosensor applications have been conducted recently [29]. Sreekanth et al. investigation on the use of MWCNT reinforced nanobiosensors in the detection Cd metal in water. Cd in particular poses a major health risk to humans and is damaging to the environment too.

In this work, a nanobiosensor was created to sense Cd metal using an electrochemical method aided by DNA. In the work, dsDNA was immobilized onto the multi-walled carbon nanotube-decorated glassy carbon electrode (GCE), which was also used to conduct the experiment. Additionally, differential pulse voltammetry (DPV) study was used to look at the detection of heavy metals. When Cd (II) ions are present, dsDNA combines with Cd to create ssDNA. The decrease peak current changes noticeably when ssDNA attaches to ethyl green (EG). Increased Cd

Table 2 List of various carbon-based nanostructures as an excellent biosensor for detecting many pharmaceutical drugs, biochemicals, bacteria, and viruses

Nanostructured electrodes	Analyte for testing	Detection of limits	Linear range (nanometer)	References
Carbon nanotubes (CNT)	Dopamine	0.82 nM	2.81 – 8.3 μM	[91]
polyethyleneimine/ (CNT)	Cardiac Troponin T	33 pg mL^{-1}	0.1^{-10} ng mL^{-1}	[33]
L-lysine/multi-walled CNT	Hydrogen peroxide	8 nM	1–3.6 μM	[30]
Ag nanofilm/ multi-walled WCNT	Zidovudine HIV drug	0.15 μM	0.37 μM–1.5 mM	[80]
Conducting polymer/ CNT	miRNA	8 fM	1–10 pM	[96]
MWCNT-functionalized polyamide 6/ poly(allylamine hydrochloride)	Dopamine	1–70 μM	0.15 μM	[66]
Gold tungsten wires coated with PEI andSWCNTs	*S. aureus* and *Escherichia coli* K-12	10^2 CFU mL^{-1}	10^2–10^5 CFU mL^{-1}	[103]
Graphene quantum dots	HBV	1 nM	10–500 nM	[102]
Ruthenium/vulcan carbon-based nanoparticles	Idarubicin	9.25×10^{-9} M	5×10^{-8} M	[49]
Multi-walled carbon nanotubes/iron(III) oxide nanoparticle–chitosan nanocomposite (MWCNT/ Fe_2O_3@chitosan NC)	Axitinib	0.904 pM	6×10^{-9} and 1×10^{-6} M	[15]

concentrations are accompanied by higher decrease peak currents. Thanks to the nanobiosensor's capability to measure Cd with a detection limit of 2–50 nm and a sensitivity of 5 nM.

Furthermore, Ballen et al. concentrated on the development of cantilever biosensors for detecting cadmium. They produced nanobiosensors based on urease/GO and researched the characteristics of biosensors for the identification of Cd in their researches. The detection limit of the urease nanobiosensor is 0.03776 ppb, while the detection limit of the GO/urease nanobiosensor is 0.01831 ppb. Taşaltn et al. investigated a nanobiosensor improved with rGO produced using an acoustic microwave-assisted technique for the propamocarb trace analysis in another work. According to the study's findings, the proposed biosensor exhibits exceptional qualities like high specificity (101.1 $\mu A \mu M^{-1}$ cm^{-2}), fast reaction (1 min), a linear dynamic range (1–5 μM), and a low LOD (0.6 μM) for pesticide [93].

In another work, Elugoke et al. developed a new electrochemical biosensor for the identification of dopamine utilizing square-wave voltammetry using a modified electrode carbon containing QDs and CuO nanostructure (SWV). According to the electrochemical results, the graphene QDs-CuO nanoparticle biosensor has a lesser LOD of 25.4 M over a broad linear range of 1–180 M [29]. Furthermore, the working principle was postulated to be based on the negatively charged oxygen-containing capabilities on the electrode surface, that was absorbed the positively charged substrate. Gaidukevic et al. used a same strategy to create a GO-based sensor for dopamine sensing in the proximity of malonic acid and additives of P_2O_5. The experimental findings indicated that the suggested biosensor has a high specificity of 28.64 AM/1 cm^2 and a reduced LOD level of 0.11 M for the identification of dopamine [31]. Furthermore, it was found that sensing mechanism of the analyte's redox reaction was altered due to the switch from reversible to irreversible. Because of the switch from a reversible to such an irreversible transition, the biological recognition of the analyte's redox reaction was altered. Furthermore, the electrochemical process occurred due to surface adsorption.

3.3 Polymer-Based Nanostructures

Polymer-based nanostructures are utilized to increase cell permeability, expand treatment application, and control and regulate dose frequency and amount since their tunable nanoscale features for the improvement of biosensor efficiency [47, 107]. Although polymer-based nanostructures have stand-alone applications, they are most commonly utilized in composite-based biosensor devices and are favored in fields like drug-delivery systems, nanomedicine, catalysts, etc. Synthetic polymer, microemulsions, nanofibers, liposomes, and polymer NPs are examples of polymer-based nanostructures utilized in biosensors. Vais et al. used an thin-film-based transducer based on electro-polymerized poly(ortho-aminophenol) (POAP) to construct a new DNA sensor for the detection of *Trichomonas vaginalis.* POAP film was designed as an electrical and chemical nanotransducer and for redox active indicator. The POAP nanotranducer demonstrated long-term stability, good repeatability, strong specificity for bacterial vaginosis, and reusability in this study. POAP film has also been claimed to have substantial promise in the creation of biosensors for the immobilization of DNA in biological field and applications [22]. Singh et al. investigated a unique L-asparaginase (L-ASP) encapsulation approach using a loader and carrier of collagen alginate nanoparticles (GANp) produced by ionic gelation [85]. In the study, the NPs were investigated as a polymer nanocarriers' fiber optic-asparagine biosensor. When there is an increase in the fluorescence property of Rhodamine 6G using L-ASP as a bio-receptor, the tested biosensor detects it. According to the scientists, the polymeric nanocomposite of GANp was efficient in L-ASP immobilization and provided a specific, sensitive, and repeatable alternative for the L-ASP biosensor, which might also be employed in leukemia patients [86].

By altering the surface electrode with PEDOT: PSS/graphene, a team effectively produced an innovative poly (3, 4-ethylenedioxythiophene)-poly (styrene sulfonate) (PEDOT:PSS) structured 3D porous laser-induced graphene-based electrical and showed pH biosensor with a high sensitivity for glucose sensing. This research highlights several benefits of conductive PEDOT: PSS for biosensor applications, including superior electrical conductivity, chemical stability, and biocompatibility [105]. The creation of electrochemical biosensors based on conducting polymers for detecting proteins and nucleic acids as COVID-19 biomarkers was also emphasized for accurate and detection and diagnosis [96]. Thus, several studies on the metal, carbon, and hybrid nanostructure are still going on to provide the information at very low level or amount such as Pico molar and lower than that may further detect the disease at the early stages.

4 Conclusion

Biosensors have been used in significant sectors including biomedical diagnostics, illness monitoring, and hazard assessment as an efficient analytical tool. The most researched form of common biosensors is electrochemical biosensors, which have the advantages of a relatively low LOD, sensitivity, ease of manufacture, and simple use. The production of nano-electrochemical sensing and nanobiosensors greatly improved biosensor performance by utilizing adaptable nanostructures. Furthermore, advancements in electronic instrumentation have led to the creation of portable, multi-functional, multiplexed, and affordable digital biosensing devices, such as lab-on-chip devices, for both in-person and at-home diagnoses, now available on the market. As a result, around 200 businesses are investing in the manufacturing and marketing of biosensors and bioelectronics. However, between 85 and 90 percent of all currently available devices are used for blood patient glucose monitoring. The challenges and issues relating to sensor stability, cost, longevity, accuracy, and precision require urgent academic-industry cooperation in order to be resolved. From a different angle, future wearable biosensing technologies will make it possible to monitor and test for a variety of analytes, including as glucose, metabolites, protein, and nucleic acids, without causing any discomfort. This will eventually allow for advanced performance and self-diagnosis. Furthermore, the integration of machine learning (artificial intelligence), algorithms, nanotechnology, and nanoelectronics may lead to significant advancements in the design and fabrication of smart nanobiosensors in the far future that will satisfy the global market for biosensors and have an impact on sensor commercialization.

References

1. Abd El-Haleem HS, Hefnawy A, Hassan RYA et al (2016) Manganese dioxide-core–shell hyperbranched chitosan (MnO_2–HBCs) nano-structured screen printed electrode for enzymatic glucose biosensors. RSC Adv 6:109185–109191. https://doi.org/10.1039/C6RA24419J
2. Ahmad R, Ahn M-S, Hahn Y-B (2017a) Fabrication of a non-enzymatic glucose sensor field-effect transistor based on vertically-oriented ZnO nanorods modified with Fe_2O_3. Electrochem Commun 77:107–111. https://doi.org/10.1016/j.elecom.2017.03.006
3. Ahmad R, Tripathy N, Ahn M-S et al (2017) Highly efficient non-enzymatic glucose sensor based on CuO modified vertically-grown ZnO Nanorods on electrode. Sci Rep 7:5715. https://doi.org/10.1038/s41598-017-06064-8
4. Alavi-Tabari SAR, Khalilzadeh MA, Karimi-Maleh H (2018) Simultaneous determination of doxorubicin and dasatinib as two breast anticancer drugs uses an amplified sensor with ionic liquid and ZnO nanoparticle. J Electroanal Chem 811:84–88. https://doi.org/10.1016/j.jelechem.2018.01.034
5. Alle M, Bandi R, Sharma G, et al (2022) Gold nanoparticles spontaneously grown on cellulose nanofibrils as a reusable nanozyme for colorimetric detection of cholesterol in human serum. Int J Biol Macromol 201:686–697. https://doi.org/10.1016/j.ijbiomac.2022.01.158
6. Alle M, Bandi R, Sharma G et al (2021) Current trends in engineered gold nanoparticles for cancer therapy BT—smart nanomaterials in biomedical applications. In: Alle M, Husen A (eds) Kim J-C. Springer International Publishing, Cham, pp 1–40
7. Aydoğdu G, Zeybek DK, Pekyardımcı Ş, Kılıç E (2013) A novel amperometric biosensor based on ZnO nanoparticles-modified carbon paste electrode for determination of glucose in human serum. Artif Cells Nanomed Biotechnol 41:332–338. https://doi.org/10.3109/21691401.2012.744994
8. Bandi R, Tummala S, Dadigala R et al (2021) Role of metal-doped carbon dots in bioimaging and cancer therapy BT—smart nanomaterials in biomedical applications. In: Alle M, Husen A (eds) Kim J-C. Springer International Publishing, Cham, pp 101–123
9. Bardajee GR, Sharifi M, Mahmoodian H (2021) Novel CMC-CdTe/ZnS QDs Nanosensor for the detection of anticancer drug Epirubicin. J Fluoresc 31:651–658. https://doi.org/10.1007/s10895-021-02687-3
10. Bertok T, Lorencova L, Chocholova E, et al (2019) Electrochemical impedance spectroscopy based biosensors: mechanistic principles, analytical examples and challenges towards commercialization for assays of protein cancer biomarkers. ChemElectroChem 6:989–1003. https://doi.org/10.1002/celc.201800848
11. Bhalla N, Jolly P, Formisano N, Estrela P (2016) Introduction to biosensors. Essays Biochem 60:1–8. https://doi.org/10.1042/EBC20150001
12. Bukkitgar SD, Shetti NP, Kulkarni RM (2018) Construction of nanoparticles composite sensor for atorvastatin and its determination in pharmaceutical and urine samples. Sensors Actuators B Chem 255:1462–1470. https://doi.org/10.1016/j.snb.2017.08.150
13. Bukkitgar SD, Shetti NP, Kulkarni RM, Doddamani MR (2016) Electro-oxidation of nimesulide at 5% barium-doped zinc oxide nanoparticle modified glassy carbon electrode. J Electroanal Chem 762:37–42. https://doi.org/10.1016/j.jelechem.2015.12.023
14. Castillo J, Gáspár S, Leth S, et al (2004) Biosensors for life quality: design, development and applications. Sensors Actuators B Chem 102:179–194. https://doi.org/10.1016/j.snb.2004.04.084
15. Cetinkaya A, Kaya SI, Şenel P et al (2022) Detection of Axitinib using Multiwalled carbon nanotube-Fe_2O_3/Chitosan nanocomposite-based electrochemical sensor and modeling with density functional theory. ACS Omega 7:34495–34505. https://doi.org/10.1021/acsomega.2c04244
16. Chauhan S, Patel K, Jain P et al (2022) Matrix metalloproteinase enzyme responsive delivery of 5-Fluorouracil using Collagen-I peptide functionalized Dendrimer-Gold nanocarrier. Drug Dev Ind Pharm 48:333–342. https://doi.org/10.1080/03639045.2022.2113404

17. Chen G, Roy I, Yang C, Prasad PN (2016) Nanochemistry and nanomedicine for nanoparticle-based diagnostics and therapy. Chem Rev 116:2826–2885. https://doi.org/10.1021/acs.chemrev.5b00148
18. Cheon J, Kang N-J, Lee S-M et al (2004) No title. J Am Chem Soc 126:1950
19. da Silva W, Ghica ME, Ajayi RF, et al (2019) Tyrosinase based amperometric biosensor for determination of tyramine in fermented food and beverages with gold nanoparticle doped poly(8-anilino-1-naphthalene sulphonic acid) modified electrode. Food Chem 282:18–26. https://doi.org/10.1016/j.foodchem.2018.12.104
20. Dadigala R, Bandi R, Alle M et al (2021) Graphene-based smart nanomaterials for Photothermal therapy BT—smart nanomaterials in biomedical applications. In: Alle M, Husen A (eds) Kim J-C. Springer International Publishing, Cham, pp 125–153
21. de Brito AR, dos Santos RN, Oliveira PC et al (2020) Development of amperometric biosensor in modified carbon paste with enzymatic preparation based on lactase immobilized on carbon nanotubes. J Food Sci Technol 57:1342–1350. https://doi.org/10.1007/s13197-019-04168-9
22. Dehdari Vais R, Heli H, Sattarahmady N, Barazesh A (2022) A novel and ultrasensitive label-free electrochemical DNA biosensor for Trichomonas vaginalis detection based on a nanostructured film of poly(ortho-aminophenol). Synth Met 287:117082. https://doi.org/10.1016/j.synthmet.2022.117082
23. Dong Y, Luo X, Liu Y, et al (2022) A disposable printed amperometric biosensor for clinical evaluation of creatinine in renal function detection. Talanta 248:123592. https://doi.org/10.1016/j.talanta.2022.123592
24. Du JZ, Du XJ, Mao CQ, Wang J (2011) Tailor-Made dual pH-sensitive polymer-doxorubicin nanoparticles for efficient anticancer drug delivery. J Am Chem Soc 133:17560–17563. https://doi.org/10.1021/ja207150n
25. Dutta P, Sharma V, Bhardwaj H et al (2022) Fabrication of electrochemical biosensor using zinc oxide Nanoflowers for the detection of uric acid. Mapan 37:585–595. https://doi.org/10.1007/s12647-022-00598-7
26. Dutta SD, Ganguly K, Bandi R, Alle M (2021) A new era of cancer treatment: carbon nanotubes as drug delivery tools BT—smart nanomaterials in biomedical applications. In: Alle M, Husen A (eds) Kim J-C. Springer International Publishing, Cham, pp 155–171
27. Eivazzadeh-Keihan R, Bahojb Noruzi E, Chidar E, et al (2022) Applications of carbon-based conductive nanomaterials in biosensors. Chem Eng J 442:136183. https://doi.org/10.1016/j.cej.2022.136183
28. El-Raheem HA, Hassan RYA, Khaled R et al (2021) New sensing platform of poly(ester-urethane)urea doped with gold nanoparticles for rapid detection of mercury ions in fish tissue. RSC Adv 11:31845–31854. https://doi.org/10.1039/D1RA03693A
29. Elugoke SE, Adekunle AS, Fayemi OE, et al (2020) Carbon-based quantum dots for electrochemical detection of monoamine neurotransmitters—review. Biosensors 10
30. Ezhil Vilian AT, Chen S-M, Lou B-S (2014) A simple strategy for the immobilization of catalase on multi-walled carbon nanotube/poly (l-lysine) biocomposite for the detection of H_2O_2 and iodate. Biosens Bioelectron 61:639–647. https://doi.org/10.1016/j.bios.2014.05.023
31. Gaidukevic J, Aukstakojyte R, Barkauskas J, et al (2022) A novel electrochemical sensor based on thermally reduced graphene oxide for the sensitive determination of dopamine. Appl Surf Sci 592:153257. https://doi.org/10.1016/j.apsusc.2022.153257
32. George Kerry R, Ukhurebor KE, Kumari S et al (2021) A comprehensive review on the applications of nano-biosensor-based approaches for non-communicable and communicable disease detection. Biomater Sci 9:3576–3602. https://doi.org/10.1039/D0BM02164D
33. Gomes-Filho SLR, Dias ACMS, Silva MMS, et al (2013) A carbon nanotube-based electrochemical immunosensor for cardiac troponin T. Microchem J 109:10–15. https://doi.org/10.1016/j.microc.2012.05.033
34. Grieshaber D, MacKenzie R, Vörös J, Reimhult E (2008) Electrochemical biosensors—sensor principles and architectures. Sensors 8:1400–1458
35. Haleem A, Javaid M, Singh RP, et al (2021) Biosensors applications in medical field: a brief review. Sensors Int 2:100100. https://doi.org/10.1016/j.sintl.2021.100100

36. Hassan RYA (2022) Advances in electrochemical nano-biosensors for biomedical and environmental applications: from current work to future perspectives. Sensors 22
37. Hassan RYA, Bilitewski U (2011) A viability assay for Candida albicans based on the electron transfer mediator 2,6-dichlorophenolindophenol. Anal Biochem 419:26–32. https://doi.org/10.1016/j.ab.2011.07.025
38. Hassan RYA, Hassan HNA, Abdel-Aziz MS, Khaled E (2014) Nanomaterials-based microbial sensor for direct electrochemical detection of Streptomyces Spp. Sensors Actuators B Chem 203:848–853. https://doi.org/10.1016/j.snb.2014.07.059
39. Hassan RYA, Mekawy MM, Ramnani P, Mulchandani A (2017a) Monitoring of microbial cell viability using nanostructured electrodes modified with Graphene/Alumina nanocomposite. Biosens Bioelectron 91:857–862. https://doi.org/10.1016/j.bios.2017.01.060
40. Hassan RYA, Sultan MA, El-Alamin MMA, et al (2017b) A disposable carbon nanotubes-screen printed electrode (CNTs-SPE) for determination of the antifungal agent Posaconazole in biological samples. Electroanalysis 29:843–849. https://doi.org/10.1002/elan.201600621
41. Holzinger M, Le Goff A, Cosnier S (2014) Nanomaterials for biosensing applications: a review. Front Chem 2:63. https://doi.org/10.3389/fchem.2014.00063
42. Houtan J, Byungjoo C, SangHyun L (2019) Application of wearable biosensors to construction sites. I: assessing workers' stress. J Constr Eng Manag 145:4019079. https://doi.org/10.1061/(ASCE)CO.1943-7862.0001729
43. Jain P, Bhagat S, Tunki L et al (2020) Serotonin-stearic acid Bioconjugate-coated completely biodegradable Mn_3O_4 Nanocuboids for hepatocellular carcinoma targeting. ACS Appl Mater Interfaces 12:10170–10182. https://doi.org/10.1021/acsami.0c00331
44. Jain P, Patel K, Jangid AK, et al (2021) Biotinylated Mn3O4 nanocuboids for targeted delivery of gemcitabine hydrochloride to breast cancer and MRI applications. Int J Pharm 606:120895. https://doi.org/10.1016/j.ijpharm.2021.120895
45. Jain P, Patel K, Jangid AK, et al (2020b) Modulating the delivery of 5-fluorouracil to human colon cancer cells using multifunctional arginine-coated manganese oxide nanocuboids with MRI properties. ACS Appl Bio Mater.https://doi.org/10.1021/acsabm.0c00780
46. Jung D-U-J, Ahmad R, Hahn Y-B (2018) Nonenzymatic flexible field-effect transistor based glucose sensor fabricated using NiO quantum dots modified ZnO nanorods. J Colloid Interface Sci 512:21–28. https://doi.org/10.1016/j.jcis.2017.10.037
47. Kankala RK, Liu CG, Yang DY et al (2020) Ultrasmall platinum nanoparticles enable deep tumor penetration and synergistic therapeutic abilities through free radical species-assisted catalysis to combat cancer multidrug resistance. Chem Eng J 383:123138. https://doi.org/10.1016/j.cej.2019.123138
48. Kashish GS, Dubey SK, Prakash R (2015) Genosensor based on a nanostructured, platinum-modified glassy carbon electrode for Listeria detection. Anal Methods 7:2616–2622. https://doi.org/10.1039/C5AY00167F
49. Kaya SI, Karadurmuş L, Cetinkaya A, et al (2022) Types of biosensors and their importance in cardiovascular applications. In: Predicting heart failure, pp 47–79
50. Kim DC, Kang DJ (2008) Molecular recognition and specific interactions for biosensing applications. Sensors 8:6605–6641
51. Kobun RBT-AFAT (ed) (2021) Chapter 7—biosensor technology to detect chemical contamination in food. Academic Press, pp 127–146
52. Kourani K, Jain P, Kumar A et al (2022) Inulin coated Mn_3O_4 nanocuboids coupled with RNA interference reverse intestinal tumorigenesis in Apc knockout murine colon cancer models. Nanomed Nanotechnol Biol Med 40:102504. https://doi.org/10.1016/j.nano.2021.102504
53. Kuzin Y, Porfireva A, Stepanova V, et al (2015) Impedimetric detection of DNA damage with the sensor based on silver nanoparticles and neutral red. Electroanalysis 27:2800–2808. https://doi.org/10.1002/elan.201500312
54. Lee W-I, Subramanian A, Mueller S et al (2022) Potentiometric biosensors based on molecular-imprinted self-assembled monolayer films for rapid detection of influenza a virus and SARS-CoV-2 spike protein. ACS Appl Nano Mater 5:5045–5055. https://doi.org/10.1021/acsanm.2c00068

55. Leva-Bueno J, Peyman SA, Millner PA (2020) A review on impedimetric immunosensors for pathogen and biomarker detection. Med Microbiol Immunol 209:343–362. https://doi.org/10.1007/s00430-020-00668-0
56. Li X, Deng Y, Tang W, et al (2018) Urban-rural disparity in cancer incidence, mortality, and survivals in Shanghai, China, during 2002 and 2015. Front Oncol 8. https://doi.org/10.3389/fonc.2018.00579
57. Locatelli C, Fagioli F, Bighi C, Garai T (1984) A critical comparison of hanging mercury drop electrode and long lasting sessile drop mercury electrode in A.C. Anodic Stripping Voltammerty Anal Lett 17:623–650. https://doi.org/10.1080/00032718408069091
58. Madianos L, Tsekenis G, Skotadis E, et al (2018) A highly sensitive impedimetric aptasensor for the selective detection of acetamiprid and atrazine based on microwires formed by platinum nanoparticles. Biosens Bioelectron 101:268–274. https://doi.org/10.1016/j.bios.2017.10.034
59. Magar HS, Hassan RYA, Mulchandani A (2021) Electrochemical impedance spectroscopy (EIS): principles, construction, and biosensing applications. Sensors 21
60. Magar HS, Magd EEA-E, Hassan RYA, Fahim AM (2022) Rapid impedimetric detection of cadmium ions using Nanocellulose/ligand/nanocomposite (CNT/Co3O4). Microchem J 182:107885. https://doi.org/10.1016/j.microc.2022.107885
61. Mahmoud A, Echabaane M, Omri K, et al (2019) Development of an impedimetric non enzymatic sensor based on ZnO and Cu doped ZnO nanoparticles for the detection of glucose. J Alloys Compd 786:960–968. https://doi.org/10.1016/j.jallcom.2019.02.060
62. Mallick S, Singh KRB, Nayak V, et al (2022) Potentialities of core@shell nanomaterials for biosensor technologies. Mater Lett 306:130912. https://doi.org/10.1016/j.matlet.2021.130912
63. Manickam A, Chevalier A, McDermott M et al (2010) A CMOS electrochemical impedance spectroscopy (EIS) biosensor array. IEEE Trans Biomed Circuits Syst 4:379–390. https://doi.org/10.1109/TBCAS.2010.2081669
64. Marie M, Manoharan A, Kuchuk A et al (2018) Vertically grown zinc oxide nanorods functionalized with ferric oxide for in vivo and non-enzymatic glucose detection. Nanotechnology 29:115501. https://doi.org/10.1088/1361-6528/aaa682
65. Mehrotra P (2016) Biosensors and their applications—a review. J Oral Biol Craniofacial Res 6:153–159. https://doi.org/10.1016/j.jobcr.2015.12.002
66. Mercante LA, Pavinatto A, Iwaki LEO et al (2015) Electrospun Polyamide 6/Poly(allylamine hydrochloride) nanofibers functionalized with carbon nanotubes for electrochemical detection of dopamine. ACS Appl Mater Interfaces 7:4784–4790. https://doi.org/10.1021/am508709c
67. Millner PA, Caygill RL, Conroy DJR, Shahidan MA (2012) 4—Impedance interrogated affinity biosensors for medical applications: novel targets and mechanistic studies. In: Higson SBT-B for MA (ed) Woodhead Publishing Series in Biomaterials. Woodhead Publishing, pp 103–134
68. Mohammad-Razdari A, Ghasemi-Varnamkhasti M, Izadi Z et al (2019) An impedimetric aptasensor for ultrasensitive detection of Penicillin G based on the use of reduced graphene oxide and gold nanoparticles. Mikrochim Acta 186:372. https://doi.org/10.1007/s00604-019-3510-x
69. Mohammadi-Manesh E, Vaezzadeh M, Saeidi M (2015) Cu- and CuO-decorated graphene as a nanosensor for H2S detection at room temperature. Surf Sci 636:36–41. https://doi.org/10.1016/j.susc.2015.02.002
70. Mustafa F, Hassan RYA, Andreescu S (2017) Multifunctional nanotechnology-enabled sensors for rapid capture and detection of pathogens. Sensors 17
71. Nadzirah S, Hashim U, Gopinath SCB, et al (2020) Titanium dioxide–mediated resistive nanobiosensor for E. coli O157:H7. Microchim Acta 187:235. https://doi.org/10.1007/s00604-020-4214-y
72. Naresh V, Lee N (2021) A review on biosensors and recent development of nanostructured materials-enabled biosensors. Sensors 21
73. Nguyen HH, Lee SH, Lee UJ, et al (2019) Immobilized enzymes in biosensor applications. Mater (Basel, Switzerland) 12.https://doi.org/10.3390/ma12010121

74. Pabbi M, Mittal SK (2017) An electrochemical algal biosensor based on silica coated ZnO quantum dots for selective determination of acephate. Anal Methods 9:1672–1680. https://doi.org/10.1039/C7AY00111H
75. Plata MR, Contento AM, Ríos A (2010) State-of-the-art of (Bio)chemical sensor developments in analytical spanish groups. Sensors 10:2511–2576
76. Plekhanova Y, Tarasov S, Kitova A, et al (2022) Modification of thermally expanded graphite and its effect on the properties of the amperometric biosensor. 3 Biotech 12:42. https://doi.org/10.1007/s13205-021-03107-w
77. Polat EO, Cetin MM, Tabak AF, et al (2022) Transducer technologies for biosensors and their wearable applications. Biosensors 12
78. Poletti F, Scidà A, Zanfrognini B, et al (2022) Graphene-paper-based electrodes on plastic and textile supports as new platforms for Amperometric Biosensing. Adv Funct Mater 32:2107941. https://doi.org/10.1002/adfm.202107941
79. Putzbach W, Ronkainen NJ (2013) Immobilization techniques in the fabrication of nanomaterial-based electrochemical biosensors: a review. Sensors 13:4811–4840
80. Rafati AA, Afraz A (2014) Amperometric sensing of anti-HIV drug zidovudine on Ag nanofilm-multiwalled carbon nanotubes modified glassy carbon electrode. Mater Sci Eng C 39:105–112. https://doi.org/10.1016/j.msec.2014.02.037
81. Rick J, Tsai M-C, Hwang BJ (2016) Biosensors incorporating bimetallic nanoparticles. Nanomaterials 6
82. Saylan Y, Erdem Ö, Ünal S, Denizli A (2019) An alternative medical diagnosis method: biosensors for virus detection. Biosensors 9
83. Shkotova L, Bohush A, Voloshina I et al (2019) Amperometric biosensor modified with platinum and palladium nanoparticles for detection of lactate concentrations in wine. SN Appl Sci 1:306. https://doi.org/10.1007/s42452-019-0315-9
84. Shul'ga AA, Soldatkin AP, El'skaya A V, et al (1994) Thin-film conductometric biosensors for glucose and urea determination. Biosens Bioelectron 9:217–223. https://doi.org/10.1016/0956-5663(94)80124-X
85. Singh A, Verma N, Kumar K (2022a) Fabrication and characterization of L-asparaginase entrapped polymeric nanoparticles for asparagine biosensor construction. Mater Today Proc 67:591–597. https://doi.org/10.1016/j.matpr.2022.05.227
86. Singh A, Verma N, Kumar K (2022) Fabrication and construction of highly sensitive polymeric nanoparticle-based electrochemical biosensor for asparagine detection. Curr Pharmacol Reports 8:62–71. https://doi.org/10.1007/s40495-021-00271-8
87. Singhal C, Pundir CS, Narang J (2017) A genosensor for detection of consensus DNA sequence of Dengue virus using ZnO/Pt-Pd nanocomposites. Biosens Bioelectron 97:75–82. https://doi.org/10.1016/j.bios.2017.05.047
88. Soldatkin OO, Peshkova VM, Saiapina OY, et al (2013) Development of conductometric biosensor array for simultaneous determination of maltose, lactose, sucrose and glucose. Talanta 115:200–207. https://doi.org/10.1016/j.talanta.2013.04.065
89. Staerz A, Weimar U, Barsan N (2016) Understanding the potential of WO3 based sensors for breath analysis. Sensors 16
90. Sun X, Sun X, Wang Q, et al (2022) Biosensors toward behavior detection in diagnosis of alzheimer's disease. Front Bioeng Biotechnol 10.https://doi.org/10.3389/fbioe.2022.1031833
91. Suzuki I, Fukuda M, Shirakawa K, et al (2013) Carbon nanotube multi-electrode array chips for noninvasive real-time measurement of dopamine, action potentials, and postsynaptic potentials. Biosens Bioelectron 49:270–275. https://doi.org/10.1016/j.bios.2013.05.023
92. Švancara I, Vytřas K, Barek J, Zima J (2001) Carbon paste electrodes in modern electroanalysis. Crit Rev Anal Chem 31:311–345. https://doi.org/10.1080/20014091076785
93. Taşaltın N, Karakuş S, Taşaltın C, Baytemir G (2022) Highly sensitive and selective rGO based Non-Enzymatic electrochemical sensor for propamocarb fungicide pesticide detection. Food Chem 372:131267. https://doi.org/10.1016/j.foodchem.2021.131267
94. Thakur MS, Ragavan KV (2013) Biosensors in food processing. J Food Sci Technol 50:625–641. https://doi.org/10.1007/s13197-012-0783-z

95. Thanh NTK, Maclean N, Mahiddine S (2014) Mechanisms of nucleation and growth of nanoparticles in solution. Chem Rev 114:7610–7630. https://doi.org/10.1021/cr400544s
96. Tran DH, Cuong HQ, Tran HT, et al (2021) A comparative study of isothermal nucleic acid amplification methods for SARS-CoV-2 detection at point-of-care. bioRxiv 2020.05.24.113423. https://doi.org/10.1101/2020.05.24.113423
97. Vidic J, Manzano M (2021) Electrochemical biosensors for rapid pathogen detection. Curr Opin Electrochem 29:100750. https://doi.org/10.1016/j.coelec.2021.100750
98. Walker NL, Roshkolaeva AB, Chapoval AI, Dick JE (2021) Recent advances in potentiometric biosensing. Curr Opin Electrochem 28:100735. https://doi.org/10.1016/j.coelec.2021.100735
99. Wang C, Liu M, Wang Z, et al (2021) Point-of-care diagnostics for infectious diseases: From methods to devices. Nano Today 37:101092. https://doi.org/10.1016/j.nantod.2021.101092
100. Wang Y, Duncan T V (2017) Nanoscale sensors for assuring the safety of food products. Curr Opin Biotechnol 44:74–86. https://doi.org/10.1016/j.copbio.2016.10.005
101. Watanabe H, Omoto S, Hoshi Y, et al (2021) Electrochemical impedance analysis on positive electrode in lithium-ion battery with galvanostatic control. J Power Sources 507:230258. https://doi.org/10.1016/j.jpowsour.2021.230258
102. Xiang Q, Huang J, Huang H et al (2018) A label-free electrochemical platform for the highly sensitive detection of hepatitis B virus DNA using graphene quantum dots. RSC Adv 8:1820–1825. https://doi.org/10.1039/C7RA11945C
103. Yamada K, Choi W, Lee I, et al (2016) Rapid detection of multiple foodborne pathogens using a nanoparticle-functionalized multi-junction biosensor. Biosens Bioelectron 77:137–143. https://doi.org/10.1016/j.bios.2015.09.030
104. Yoon J, Lim J, Shin M, et al (2022) Recent progress in nanomaterial-based bioelectronic devices for biocomputing system. Biosens Bioelectron 212:114427. https://doi.org/10.1016/j.bios.2022.114427
105. Zahed MA, Barman SC, Das PS, et al (2020) Highly flexible and conductive poly (3, 4-ethylene dioxythiophene)-poly (styrene sulfonate) anchored 3-dimensional porous graphene network-based electrochemical biosensor for glucose and pH detection in human perspiration. Biosens Bioelectron 160:112220. https://doi.org/10.1016/j.bios.2020.112220
106. Zhang Y, Wei Q (2016) The role of nanomaterials in electroanalytical biosensors: A mini review. J Electroanal Chem 781:401–409. https://doi.org/10.1016/j.jelechem.2016.09.011
107. Zhang Z, Li Q, Du X, Liu M (2020) Application of electrochemical biosensors in tumor cell detection. Thorac Cancer 11:840–850. https://doi.org/10.1111/1759-7714.13353
108. Zhou C, Xu L, Song J et al (2014) Ultrasensitive non-enzymatic glucose sensor based on three-dimensional network of ZnO–CuO hierarchical nanocomposites by electrospinning. Sci Rep 4:7382. https://doi.org/10.1038/srep07382
109. Zhou Z, Liu X, Zhu D, et al (2017) Nonviral cancer gene therapy: delivery cascade and vector nanoproperty integration. Adv Drug Deliv Rev 115:115–154. https://doi.org/10.1016/j.addr.2017.07.021

State of the Art in Integrated Biosensors for Organ-on-a-Chip Applications

Tanima Dey, Priya Mitra, Binita Chakraborty, Arka Sanyal, Aditi Acharjee, Anushikha Ghosh, and Dindyal Mandal

Abstract Organ-on-a-chip (OoaC) is a potential emerging technology in the healthcare business based on bioengineered tissue constructions combined with microfluidics. The chip's microenvironment mimics tissue interfaces and mechanical stimulation of the organ, imitating its form and function using a synergy of cell biology, engineering, and biomaterial technologies. In this chapter, we discussed the sorts of biosensors required for various OoaC applications. OoaC critically monitors tissue structures, functionality, and metabolic activity of the cells to verify the cellular behavior and environments which mimic in vivo physiology. This chapter introduces the notion of OoaC about several organs. Application of OoaC in disease modeling concerning different organs, drug discovery and development, individualized precision medicine, organ transplantation, and cancer diagnosis and treatment will revolutionize the healthcare and bioengineering industries. In addition, we addressed the present challenges and prospects for OoaC applications.

Keywords Organ-on-a-chip · Biosensors · Bioengineering · Microfluidics · Drug discovery and development · Precision medicine · Cancer diagnosis and treatment

Abbreviations

ADME	Absorption, distribution, metabolism, and excretion
BBB	Blood-brain-barrier
BMOC	Bone marrow-on-a-chip
BOC	Brain-on-a-chip
CF	Cystic fibrosis

Equal first author contribution: Tanima Dey and Priya Mitra.

T. Dey · P. Mitra · B. Chakraborty · A. Sanyal · A. Acharjee · A. Ghosh · D. Mandal (✉)
KIIT School of Biotechnology, Kalinga Institute of Industrial Technology (KIIT-DU), Bhubaneswar, Odisha, India
e-mail: dmandal@kiitbiotech.ac.in

A. Madhusudhan et al. (eds.), *Functional Smart Nanomaterials and Their Theranostics Approaches*, Smart Nanomaterials Technology,
https://doi.org/10.1007/978-981-99-6597-7_10

CFTR	CF transmembrane conductance regulator
CLLE	*Curcuma longa* leaf extract
COPD	Chronic obstructive pulmonary disease
DLM	Decellularized liver matrices
DO	Dissolved oxygen
EC	Electrochemical
ECIS	Electrical cell-substrate impedance sensing
GelMA	Gelatin methacryloyl
GOC	Gut-on-a-chip
HSPC	Hematopoietic stem cells and progenitor cells
Hz	Hertz
H_2O_2	Hydrogen peroxide
IL-6	Interleukin 6
IOC	Intestine-on-a-chip
iPSCs	Induced pluripotent stem cells
LED	Light-emitting diode
LOD	Limits of detection
MEA	Multi-electrode arrays
NIRS	Near-infrared spectroscopy
OoaC	Organ-on-a-chip
PDMS	Polydimethylsiloxane
POC	Pancreas-on-a-chip
TEER	Transepithelial/transendothelial electrical resistance
TGF-β1	Transforming growth factor—β1
TNf-α	Tumor necrosis factor—α
2D	2-Dimensional
3D	3-Dimensional

1 Introduction

An "organ-on-a-chip" (OoaC) is a bioengineered microfluidic platform that replicates particular attributes, such as the morphology, cellular organization, and functions of in vivo organs and tissues of human [1]. An OoaC with numerous compartments connected by microfluidic channels reflects the multi-cellular nature of the smallest functional unit of a complex organ [2]. By observing different types of cultivated tissue in real-time, OoaC systems typically simulate the dynamics, physiological reactions, and functioning of organs. Four primary components of an integrated OoaC (Fig. 1) include a microfluidic component, cells/tissues cultured in chip, stimulus to develop the tissues, and sensors to interprete results. Biosensors coupled with OoaCs are potential alternatives to monitor simulated human physiology in a less invasive way, thereby improving the functional relevance of in vivo counterparts of OoaC models. Figure 2 depicts the flow of the notion of developing an OoaC.

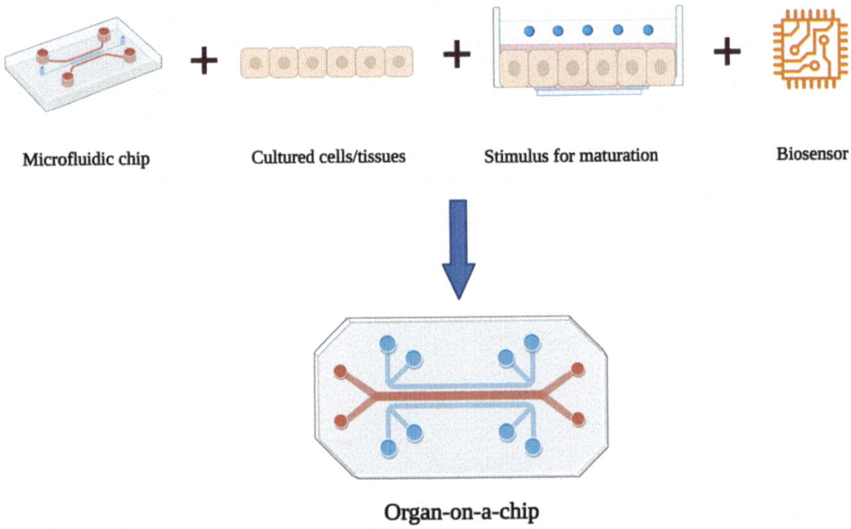

Fig. 1 Basic parts of OoaC are illustrated. OoaC consists of microfluidic chip, cultured cell/tissues, stimulus for maturation, and integration of biosensors for result interpretation. [Created with Bio Render.com]

Microfabrication techniques used for OoaC development have been discussed in Table 1. OoaC systems, combining microfluidics and 3D cell culture concepts [3, 4], first surfaced in 2003, and they mimic in vitro interactions between cells or between cells and their environments [5, 6]. The properties of the cultured cells/tissues in OoaC must be carefully controlled as they directly affect the survival, proliferation, organization, differentiation, and function of the cells. As a result, it is crucial to gauge the physical conditions and cellular activities of the cultivated cells. Furthermore, employing biosensors in conjunction with OoaC, the response to outside stimuli may be monitored. In preclinical testing, OoaCs can take the place of conventional 2D cell culture and animal models [7]. Given that OoaC models may be used to specifically predict pharmaceutical reactions, stem cells provide vast potential for drug discovery and personalized medicine applications [8]. Various applications of different types of OoaC have been shown in Fig. 3.

This chapter concentrates on cutting-edge technology to produce OoaC. Initially, we briefly discuss various biosensing techniques that are particularly relevant in monitoring physical parameters of microenvironment, cellular activities, and response to external stimuli within OoaC models. The advantages and limitations associated with different types of OoaC have been summarized in Table 2. The primary benefit of OoaCs is the ability to model the physiology of the human organ, which revolutionize biological research like cancer treatment, and transplantation of organ. Furthermore, by adopting OoaCs, the shortcomings of traditional drug development and screening techniques will be considerably reduced, allowing the production of efficient personalized medicines. OoaCs can also be employed in drug

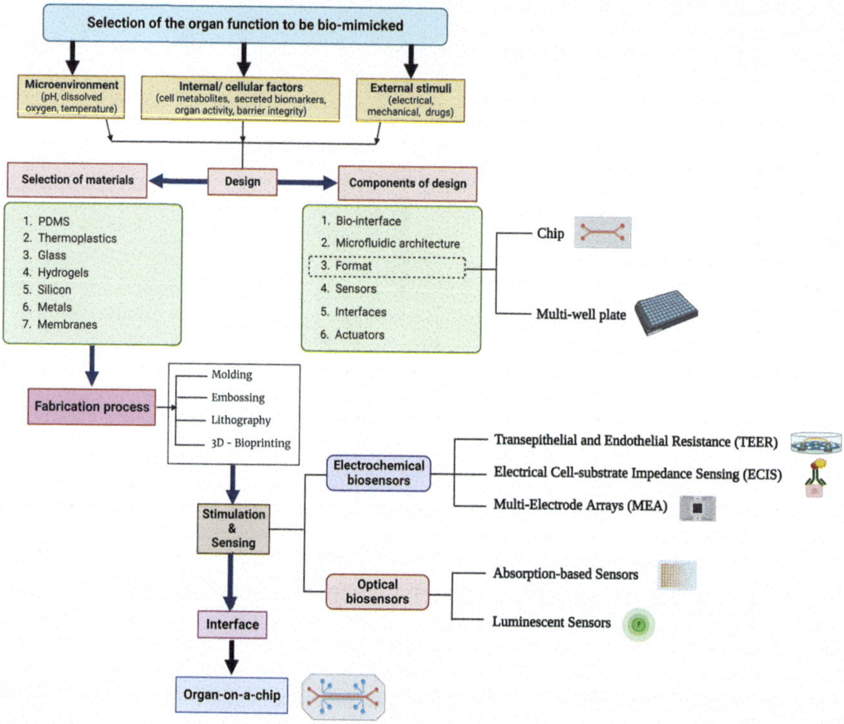

Fig. 2 Concept of designing OoaC based on the function to be bio-mimicked. [Created with Bio Render.com]

absorption, distribution, metabolism, excretion (ADME), and toxicity research and in vivo pharmacokinetics and pharmacodynamics [19]. Finally, we discussed key hurdles and probable future scenarios of OoaC applications for greater avenues such as body-on-a-chip surveillance, as well as the integration of microrobotics and artificial intelligence in OoaC devices.

2 Types of Biosensors for OoaC Applications

A biosensor is defined as a device that generates a quantifiable signal proportional to a target analyte's concentration, derived from biological interactions [50]. Recently, biosensor research and development has flourished extensively due to its simplicity, rapidity, inexpensiveness, highly sensitivity, and selectivity. It contributes significantly in personalized medicine and ultra-sensitive point-of-care detection of disease markers. Concerning OoaC applications, two types of biosensors have been extensively researched and shown limited yet promising breakthroughs: electrochemical biosensors and optical biosensors.

Table 1 Different methods of fabrication used for various OoaC development

Methods of fabrication	Materials used	Organs employed	Advantages	Disadvantages	Refs.
Soft lithography (fabricated by combining photo-lithography and molding)	Elastomers like PDMS	Brain-on-a-chip, kidney-on-a-chip, gut-and-liver-on-a-chip	– Quick prototyping – Inexpensive	– Dependency on other lithographic processes	[9, 10]
Hot embossing (master mold fabrication)	Thermoplastic materials like PMMA	–	– Inexpensive – Suitable and flexible method of microfabrication – Polymeric microstructure which has high aspect ratio	Precise controlling of the temperature	[11, 12]
Injection molding (master mold fabrication)	Polymers	Human umbilical endothelial cells (HUVEC), lung fibroblasts	– High-precision microfabrication	Need to control injection rate, temperature, and pressure	[13, 14]
3D printing (“layer-by-layer approach”)	PDMS	Liver, nervous system	– Rapid prototyping – Direct fabrication – Precise control	Smallest dimensions, not compatible with all materials	[15, 16]
Creative methods (used in preliminary experimentations)	Polymers	BBB-on-chip, intestine-on-chip,	– Easy implementation – Inexpensive	Large uncertain geometry of materials	[17, 18]

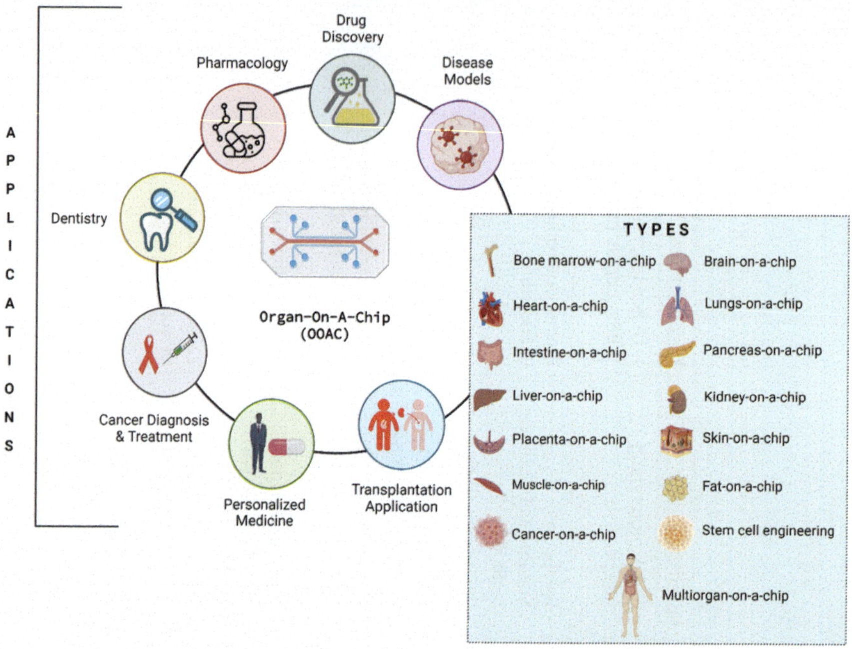

Fig. 3 Wide range applications of different types of OoaC. [Created with BioRender.com]

2.1 *Electrochemical Biosensors*

Electrochemical (EC) biosensors are analytical devices that use their unique binding properties and biocatalytic activity [51, 52] to identify and convert a variety of biorecognition components such as enzymes, antibodies, nucleic acids, aptamers, cells, microbes, and biochemical events into electrical signals such as current, voltage, impedance, and so on. An immunosensor detects binding events of antigen–antibody in bioelectrochemical processes by evaluating measurable current (amperometric), detectable potential accumulation (potentiometric), or quantitatively changing the conductive characteristics of material between electrodes (conductometric) [53]. Low limits of detection (LOD) and large values for dynamic precision can be obtained by integrating electrochemical biosensors in OoaC models [54]. Amperometry and voltammetry-based EC biosensors have been widely employed for real-time monitoring of the physical properties of the microenvironment and identifying metabolic indicators in OoaC models [55]. Various electrochemical sensing techniques have been used in the context of OoaC applications.

Table 2 Advantages and disadvantages of different OoaC systems

Organ-on-a-chip	Applications	Advantages	Disadvantages	Refs.
Bone marrow-on-a-chip	– Used to process the multi-potent hematopoietic stem cells and progenitor cells (HSPC) – Enables researchers to evaluate the effectiveness of drugs developed	– Preserve the developed bone marrow – Keep track of blood cell populations long after radiation damage – Tests immunotoxicity and immunogenicity in long-term culture with repeated antigen stimulation	– Does not completely represent physiological bone marrow – Unable to maintain for more than 14 days	[20]
Brain-on-a-chip	– Assess the neurotoxicity of any drug – Studies the mechanism of neurological disorders such as Alzheimer's disease – Studies neuronal cell migration in response to chemotactic factors	– Limitations of existing 2D and 3D models have been addressed	– Difficult to simulate exact brain since it has a complex network of interconnected cells, including astrocytes, microglia, neurons, and oligodendrons – Non-standard protocol – Complex designs for manufacturing – Expensive fabrication – Unable to model cerebrovascular interface	[21, 22]
Heart-on-a-chip	– Monitors effects of drugs on heart cells – Studies various cardiac diseases – Helps in drug screening and testing	– Real-time monitoring is possible	– Challenging to replicate physiological environment made up of cardiomyocytes with such complicated features of polarization and electric impulses	[23, 24]

(continued)

Table 2 (continued)

Organ-on-a-chip	Applications	Advantages	Disadvantages	Refs.
BBB-on-a-chip	– Comprehends numerous methods of drug delivery – Investigated nanoparticle distribution of and cellular response	– Continuously monitors inflammation and cell damage – Identifies markers associated with neurodegenerative diseases	– Limited scalability – Complex fabrication	[25, 26]
Lungs-on-a-chip	– Simulates lung diseases such as asthma, lung cancer, pulmonary fibrosis, etc	– Studies the effects of environmental toxins, absorption of therapeutics – Tests safety and efficacy of novel drugs	– Functional period of lung-on-a-chip is only 4 weeks – Limits modeling of some chronic disease condition, such as COPD	[27– 29]
Intestine-on-a-chip and gut-on-a-chip	– Aids in understanding intestinal inflammation and the screening of anti-inflammatory substances – Investigates microbiome, intestinal pathophysiology	– Helps personalized medicine development – Helps in microbiome study – Enterovirus infection is well studied	– Adsorption of lipophilic compounds onto device's walls – Difficulty in as permeability limits the development of IOC, GOC	[30–32]
Pancreas-on-a-chip	– Studies diabetic and cystic fibrosis-related diseases	– Real-time evaluation platform for assessing islet potency and quality	– Difficulty in Islet culturing – Difficult to maintain homogeneity	[35, 34]
Liver-on-a-chip	– Studies drug and toxin metabolisms	– Stable – Provides an improved model for anticancer drug pharmacological research	– Low throughput – Cannot simulate liver-specific functions	[35–[37]
Kidney-on-a-chip	– Evaluates nephrotoxicity – Design therapeutic – Study kidney diseases	– Real-time monitoring – Quantitative analysis of various biological processes – Study effect of drugs	– Limited life span of cells on-chip – Difficult to optimize biomaterial used	[38, 39]

(continued)

Table 2 (continued)

Organ-on-a-chip	Applications	Advantages	Disadvantages	Refs.
Placenta-on-a-chip	– Used numerous in vivo and ex vivo studies	– Demonstrates a functioning placenta	– Difficult to design – Sometimes provide inconsistent results and inaccurate predictions	[40–42]
Skin-on-a-chip	– Aid in understanding numerous immune response mechanisms that take place in human skin – Tests drug efficacy and toxicity	– Portable – Inexpensive – High throughput – 3D fabrication used	– Incapable to recapitulate heterogenous nature of human skin	[43]
Muscle-on-a-chip	– Monitors drug effect on muscle	– Assess the contractility of both smooth and striated muscles at the same time through an automated projection tracking system	– Decreased fabrication efficiency – Planar vascular geometry with the constrained resolution is an issue	[44, 45]
Fat-on-a-chip	– Examines relationship between immune cells and adipocytes	– Device's membrane separated media channel and cellular chamber – Protects cells from shear stress	– Uncontrolled parameters – Maintains homeostasis by compartmentalization	[46]
Cancer-on-a-chip	– Diagnoses and treats cancer – Studies drug toxicity – Manufacturing personalized cancer models	– Replicates tumor microenvironments – Helps to study interaction of cancer and immune response to it – Studies drug efficacy and personalized medicine	– Limited recombination ability	[47]
Stem cells in OoaC	– Used in stem cell research	– Embryonic stem cells, iPSCs, ASCs can be differentiated and integrated into microfluidic chips	– Mesenchymal stem cells are less useful in this chip	[48, 49]

2.1.1 Transepithelial/Transendothelial Resistance

The resistance across probes on either side of a selectively permeable membrane, when a biological barrier is created by developing endothelium or epithelial cells, is known as transepithelial/transendothelial resistance (TEER). It is the most prevalent kind of electrical sensing in OoaC applications. This is a label-free, fast, and extremely sensitive technique that measures resistance/impedance across a wide frequency range [56]. A high TEER value indicates a tight junction formation. It helps to quantitatively assess the integrity of tight junction dynamics in endothelial and epithelial monolayer cell culture models, ensuring effective in vivo permeability tests [57]. It is vital to monitor barrier function and integrity while developing endothelium or epithelial layers to simulate functioning organoid cell cultures. Such approach has been used to describe OoaC models containing biological barrier including blood–brain-barrier (BBB), the air–liquid interface in lung-on-a-chip models, and the intestinal epithelial barrier in gut-on-a-chip models [58].

2.1.2 Electrical Cell-Substrate Impedance Sensing

Electrical cell-substrate impedance sensing (ECIS) is described by Giaever and Keese as cell impedance and impedance arising from cell-to-substrate contact that may be monitored by designed probes [59]. The ECIS technology measures the impedance of cells inside the adherent cell layers in vitro. Between the electrodes, a continuous tiny alternating current is provided, and the voltage across is measured where the cell membrane provides constant resistance. Measuring cellular impedance allows the automated investigation of parameters such as cell attachment, growth, morphology, function, and motility [60]. It involves a system in which the complex electric impedance of cell is monitored in a time-resolved manner over a broad range of frequency ranging from a few Hz to kHz, or even MHz. Impedance is inversely proportional to electrode size. Absolute impedance is often measured at a specified frequency where the impedance change recorded over the course of the experiment is large enough to provide a qualitative assessment of cell activity fluctuations [61]. Based on the frequency, cell-based barrier models are observed and changes in cell membrane capacity, conductivity, and resistance generated by tight junctions or cellular motions can be studied.

2.1.3 Multi-electrode Arrays

Multi-electrode arrays (MEA) are devices that include a large number of microelectrodes, acting as neural interfaces connecting neurons and electronic circuit to acquire or distribute neural signals. The potential change due to depolarization and repolarization of neurons' and cardiomyocytes' membrane which are electrically active is measured by voltage-detecting probes. MEAs are sets of separated microelectrodes used for mapping of field potentials or to record the presence of voltage peaks over a given threshold. MEAs are commonly employed to study electrogenic cell connections, physiology, and in vitro neuronal and muscular models, providing a novel electrical interface [62]. Cells may be cultured directly on the MEA substrate in vitro. It is possible to study both high-frequency potentials produced by individual cells and low-frequency variations reflecting due to cellular function [63]. MEA's potential to record and excite vast populations of excitable cells without physical injury is considered to be a significant advantage [64]. After being exposed to common drug, metabolites formed by hepatic metabolism, heart beating was studied utilizing a MEA for potentiometric measurements in conjunction with cantilever array [65]. MEA helps in continuous monitoring of dissolved oxygen (DO) concentrations in cell culture based on amperometric oxygen reduction [66]. Citing the feasibility of microfluidic OoaCs as alternative to standard cell culture platforms, there has recently been a spike in interest in combining oxygen-sensing devices with microfluidics for cell culture applications.

2.2 *Optical Biosensors*

Optical biosensors can detect biological factors by measuring changes in various light properties such as resonance, reflectance, polarization, luminescence, absorption, refractive index, scattering and plasmon effect, due to a bioreceptor-analyte interaction. Besides simple integration with microfluidics, label-free and minimum intrusive nature, the lack of contact between the sensing device and the detector is a significant advantage for all optical sensors. They are used to monitor the physical microenvironment of cells and also used in point-of-care devices.

2.2.1 Absorption-Based Sensor

Being one of the earliest types of biosensors, it employed fiber channels to transport light to and from the area around the fiber tip where the inherent spectroscopic characteristics of the analyte matrix were investigated. This is an ideal approach for applications where measurements must be taken over a lengthy period of time [67]. The technology is extremely effective and conceptually simple, however, the scope of applications it can handle might be limited. Indicator pigments are used in absorption assays because they change their absorption spectra in the presence of an analyte. According to the Beer-Lambert Law, the analyte concentration, optical path length along the microfluidic channel, and dye molar absorption all affect the measured absorption. The tiny size of miniature OoaC systems complicates the implementation of an absorption sensor. However, these sensors can be used to estimate the pH of cell culture medium by measuring light absorption of the phenol red dye. It detects pH changes in culture medium to assess cell health indicating yellow at 6.2 acidic pH which turns to red at 8.2 basic pH and gradually to a bright pink color when the pH exceeds 8.2 [68].

2.2.2 Luminescent Sensor

Luminescent sensor is the most common type of optical biosensor. It consists of a luminous dye that is sensitive to the target molecule and polymeric matrix that houses this dye. Because of their lower LOD, lower cost, and ease of manufacture, these sensors have been used more than electrochemical methods for dissolved oxygen (DO) monitoring in OoaC devices. In order to detect DO optically, the photoluminescence quenching effect of oxygen depends on the partial pressure of oxygen and luminophores' intensity [69]. Ruthenium- and metalloporphyrin-based fluorophores have high quenching constants, hence useful in these kinds of sensors. Luminescent intensity may be evaluated spatially using a typical fluorescence microscope. Intensity measurements, on the other hand, are susceptible to inaccuracies induced by various factors. Since lifetime is an inherent feature of the indicator molecule, measurements may be obtained by pulsed excitation utilizing single photon counting or phase modulation and are therefore more accurate [70].

3 Biosensors for Monitoring Micro-physiological Systems

Micro-physiological systems on OoaC device aim to restate the functional components of human tissues/organs in vitro. OoaC progressed over time, but sensor-integrated OoaCs are needed to provide real-time information about the metabolic activity and viability of the tissue/organ besides monitoring the physical and biological parameters.

3.1 Physical Parameters

The microenvironment of organ-on-a-chip technologies is crucial for cellular viability, maturation, and differentiation. Continuous physical parameter monitoring allows physiological cell activity maintenance in cultured organs and tissues within OoaC devices [71]. In OoaC models, a number of physical properties of the microenvironment like pH, temperature, DO, carbon dioxide, surface tension, humidity, osmolarity, mechanical strength, nutrient content, flow rates, etc. have been monitored [72].

The real-time DO monitoring is crucial to monitor cellular activity because mitochondrial respiration might be lost in a hypoxic milieu affecting cell survival [55]. pH is the measure of efficient functioning of various organs, indicator of many diseases including cancer as well as the pH-dependent action of certain nanocarriers. Light addressable potentiometric sensors (LAPS), metal oxide-based potentiometric sensors, ion-selective field effect transistors (ISFET), and optical pH sensors are commonly used to monitor pH [73]. For instance, Mattes et al. created a near-infrared spectroscopy-based biosensor to track pH and osmolality in a bioreactor cultured with cells loaded on microcarriers because these two parameters can influence cell survival and development by controlling nutrients and water transport through the cell membrane [55]. As temperature is also a crucial parameter to be considered for long period of culturing and testing in OoaC, Yu et al. created a microthermal sensor with a control software at 37 ± 0.3 °C and a platinum-based grid microheater that was built on glass. Zhang et al. built a multi-sensor-integrated multi-OoaC system in which they reported a fully integrated multiplexed biophysical sensing platform to monitor oxygen, temperature, and pH in a heart-liver-on-chip model for more than five days [74]. Such sensors for long period of monitoring of the mentioned physical characteristics within OoaC devices have a significant amount of space for implementation and enhancement.

3.2 *Biological Parameters*

Different microsensors have been incorporated to determine the metabolic functions of cultured cells. Here, we will discuss OoaC-integrated biosensors for measuring oxygen, glucose, lactate, cytokines, and other metabolites.

Oxygen

Oxygen is an essential parameter in cell metabolism and real-time monitoring of oxygen levels in OoaCs can help to maintain oxygen distribution in tissues/organs. To detect absolute oxygen in OoaC devices, two types of sensors are typically used: optical and electrochemical sensors. Based on the reduction of the oxygen molecules, electrochemical sensors are generally preferred for online monitoring due to their robustness and reproducibility, whereas optical sensors are preferred in cases of low oxygen levels [72].

Glucose and lactate

Cells use glucose as their primary energy source, which is converted into pyruvate in presence of oxygen or lactate in the anaerobic circumstances. Thus, increase in lactate concentrations indicates stress and dysfunctions within cells. This leads to the integration of electrochemical biosensors into OoaC devices. Cellular enzymes likely glucose and lactate oxidase are immobilized on a membrane directly to catalyze glucose or lactate, producing H_2O_2 which provides a direct correlation with glucose or lactate concentration [66, 75].

Cytokines and other metabolites

During immune responses, cytokines control different cellular functions like migration, proliferation, cell activation, and play crucial roles in cell–cell signaling and cancer progression. The most widely used in situ approach for small proteins and cytokines detection on-chip is immunoassays such as bead-based assays, aptamer-based assays, and antigen–antibody assays. Ortega et al. established a polydimethylsiloxane-based muscle-on-a-chip device for local inspection of cytokines IL-6 and TNf-α produced by murine skeletal myoblasts [76]. Son et al. have designed fluorescent microbead-based devices to measure TGF-β1 and hepatocyte growth factor secretion levels across primary hepatocytes over seven days [77]. A specific microfluidic device allows for local cytokine assessment from cells isolated from sensor channels by a porous hydrogel layer [66].

4 Organ-on-a-Chip

4.1 Bone Marrow-on-a-Chip (BMOC)

The hematopoietic process is maintained by a complex network of biological, chemical, physical, and structural signals in the bone marrow microenvironment. OoaC devices have been created to meet the demand for in vitro systems that replicate the activities that take place in the bone marrow microenvironment [78]. To address the functional gap between in vitro and in vivo methods, a BMOC culture system was produced which consists of synthetic bone and living marrow. The bone marrow developed on microfluidic chip reacts to detrimental radiation presentation accompanied by composition treatment which aids in recovering blood cells to mimic living bone marrow. The microdevice's ability to preserve the developed bone marrow and keep track of blood cell populations long after radiation damage has enabled researchers to evaluate the effectiveness of experimental drugs. Recently, a BMOC microfluidic system was developed to activate the living hematopoietic bone marrow culture, and it supports immunotoxicity and immunogenicity testing in long-term culture with repeated antigen stimulation [79]. In a case study, Torisawa et al. developed a BMOC device by inserting a scaffold into a mouse and enabling the mouse cells to colonize it. The construct was then removed and infused with culture media in a microfluidic chip to assist the preservation of primitive hematopoietic cells [20]. HSPCs are retained in the engineered bone marrow for a week providing an innovative technique for analysis of drug responses and toxicities in bone marrow.

4.2 Brain-on-a-Chip (BOC)

Brain controls memory, emotion, thought, motor skills, breathing, vision, touch, temperature, hunger, and other process that regulate our body. Besides the study of neurotoxicity due to various drugs, the urge to study the mechanism of neuro-diseases like Parkinson's disease, Alzheimer's, multiple sclerosis, nerve injuries, etc. remains intriguing. BOC models were designed. Human cells and dynamic systems are used in BOCs to create micro-physiological platforms which exhibit unique brain tissues and particular functions. BOC integrates non-neuronal cell like glia and vasculature, neuronal chambers, biomaterial scaffolds, and static or dynamic fluid conditions. Recently, Yue and his co-workers developed the 3D hydrogel-based multi-component BOC model which helps in screening for drug development [22]. From a case study, we see that Park et al. developed a BOC device for Alzheimer's studies. In addition, Park et al. confirmed this BOC as a cytotoxicity testing and drug screening tool [21].

BBB-on-a-Chip

Blood–brain barrier (BBB) plays crucial role in brain homeostasis maintenance and prevents toxic substances to enter the brain. BBB's functionality is hampered in neurodegenerative diseases. A practical model of BBB is necessary to understand the numerous methods associated with drug delivery applications. Ahn et al. recently used a BBB-on-a-chip device to investigate the distribution of nanoparticles and cellular response [25]. This model has two layers: the upper vascular layer consists of human brain microvascular endothelial cells and the lower perivascular layer consists of human astrocytes in matrigel and human brain vascular pericytes separated by a membrane. Despite the numerous challenges, BBB-on-a-chip has been commercialized. Furthermore, new BBB-on-a-chip models are expected to enable continuous monitoring of inflammation and cell damage markers associated with neurodegenerative diseases [26].

4.3 Heart-on-a-Chip

The heart is the center of circulation system, and cardiovascular diseases are one of the leading causes of human death. Heart-on-a-chip is a microfluidic device reproducing the heart mechanisms to test drug efficacy and its effects on heart cells. This heart-on-a-chip is used for drug screening and other critical applications due to significant events like recapitulation of the microenvironment via the introduction of cyclic stretch, anisotropic arrangement of cells, electrical signals, fluidic flow, and incorporation of sensors to evaluate stress response, calcium transients, cardiomyocyte contraction, electrical action potential, etc. Xiao et al. built a device to study cardiac pharmacology by inserting a polytetrafluoroethylene tube into a bioreactor that provides contact conduction to bind and expand cardiomyocytes. To solve monitoring issues, 3D printing methods were used by Lind et al. where strain gauge sensors were embedded in the heart-on-a-chip device using piezo-resistive biocompatible high-conductance soft ink [23]. A biosensor-based functional heart-on-a-device was built by Marsano et al. which mimicked the mechanical and physiological characteristics of cardiomyocytes, aids in real-time monitoring of heart activities, and is useful for cardiovascular investigations [24].

4.4 Lungs-on-a-Chip

The COVID-19 pandemic has increased the importance of pulmonary research because it is related to severe lung injuries. Lung-on-a-chip devices mimic the physiological and mechanical environment of a human lung offering a unique in vitro approach for drug screening by combining diverse microfabrication techniques with tissue engineering. Lung-on-a-chip replicated the breathing mechanism. Huh et al. developed a lung-on-a-chip model by dividing this chip into areas separated by a polydimethylsiloxane (PDMS) membrane with an extracellular environment utilizing soft lithography [27]. The lower region consists of human pulmonary microvascular endothelial cells and the upper region has alveolar epithelial cells, which imitate the alveolar capillary-barrier. Stucki et al. developed a lung-on-a-chip mimicking the lung parenchyma in 2015 [28]. A microfluidic chip was used by Xu et al. to imitate the lung cancer microenvironment consisting of primary cancer cells and cancer cell lines, and various chemotherapeutic medications were tested [29]. The less functional period of the chip limits modeling of some chronic conditions, such as chronic obstructive pulmonary disease (COPD) [80] including asthma, chronic bronchitis, etc.

4.5 Intestine-on-a-Chip and Gut-on-a-Chip

The small intestine consisting of muscles, epithelium, and submucosal layers helps in digestion and drug-nutrient absorption. To study intestinal health and disease, microfluidic intestine-on-a-chip (IOC) has been built. The development of IOC can provide us with important information's about absorption, permeation, oral drug metabolism, and drug screening. Sung et al. developed the first 3D hydrogel to imitate the human intestinal villi [81]. Jalili et al. created an IOC in which healthy aerobic and anaerobic human gut microbiota were co-cultured with living human intestinal epithelium to detect real-time oxygen concentration [82]. Kim et al. created bionic devices, implants that used electrical impulses to restore lost biological functions, to aid researchers in understanding the origins of intestinal disorders and finding effective treatments [83]. Additionally, IOC aids in comprehending the potential applications in personalized medicine [80]. The 3D in vitro IOC model developed by Gijzen et al. comprises immune-competent and epithelial cells. By examining changes in cytokine release and barrier function, they demonstrated that specific chemicals may either prevent or promote intestinal inflammation, which aids in understanding intestinal inflammation and the screening of anti-inflammatory substances [30].

Gut-on-a-chip (GOC) offers an excellent in vitro platform for studying human gut physiology and pathophysiology [31]. Investigation of microbiome commitments to intestinal pathophysiology, together with the disease mechanisms study, can be performed in a measured background. A GOC device made by Villenave et al. recapitulates tissue's condition as a perfect model so that enterovirus infection can be investigated properly [84]. Some strategies have been developed to address the disadvantages of using PDMS in a GOC model [32].

4.6 Pancreas-on-a-Chip

Pancreas-on-a-chip (POC), which primarily refers to the analysis of the pancreatic endocrine component on a microfluidic device, may be implemented as a traditional and real-time evaluation platform for assessing islet potency and quality. Diabetic and cystic fibrosis-related diseases can be studied using POC models. Cystic fibrosis (CF) is modeled in one of the POC prototypes using pancreatic ductal epithelial cells and islet cells co-cultured in a microfluidic device segregated by a thin porous barrier. The system determined the functional connection of islets cells with the ductal cells, as well as the regulatory function of CF transmembrane conductance regulator (CFTR) in the production of insulin [34]. The microfluidic chip designed by Zbinden and colleagues traps individual pseudo-islets derived from conditionally immortalized EndoC-bH3 cell lines using the hydrodynamic trapping technique [85]. Low-shear perfused pancreas-on-a-chip analyzes the glucose response and endocrine activities via the incorporation of on-chip Raman analysis. Native islets of the pancreas are grouped and contain a variety of endocrine cells that communicate through both endocrine and autocrine signaling [86]. These capillary pathways encompassed by these islets play a crucial role in maintaining glucose homeostasis [87]. Keeping the maintenance of homogeneity as primary concern in the shape of spheroids for improving cellular interactions, Jun et al. have developed a microfluidic device with concave microarrays. Osmotic micropumps were used for lowering flow rates and to prevent shear-induced cellular injury without affecting the supply of oxygen and nutrients [88]. Due to the significant advancements of POC, it is used as a universal tool to study islet quality, and it replaced animal testing in the creation of diabetes interventions.

4.7 Liver-on-a-Chip

The liver carries out numerous metabolic processes crucial for the body's production of necessary compounds as well as the removal of harmful and toxic substances such urea, glucose, lipids, proteins, carbohydrates, and cytochrome P450. The hepatic system is the primary location of drug and toxin metabolism [89]. The first liver-on-a-chip system created by Kane et al. comprised microfluidic openings in which rat hepatocytic cells and 3T3-J2 fibroblasts were co-cultured to resemble an airway interface [36]. Lee et al. created a chip that replicated cultured primary hepatocytes and endothelial interstitial structure using media perfused outside the gap [90]. Hegde et al. created a two-layer chip with porous polyethylene terephthalate (PET) membranes to separate the channels and continuously perfused rat primary hepatocytes that had been coated in collagen and fibronectin into the lower channel via the upper chamber [35]. Isolated hepatocytes cultivated in 3D exhibit increased liver functions and retain their differentiated characteristics for a prolonged time [91]. Decellularized liver matrixes (DLM) integrated with gelatinmethacryloyl (GelMA) were developed by Lu et al. to create biomimetic liver tumors that mimic the 3D tumor microenvironment [37]. This technique provides an improved disease model for a variety of upcoming anticancer drug pharmacological research studies.

4.8 Kidney-on-a-Chip

The need for drug screening systems is emphasized due to the permanent loss of renal filtration (comprised of nephrons) induced by drug toxicity. Microfluidics has the ability to mimic the fluid environment that promotes cellular growth while supporting the porous membranes to retain cell polarity. To imitate renal filtration, the first multi-layered microfluidic device was designed by Jang et al. using murine kidney medullary collecting duct cells. In response to hormone stimulation, the device's biomimetic environment promoted cytoskeletal remodeling and molecular transport improving the polarity of the inner medullary collecting duct [38]. In 2013, they cultured human primary renal epithelial cells using the same microfluidic apparatus for drug transport and toxicity studies [92]. Musah et al. demonstrated techniques for inducing podocytes derived from pluripotent stem cells in organ culture devices that make human glomerular chips [93]. These mimic the glomerular capillary wall's structure and function, which could not be achieved with earlier techniques. The chip aids to evaluate nephrotoxicity, design therapeutics, conduct regenerative medicine, and study kidney diseases. Human proximal tubules and glomeruli were used by Sakolish et al. [39] to create a reusable microfluidic chip that allowed renal epithelial cells to develop in various environments. The kidney-on-a-chip has also been utilized to simulate renal failure caused by viral infection to investigate potential pathogenesis [94]. The pseudorabies virus infection alters microvilli, breaks the barrier, and renders sodium reabsorption dysfunction. The barrier structure and sodium reabsorption can be recapitulated by the distal tubule-on-a-chip [95]. The kidney-on-a-chip technology simulates physiological influences and tracks drug-induced renal failure and its biological effects.

4.9 Placenta-on-a-Chip

Despite the advancements in OoaC technology, development of human placenta on microfluidic device is more challenging than other human organs. The human placenta develops after conception as a temporary organ that binds mothers and fetuses to provide nourishment, remove waste, exchange gases, and prevent internal infections. To understand the mechanism of placental transport, numerous in vivo and ex vivo studies were conducted. Because of differences in the functional properties including nutrient distribution, blood circulation, permeability, and metabolic activity in the human or animal placental membrane, it is assumed to be the cause of inconsistent and inaccurate projections in animal models. The placental trophoblast and fetal capillary endothelium are the responsible distinct cell regimes that separate the maternal blood from the fetal circulation during pregnancy. Endothelial and trophoblast cells with a membrane covered in highly concentrated collagen are used to depict the placental barrier to investigate how the membrane influences glucose transport across endothelial-trophoblast barriers [40, 41]. To simulate the same, Blundell et al. developed a microfluidic platform that mimics placenta in vitro [96]. The focus of placenta research has thus been changed to "organ-on-a-chip" technology to better understand how placenta works [97].

4.9.1 Skin-on-a-Chip

Skin is the largest organ made up of two layers: dermis and epidermis. Human skin models are needed to examine dermal pharmacokinetics and pharmacodynamics, understand the causes of dermatological illnesses, pollutant exposure-induced toxicity, develop innovative therapies for skin diseases, and optimize medication delivery. Various skin-on-a-chip models have been designed recently to test the effects of UV radiation [98] and drugs [99] as a substitute for animal models. Jeon et al. created a microfluidic skin model that included three layers that mimicked the dermal, epidermal, and endothelial equivalents. Two transwell membranes were sandwiched between patterned PDMS chips to construct the device, which is used to analyze a variety of test chemicals, and a dual parameter model includes the assay for cellular viability. For simulating epidermal, dermal, and endothelial compartments, HaCaT, Hs27, and HUVEC cells were used respectively [100]. A microfluidic device, consisting of RBC filter, blood loading channel, a skin column loading channel, and a migration channel, designed by Kim et al. studied the neutrophil response to a skin infection caused by bacteria. Neutrophil migration is a significant indicator of skin and soft tissue infections [101]. The antiaging effect of *Curcuma longa* leaf extract (CLLE) was examined by Kim et al. on a gravity-driven skin-on-a-chip platform without a pump. A porous membrane divided the two layers of PDMS chips that made up the device, the upper layer contained a medium chamber and a culture chamber, whereas the lower layer contained microfluidic channel. The skin-on-a-chip system evaluates the effectiveness of CLLE suggesting potential use in cosmetics [102]. The study on effectiveness of coenzyme Q10 was carried out by the same team using the pump-less skin-on-a-chip method. The chip showed enhanced epidermal layer thickness while increasing coenzyme Q10 concentration and increased filaggrin expression [43].

4.9.2 Muscle-on-a-Chip

Numerous OoaC models have been designed to monitor and evaluate the toxicity of drug on muscles. Accurate information on muscle contractibility is necessary to create safer medications [44]. Based on the thin-film technique, Grosberg et al. created a muscle-on-a-chip that had a thin elastic film with 2D-designed muscle tissue on one side. The same apparatus could assess the contractility of both smooth and striated muscles at the same time through an automated projection tracking system [45]. Microfluidic-based tissue engineering equipment can be created using a variety of methods: rapid prototyping [103], soft lithography [104], and microcell printing [105]. Using a temperature-sensitive template, Wan et al. created skeletal-muscle-on-a-chip that exhibited 3D vasculature in synthetic extracellular matrix [106]. The mechanobiology and geometrical characteristics are essential for the development of muscle-on-a-chip [107]. The geometric features of the chip could be integrated due to the sacrificial template constructed from a polyester wax and paraffin wax mixture, which also enabled the tissues to respond to mechanical stimuli by myogenesis [106].

4.9.3 Fat-on-a-Chip

Highly reactive adipose tissues store and emit energy in response to hormonal and energetic signals. Traditional cell culture systems for fat models have limitations due to a lack of control over parameters like fluid volume ratio. Since adipose cells possess endocrine functions, compartmentalization is also required to maintain homeostasis [46]. Tanataweethum et al. created a dual-channel microfluidic device with a permeable membrane to cultivate and retain primary mouse adipose cells for several weeks, which clarified the influence of dynamic environment on adipocyte differentiation. The device's membrane keeps the media channel and cellular chamber separated, protecting the cells from shear-induced injury [108]. To examine the relationship between immune cells and adipocytes in human diabetic model, Liu et al. built co-culture system of monocytes and adipocytes (U937) on microfluidic platform. Insulin resistance was observed in adipocytes when co-cultured with U937 [109].

4.9.4 Cancer-on-a-Chip

Microfluidic chips are more suitable for cancer diagnosis and treatment. Toxicity studies that are carried out in 2D and 3D culture or in vivo cannot replicate major important facets of cancer cells. Cells on-a-chip tracks cell–cell interactions in bone formation, immune response, neuronal signaling, tumor growth, and other pathological conditions [110]. For example, such a chip can be used to examine B- and T-cell repertoires to provide antibody screening, improve cancer immunotherapy, and the dynamic relationships between tumor cells, immune cells, and local tissue. OoaC systems provide replicative tumor microenvironments, that can be controlled and monitored, indicating significant cancer therapy studies and developing personalized medicines. Besides metastasis and phenotype studies on cancerous cells, cancer-on-a-chip platforms are valuable for evaluating cancer drugs and manufacturing personalized cancer models. Combining cancer and the immune system on microfluidic chips would offer insight on immune system responses, metastatic processes, and tumor states, thereby minimizing cancer mortality and expenses [47].

4.9.5 Stem Cells in OoaC

A stem cell is defined as any self-renewing cell with the ability to develop into one or more specialized cell types. Mesenchymal stem cells are pluripotent stem cells having limited growing abilities and lack consistent derivation protocols and clear biological responses, hence are less useful in OoaC designing. As with cell lines and primary cells, embryonic stem cells, induced pluripotent stem cells (iPSCs), and adult stem cells (ASCs) may be differentiated and inserted onto microfluidic chips. Organs like lungs and liver can be cultured from stem cells [49] and integrated onto chips to understand the organ-associated-diseases better. Kane et al. created the first liver-based system, which replicated an airway interface by using microfluidic holes to co-culture 3T3-J2 fibroblasts and rat liver cells. This could continuously synthesize albumin and undergo metabolism [36]. In another way, electric field gradients were used to pattern the cells onto circular PDMS which simulated hepatic lobule structure [111]. Biomimetic liver tumors were designed by integrating DLM with GelMA to mirror the 3D tumor microenvironment. Xu et al. developed microfluidic platform to simulate the lung tumor microenvironment and evaluate several chemotherapeutic drugs using cancer cell lines and cancer cells [29].

4.9.6 Multi-organ-on-a-Chip

The complexities of inter-tissue interactions, functional changes, and integrity of organ function are not accurately reflected by single OoaC. Multi-organ-on-a-chip integrates several organs concurrently, drawing research interest [112]. Multi-organ-on-a-chip is achieved by simultaneously culturing cells from several organs and tissues that are connected by channels, "bionic blood vessels," allowing the interactions to form a system [113]. The organs simulated chambers are connected by fluidic networks in semi-static systems using Transwell ®-based [114] tissue implants. Individual OoaC platforms are connected via flexible microchannels [115]. Even though the idea of multiple-organ on-a-chip is still in its infancy, significant advancements have been made which is discussed in detail in Table 3, including the design of two [116, 117], three [118], four [119, 120], and ten [121] organs on a chip. Van et al. first combined intestines and liver in a microfluidic platform in 2010 [117], demonstrating the chip's application in organ interactions, such as the regulation of synthesis of bile acid. This technique provided in vitro research possibilities and insights into

Table 3 Study of different multi-organ-on-a-chip systems

Multi-organ-on-a-chip (MOC)	Pumping system	Impacts on	Preclinical model	Remarks	Refs.
HUMIMIC Chip2 (Human epidermis models, and HepaRG-stellate spheroids)	On-chip micropump	Skin-liver axis	In vitro	– Gives information on a metabolic fate of drugs following topical and systemic administration – Supports two different culture modes	[124]
Human gut-on-a-chip and liver-on-a-chip	Static, single-pass, recirculating	Gut-liver axis	In vitro	– Diagnostic tool for evaluation of the onset of drug-induced NAFLD – Gut epithelial cells (Caco-2) and the liver cells (HepG2) are cultured in two separate layers – Transwell-based model with hepatocytes below and Caco-2 cells inside the transwell – Enabled pharmacokinetic studies	[116, 125, 126]

(continued)

Table 3 (continued)

Multi-organ-on-a-chip (MOC)	Pumping system	Impacts on	Preclinical model	Remarks	Refs.
Human pancreatic islets and liver spheroids-on-a-chip	Recirculating, on-chip micropump	Liver-pancreas	Ex vivo	– Studies of insulin and glucose regulation-based pancreatic islet-liver cross-talk	[127]
Liver cells with different cytochrome P450 (CYP) subtypes and glioblastoma multiforme (GBM) brain cancer-on-a-chip	Peristaltic pump	Liver-cancer	In vitro	– Evaluates the cytotoxicity of anticancer drugs – 3D cultures are more important than 2D cultures in drug dosage studies	[128]
IdMOC co-culture system for metabolically competent human primary hepatocytes, and metabolically incompetent mouse 3T3 fibroblasts cell line	–	Liver-fibroblast	In vitro	– Classifies metabolism-based toxicants – Used for drug-drug interactions, drug metabolism, and hepatotoxicity studies – Toxicity increases when metabolically active hepatocytes are present	[129]

(continued)

Table 3 (continued)

Multi-organ-on-a-chip (MOC)	Pumping system	Impacts on	Preclinical model	Remarks	Refs.
Liver (hepatocytes), kidney (kidney cortical cells), lung (small airway epithelial cells), central nervous system (astrocytes), blood vessels (aortic endothelial cells), MCF-7 human breast adenocarcinoma cells	–	Liver-kidney-lung-neural vasculature-cancer	In vitro	– Individual organs cultured are in shallow compartments – They connected via media within a larger compartment – Aids in drug toxicity study – Cytotoxicity of tamoxifen is compared in different human organs	[130]
Human organoids liver, and heart	Microfluidics-controlling recirculating pump	Liver-heart and liver-cancer-heart	In vitro	– Automated, in situ continual monitoring – Demonstrates long-term monitoring of chronic drug responses – Short-term evaluation of toxicity	[74]

(continued)

Table 3 (continued)

Multi-organ-on-a-chip (MOC)	Pumping system	Impacts on	Preclinical model	Remarks	Refs.
Four human cell types of liver, lung, kidney, and the adipose tissue on-a-chip	Peristaltic pump	Liver-lung-kidney-fat	In vitro	– Developed common media capable of supporting all organ systems – Minimal communication across cell culture compartments – Concurrent cultivation of four different human cell types (C3A, A549, HK-2 and HPA)	[131]
Cell lines, primary cells and cells derived from iPSC for cardiomyocytes, skeletal muscle, liver, and neuron	Gravity-driven flow	Cardiac-muscle-neuronal-liver	In vitro	– Uses serum-free media, electrical and mechanical readouts – Evaluates multi-organ toxicity – Presents phenotypic culture model	[120]
μOrgano system - multiple heart chips in series	Diffusive transport	Heart-heart	In vitro	– Plug-and-play interfaces enable linear integration of modular OoaCs	[132]
Cell lines, primary cells of liver, kidney, intestine, epidermis	Pulsatile medium perfusion, single-pass	Intestine-liver-skin-kidney	In vitro	– Used in drug discovery	[133]

organ-organ interactions. A high-throughput multi-organ-on-a-chip system developed on a microplate-sized pneumatic pressure-driven medium circulation platform was reported by Satoh et al. [122]. This platform is useful as a research tool for finding new drugs. An innovative combination of laser technology was developed by Lantada et al. [123]. The efficiency of the method was confirmed by the analysis of human mesenchymal stem cells, and the resulting chip was transparent, making imaging processes easier. Recently, rapid advancement in OoaC and multi-organ-on-a-chip improved our understanding of all the major organs.

5 Applications

5.1 Disease Models

Lung models previously have been studied from animal models, in vitro 2D cell models, 3D spheroids, and human organoid models. However, these are not capable of analyzing engagement of circulating immune cells in the presence of active fluid flow. Being the first organ-on-a-chip system, lung-on-a-chip is a multifunctional microexperimental platform reproducing human alveolar–capillary interface to mimic organ-level lung function. A lung-on-a-chip model created by Huh et al. had human pulmonary microvascular endothelial cells in the lower sections and alveolar epithelial cells in the upper regions, separated by a 10 μm PDMS membrane having extracellular matrix [27]. A second-generation alveolar array tries to mimic the alveolar network, and its biochemical characteristics to investigate human non-small cell lung cancer-on-a-chip [134]. A three-channel 3D LOC model is being used to investigate the toxic effect of TiO_2/ZnO nanoparticles on lungs. The purpose of multi-organ lung cancer metastasis-on-a-chip is to recreate the in vivo cancer metastatic milieu [135].

Most of the drugs fail to cross BBB and do not treat the neurodegenerative diseases. The BBB endothelial cell layer's permeability and cellular activity can be determined by integrating TEER biosensor in BBB-on-a-chip. Integration of biosensor helps to monitor a wide range of analytes and biomarkers for personalized disease research or pharmacological testing. Astrocytes are critical in the creation of BBB-on-a-chip because they regulate expression of proteins and endothelial differentiation [136]. Recent BBB-on-a-chip device utilized human mesenchymal stem cells to simulate more realistic physiological barrier [137]. Another study includes human BBB 3D self-organized microvascular design to validate barrier functioning and cell behavior, consisting of hiPSC produced endothelial cells, primary pericytes, astrocytes, and hiPSC-derived BBB microchannel [138].

5.2 Drug Discovery

As the drug toxicity pathways are different for humans than the animal models, 48% of preclinical trials of drug shows the adverse reactions [139]. The main goal of creating an OoaC model for drug development is to produce a translational in vitro model and imitate in vivo physiology. Currently, the use of OoaC models is limited to preclinical drug discovery, with the data limited to internal research reports. The OoaC devices should have a detailed drug testing procedure to gain regulatory acceptance. The industry has established a list of reference drugs containing data on their toxicology and metabolism to validate liver-on-chip models [140]. Multi-organ models such as having a combination of four or more organs provide an interesting proof-of-concept studies, but their significance to drug development is to be determined [121]. Drugs like diclofenac have been tested on liver-on-a-chip devices to check its toxic effects in human liver. The lung-on-a-chip model allows for a more detailed analysis of antifibrotic medicines than is achievable in vivo, such as the inhibitory effect of nintedanib on neo-vascularization [141]. Although OoaC models are still in their early phases of drug discovery, it has been a pioneer in drug toxicology.

5.3 Pharmacology

OoaC models were used in drug toxicity assessment due to the limitations of employing animal models or cell cultures. In liver-on-a-chip model, Beart et al. showed germ cell toxicity of antineoplastic pro-drug cyclophosphamide after getting bioactivated in liver. It boosts the activity of a particular cytochrome in liver spheroids while decreasing the germ cells in testicular organoids [142]. Hepatocytes and nonparenchymal cells are used in a model of immunocompromised liver-on-a-chip by Sarkar et al. to show liver injury caused by diclofenac [139]. Cohen et al. examined the mechanism underlying cyclosporine and cisplatin's hepatotoxic effects using kidney-on-a-chip model [143]. By using microfluidic devices into preclinical research, drug ADME studies and toxicity screening can be accelerated [7]. Khoo et al. designed a microfluidic device containing patient-derived circulatory tumor cells to understand the therapeutic effects [144].

5.4 Cancer Diagnosis and Treatment

Tumor-on-a-chip is an OoaC technology that replaces healthy cells with malignant cells and is very versatile for theragnostic applications [145]. Tumor-on-a-chip allows researchers for investigating the effectiveness of smart drug nanocarriers to minimize chemotherapy cost while increasing its efficacy. In 2005, a simple microfluidic device was made by Farokhzard and co-workers having monolayer of cancer cells, to study the dynamic interaction of nanoparticles with different cancer cells. A study using gold nanoparticles found that the diameter of the nanoparticles affects their penetration into the tumor and targeting the receptor can increase retention [146, 147]. To correctly maintain a tumor environment, OoaC systems require detectors to measure pH, oxygen, and temperature [148]. Ion-selective field effect transistors, light addressable potentiometric sensors, metal oxide-based potentiometric sensors such as iridium oxide, tungsten oxide, or ruthenium oxide, or optical pH sensors are used for pH monitoring [149]. Zhang et al. developed a multi-sensor-integrated OoaC with a reliable temperature sensor that was tested in a benchtop incubator for seven days [74]. H_2O_2 is a reactive oxygen species that plays a vital part in biological processes and is measured directly to monitor cell growth in case of cancer. As a result, OoaC offers a higher possibility of detecting cancer and treating it successfully.

5.5 *Personalized Medicine*

Personalized medicine is an advanced therapeutic technique that uses genetic profile of an individual for disease diagnosis and treatment. Some methods of individualized cancer treatment are gene sequencing of solid tumors and identification of crucial molecular targets. Precision medicine is being used to treat other disorders like hepatitis B, HIV/AIDS and other chronic ailments. Patients' blood samples and health data can be utilized in personalized OoaC platforms [150]. COPD and asthma were studied by Benam and co-workers using an airway-on-a-chip model. This device was effective in modeling patient-specific issues on-a-chip [151]. Personalized patient-derived multiple-organ-on-chip models have some criteria to meet. Firstly, they should be able to portray the complexity of a patient's tissues while being therapeutically relevant. Next, personalized treatment can only be developed by integrating various "personalized organs" to allow tissue interaction. Third, they must have parallel processing capabilities in order to enable for high-throughput screening of personalized therapeutic medications, which is mainly lacking in present systems. Some commercial technologies, such as CNBio's OoaC plates with automated fluid management, AIM Biotech 3D cell culture chips, and 384-well OrganoPlate from MIMETAS, provide some degree of parallelization [152].

5.6 *Transplantation Application*

OoaCsystems replicate organ function in a similar physiological environment and provide information on the implementation and consequences of transplantation applications. A pancreas-on-a-chip has been built, enabling microfluidic platforms for islets to utilize this knowledge for clinical transplantation [153]. All transplantation protocols revolve around immune compatibility. Although islet encapsulation or iPSCs abolish the requirement for immune suppression, it is documented that transplantation of iPSCs into recipients can trigger immunological responses [154]. A simplified animal model for multiple organ transplantation—liver, kidney, spleen, and intestine—was also established employing donor procedures such as nephrectomy, hepatectomy, splenectomy, and jejunum segment surgery. During multi-organ transplant surgeries, the omentum was transplanted with chips of each organ. This approach could be appropriate for the simultaneous transplantation of four organs into a single recipient. The big momentum in the aforementioned model has the advantage of maintaining its anatomical position after surgery without causing any local circulation problems [155]. During the organ harvesting and isolation process, a number of variables have negative impacts on the viability of the cells provoking a post-transplant obligation to achieve reliance. It may be possible to assess cell quality post-isolation more accurately if there are standard techniques [156]. Regardless of breakthroughs in the isolation process, the proportion of cells used for research and clinical transplantation is limited because of less eligible donors for the isolation approach. The functionality of OoaC technology has recently increased. Considering its start in this arena, microfluidic engineering has produced unique and efficient chip platforms [157].

5.7 *Dentistry*

According to the most recent data, more than 3.5 billion people worldwide experience oral disorders, while malignancies in the oral cavity are one of the most prevalent types of cancer. Severe periodontal diseases lead to tooth deterioration, which is expected to affect millions of people, primarily the elderly [158]. The distinct configuration of the tooth is pointed out as an obstacle, making the treatment approach more challenging. The cultured dentine cells have been integrated within a microfluidic chip to construct an OoaC model; specifically, a tooth-on-a-chip. It is constructed of PDMS that imitates the characteristics of dentin-pulp interface. Stem cells from the apical papilla were grown in an odontogenic environment and implanted onto the tooth surfaces with continuous monitoring under live-cell microscopy to assess pulp cellular responses to the materials on-chip [159]. Characterization of mucosal plasticity and responsiveness of epithelial and subepithelial layers to the oral environment was also assessed for a lateral arrangement of undifferentiated gingival fibroblasts and keratinocytes containing oral mucosa-on-a-chip [160]. The oral bacteria *Streptococcus* mutants and the dental monomer 2-hydroxyethyl methacrylate were introduced individually to the mucosal construct, where the cells were imaged using epifluorescence microscopy to verify layer characteristics. The oral mucosa-on-a-chip is an useful tool for the study of the histological representations of oral mucosa's interactions with both bacteria and biomaterials by reducing transepithelial electrical resistance and mucosal cell viability to bacterial exposure and for treatment of dental monomer. Another microfluidic chip was created utilizing soft lithography for odontoblast culture, with the microchannels serving as dentin tubule mimics [161].

With the use of this technology, odontoblast processes were effectively stimulated to proliferate from cell bodies. This microfluidic chip, however, might be helpful for future researches on pathology and physiology of the odontoblast process as well as the development of dentin hypersensitivity treatments.

6 Challenges and Outlook

With consideration of the breakthrough innovations in the domains of sensors and OoaC systems, attempts to incorporate sensors in OoaCs are still in their infancy. The objective of this state-of-the-art technique is to recreate functional tissue models, aiming to imitate in vivo models. Both physical and biochemical sensors are essential to satisfy these criteria respectively, which demanded incorporating both sensing platforms in close proximity in the OoaC systems to mimic the human tissue/organ. Despite the advantages, the major shortcoming of OoaCs has not yet been adequately addressed. The quick saturation tendency of biosensors limits its application for one-time use only, which necessitates further research for regeneration methods to enable multiple-time usage of these biosensors for ongoing monitoring of biomarkers. Other difficulties with biosensors, including reliability, scalability, and biofouling, are needed to be resolved to monitor biomarkers in real-time, which would be ideal to have a single multi-sensor system for the automated physical and biological sensing operation [94]. While measuring cellular and tissue reactions, the right output from the electrochemical and optical sensors is largely dependent on the material selection, fabrication, and chip assembly [162]. For the establishment of a sensor-integrated OoaC platform, additional domains including material science, microfabrication, synthetic chemistry, and biosensors must need to evolve. For more precise and reliable measurements of physiological indicators, highly sensitive and specific microsensors are needed. The pursuit for even more insightful systems has been sparked by the fact that neither of the OoaCs created till now can completely recreate the function of an organ. However, the present limitations and upcoming potential of OoaC have been summarized in Fig. 4. Additionally, there are limiting aspects of iPSC research due to their reproducibility [163]. Automated fluid management is another requirement since the fluidic system calls for highly accurate chemical extraction and injection. A body-on-a-chip concept eventually requires to be refined more in order to assess the overall system, including the method of cell migration, the interactions between various organs, and the process of communication between them.

Nevertheless, despite all the limitations, OoaCs do have a bright future. Microbiorobotics may be used synergistically with an OoaC for multiple activities, including drug administration, real-time monitoring, and so on [164]. New open-source software and improved neural network models have made it possible for OoaCs and artificial intelligence technology to handle exceedingly labor-intensive

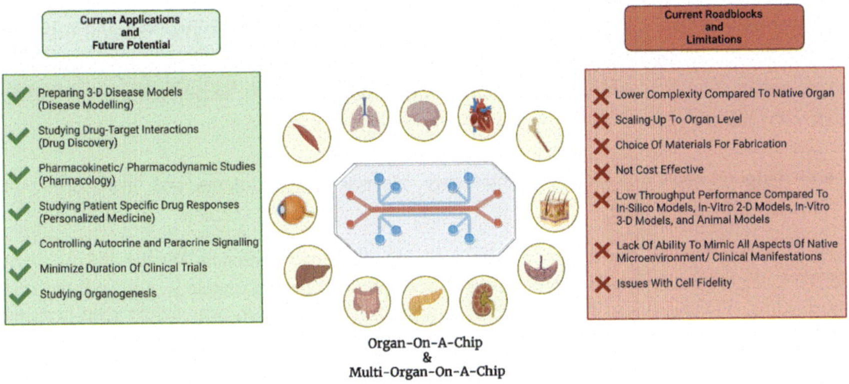

Fig. 4 Summary of limitations and outlooks of OoaC. The figure has been recreated with permission [165]

data, improving automation and data processing. In the near future, it is anticipated that OoaC and AI-integrated systems will advance to extend the realm of their application.

7 Conclusion

We have reviewed recent advancements in OoaC technology and its various applications in the field of dentistry, personalized medicines, organ transplantation, and cancer research. Although the use of OoaC technology in drug discovery is still in its infancy, its potential to forecast therapeutic efficacy and safety could have a big impact on the entire process. By precisely mimicking the intricate physiology, mechanics, and biochemistry of human tissues, OoaC models have ability to examine intricate functional characteristics that are outside the purview of conventional in vitro cultures. These models' qualification and regulatory acceptability should make widespread deployment easier while also reducing the risk associated with the pharmaceutical industry's decision to use them. The development of OoaC is favorably supported by microfluidic chips. Its development has drawn research interest from all across the world, and significant scientific progress has been accomplished. A number of different OoaCs have been designed and developed over the years. Numerous human organs have been researched. The end outcome of OoaC will be a "human-on-a-chip" by incorporating many organs onto a single chip and creating complicated multi-organ chip platform. Despite the high hopes, developing an OoaC is not simple and involves overcoming a number of challenges. OoaCs cannot yet be used widely due to the high cost of manufacture and experimental implementation; hence, components must be inexpensive and simple to dispose. The more expensive parts ought to be reusable. The chip's functionality may be hampered by sample collection, which might affect the levels of different metabolites. Thus, highly sensitive and specific microsensors are required for more precise and reliable measurements of physiological indicators. Further study is anticipated to eliminate the necessity for animal testing, enabling all experiments to be performed on a chip as in vivo labs.

Acknowledgements The authors acknowledge respective departments and institutions for providing facilities and support.

Author's Contribution The manuscript was conceptualized and written by TD and PM. BC, AA, and AG collected the information and analyzed it. All authors participated in writing, editing, and proofreading. TD and AS worked on the illustrations. The manuscript was finally modified and supervised by DDM for the submission.

References

1. Danku AE, Dulf EH, Braicu C, Jurj A, Berindan-Neagoe I (2022) Organ-on-a-chip: a survey of technical results and problems. Front Bioeng Biotechnol 10:840674. https://doi.org/10.3389/fbioe.2022.840674
2. Kirby BJ (2010) Micro- and nanoscale fluid mechanics: transport in microfluidic devices. Cambridge University Press, Cambridge
3. Ahadian S, Civitarese R, Bannerman D, Mohammadi MH, Lu R, Wang E, et al (2018) Organ-on-a-chip platforms: a convergence of advanced materials, cells, and microscale technologies. Adv Healthc Mater 7(2). https://doi.org/10.1002/adhm.201700506
4. Huh D, Hamilton GA, Ingber DE (2011) From 3D cell culture to organs-on-chips. Trends Cell Biol 21(12):745–754. https://doi.org/10.1016/j.tcb.2011.09.005
5. Park TH, Shuler ML (2003) Integration of cell culture and microfabrication technology. Biotechnol Prog 19(2):243–253. https://doi.org/10.1021/bp020143k
6. Sun B, Jiang J, Shi N, Xu W (2016) Application of microfluidics technology in chemical engineering for enhanced safety. Proc Saf Prog 35(4):365–373
7. Esch EW, Bahinski A, Huh D (2015) Organs-on-chips at the frontiers of drug discovery. Nat Rev Drug Discov 14(4):248–260. https://doi.org/10.1038/nrd4539
8. Ebert AD, Liang P, Wu JC (2012) Induced pluripotent stem cells as a disease modeling and drug screening platform. J Cardiovasc Pharmacol 60(4):408–416. https://doi.org/10.1097/FJC.0b013e318247f642
9. Ishida S (2018) Organs-on-a-chip: current applications and consideration points for in vitro ADME-Tox studies. Drug Metab Pharmacokinet 33(1):49–54
10. Qin D, Xia Y, Whitesides GM (2010) Soft lithography for micro- and nanoscale patterning. Nat Protoc 5(3):491–502. https://doi.org/10.1038/nprot.2009.234
11. Sticker D, Rothbauer M, Lechner S, Hehenberger M-T, Ertl P (2015) Multi-layered, membrane-integrated microfluidics based on replica molding of a thiol–ene epoxy thermoset for organ-on-a-chip applications. Lab Chip 15(24):4542–4554. https://doi.org/10.1039/C5LC01028D
12. Kricka LJ, Fortina P, Panaro NJ, Wilding P, Alonso-Amigo G, Becker H (2002) Fabrication of plastic microchips by hot embossing. Lab Chip 2(1):1–4. https://doi.org/10.1039/B109775J
13. Ng SH, Tjeung RT, Wang Z (2006) Hot embossing on polymethyl methacrylate.In: Paper presented at the 2006 8th electronics packaging technology conference
14. Chen C-S, Chen S-C, Liao W-H, Chien R-D, Lin S-H (2010) Micro injection molding of a micro-fluidic platform. Int Commun Heat Mass Transfer 37(9):1290–1294. https://doi.org/10.1016/j.icheatmasstransfer.2010.06.032

15. Lee Y, Choi JW, Yu J, Park D, Ha J, Son K, Jeon NL (2018) Microfluidics within a well: an injection-molded plastic array 3D culture platform. Lab Chip 18(16):2433–2440. https://doi.org/10.1039/C8LC00336J
16. Chang R, Emami K, Wu H, Sun W (2010) Biofabrication of a three-dimensional liver micro-organ as an in vitro drug metabolism model. Biofabrication 2(4):045004. https://doi.org/10.1088/1758-5082/2/4/045004
17. Johnson BN, Lancaster KZ, Hogue IB, Meng F, Kong YL, Enquist LW, McAlpine MC (2016) 3D printed nervous system on a chip. Lab Chip 16(8):1393–1400. https://doi.org/10.1039/C5LC01270H
18. Sooriyaarachchi D, Zhou Y, Maharubin S, Tan GZ (2020) Microtube-embedded microfluidic devices for potential applications in blood brain barrier research. Procedia Manuf 48:294–301. https://doi.org/10.1016/j.promfg.2020.05.050
19. Winkler TE, Feil M, Stronkman EFGJ, Matthiesen I, Herland A (2020) Low-cost microphysiological systems: feasibility study of a tape-based barrier-on-chip for small intestine modeling. Lab Chip 20(7):1212–1226. https://doi.org/10.1039/D0LC00009D
20. Lowe CR (1984) Biosensors. Trends Biotechnol 2(3):59–65. https://doi.org/10.1016/0167-7799(84)90011-8
21. D'Orazio P (2003) Biosensors in clinical chemistry. Clin Chim Acta 334(1):41–69. https://doi.org/10.1016/S0009-8981(03)00241-9
22. Schöning MJ, Poghossian A (2002) Recent advances in biologically sensitive field-effect transistors (BioFETs). Analyst 127(9):1137–1151. https://doi.org/10.1039/B204444G
23. Chaubey A, Malhotra BD (2002) Mediated biosensors. Biosens Bioelectron 17(6):441–456. https://doi.org/10.1016/S0956-5663(01)00313-X
24. Karimi-Maleh H, Karimi F, Alizadeh M, Sanati AL (2020) Electrochemical sensors, a bright future in the fabrication of portable kits in analytical systems. Chem Rec 20(7):682–692. https://doi.org/10.1002/tcr.201900092
25. Zhu Y, Mandal K, Hernandez AL, Kawakita S, Huang W, Bandaru P, Khademhosseini A (2021) State of the art in integrated biosensors for organ-on-a-chip applications. Curr Opin Biomed Eng 19:100309. https://doi.org/10.1016/j.cobme.2021.100309
26. Srinivasan B, Kolli AR, Esch MB, Abaci HE, Shuler ML, Hickman JJ (2015) TEER measurement techniques for in vitro barrier model systems. J Lab Autom 20(2):107–126. https://doi.org/10.1177/2211068214561025
27. Odijk M, van der Meer AD, Levner D, Kim HJ, van der Helm MW, Segerink LI, van den Berg A (2015) Measuring direct current trans-epithelial electrical resistance in organ-on-a-chip microsystems. Lab Chip 15(3):745–752. https://doi.org/10.1039/C4LC01219D
28. Vatine GD, Barrile R, Workman MJ, Sances S, Barriga BK, Rahnama M, Svendsen CN (2019) Human iPSC-derived blood-brain barrier chips enable disease modeling and personalized medicine applications. Cell Stem Cell 24(6):995-1005.e1006. https://doi.org/10.1016/j.stem.2019.05.011
29. Giaever I, Keese CR (1993) A morphological biosensor for mammalian cells. Nature 366(6455):591–592. https://doi.org/10.1038/366591a0
30. Au-Szulcek R, Au-Bogaard HJ, Au- van Nieuw Amerongen GP (2014) Electric cell-substrate impedance sensing for the quantification of endothelial proliferation, barrier function, and motility. JoVE 85:e51300. https://doi.org/10.3791/51300
31. An Y, Jin T, Zhang F, He P (2019) Electric cell-substrate impedance sensing (ECIS) for profiling cytotoxicity of cigarette smoke. J Electroanal Chem 834:180–186. https://doi.org/10.1016/j.jelechem.2018.12.047
32. Spira ME, Hai A (2013) Multi-electrode array technologies for neuroscience and cardiology. Nat Nanotechnol 8(2):83–94. https://doi.org/10.1038/nnano.2012.265
33. Maoz BM, Herland A, Henry OYF, Leineweber WD, Yadid M, Doyle J, Ingber DE (2017) Organs-on-Chips with combined multi-electrode array and transepithelial electrical resistance measurement capabilities. Lab Chip 17(13):2294–2302. https://doi.org/10.1039/C7LC00412E

34. Egert U, Schlosshauer B, Fennrich S, Nisch W, Fejtl M, Knott T, Hämmerle H (1998) A novel organotypic long-term culture of the rat hippocampus on substrate-integrated multielectrode arrays. Brain Res Protoc 2(4):229–242. https://doi.org/10.1016/S1385-299X(98)00013-0
35. Boudou T, Legant WR, Mu A, Borochin MA, Thavandiran N, Radisic M, Zandstra PW, Epstein JA, Margulies KB, Chen CS (2012) A microfabricated platform to measure and manipulate the mechanics of engineered cardiac microtissues. Tissue Eng Part A 18(9–10):910–919. https://doi.org/10.1089/ten.tea.2011.0341
36. Ferrari E, Palma C, Vesentini S, Occhetta P, Rasponi M (2020) Integrating biosensors in organs-on-chip devices: a perspective on current strategies to monitor microphysiological systems. Biosensors 10(9):110
37. Burgess LW (1995) Absorption-based sensors. Sens Actuat B Chem 29(1):10–15. https://doi.org/10.1016/0925-4005(95)01657-0
38. Mousavi Shaegh SA, De Ferrari F, Zhang YS, Nabavinia M, Binth Mohammad N, Ryan J, Dokmeci MR (2016) A microfluidic optical platform for real-time monitoring of pH and oxygen in microfluidic bioreactors and organ-on-chip devices. Biomicrofluidics 10(4):044111. https://doi.org/10.1063/1.4955155
39. Azimzadeh M, Khashayar P, Amereh M, Tasnim N, Hoorfar M, Akbari M (2022) Microfluidic-based oxygen (O2) sensors for on-chip monitoring of cell, tissue organ metabolism. Biosensors 12(1):6
40. Fuchs S, Johansson S, Tjell AØ, Werr G, Mayr T, Tenje M (2021) In-line analysis of organ-on-chip systems with sensors: integration, fabrication, challenges, and potential. ACS Biomater Sci Eng 7(7):2926–2948. https://doi.org/10.1021/acsbiomaterials.0c01110
41. Clarke GA, Hartse BX, Niaraki Asli AE, Taghavimehr M, Hashemi N, Abbasi Shirsavar M, Hashemi NN (2021) Advancement of sensor integrated organ-on-chip devices. Sensors 21(4):1367
42. Kilic T, Navaee F, Stradolini F, Renaud P, Carrara S (2018) Organs-on-chip monitoring: sensors and other strategies. Microphysiol Syst 2
43. Rodrigues RO, Sousa PC, Gaspar J, Bañobre-López M, Lima R, Minas G (2020) Organ-on-a-chip: a preclinical microfluidic platform for the progress of nanomedicine. Small 16(51):2003517. https://doi.org/10.1002/smll.202003517
44. Zhang YS, Aleman J, Shin SR, Kilic T, Kim D, Mousavi Shaegh SA, Khademhosseini A (2017) Multisensor-integrated organs-on-chips platform for automated and continual in situ monitoring of organoid behaviors. Proc Natl Acad Sci 114(12):E2293–E2302. https://doi.org/10.1073/pnas.1612906114
45. Alle M, Park SC, Bandi R, Lee S-H, Kim J-C (2021) Rapid in-situ growth of gold nanoparticles on cationic cellulose nanofibrils: recyclable nanozyme for the colorimetric glucose detection. Carbohyd Polym 253:117239. https://doi.org/10.1016/j.carbpol.2020.117239
46. Ortega MA, Fernández-Garibay X, Castaño AG, De Chiara F, Hernández-Albors A, Balaguer-Trias J, Ramón-Azcón J (2019) Muscle-on-a-chip with an on-site multiplexed biosensing system for in situ monitoring of secreted IL-6 and TNF-α. Lab Chip 19(15):2568–2580. https://doi.org/10.1039/C9LC00285E
47. Son KJ, Gheibi P, Stybayeva G, Rahimian A, Revzin A (2017) Detecting cell-secreted growth factors in microfluidic devices using bead-based biosensors. Microsyst Nanoeng 3(1):17025. https://doi.org/10.1038/micronano.2017.25
48. Zheng Y, Chen J, Craven M, Choi NW, Totorica S, Diaz-Santana A, Stroock AD (2012) In vitro microvessels for the study of angiogenesis and thrombosis. Proc Natl Acad Sci 109(24):9342–9347. https://doi.org/10.1073/pnas.1201240109
49. Kim J, Lee H, Selimović Š, Gauvin R, Bae H (2015) Organ-on-a-chip: development and clinical prospects toward toxicity assessment with an emphasis on bone marrow. Drug Saf 38(5):409–418. https://doi.org/10.1007/s40264-015-0284-x
50. Torisawa Y-S, Spina CS, Mammoto T, Mammoto A, Weaver JC, Tat T, Ingber DE (2014) Bone marrow–on–a–chip replicates hematopoietic niche physiology in vitro. Nat Methods 11(6):663–669. https://doi.org/10.1038/nmeth.2938

51. Yue H, Xie K, Ji X, Xu B, Wang C, Shi P (2020) Vascularized neural constructs for ex-vivo reconstitution of blood-brain barrier function. Biomaterials 245:119980. https://doi.org/10.1016/j.biomaterials.2020.119980
52. Park J, Lee BK, Jeong GS, Hyun JK, Lee CJ, Lee S-H (2015) Three-dimensional brain-on-a-chip with an interstitial level of flow and its application as an in vitro model of Alzheimer's disease. Lab Chip 15(1):141–150. https://doi.org/10.1039/C4LC00962B
53. Ahn SI, Sei YJ, Park H-J, Kim J, Ryu Y, Choi JJ, Kim Y (2020) Microengineered human blood–brain barrier platform for understanding nanoparticle transport mechanisms. Nat Commun 11(1):175. https://doi.org/10.1038/s41467-019-13896-7
54. Mir M, Palma-Florez S, Lagunas A, López-Martínez MJ, Samitier J (2022) Biosensors integration in blood-brain barrier-on-a-chip: emerging platform for monitoring neurodegenerative diseases. ACS Sens 7(5):1237–1247. https://doi.org/10.1021/acssensors.2c00333
55. Lind JU, Busbee TA, Valentine AD, Pasqualini FS, Yuan H, Yadid M, Parker KK (2017) Instrumented cardiac microphysiological devices via multimaterial three-dimensional printing. Nat Mater 16(3):303–308. https://doi.org/10.1038/nmat4782
56. Marsano A, Conficconi C, Lemme M, Occhetta P, Gaudiello E, Votta E, Rasponi M (2016) Beating heart on a chip: a novel microfluidic platform to generate functional 3D cardiac microtissues. Lab Chip 16(3):599–610. https://doi.org/10.1039/C5LC01356A
57. Huh D, Matthews BD, Mammoto A, Montoya-Zavala M, Hsin HY, Ingber DE (2010) Reconstituting organ-level lung functions on a chip. Science 328(5986):1662–1668. https://doi.org/10.1126/science.1188302
58. Stucki AO, Stucki JD, Hall SRR, Felder M, Mermoud Y, Schmid RA, Guenat OT (2015) A lung-on-a-chip array with an integrated bio-inspired respiration mechanism. Lab Chip 15(5):1302–1310. https://doi.org/10.1039/C4LC01252F
59. Xu Z, Gao Y, Hao Y, Li E, Wang Y, Zhang J, Wang Q (2013) Application of a microfluidic chip-based 3D co-culture to test drug sensitivity for individualized treatment of lung cancer. Biomaterials 34(16):4109–4117. https://doi.org/10.1016/j.biomaterials.2013.02.045
60. Wu Q, Liu J, Wang X, Feng L, Wu J, Zhu X, Gong X (2020) Organ-on-a-chip: recent breakthroughs and future prospects. Biomed Eng Online 19(1):9. https://doi.org/10.1186/s12938-020-0752-0
61. Kim SH, Lee JW, Choi I, Kim Y-C, Lee JB, Sung JH (2013) A microfluidic device with 3-D hydrogel villi scaffold to simulate intestinal absorption. J Nanosci Nanotechnol 13(11):7220–7228. https://doi.org/10.1166/jnn.2013.8088
62. Jalili-Firoozinezhad S, Gazzaniga FS, Calamari EL, Camacho DM, Fadel CW, Bein A, Ingber DE (2019) A complex human gut microbiome cultured in an anaerobic intestine-on-a-chip. Nat Biomed Eng 3(7):520–531. https://doi.org/10.1038/s41551-019-0397-0
63. Kim WS, Paik J (2021) Soft bionic sensors and actuators. Adv Intel Syst 3(3):2100003. https://doi.org/10.1002/aisy.202100003
64. Gijzen L, Marescotti D, Raineri E, Nicolas A, Lanz HL, Guerrera D, Kurek D (2020) An intestine-on-a-chip model of plug-and-play modularity to study inflammatory processes. SLAS Technol 25(6):585–597. https://doi.org/10.1177/2472630320924999
65. Kim HJ, Huh D, Hamilton G, Ingber DE (2012) Human gut-on-a-chip inhabited by microbial flora that experiences intestinal peristalsis-like motions and flow. Lab Chip 12(12):2165–2174. https://doi.org/10.1039/C2LC40074J
66. Villenave R, Wales SQ, Hamkins-Indik T, Papafragkou E, Weaver JC, Ferrante TC, Ingber DE (2017) Human gut-on-a-chip supports polarized infection of coxsackie b1 virus in vitro. PLoS ONE 12(2):e0169412. https://doi.org/10.1371/journal.pone.0169412
67. Toepke MW, Beebe DJ (2006) PDMS absorption of small molecules and consequences in microfluidic applications. Lab Chip 6(12):1484–1486. https://doi.org/10.1039/B612140C
68. Shik Mun K, Arora K, Huang Y, Yang F, Yarlagadda S, Ramananda Y, Naren AP (2019) Patient-derived pancreas-on-a-chip to model cystic fibrosis-related disorders. Nat Commun 10(1):3124. https://doi.org/10.1038/s41467-019-11178-w
69. Zbinden A, Marzi J, Schlünder K, Probst C, Urbanczyk M, Black S, Loskill P (2020) Non-invasive marker-independent high content analysis of a microphysiological human pancreas-on-a-chip model. Matrix Biol 85–86:205–220. https://doi.org/10.1016/j.matbio.2019.06.008

70. Caicedo A (2013) Paracrine and autocrine interactions in the human islet: more than meets the eye. Semin Cell Dev Biol 24(1):11–21. https://doi.org/10.1016/j.semcdb.2012.09.007
71. Eberhard D, Kragl M, Lammert E (2010) 'Giving and taking': endothelial and β-cells in the islets of Langerhans. Trends Endocrinol Metab 21(8):457–463. https://doi.org/10.1016/j.tem.2010.03.003
72. Jun Y, Lee J, Choi S, Yang JH, Sander M, Chung S, Lee S-H (2019) In vivo–mimicking microfluidic perfusion culture of pancreatic islet spheroids. Sci Adv 5(11):eaax4520. https://doi.org/10.1126/sciadv.aax4520
73. Mccuskey RS (2008) The hepatic microvascular system in health and its response to toxicants. Anat Rec 291(6):661–671. https://doi.org/10.1002/ar.20663
74. Kane BJ, Zinner MJ, Yarmush ML, Toner M (2006) Liver-specific functional studies in a microfluidic array of primary mammalian hepatocytes. Anal Chem 78(13):4291–4298. https://doi.org/10.1021/ac051856v
75. Lee PJ, Hung PJ, Lee LP (2007) An artificial liver sinusoid with a microfluidic endothelial-like barrier for primary hepatocyte culture. Biotechnol Bioeng 97(5):1340–1346. https://doi.org/10.1002/bit.21360
76. Hegde M, Jindal R, Bhushan A, Bale SS, McCarty WJ, Golberg I, Yarmush ML (2014) Dynamic interplay of flow and collagen stabilizes primary hepatocytes culture in a microfluidic platform. Lab Chip 14(12):2033–2039. https://doi.org/10.1039/C4LC00071D
77. Bhise NS, Manoharan V, Massa S, Tamayol A, Ghaderi M, Miscuglio M, Khademhosseini A (2016) A liver-on-a-chip platform with bioprinted hepatic spheroids. Biofabrication 8(1):014101. https://doi.org/10.1088/1758-5090/8/1/014101
78. Lu S, Cuzzucoli F, Jiang J, Liang L-G, Wang Y, Kong M, Wang S (2018) Development of a biomimetic liver tumor-on-a-chip model based on decellularized liver matrix for toxicity testing. Lab Chip 18(22):3379–3392. https://doi.org/10.1039/C8LC00852C
79. Jang K-J, Suh K-Y (2010) A multi-layer microfluidic device for efficient culture and analysis of renal tubular cells. Lab Chip 10(1):36–42. https://doi.org/10.1039/B907515A
80. Jang K-J, Mehr AP, Hamilton GA, McPartlin LA, Chung S, Suh K-Y, Ingber DE (2013) Human kidney proximal tubule-on-a-chip for drug transport and nephrotoxicity assessment. Integr Biol 5(9):1119–1129. https://doi.org/10.1039/c3ib40049b
81. Musah S, Dimitrakakis N, Camacho DM, Church GM, Ingber DE (2018) Directed differentiation of human induced pluripotent stem cells into mature kidney podocytes and establishment of a glomerulus chip. Nat Protoc 13(7):1662–1685. https://doi.org/10.1038/s41596-018-0007-8
82. Sakolish CM, Philip B, Mahler GJ (2019) A human proximal tubule-on-a-chip to study renal disease and toxicity. Biomicrofluidics 13(1):014107. https://doi.org/10.1063/1.5083138
83. Koyilot MC, Natarajan P, Hunt CR, Sivarajkumar S, Roy R, Joglekar S, Yadav KK (2022) Breakthroughs and applications of organ-on-a-chip technology. Cells 11(11):1828
84. Wang J, Wang C, Xu N, Liu Z-F, Pang D-W, Zhang Z-L (2019) A virus-induced kidney disease model based on organ-on-a-chip: pathogenesis exploration of virus-related renal dysfunctions. Biomaterials 219:119367. https://doi.org/10.1016/j.biomaterials.2019.119367
85. Blundell C, Yi Y-S, Ma L, Tess ER, Farrell MJ, Georgescu A, et al (2018a) Placenta-on-a-chip: placental drug transport-on-a-chip: a microengineered in vitro model of transporter-mediated drug efflux in the human placental barrier. Adv Healthc Mater 7(2):1870008. https://doi.org/10.1002/adhm.201870008
86. Blundell C, Yi Y-S, Ma L, Tess ER, Farrell MJ, Georgescu A, Huh D (2018b) Placental drug transport-on-a-chip: a microengineered in vitro model of transporter-mediated drug efflux in the human placental barrier. Adv Healthc Mater 7(2):1700786. https://doi.org/10.1002/adhm.201700786
87. Lee JS, Romero R, Han YM, Kim HC, Kim CJ, Hong J-S, Huh D (2016) Placenta-on-a-chip: a novel platform to study the biology of the human placenta. J Matern Fetal Neonatal Med 29(7):1046–1054. https://doi.org/10.3109/14767058.2015.1038518
88. Cheng Y-C, Tsao C-W, Chiang M-Z, Chung C-A, Chien C-C, Hu W-W, Li C (2015) Microfluidic platform for human placenta-derived multipotent stem cells culture and applied for

enhanced neuronal differentiation. Microfluid Nanofluid 18(4):587–598. https://doi.org/10.1007/s10404-014-1455-2
89. Ramadan Q, Ting FCW (2016) In vitro micro-physiological immune-competent model of the human skin. Lab Chip 16(10):1899–1908. https://doi.org/10.1039/C6LC00229C
90. Wufuer M, Lee G, Hur W, Jeon B, Kim BJ, Choi TH, Lee S (2016) Skin-on-a-chip model simulating inflammation, edema and drug-based treatment. Sci Rep 6(1):37471. https://doi.org/10.1038/srep37471
91. Jeon B, Lee G, Wufuer M, Huang Y, Choi Y, Kim S, Choi TH (2020) Enhanced predictive capacity using dual-parameter chip model that simulates physiological skin irritation. Toxicol In Vitro 68:104955. https://doi.org/10.1016/j.tiv.2020.104955
92. Kim JJ, Ellett F, Thomas CN, Jalali F, Anderson RR, Irimia D, Raff AB (2019) A microscale, full-thickness, human skin on a chip assay simulating neutrophil responses to skin infection and antibiotic treatments. Lab Chip 19(18):3094–3103. https://doi.org/10.1039/C9LC00399A
93. Kim J, Kim K, Sung GY (2020) Coenzyme Q10 efficacy test for human skin equivalents using a pumpless skin-on-a-chip system. Int J Mol Sci 21(22):8475
94. Butler L, Cros C, Oldman KL, Harmer AR, Pointon A, Pollard CE, Abi-Gerges N (2015) Enhanced characterization of contractility in cardiomyocytes during early drug safety assessment. Toxicol Sci 145(2):396–406. https://doi.org/10.1093/toxsci/kfv062
95. Grosberg A, Nesmith AP, Goss JA, Brigham MD, McCain ML, Parker KK (2012) Muscle on a chip: in vitro contractility assays for smooth and striated muscle. J Pharmacol Toxicol Methods 65(3):126–135. https://doi.org/10.1016/j.vascn.2012.04.001
96. Yeong W-Y, Chua C-K, Leong K-F, Chandrasekaran M (2004) Rapid prototyping in tissue engineering: challenges and potential. Trends Biotechnol 22(12):643–652. https://doi.org/10.1016/j.tibtech.2004.10.004
97. Kim P, Kwon KW, Park MC, Lee SH, Kim SM, Suh KY (2008) Soft lithography for microfluidics: a review
98. Lee H, Koo Y, Yeo M, Kim S, Kim GH (2017) Recent cell printing systems for tissue engineering. 3(1):15. https://doi.org/10.18063/ijb.2017.01.004
99. Wan L, Flegle J, Ozdoganlar B, LeDuc PR (2020) Toward vasculature in skeletal muscle-on-a-chip through thermo-responsive sacrificial templates. Micromachines 11(10):907
100. Duffy RM, Sun Y, Feinberg AW (2016) Understanding the role of ECM protein composition and geometric micropatterning for engineering human skeletal muscle. Ann Biomed Eng 44(6):2076–2089. https://doi.org/10.1007/s10439-016-1592-8
101. Wang RY, Abbott RD, Zieba A, Borowsky FE, Kaplan DL (2017) Development of a three-dimensional adipose tissue model for studying embryonic exposures to obesogenic chemicals. Ann Biomed Eng 45(7):1807–1818. https://doi.org/10.1007/s10439-016-1752-x
102. Tanataweethum N, Zelaya A, Yang F, Cohen RN, Brey EM, Bhushan A (2018) Establishment and characterization of a primary murine adipose tissue-chip. Biotechnol Bioeng 115(8):1979–1987. https://doi.org/10.1002/bit.26711
103. Liu Y, Kongsuphol P, Chiam SY, Zhang QX, Gourikutty SBN, Saha S, Ramadan Q (2019) Adipose-on-a-chip: a dynamic microphysiological in vitro model of the human adipose for immune-metabolic analysis in type II diabetes. Lab Chip 19(2):241–253. https://doi.org/10.1039/C8LC00481A
104. El-Ali J, Sorger PK, Jensen KF (2006) Cells on chips. Nature 442(7101):403–411. https://doi.org/10.1038/nature05063
105. Choi Y, Hyun E, Seo J, Blundell C, Kim HC, Lee E, Huh D (2015) A microengineered pathophysiological model of early-stage breast cancer. Lab Chip 15(16):3350–3357. https://doi.org/10.1039/C5LC00514K
106. Wnorowski A, Yang H, Wu JC (2019) Progress, obstacles, and limitations in the use of stem cells in organ-on-a-chip models. Adv Drug Deliv Rev 140:3–11. https://doi.org/10.1016/j.addr.2018.06.001
107. Ho C-T, Lin R-Z, Chen R-J, Chin C-K, Gong S-E, Chang H-Y, Liu C-H (2013) Liver-cell patterning lab chip: mimicking the morphology of liver lobule tissue. Lab Chip 13(18):3578–3587. https://doi.org/10.1039/C3LC50402F

108. Castiello FR, Heileman K, Tabrizian M (2016) Microfluidic perfusion systems for secretion fingerprint analysis of pancreatic islets: applications, challenges and opportunities. Lab Chip 16(3):409–431. https://doi.org/10.1039/C5LC01046B
109. Chang CJ, Yen ML, Chen YC, Chien CC, Huang HI, Bai CH, Yen BL (2006) Placenta-derived multipotent cells exhibit immunosuppressive properties that are enhanced in the presence of interferon-gamma. Stem Cells 24(11):2466–2477. https://doi.org/10.1634/stemcells.2006-0071
110. Qian T, Shusta EV, Palecek SP (2015) Advances in microfluidic platforms for analyzing and regulating human pluripotent stem cells. Curr Opin Genet Dev 34:54–60. https://doi.org/10.1016/j.gde.2015.07.007
111. Lee SH, Sung JH (2017) Microtechnology-based multi-organ models. Bioengineering 4(2):46
112. Zhang B, Montgomery M, Chamberlain MD, Ogawa S, Korolj A, Pahnke A, Radisic M (2016) Biodegradable scaffold with built-in vasculature for organ-on-a-chip engineering and direct surgical anastomosis. Nat Mater 15(6):669–678. https://doi.org/10.1038/nmat4570
113. Wagner I, Materne E-M, Brincker S, Süßbier U, Frädrich C, Busek M, Marx U (2013) A dynamic multi-organ-chip for long-term cultivation and substance testing proven by 3D human liver and skin tissue co-culture. Lab Chip 13(18):3538–3547. https://doi.org/10.1039/C3LC50234A
114. Rogal J, Probst C, Loskill P (2017) Integration concepts for multi-organ chips: how to maintain flexibility?! Fut Sci OA 3(2):FSO180. https://doi.org/10.4155/fsoa-2016-0092
115. Tsamandouras N, Chen WLK, Edington CD, Stokes CL, Griffith LG, Cirit M (2017) Integrated gut and liver microphysiological systems for quantitative in vitro pharmacokinetic studies. AAPS J 19(5):1499–1512. https://doi.org/10.1208/s12248-017-0122-4
116. van Midwoud PM, Merema MT, Verpoorte E, Groothuis GMM (2010) A microfluidic approach for in vitro assessment of interorgan interactions in drug metabolism using intestinal and liver slices. Lab Chip 10(20):2778–2786. https://doi.org/10.1039/C0LC00043D
117. Skardal A, Murphy SV, Devarasetty M, Mead I, Kang H-W, Seol Y-J, Atala A (2017) Multi-tissue interactions in an integrated three-tissue organ-on-a-chip platform. Sci Rep 7(1):8837. https://doi.org/10.1038/s41598-017-08879-x
118. Maschmeyer I, Lorenz AK, Schimek K, Hasenberg T, Ramme AP, Hübner J, Marx U (2015) A four-organ-chip for interconnected long-term co-culture of human intestine, liver, skin and kidney equivalents. Lab Chip 15(12):2688–2699. https://doi.org/10.1039/C5LC00392J
119. Oleaga C, Bernabini C, Smith AST, Srinivasan B, Jackson M, McLamb W, Hickman JJ (2016) Multi-organ toxicity demonstration in a functional human in vitro system composed of four organs. Sci Rep 6(1):20030. https://doi.org/10.1038/srep20030
120. Edington CD, Chen WLK, Geishecker E, Kassis T, Soenksen LR, Bhushan BM, Griffith LG (2018) Interconnected microphysiological systems for quantitative biology and pharmacology studies. Sci Rep 8(1):4530. https://doi.org/10.1038/s41598-018-22749-0
121. Satoh T, Sugiura S, Shin K, Onuki-Nagasaki R, Ishida S, Kikuchi K, Kanamori T (2018) A multi-throughput multi-organ-on-a-chip system on a plate formatted pneumatic pressure-driven medium circulation platform. Lab Chip 18(1):115–125. https://doi.org/10.1039/C7LC00952F
122. Díaz Lantada A, Pfleging W, Besser H, Guttmann M, Wissmann M, Plewa K, García-Ruíz JP (2018) Research on the methods for the mass production of multi-scale organs-on-chips. Polymers 10(11):1238
123. Kühnl J, Tao TP, Brandmair K, Gerlach S, Rings T, Müller-Vieira U, Maschmeyer I (2021) Characterization of application scenario-dependent pharmacokinetics and pharmacodynamic properties of permethrin and hyperforin in a dynamic skin and liver multi-organ-chip model. Toxicology 448:152637. https://doi.org/10.1016/j.tox.2020.152637
124. Esch MB, Ueno H, Applegate DR, Shuler ML (2016) Modular, pumpless body-on-a-chip platform for the co-culture of GI tract epithelium and 3D primary liver tissue. Lab Chip 16(14):2719–2729. https://doi.org/10.1039/C6LC00461J
125. Lau YY, Chen YH, Liu TT, Li C, Cui X, White RE, Cheng KC (2004) Evaluation of a novel in vitro Caco-2 hepatocyte hybrid system for predicting in vivo oral bioavailability. Drug Metab Dispos 32(9):937–942

126. Bauer S, Wennberg Huldt C, Kanebratt KP, Durieux I, Gunne D, Andersson S, Andersson TB (2017) Functional coupling of human pancreatic islets and liver spheroids on-a-chip: towards a novel human ex vivo type 2 diabetes model. Sci Rep 7(1):14620. https://doi.org/10.1038/s41598-017-14815-w
127. Ma L, Barker J, Zhou C, Li W, Zhang J, Lin B, Honkakoski P (2012) Towards personalized medicine with a three-dimensional micro-scale perfusion-based two-chamber tissue model system. Biomaterials 33(17):4353–4361. https://doi.org/10.1016/j.biomaterials.2012.02.054
128. Li AP, Uzgare A, LaForge YS (2012) Definition of metabolism-dependent xenobiotic toxicity with co-cultures of human hepatocytes and mouse 3T3 fibroblasts in the novel integrated discrete multiple organ co-culture (IdMOC) experimental system: results with model toxicants aflatoxin B1, cyclophosphamide and tamoxifen. Chem Biol Interact 199(1):1–8. https://doi.org/10.1016/j.cbi.2012.05.003
129. Li AP, Bode C, Sakai Y (2004) A novel in vitro system, the integrated discrete multiple organ cell culture (IdMOC) system, for the evaluation of human drug toxicity: comparative cytotoxicity of tamoxifen towards normal human cells from five major organs and MCF-7 adenocarcinoma breast cancer cells. Chem Biol Interact 150(1):129–136. https://doi.org/10.1016/j.cbi.2004.09.010
130. Zhang C, Zhao Z, Abdul Rahim NA, van Noort D, Yu H (2009) Towards a human-on-chip: culturing multiple cell types on a chip with compartmentalized microenvironments. Lab Chip 9(22):3185–3192. https://doi.org/10.1039/B915147H
131. Loskill P, Marcus SG, Mathur A, Reese WM, Healy KE (2015) μOrgano: a Lego®-Like plug & play system for modular multi-organ-chips. PLoS ONE 10(10):e0139587. https://doi.org/10.1371/journal.pone.0139587
132. Villasante A, Vunjak-Novakovic G (2015) Tissue-engineered models of human tumors for cancer research. Expert Opin Drug Discov 10(3):257–268. https://doi.org/10.1517/17460441.2015.1009442
133. Hassell BA, Goyal G, Lee E, Sontheimer-Phelps A, Levy O, Chen CS, Ingber DE (2017) Human organ chip models recapitulate orthotopic lung cancer growth, therapeutic responses, and tumor dormancy in vitro. Cell Rep 21(2):508–516. https://doi.org/10.1016/j.celrep.2017.09.043
134. Xu Z, Li E, Guo Z, Yu R, Hao H, Xu Y, Wang Q (2016) Design and construction of a multi-organ microfluidic chip mimicking the in vivo microenvironment of lung cancer metastasis. ACS Appl Mater Interfaces 8(39):25840–25847. https://doi.org/10.1021/acsami.6b08746
135. Janzer RC, Raff MC (1987) Astrocytes induce blood–brain barrier properties in endothelial cells. Nature 325(6101):253–257. https://doi.org/10.1038/325253a0
136. Ugolini GS, Occhetta P, Saccani A, Re F, Krol S, Rasponi M, Redaelli A (2018) Design and validation of a microfluidic device for blood–brain barrier monitoring and transport studies. J Micromech Microeng 28
137. Wu Y-Y, Chiu F-L, Yeh C-S, Kuo H-C (2019) Opportunities and challenges for the use of induced pluripotent stem cells in modelling neurodegenerative disease. Open Biol 9(1):180177. https://doi.org/10.1098/rsob.180177
138. Sarkar U, Ravindra KC, Large E, Young CL, Rivera-Burgos D, Yu J, Tannenbaum SR (2017) Integrated assessment of diclofenac biotransformation, pharmacokinetics, and omics-based toxicity in a three-dimensional human liver-immunocompetent coculture system. Drug Metab Dispos 45(7):855–866. https://doi.org/10.1124/dmd.116.074005
139. Baudy AR, Otieno MA, Hewitt P, Gan J, Roth A, Keller D, Proctor WR (2020) Liver microphysiological systems development guidelines for safety risk assessment in the pharmaceutical industry. Lab Chip 20(2):215–225. https://doi.org/10.1039/C9LC00768G
140. Zeinali S, Bichsel CA, Hobi N, Funke M, Marti TM, Schmid RA, Geiser T (2018) Human microvasculature-on-a chip: anti-neovasculogenic effect of nintedanib in vitro. Angiogenesis 21(4):861–871. https://doi.org/10.1007/s10456-018-9631-8
141. Baert Y, Ruetschle I, Cools W, Oehme A, Lorenz A, Marx U, Maschmeyer I (2020) A multi-organ-chip co-culture of liver and testis equivalents: a first step toward a systemic male reprotoxicity model. Hum Reprod 35(5):1029–1044. https://doi.org/10.1093/humrep/deaa057

142. Cohen A, Ioannidis K, Ehrlich A, Regenbaum S, Cohen M, Ayyash M, Nahmias Y (2021) Mechanism and reversal of drug-induced nephrotoxicity on a chip. Sci Transl Med 13(582):eabd6299. https://doi.org/10.1126/scitranslmed.abd6299
143. Khoo BL, Grenci G, Lim YB, Lee SC, Han J, Lim CT (2018) Expansion of patient-derived circulating tumor cells from liquid biopsies using a CTC microfluidic culture device. Nat Protoc 13(1):34–58. https://doi.org/10.1038/nprot.2017.125
144. Zhang YS, Zhang Y-N, Zhang W (2017) Cancer-on-a-chip systems at the frontier of nanomedicine. Drug Discov Today 22(9):1392–1399. https://doi.org/10.1016/j.drudis.2017.03.011
145. Albanese A, Lam AK, Sykes EA, Rocheleau JV, Chan WCW (2013) Tumour-on-a-chip provides an optical window into nanoparticle tissue transport. Nat Commun 4(1):2718. https://doi.org/10.1038/ncomms3718
146. Alle M, Sharma G, Lee S-H, Kim J-C (2022) Next-generation engineered nanogold for multimodal cancer therapy and imaging: a clinical perspectives. J Nanobiotechnol 20(1):222. https://doi.org/10.1186/s12951-022-01402-z
147. Alle M, Kim TH, Park SH, Lee S-H, Kim J-C (2020) Doxorubicin-carboxymethyl xanthan gum capped gold nanoparticles: microwave synthesis, characterization, and anti-cancer activity. Carbohyd Polym 229:115511. https://doi.org/10.1016/j.carbpol.2019.115511
148. Gotor R, Ashokkumar P, Hecht M, Keil K, Rurack K (2017) Optical pH sensor covering the range from pH 0–14 compatible with mobile-device readout and based on a set of rationally designed indicator dyes. Anal Chem 89(16):8437–8444. https://doi.org/10.1021/acs.analchem.7b01903
149. van den Berg A, Mummery CL, Passier R, van der Meer AD (2019) Personalised organs-on-chips: functional testing for precision medicine. Lab Chip 19(2):198–205. https://doi.org/10.1039/C8LC00827B
150. Benam KH, Villenave R, Lucchesi C, Varone A, Hubeau C, Lee H-H, Ingber DE (2016) Small airway-on-a-chip enables analysis of human lung inflammation and drug responses in vitro. Nat Methods 13(2):151–157. https://doi.org/10.1038/nmeth.3697
151. de Sousa e Melo F, Kurtova AV, Harnoss JM, Kljavin N, Hoeck JD, Hung J, et al (2017) A distinct role for Lgr5+ stem cells in primary and metastatic colon cancer. Nature 543(7647):676–680. https://doi.org/10.1038/nature21713
152. Abadpour S, Aizenshtadt A, Olsen PA, Shoji K, Wilson SR, Krauss S, Scholz H (2020) Pancreas-on-a-chip technology for transplantation applications. Curr DiabRep 20(12):72. https://doi.org/10.1007/s11892-020-01357-1
153. Ye L, Li L, Wan B, Yang M, Hong J, Gu W, Ning G (2017) Immune response after autologous hematopoietic stem cell transplantation in type 1 diabetes mellitus. Stem Cell Res Ther 8(1):90. https://doi.org/10.1186/s13287-017-0542-1
154. Brath E, Miko I, Kovacs J, Toth FF, Fachet J, Furka I (2003) Multiorgan transplantation with a new organ-chip technique in mice: Preliminary histological data. Microsurgery 23(5):466–469. https://doi.org/10.1002/micr.10169
155. Marx U, Andersson TB, Bahinski A, Beilmann M, Beken S, Cassee FR, et al (2016) Biology-inspired microphysiological system approaches to solve the prediction dilemma of substance testing. Altex 33(3):272–321. https://doi.org/10.14573/altex.1603161
156. Halldorsson S, Lucumi E, Gómez-Sjöberg R, Fleming RMT (2015) Advantages and challenges of microfluidic cell culture in polydimethylsiloxane devices. Biosens Bioelectron 63:218–231. https://doi.org/10.1016/j.bios.2014.07.029
157. Peres MA, Macpherson LMD, Weyant RJ, Daly B, Venturelli R, Mathur MR, Watt RG (2019) Oral diseases: a global public health challenge. The Lancet 394(10194):249–260. https://doi.org/10.1016/S0140-6736(19)31146-8
158. França CM, Tahayeri A, Rodrigues NS, Ferdosian S, Puppin Rontani RM, Sereda G, Bertassoni LE (2020) The tooth on-a-chip: a microphysiologic model system mimicking the biologic interface of the tooth with biomaterials. Lab Chip 20(2):405–413. https://doi.org/10.1039/C9LC00915A

159. Rahimi C, Rahimi B, Padova D, Rooholghodos SA, Bienek DR, Luo X, Raub CB (2018) Oral mucosa-on-a-chip to assess layer-specific responses to bacteria and dental materials. Biomicrofluidics 12(5):054106. https://doi.org/10.1063/1.5048938
160. Niu L, Zhang H, Liu Y, Wang Y, Li A, Liu R, Yang Q (2019) Microfluidic chip for odontoblasts in vitro. ACS Biomater Sci Eng 5(9):4844–4851. https://doi.org/10.1021/acsbiomaterials.9b00743
161. Capulli AK, Tian K, Mehandru N, Bukhta A, Choudhury SF, Suchyta M, Parker KK (2014) Approaching the in vitro clinical trial: engineering organs on chips. Lab Chip 14(17):3181–3186. https://doi.org/10.1039/C4LC00276H
162. Rusu E, Necula LG, Neagu AI, Alecu M, Stan C, Albulescu R, Tanase CP (2016) Current status of stem cell therapy: opportunities and limitations. Turk J Biol 40(5):955–967
163. Fuller HC, Wei T-Y, Behrens MR, Ruder WC (2020) The future application of organ-on-a-chip technologies as proving grounds for microbioRobots. Micromachines 11(10):947
164. Jalili-Firoozinezhad S, Miranda CC, Cabral JMS (2021) Modeling the Human Body on Microfluidic Chips. Trends Biotechnol 39(8):838–852. https://doi.org/10.1016/j.tibtech.2021.01.004
165. Kim K, Jeon HM, Choi KC, Sung GY (2020) Testing the effectiveness of curcuma longa leaf extract on a skin equivalent using a pumpless skin-on-a-chip model. Int J Mol Sci 21(11):3898

Nanotherapeutics for Rheumatoid Arthritis Therapy

Poonam Jain, K. Gireesh Babu, Alle Madhusudhan, and Sashikantha Reddy Pulikallu

Abstract Rheumatoid arthritis is an autoimmune disease that chronically depromotes the value of life, because of the destruction of bone and cartilage, swelling comes in the joint along with the pain and affects nearly 0.5 to 1% of the world's population. The cause of RhA remained unknown, however, the complex formation or interaction of the body's immune system including cytokines and effector cells, activation of effector cells responsible for pathogenicity. The treatment involves various traditional approaches such as DMARDs, anti-TNF-α treatment such as infliximab, adalimumab, etanercept, certolizumab, and golimumab have shown interesting results; however, the dose frequency for longer time of treatment causes several adverse effects. However, nanotechnology-based drug delivery now plays an important role in delivery of the many therapeutic agents, while several metal-based nanoparticle systems itself cause anti-inflammatory activities. Hence, in this book chapter we are providing detailed knowledge of Rheumatoid arthritis and its causes, traditional drugs and nanodrug and metal-based nanodrug delivery system which have been shown excellent in the in vitro and in vivo studies, therefore evading systemic and unpleasant effects.

Keywords Rheumatoid arthritis · DMARDs · Cartilage destruction · Nanotechnology · Metal-based nanoparticle

P. Jain · K. G. Babu (✉)
Department of Life Sciences, Parul Institute of Applied Sciences, Parul University, Limda, Vadodara, India
e-mail: gireesh.k18684@paruluniversity.ac.in

A. Madhusudhan (✉)
Department of Chemistry, The University of Memphis, Memphis, TN 38152, USA
e-mail: allemadhusudhan@gmail.com

S. R. Pulikallu
St. Jude Children's Research Hospital, Memphis, TN, USA

A. Madhusudhan et al. (eds.), *Functional Smart Nanomaterials and Their Theranostics Approaches*, Smart Nanomaterials Technology,
https://doi.org/10.1007/978-981-99-6597-7_11

1 Introduction

Rheumatoid arthritis (RhA) causes inflammation of the synovial joints chronically and finally causes damage and deformation of the joints. Although RhA is not often a deadly health problem, its complications, such as cardiac and lung problems associated with the mortality [2, 15, 26]. Some common situations like Carpal tunnel syndrome, tendon rupture, cervical myelopathy, and joint degeneration are all frequent disorders in RhA-suffered patients [9, 42]. In addition to this, there are some side effects such as Sjögren's syndrome, pericarditis, and/or sexual disturbance are other inflammatory side effects [12, 40].

It is more defined than other autoimmune disorders since it may lead to numerous in women, which suggests a possible involvement for sex hormones. A 2–3:1 ratio of women to men experiences the clinical manifestation of RhA between the ages of 18 and 38 [4, 34]. Many symptoms of the RhA are the chronic inflammation that shows the involvement of T lymphocyte in RhA blood-derived cells, primarily memory T cells, macro-infiltration and synovial joint inflammation is not well known. Absence of T cell responses, presence of plasma cells showing signs of proliferation, and limited activation of T cell-derived cytokines all indicate a particular pattern of immune system activity.

In the later stages, this disease starts invasion and results in the gradual destruction of gusted that T lymphocytes may not be necessary in maintaining cartilage and bone [18, 25]. Hence, the possibility may be due to synovial tissue which is primarily carried by cytokine induction of destructive agents intended to reduce T cell role in late-stage enzymes, primarily matrix metalloproteinases. Hence, several studies and trials support the reason for the RhA in the patients. As per the World Health Organization reports, up to the one-decade onset, in the developed and urban countries about 50% of patients are unable to maintain a full-time employment, probably owing to the impairment that results. As per the data, the yearly incident rate, the prevalence of RA is 0.15–0.6, one thousand people in Europe mainly in the northern side and the USA, and it affects 0.5–1.0% of people globally [18]. Besides the several factor, female gender, age, tobacco use, smoking, exposure of element silica, and obesity are also some of the other risk factors that causes RhA worldwide chronic rheumatic conditions (Available online: http://www.who.int/chp/topics/rheumatic/en/).

Moreover, symptomatic synovitis and the onset and spread of autoimmunity against changed auto-proteins, which can happen years before clinical symptoms appear, are all signs of disease progression [32].

2 Pathogenic Physiology of Rheumatoid Arthritis

Several immune modulators such as cytokines, interleukins, and signaling pathways contribute to the pathogenicity of RhA. Hence, joint inflammation and damage start with a complex interaction of immune modulators that mainly begin in synovial membrane (Fig. 1). Furthermore, owing to occurrence or lack of anti-citrullinated protein antibodies, there are two main subcategories of RA (ACPAs). Citrullination is a post-translational changes that is carried out by the calcium-dependent enzyme peptidyl-arginine-deiminase (PAD) that alters an arginine molecule into a polar but neutral citrulline. About 67% of patients show ACPAs symptoms and acted as first diagnosis parameters at their early stage. In comparison with ACPA-negative fraction of RhA, the ACPA-positive one exhibits a more aggressive clinical profile. Hence, anti-citrullinated protein antibodies (ACPA), rheumatoid factors (RFs), and pro-inflammatory cytokines are active protein physiologically secreted by B-cells in support of RhA. It has been studied that ACPA-negative RA shows varied genetic association patterns. Methotrexate (MTX) and rituximab are some of the drugs for the ACPA-negative groups of RA.

Therefore, immune system including various B-cells, T-cells, and macrophages are major in the pathophysiology of RhA since it is an autoimmune condition. The synovium or the peripheral blood may have such cells. The activation of T-cell is also governed by B-cells by the making of costimulatory molecules. Hence, the primary role is to trigger fibroblasts and macrophages, and damage tissues. The triggers immune cells further release variety of cytokines and chemokines that are again associated with joint pain and damage. Thus, in this section, the pathogenesis of RA

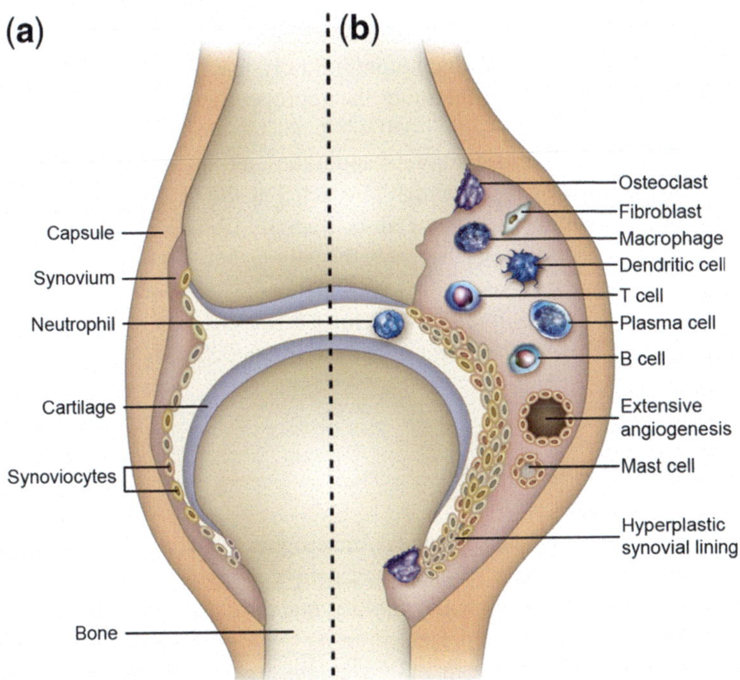

Fig. 1 Rheumatoid arthritis and accumulation of inflammatory immune cells in the synovial tissue in the joints

will be described by B lymphocyte, T lymphocyte and macrophages and another immune cells.

2.1 B Cells or Lymphocytes

B-cells play a significant role in adaptive immunity, however, they also help as one of the fundamental causes of RhA onset. B-cells that are auto-reactive are those that recognize host antigens and then continue to destroy the corresponding tissues. In the bone marrow, auto-reactive B-cells are often removed by repairing mechanisms mostly when the B-cells grow from early immature to mature or before they mature into naive B-cells. Two immune factor-based checkpoints exist. The first a) the central tolerance checkpoints that are controlled by B-cell GFs and further control B-cell receptor (BCR) and toll-like receptor (TLR) signaling. The second b) deals with peripheral B-cell tolerance, which is made up of extrinsic B-cell components such as regulatory T cells (Tregs) and serum B-cell activation factor (BAFF).

Hence, the person suffered from RhA shows the defect of these above checkpoints and synthesis of auto-reactive matured naïve B-cells on a larger scale. For example, studies have shown the production of auto-reactive B-cells in the peripheral with a 3–5 times increase than the healthy person. The possible reason behind this is the mutation in the BCR responsive gene, i.e., PTPN22 gene that increases the number of mature naïve B-cells as well as which represent poly-reactive and human epithelial (HEP-2)-reactive antibodies. Furthermore, the improper functioning in the peripheral checkpoint causes fault in Tregs and B-cell resistance for apoptosis. The auto-reactive B-cells also function in the production of antibodies automatically. Hence, the RFs and ACPA are two antibody groups that also acted as the best diagnostic markers. The most common cytokines such as TNF-α, IL-6, IL-12, IL-23, and IL-1α usually confined in the auto-reactive cells that show the pathogenesis in RhA cells that cause bone damage, inflammation as well as dysfunction in immune cells. In addition to these causes and reasons, the construction of RANKL via memory cells also associated with bone resorption effect is carried by osteoclasts (OCs).

2.2 T-Cells or Lymphocytes

However, T-cells could also be activated by macrophages and dendritic cells (DCs). Now, reports also proven that RhA is caused by CD4 + T-cells connected with leukocyte antigen or major histocompatibility mainly class II content, CD28, co-stimulatory molecules that are represented on APCs. Moreover, these type of interaction lead to onset of downstream PI3K signaling pathway that helps in CD4 + cells maturation that stimulate on set of naïve CD8 + cells and cause inflammation and chronically increases the chronic autoimmune response. The particular

MHC-II alleles, HLA-DR4 show the same amino acid motifs in the third hypervariable region of DRB-chain and are associated with CD4 + T-cells-based aggressively causes RhA chronic bone inflammation that is further related to systemic morbidities like vasculitis and acute coronary syndrome. Despite this, CD4 + T-helper was also found as causative reason for RhA inflammation and autoimmune disease. In RhA, the high activation of Type 1 T-helper cells causes the secretion of pro-inflammatory cytokines. For example, IFN-gamma (IFN-γ), IL-2 and TNF-α. In addition to this, Type 1 T-helper cells also secrete IFN-gamma IL-2 and TNF-α and stimulate macrophages to act as an APC to present MHC-II molecules to the T-cells. Next, the type of CD4 + type 2 T-helper (Th2) cells show essential roles in B-cell activation and immunoglobulin (Ig) class moving to IgE along with secretion of IL-4 and IL-5. Some other helper T-cells primarily secrete IL-17 that triggers the yielding of pro-inflammatory cytokines, chemokines, and matrix metalloproteinases (MMPs) and causes RhA pathogenesis.

The other immune cells like mast cells and macrophages secret the main source of IL-17 along with helper T-17 cells, while IL-17 boosts the yielding of vascular endothelial growth factor-A (VEGF-A), IL-6, IL-8, MMP-1, and MMP-3 in RhA synovial fibroblasts. In the synovial tissues and fluid, Th-9 cells secrete IL-9 and extend the neutrophils survival rate, stimulated the making of MMP-9, and facilitated Th17 cell differentiation.

Although the prevention of autoimmunity is occurred by CD4 + CD25 + Tregs cells via suppressing auto-reactive lymphocytes facilitated by IL-10 and transforming growth factor-beta (TGF-β), but failed to suppress TNF-α and IL-6, pro-inflammatory cytokines that are secreted from activated T-cells and monocytes and results in the presence of Tregs cells in the RhA's patients blood. One another T-cells, i.e., T follicular helper (Tfh) combine with B-cells mainly within the follicles of secondary lymphoid organs as well as tissues of inflamed peripheral specifically by extremely expressing Bcl6, CXCR5, CXCL13, IL-21, PD-1, and ICOS. Hence, overproduction of the expressive B-cell-helper phenotype causes RhA pathogenesis.

Altogether, multiple T-cells and the respective effector pathways cause RhA through chiefly facilitating the chronic inflammatory method.

2.3 Macrophages

In the normal conditions, the synovial membrane is acellular, where fibroblast stromal tissues and blood vessels together with macrophages. Hence, the dendritic-like cells are found majorly in the perivascular areas of the sub-lining layer. Among several immune cells that are involved in RhA, macrophages are found in hyper-condition in synovial tissues, fluid and control the pro-inflammatory cytokines secretion and damage enzymes. The results of the secretion are relatively caused inflammatory responses and joint destruction over the time of period. Hence, extreme discomfort and loss of joint function are frequently the results of inflammation, which causes pannus development, and the breakdown of cartilage and bone. Next, the macrophages also produce cytokines and mediate several activities such as lymphocyte maturation, damage in bones, angiogenesis, and erosion of joints. Furthermore, effector T-cells are produced because of the M-mediated T-cell activation, as well as pro-inflammatory mediators such as IL-1, IL-1, and MMPs that help the autoimmune disease, i.e., RA. The pro-inflammatory phenotype (M1) can change into an anti-inflammatory phenotype because of the switchable nature of macrophages (M2). These cells basically act as professional antigen-presenting cells (APCs) that generate T-cell responses and are associated significantly with joint destruction. Its obvious increase in number in the inflammatory synovial membrane and at the cartilage-pannus interface emphasizes their crucial relevance. Therefore, many cytokines and chemokines, such as TNF, IL-1, IL-6, IL-10, IL-12, IL18, IL-15, IL-10, GM-CSF, M-CSF, and TGF, remain expressed in the synovium of people with RhA, and they are responsible for the chronicity of inflammation and illness. Notably, the synovial membrane and the cartilage-pannus interface are where pathogenic TNF is mostly produced by monocytes and macrophages. In vivo treatment of a monoclonal antibody against human TNF significantly halted the evolution of the illness, according to preliminary research using a new model expressing a modified TNF gene. Transgenic mice developed chronic inflammatory polyarthritis.

Despite the TNF, another proinflammatory mediator, IL-1 cytokine, enhanced in RhA synovial membrane and closely connected to inflammation of joint. It is mostly produced by CD14 + macrophages. Its synthesis is controlled by TNF. MMPs, iNOS, and RANKL are all powerfully induced by IL-1, among many other things. On the other hand, only a small portion of IL-6 is generated by macrophages or lymphocytes, with the majority originating from synovial fibroblasts.

Proinflammatory fluids generated through macrophages and IL-15, an IL2-like cytokine, and IL-18, an IL1-like cytokine, are essential for the control of T lymphocyte activity. It is proven from some studies that IL-15 and IL-18, which have both been found in tissues of synovial of RhA patients and in arthritic models at highly higher levels, maintain Th1 responses in RhA joints. NF-κB is a transcription factor and as it controls the presence of several pro-inflammatory cytokines, such as TNF, IL-1, and IL-6, without altering the expression of anti-inflammatory cytokines like IL-10, IL-11, and IL-1ra, NF-κB activation is crucial for the pathophysiology of RhA. Interestingly, RhA fibroblasts secrete receptor activator of NF-κB ligand (RANKL) and macrophage colony-stimulating factor 1 (M-CSF, also known as CSF-1) promoting the formation of osteoclasts and subsequent bone resorption. Lastly, there is proof that synovial fibroblasts may serve as unusual antigen-presenting cells that intensify local adaptive immune responses, adding to the pathogenesis of illness.

2.4 *Natural Killer Cell, Dendritic, and Mast Cells as Causative Agent for RhA Pathogenesis*

Human body has several immune cells, for example, B-cells, T-cells and macrophages, and memory cells. Among these, some other cells, for instance, mast cells, dendritic cells as well as natural killers, play critical role in the pathogenicity of RhA.

It is well known that mast cells are localized in the synovial tissue and main causative agent of irritation and inflammation also investigated as a possible RhA-Synovial biomarker. Tissues of synovial joint of RhA patients possess the increased amount of activated DCs. DCs function a critical role in together stimulating joint inflammation and sustaining the pro-inflammatory milieu of synovial linings. The diseased function of NK cells, despite this, significant innate immune system cell type are documented in RhA. Inflamed joints were shown to have an overexpression of CD56 + NK cells, which generated more IFN-gamma than NK cells from peripheral blood. Hence, these above-described cells warranted decipher for the multifaceted role of numerous immune cells and other related constituents in RhA pathogenesis.

3 Conventional and FDA-Approved Drugs for the Treatment of Rheumatoid Arthritis

As mentioned in Sects. 1 and 2 of the book chapter, the RhA is an autoimmune disease in which patients suffered chronically by inflammatory, destructive joint disease due to which approximately 0.4–1% of the whole population is significantly disabled and facing the death-related problems. RhA has a prevalence of about 0.3%–4.0% worldwide but still a big for global public health challenges. Current treatment yet not has able to able to understand the etiology of RhA and that is the main reason RhA-based autoimmune disease yet could not cured entirely. Hence, critically in the current state for the quantitative analysis, many research is being focused on the B, T, dendritic cells, macrophages, and also NK-related RhA pathogenesis. Due to Carpal tunnel syndrome, tendon rupture, cervical myelopathy, and joint degeneration are all frequent disorders in RhA patients. Sjögren's syndrome, pericarditis, pleuritis, scleritis, vasculitis, and reproductive dysfunction are other inflammatory side effects. Hence, due non-acute nature of RhA, the treatment mainly focus on pain reliever, prevention joint malfunctioning and recover or maintain the biological activity of joint and try to alleviate the upcoming RhA-related syndromes.

3.1 Specific FDA Approved Medicine for Rheumatoid Arthritis

The primary line of therapy for RhA at the moment is Disease-modifying antirheumatic drugs (DMARDs). DMARDs work by modifying many components of inflammatory and immunological responses that result in the clinical symptoms of RhA. The MTX followed by others medications such as hydroxychloroquine, leflunomide, sulfasalazine, azathioprine and cyclosporine are common medication in the process of DMARDs. About 50% patients experiencing radiographically evident destruction in the joints within two years of the disease's inception, on time treatment, it has been advisable that DMARDs to be given to patients within the first quarter of year of the pathogenesis with proven RhA that takes two-5 months for the maximum effects, where anti-inflammatory drugs which are non-steroidal in nature and glucocorticoids could administered to slow down the pain.

Despite this, TNF inhibitors-based treatments have been utilized which are usually based on the TNF targeting and blocking mechanism that mainly check the inflammation signaling to the RhA. There are some most popular medications as anti-TNF-a treatment for RhA persons include infliximab, adalimumab, etanercept, certolizumab, and golimumab. CD20 on the B cells (Rituximab) is recommended immune modulators when patient does not react well to DMARDs and anti-TNF treatment. Along with this, Sarilumab and Tocilizumab are another RhA therapeutic agent and an antagonist for IL-6 receptor that act as immune modulator on the B-cells. Likewise, T-cells-based RhA treatment showed the inhibition to lower down the T-cell-mediated killing of 'own' cells which is the chief reason cause of all autoimmune disorders. In this, the treatment process mainly depends on the possible depletion of B-cell and T-cells. The RhA treatments target cell membrane-connected and soluble factors (mostly cytokines) such as GM-CSF, BAFF, and TNF, IL-6 receptors and CD20. Some anti-TNF agents have actively shown the ability to cause apoptosis in immune cells mainly in both synovial fluid and peripheral blood in vivo. Thus, anti-TNF agent reduces the infiltration process of granulocytes and macrophages, and IL-8 and MCP-1 expression in the RhA synovium [259]. Down-regulation of IL-1 along with some significant pro-inflammatory mediators, for example, GM-CSF, IL-6, and chemokines of inflammation in RhA. There are numerous anti-TNF agents have been approved including adalimumab, infliximab, etanercept, certolizumab, and golimumab for the RhA therapy. Now, anti-RA drugs such as tofacitinib and baricitinib got approval in 2012 and 2018, respectively, preventing the phosphorylation of Janus kinase (JAK) and functioning a significant role in stopping inflammatory action and joint pathogenesis. Recently, ruxolitinib and tofacitinib are proven and approved in the USA as JAK inhibitors. Functionally, these JAK inhibitors stop up the activation, via phosphorylation of JAK; therefore, the intracellular inflammatory process also gets blocked. This group of drugs is used sometimes with the combination of MTX for biological therapies or treatment of RhA. Targeting macrophages has been shown to put the illness into remission due to their significant involvement in RhA, particularly by reducing inflammation and bone degradation. In the case of macrophage and mast cell-based therapy, a macrophage's ability to transition between pro- and anti-inflammatory phenotypes (M1 and M2) is known as switchability. IL-1 receptor antagonist known as Anakinra is utilized for the RhA treatment, but less effective than the other biologics due to its lower effectiveness. Recently, anti-IL antibody-based treatments have shown promising outcomes for the RhA treatment via in vivo mice model studies. For example, great results were published by working on mouse models where the effect of anti-IL-12 Ab showed depletion of arthritis. In another example, a human anti-IL-15 monoclonal Ab, namely HuMax-IL15, was explored successfully both in vitro and in a phase I/II study of RhA patients. In one study, IL-18 bp is a natural inhibitor to IL-18, pausing synovial irritation and destruction in a collagen-induced arthritis (CIA) model.

Along with the mentioned Ab, another kind of neutralizing Ab is on the developing stage against GM-CSF, M-CSF, or MCSF receptor, i.e., Mavrilimumab and phase II clinical trials, and has shown effeteness in the treatment of RhA. Small interfering RNA (siRNA), anti-TNF, and nanosystems are further approaches that target macrophages. Perhaps, anti-rheumatic medications might be created from these medicinal compounds. Therefore, therapy options for RhA that are presently being investigated in clinical trials as well as the current medicines that target the human immune system are covered in this section.

4 Drawback of the Conventional RhA Therapy

However, it is important to note the DMARDs' significant adverse effects. For instance, using biological DMARDs like anti-tumor necrosis factor (TNF) treatment increases the chance of contracting TB. Therefore, it is recommended that patients receiving these treatments promptly get screened for TB. Leflunomide with MTX usage frequently results in adverse reactions including rashes, upset stomach, diarrhea, and abnormal liver enzymes. While kidney damage is a typical side effect of cyclosporine, azathioprine's most common adverse effects include nausea, vomiting, and occasionally stomach discomfort and diarrhea. Anti-TNF medication has also been linked to a higher chance of developing TB, much as DMARDs. The more costly JAK inhibitor medications, such as ruxolitinib and tofacitinib, are typically prescribed to patients who do not react to MTX or other traditional DMARDs.

5 Importance and Designing of Nanomedicine-Based Drug Delivery in Rheumatoid Arthritis

However, the DMARDs have shown the effect in the RhA disease control via disease-modifying anti-rheumatoid drugs (DMARDs) that eliminated the signs, enhances bodily function and lower down the maturation of immune cells along with excess secretion. Along with DMARDS, some other medications such as analgesics, corticosteroids, and non-steroidal anti-inflammatory drugs (NSAIDs) were utilized for the clinical application that showed reduction in RhA disease. The concept of treat to target was proposed by Smolen et al. [32] and proven as extremely efficient in a wide range of treatment modalities. In one clinical trial using among RhA patients, baricitinib is an oral, reversible, and selective Janus kinase (JAK)1/JAK2 inhibitor is frequently utilized. However, due to the likelihood of unusual adverse events of special interest (AESI), it is necessary to regularly monitor and analyze the changing long-term safety profile. For example, malignancies and major adverse cardiovascular events (MACE) [33]. Numerous impairments to the practical application of treat-to-target principles in RhA include drug toxicity, intolerance, and complications with follow-up visits. Hence, to represent the non-effect behavior or longer time of treatment of traditional drugs, currently researchers are focusing on the growth of nanoparticles-based formulation for the RhA treatment. Because the nanoparticles are shown excellent stability, targeting effect, bioavailability and prolonged effect and bioimaging property, the nanomedicines are proposed and many studies have been performed in vitro and in vivo studies. Further, it has been verified that nanoformulation also decreases the side effect of drug and expands the bioavailability at diseased sites, simultaneously reduced the malignancies and MACE-related toxicity effect. Several anti-inflammatory therapeutic drugs, for example, glucocorticoids, NSAIDS, DMARDs, etc., are well-known therapeutic agents for RhA treatment but simultaneously show some side effects like fatigue, nausea, transaminase, and cytopenia. After administration of GCs, infection and hyperglycemia and weight gain osteoporosis and necrosis are another major side effects. Although drugs such as MTX and GCs are still used for the curing of the diseases, meanwhile, utilization of nanocarrier improves the drug efficacy and reduces the side effects. By definition, any substance which having 100 nm scale, at least one dimension, which shows the property of high surface area-to-volume ratio. The nanotherapeutic nanocarrier is broadly classified as a. polymeric, solid lipid nanoparticles, and dendrimers and b. inorganic nanoparticles, c. carbon-based nanoparticles, and their quantum dots. Among them, inorganic nanoparticles such as gold, silver, superparamagnetic iron oxide, manganese oxide, zinc oxide, titanium oxide nanoparticles, and carbon-based quantum dots show imaging properties due to which they are considered as therapeutic nanoparticles but show toxicity issue at some cases. However, polymeric nanoparticles could efficiently load the drug in the capsular structure and drug release in a sustained level and are considered as highly nanomaterials. Nevertheless, nanodrug delivery systems increase blood circulation time, passively convey drug retention duration, minimize dose frequency, and exhibit therapeutic effectiveness at low concentrations.

Based on three different classifications of drug design, for example, nanodrug delivery systems (NDDS), nanodrugs, and enhanced nanotherapy, this book chapter will provide an overview of recent research on nanomedicine for RhA. We seek to shed insight on the viability of nanomedicine in RhA therapy and provide novel approaches to the clinical problem by synthesizing preclinical studies on diverse nanomedicine designs.

5.1 MTX as Conventional Drug Delivery via Nanodrug Delivery System

MTX works as an anti-folate metabolite by interfering in the DNA replication, repair, and synthesis. For clinical case, MTX provides anti-inflammatory effects as well as anti-proliferative effects. MTX is also the preferred strategy for RhA therapy as per the RhA management guidelines and recommendations issued by the ACR and the EULAR. Nevertheless, over 20–30% of RhA-affected person stop using MTX in their first year because they cannot tolerate the adverse effects brought on by the MTX medication, even though the danger of adverse effect may persist for half-decade. Due to MTX-based treatment, the patient also suffered from hematotoxicity, nephrotoxicity, and hepatotoxicity. Other, less common symptoms include respiratory and gastrointestinal problems due to the toxic effect of MTX-based RhA treatment. Due to a less plasma half-life and lesser drug concentration in the diseased site as injected MTX is eliminated from body via kidney, which also limit their use for the RhA treatment. Nanodrug delivery system, however, improves the blood circulation time, retention time of passively transported drugs, reduces the dosing frequency and at low concentration show therapeutic efficacy. For example, liposomal MTX formulation was developed by William et al. [36], for the adjuvant-induced arthritis in rats to improve the efficacy of MTX in the joint. In another study, synovial tissues are treated by liposomal–MTX lowered down the transcript levels of inflammatory factors such as IL-1 and IL-6 than pure MTX [36]. Furthermore, the albumin-based carrier also acts as biomimetic agent and reduces the risk of unnecessary elimination from the body and cytotoxicity. Moreover, the increment in the metabolic rate also makes necessities of albumin for arthritis joints, and therefore the albumin improves the transportation and delivery of drug. In this concept, one study showed the development of MTX-human serum albumin nanoparticles that showed excellent MTX accumulation and prolonged the retention time in the inflammatory area of joints than free drugs [19]. The targeted nanoparticles like mannose functionalized human serum albumin nanoparticles were developed for loading MTX which enables the reduction of serum inflammatory markers, joint swelling, and bone erosion via improving the neutrophil uptake [22, 44].

5.2 Glucocorticoids as Conventional Drug Delivery via Nanodrug Delivery System

Glucocorticoids belong to drug that shows anti-inflammatory effects and advises as adjunctive therapy. Owing to adverse events caused by GCs, these drugs have been loaded onto the nanosystem for active and passive delivery or via simple delivery strategy. Owing to stable nature of GCs-loaded nanoformulation mitigates the arthritis-based symptoms.

For example, Lorscheider et al. successfully designed dexamethasone (Dex) palmitate nanoparticles (NPs), which aggregated in arthritic joints. These NPs passively acted as anti-inflammatory molecules, and it was suggested that the controlled release of Dex reduced inflammation without causing side effects. Next, to eliminate the large dosage and frequent dosing, several studies have applied to develop and avoid the bad impact of GCs. Hence, a designed pH-sensitive acetone-based ketal-linked prodrug of Dex nanoparticles was synthesized and showed higher joint accumulation in the CIA rat model. Similarly, in order to target macrophage, another teams have developed folic acid (FA) functionalized Dex/Oxi-α cyclodextrin NPs which significantly eliminated the RA-related cytokines via pH-responsive drug release, including TNF-α, B cell activating factor. Similarly, cyclodextrin nanoparticles-based drug delivery system was developed for the carrying the dexamethasone sodium phosphate (DSP) that increased the chances of nanoparticles' extravasate into the inflammation site. The in vivo study in the adjuvant-induced arthritis (AIA) rat model, where DSP-CD NPs significantly in arthritic score, paw thickness, and blood levels of inflammatory cytokines. Studies evaluating adverse effects show a considerable decrease in the related negative effects on body weight, blood sugar levels, renal impairment, and hematological abnormalities when compared to commercial formulations [13].

5.3 *NSAIDs as Conventional Drug Delivery via Nanodrug Delivery System*

The majority of RhA patients are treated by NSAIDs as a classic anti-inflammatory and analgesic medication, especially for symptomatic treatment. NSAIDs basically worked by blocking the cyclooxygenase enzyme activity. However, some consequent complications such as fluctuation in blood pressure, stroke, and upper gastrointestinal as well as acute kidney ischemia-related are making obstacles to the patient's medications. Thus, nanocarrier-based drug delivery reduces the chance of adverse effects and boosts the medications. For example, COX-2 inhibitors like Celecoxib are used as NSAIDs for the treatment of RhA.

5.4 *TNF-A Inhibitors as Conventional Drug Delivery via Nanodrug Delivery System*

The targeting of TNF-α-based drug delivery is another way to cure the RhA disease. Attributing to adverse effects of many drugs restricts their usage and shows the necessities of well-designed formulations. It has been found that patients face a higher risk of tuberculosis and related-histoplasmosis, if the patients' therapy is based on TNF-α inhibitors. Likewise, elder patients show a high infection risk of herpes zoster and infection in the upper respiratory tract during Janus kinase (JAK) inhibitors-based therapy [13]. Candidiasis and hypogammaglobulinemia are other infections during IL-17 targeting medications and rituximab in the patients, respectively.

However, nanodrug delivery system reduces drug-related risk associated with infection and adverse effect. For example, AuNPs for drug delivery system were used for etanercept delivery and monitored inflammation in a rat arthritis model. Functionally, etanercept is a protein that binds to TNF receptors. However, etanercept nanomedicine has received less attention in preclinical RhA research. In a recent study, Sun et al. worked on molybdenum disulfide nanoflowers to deliver the drug etanercept to injured spinal cord tissue. For better conjugation and delivery, the antibody and nanocarrier are also exploring nowadays to improve the stability of therapeutic agents. Hence, infliximab-carbosilane dendrimers conjugate was prepared to increase the infliximab specificity and affinity toward TNF-α receptors. As carbosilane dendrimers are hydrophobic in nature and explored in an intestinal epithelial barrier model via in vitro studies. Likewise, for targeted drug delivery AuNPs were synthesized and functionalized using hyaluronic acid and Tocilizumab was conjugated for RhA treatment.

One another team synthesized polymeric, i.e., poly (lactic-co-glycolic acid NPs for successful delivery of rapamycin and adalimumab (anti-TNFa Ab) in order to treatment of RhA with overcoming immunological aggressiveness via inhibiting antigen-specific CD4 + T cell and B cell activation, and induction tolerogenic dendritic cells and improved the pharmacokinetics of adalimumab. Hence, still the preclinical studies and data should be explored to clarify that whether nanodrug delivery system could overcome the side effect of delivered drugs without causing immune cells and related antibodies.

6 Biocompatible Nanostructured-Based Treatment Approaches and Key Properties

Nanodrug delivery system is booming rapidly due to ease of modification, passively and actively transport mechanism and biocompatibility further showing enhancement in the therapeutic efficacy at low dosing frequency. In addition to drug delivery, nanovehicle is also attributed to thermal property and so far used in phototherapy. The imaging property is another property of some of the nanoparticles, and hence, diagnosis and therapy both make them theranostic nanomaterials. Thus, over the past decades, photothermal therapy and photodynamic therapy based on near-infrared (NIR) resonant, RhA is extensively diagnosed and treated at the same time. Hence, in this section, we will discuss the multitasking nanovehicle that combines drug/gene delivery, phototherapy, and contrast agents (CAs) may have beneficial impacts on RA treatment and diagnosis. This section 6.0 will also emphasize on gold NPs (AuNPs), silver NPs (AgNPs), reactive oxygen/nitrogen species (ROS/RNS) scavenging NPs, and biomimetic NPs, polymeric nanoparticles, and carbon-based nanovehicles and dendrimers.

6.1 Gold Nanoparticles (AuNPs)

AuNPs exhibit distinctive optical, surface plasmon resonance, and electrical properties and also customizable surfaces with functional ligands. The most famous applications for AuNPs include sensitivity testing, thermal ablation, the development of radiotherapy, and the transfer of genes and therapeutic drugs. It has been revealed that Au ion, i.e., zero-valent gold (Au^0), is act as therapeutic agent and biocompatible in nature. It has been found that AuNPs show anti-inflammatory property; hence for the first time, citrate-stabilized AuNPs were down-regulated the extracellular IL-1β that generally induces inflammation. Further, property of Au nanocages is tendency to block cytokine. In the CIA rat model, AuNPs administration dramatically reduces TNF-, IL-1, COX-2, and NF-B transcription. Khan et al. attribute AuNPs' anti-arthritic properties to their ability to synthesize inflammatory cells from infiltrating joints. Despite this, Au cluster with glutathione ($Au_{29}SG_{27}$) named as GA was synthesized to stop up the damage in bone and inflammatory activity in CIA in rats. The results reflect the dose-dependent effectiveness of GA in reducing 30-day toxicity. Continuous intraperitoneal injection of GA at a dose of 5 mg/kg (Au content) significantly improved the symptoms of RhA, suggesting that even at a lower dose, GA can suppress inflammation and mitigate bone damage to some extent [38].

In another study, generation 5 (G5) poly(amidoamine) dendrimers-encapsulated AuNPs were utilized as an excellent nanocarrier for the delivery of alpha-tocopheryl succinate (α-TOS) and anti-inflammatory anti-TNF-α siRNA to macrophage cells. Here, the reason for the delivery of an antioxidant alpha-tocopheryl succinate (α-TOS) was to reduce the reactive oxygen species that generally boost the RhA pathogenic process. Hence, G5 PAMAM dendrimer is serially decorated with 1,3-propane sultone (1,3-PS), α-TOS via a polyethylene glycol (PEG) spacer, and FA, and then entrapped with AuNPs, enabling serum-enhanced siRNA distribution to M1-type macrophage cells, where in vivo results showed effect in inflammatory cytokines down-regulation of RA lesion and bone erosions in collagen-induced arthritis mouse model [16]. In one more example, folate receptors targeted solid-phased Au NPs (~103 nm) were functionalized with folic acid for delivery of MTX. The whole formulation is successfully prolonging blood circulation time of delivery of the MTX and exhibited significant inhibitory action against human fibroblast-like synoviocytes of RhA along with the enhancement of MTX in accumulation in inflamed joints ($p < 0.01$) without causing any other toxicity to normal cells [17].

Innovative dendrimer-entrapped AuNPs that delivers TNF-siRNA and alpha-tocopheryl succinate (an antioxidant) have been developed to generate multifunctional nanoplatforms for the effective treatment of RhA. In order to regulate the development of RhA from several angles, Lee et al. presented a flexible therapy plan. The team has developed a noncomplex made of hyaluronate, monoclonal antibody against the human IL-6 receptor, and anti-angiogenic and anti-inflammatory AuNPs (cartilage-protective and lubricant effects) [16].

Despite their inherent anti-inflammatory abilities, AuNPs have the potential to be used as a drug delivery system for traditional medications to lessen their adverse side effects relevance to secondary adrenal insufficiency, bleeding in gastrointestinal, Cushing's syndrome, other bone-related diseases like osteoporosis, and carbohydrate metabolism disorder like hyperglycemia in diabetic patients. For example, as shown in Fig. 2 triamcinolone –Au NPs complex was developed in order to check the pro-inflammatory responses and improve the anti-inflammatory responses of fibroblast-like synoviocytes and macrophages via repolarization of macrophages from the M1 to the M2 phenotype. Hence, analyses performed in vitro, ex vivo, and in vivo on human, mouse, and human samples, respectively, amply showed that triamcinolone-AuNPs complex efficiently controlled the production of both pro- and anti-inflammatory cytokines in fibroblast-like synoviocytes and efficiently repolarized macrophage activity in the inflamed synovium as compared to triamcinolone alone [27].

In general, compared to larger size particles, smaller size nanoparticles have a stronger tendency to pass the stratum corneum. Hence, in view of this, a research performed on the effect of particle size on drug permeation where small size AuNPs (22 nm) had superior penetration behavior than those with larger size of AuNPs (100–200 nm) through the stratum corneum of the rat's thick-skinned hind paw [29]. Moreover, MTX-loaded polyglycolic acid AuNPs that successfully underwent photothermal reaction show the ability of multi-responsive systems in reducing inflammatory activity and enhancing the therapeutic impact of arthritis. The capacity of carriers to respond to several stimuli at various levels allows for the unique controlled release of commodities. This is most notable for the ability of carriers to respond to multiple orthogonal stimuli at different levels of change [6].

Only after clarifying the numerous physicochemical features of each unique AuNPs and determining the corresponding kinetics and interaction of these

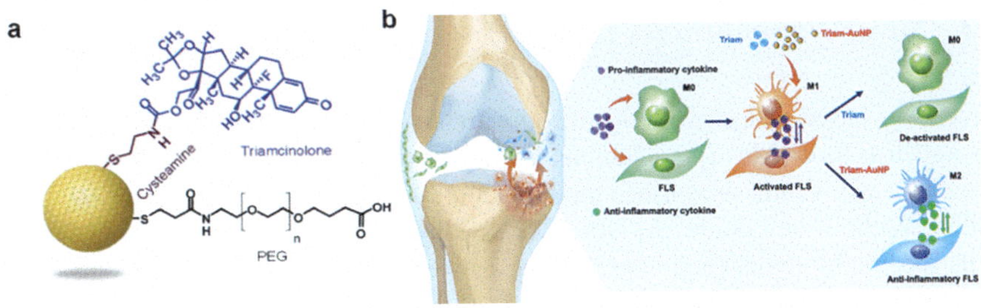

Fig. 2 Representation of the chemical bonding of triamcinolone-AuNPs complex. Cysteamine was used to link PEGylated AuNPs and triamcinolone b. Increase in the anti-inflammatory response of fibroblast-like synoviocytes by the nanodrug which shows fibroblast-like synoviocytes activation by M1 macrophages and deactivation of the activated fibroblast-like synoviocytes by alone drug Triam and drug-AuNPs. (This figure is adapted from Jun Young [27])

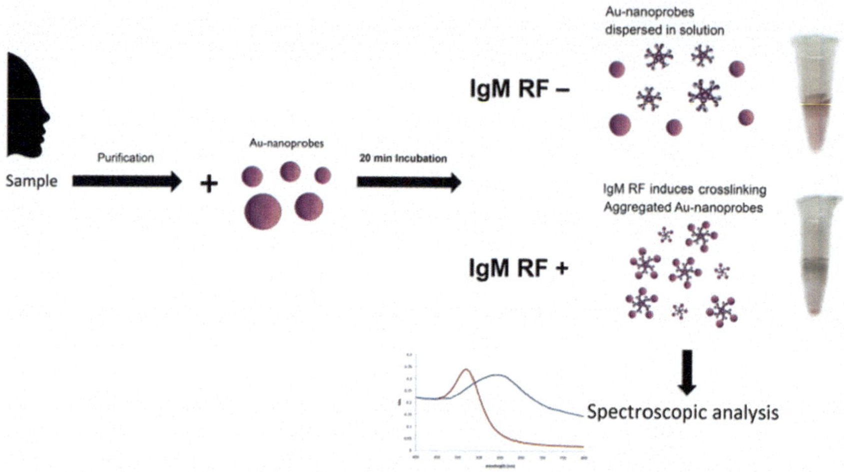

Fig. 3 Schematic illustration of the Au-nanoprobe detection process and of early rheumatoid arthritis biomarker, i.e., IgM RF with the readout coming from the colorimetric output. Inducing Au nanoprobe aggregation, which in turn causes a shift to the LSPR peak that can be seen by UV–vis spectroscopy, is one of the ways that IgM-RF selectively binds to the nanoconjugates. Actual images of genuine detection samples are shown on the right to demonstrate how the human eye can distinguish between colors (This figure is adapted from Veigas et al. [35])

compounds with target tissues in animals can knowledge on an efficient and safe application of AuNPs be developed. In another study, researchers developed low-cost Au nanoprobes for the detection and quantification of rheumatoid factor markers. The result shows, in Fig. 3, the change in the color from red to purple due to Au nanoprobes cross-linked in the presence of pentameric immunoglobin factor M as rheumatoid factor with a value, limit of detection (LOD) of 4.15 UA mL^{-1} IgM RF (clinical threshold is set for 20 IU mL^{-1}) [35].

6.2 *Silver Nanoparticles (AgNPs)*

Silver(Ag)-based nanostructures are commercialized years ago for various applications such as food industry, packaging, textiles biomedical application, and so on due to their ease of synthesis, surface modification, functionalization, and drug delivery ability. Without any doubt, AgNPs have already gained so much attention owing to its specific biological activities, such as anti-microbial properties, wound healing effect, and antioxidant activity. In recent years, the anti-inflammatory action and delivery of related drugs for the treatment of RhA have garnered significant attention in the biomedical field, along with several other diseases. For instance, in one study, an anti-inflammatory activity could promote M2 polarization and was observed by AgNPs. Despite these AgNPs-based applications, the ability of anti-inflammatory effect of AgNPs have also been encouraging due to the use of low dose and better effects without causing toxicity toward the normal cells. For example, folic acid functionalized PEGylated AgNPs were delivered that were successfully uptake by the M1 macrophages. The results showed induction of apoptosis and remove ROS to provide M2 polarization via disintegration and release of AgNPs into Ag ions in reaction to intracellular glutathione [37].

In another example, nanocargo was developed using gum acacia stabilized AgNPs for the delivery and anti-arthritic efficacy of hesperidin. The results exhibited minimal arthritic symptoms with mild to moderate tissue swelling, reduction in degenerative changes along with mild articular changes studied in in adjuvant-induced arthritic rats [30].

In another study, natural substance, i.e., *Commiphora mukul* (guggul or G), was used to synthesize AgNPs in order to avoid unwanted cytotoxicity and environment concerns. The plant extract of guggul acted as a reducing agent and stabilized AgNPs, demonstrating excellent anti-arthritic potential and potent anti-inflammatory action in the AIA rat model. Also, treatment with guggul-AgNPs effectively restored the activity of the antioxidant enzymes superoxide dismutase and catalase when compared to arthritic rats, demonstrating the effectiveness of synthetic guggul-AgNPs in reducing the oxidative stress related to RhA. Eventually, after treatment with guggul-AgNPs, histological analysis represents that the infiltration of inflammatory cells in the tissue of the ankle joint was significantly reduced [31]. In another example, for the improving cellular absorption, liposomal formulation has been developed and entrapped AgNPs on macrophages that are further improved by caspase-dependent apoptosis activation [39]. Despite self-anti-inflammatory action and nanocarrier duty, AgNPs have been also found as bacterial eradicating agents against RhA-based bacteria. For example, in those patients who are suffering from RhA, *Proteus* infections may serve as a trigger for rheumatoid arthritis (RA) as evidence by the presence of Proteus species in isolation of Proteus from urine. Another other evidence such as Proteus antibody increase in RhA patient sera and lethal effects of these serum antibodies on joint tissues containing cross-reactive Proteus antigens. Hence, the effectiveness of AgNPs for cancer treatment agent, the little advancement of bacterial resistance, and the absence of hazardous side effects are some benefits. In concept of this, reduced risk of bone damage from Proteus-associated cross-reactive immune responses may result from treating infections brought on by this bacterium in the early stages of RA. For example, AgNPs/ $Ag+$ solution was observed as highly effective where AgNPs (LR-049) was given orally for one week before volunteers who had not yet received DMARD medication showed a substantial reduction in Proteus load in the lower intestines (by 90%) and bladder (by 87%) without causing any adverse effect toward normal cells [7].

6.3 *Copper Nanoparticles (Cu NPs)*

The copper (Cu)-based NPs have extensively receiving a great deal of interest in biomedical applications. It is known as essential biological element to fulfill various biological mechanisms. Like other nanoparticles, CuO NPs with different sizes and morphologies are synthesized by different methods and play important role for drug delivery, etc. It has been found that the green technique is the ideal way to synthesize these nanoparticles since it may remove the risk of toxicity problems and neurodegenerative illnesses. Eco-friendly methods have also garnered more attention when producing metallic NPs. CuO NPs have demonstrated their therapeutic benefit against the inflammatory illness RhA, as well as other diseases. For example, CuO NPs were synthesized using *Rabdosia rubescens* and explored the anti-arthritic potential against CFA-stimulated arthritis in rats. The treatment using metallic plant extract-based CuO NPs showed enhanced the antioxidant enzymes and reduced pro-inflammatory markers, while the inflammatory cell penetration and bone damage were also prevented as investigated by histopathological studies. Further, the results also represented the considerable diminishing of RBCs, Hb, and WBCs status in CFA-triggered arthritic rats as compared to control conversely body weight increased along with paw volume severely enlargement with appreciably decrease the immune organ indexes [43].

In advanced studies, unique Cu7.2S4 NPs were developed for treating RhA via generating photothermal therapy (PTT) and photodynamic therapy (PDT) where Cu has served as photothermal agent. Here Cu7.2S4 NPs, a Cu-based nanomaterial, demonstrated combined NIR (808 nm, 1 W cm^2) irradiation that not only improved bone preservation, including increased bone mineral density (BMD) and bone volume/total volume, but also inhibited inflamed synovial invasion, degradation of cartilage, and production of proinflammatory cytokines in vivo. Despite the cure of RhA pathogenesis, the same metallic nanoparticles effectively destroyed clinical pathogenic *Staphylococcus aureus* and *Escherichia coli* and prevented pathogenic bacterial infection in the course of intra-articular injection [21].

6.4 *Selenium Nanoparticles (SeNPs)*

Selenium (Se) is a vital micronutrient trace element with significant importance for healthy human functioning as well as for a number of pathophysiological disorders. Studies have indicated that blood levels of Se, an essential component of antioxidant enzymes, are often fluctuating in people with RhA. It has been found that inorganic Se exhibits in their four oxidation state, namely selenite (Se^{+4}), selenate (Se^{+6}), elemental Se (Se^0), and selenide (Se^{-2}). In the biological system, they are capable of being transformed into organic and bioavailable forms; for example, selenocysteine is the 21st amino acid and selenomethionine, is a Se-bioconjugated natural methionine amino acid through biological processes. Moreover, 25 selenoproteins have been identified in humans. Hence, despite its several biological properties, low Se levels have been linked to an increased chance of developing RhA, according to epidemiological studies, underscoring the importance of antioxidants in halting the spread and maintenance of the condition. Furthermore, currently, Se-based NPs were examined for the therapeutic benefits in the numerous oxidative stress and inflammation-mediated diseases, like arthritis, cancer, diabetes, and nephropathy, in recent years. Thus, in one study, SeNPs enhanced the antioxidant enzyme levels in various organs, expressively improved adjuvant-induced arthritis, and may be observed as a feasible supplemental medication for RhA pathogenesis [23]. In another example, SeNPs were developed and surface modified using polyethylene glycol, and functionalized using RGD and ruthenium complexes, i.e., $Ru(Phen)_2(4idip)](ClO_4)_2$. The developed nanosystem showed multitasking activities such as trigger to generate nitric oxide (NO) in cells via Ru. The results showed Se@RuNPs target the extensive neovascular network of inflammatory areas to produce NO, encourage HUVEC death, and prevent the development of new blood vessels in the surrounding tissue. As shown in Fig. 4. Further, the induction of autophagy promoted by NO via modulating signaling pathways of AMPKa and mTOR, increased the flux of autophagy, inhibited the activity of NF-kB-p65, which modulating the levels of inflammatory cytokines. Moreover, effectively reduced synovitis, cartilage corrosion, and inflammatory cytokine expression levels were observed when treated using Se@RuNPs and suggested its satisfactory therapeutic effects [20].

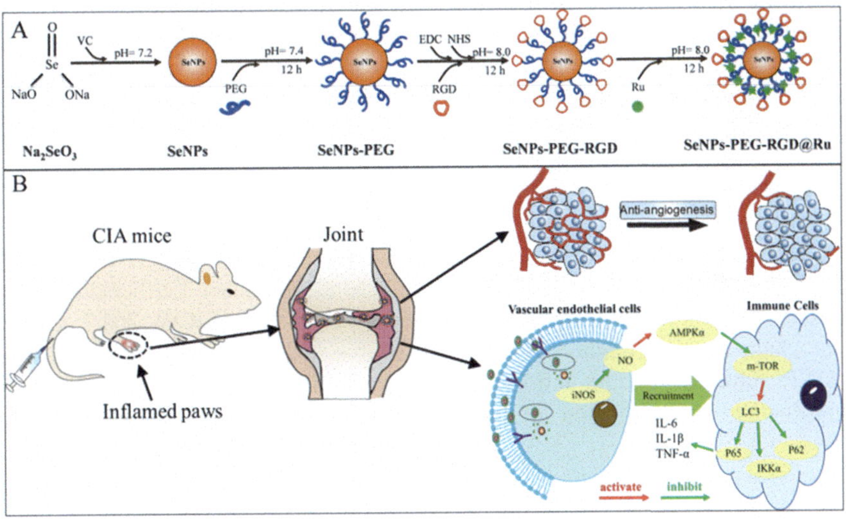

Fig. 4 Scheme 1 **a** Synthetic of Se@RuNPs. **b** The anti-angiogenic effect of Se@RuNPs with the mechanism of Se@RuNPs inducing NO to recruit immune cells and regulate inflammatory response. (This figure is adapted from Liu et al. [20])

Recently, biogenic SeNPs derived from plants like *Trachyspermum ammi* and *Foeniculum vulgare* Mill were recently demonstrated to have anti-arthritic and antioxidant activities in a research utilizing an animal model of arthritis [3, 28].

6.5 Zinc Oxide Nanoparticle (ZnONPs)

Zinc oxide (ZnO) nanostructures in the current situation gained excellent attention due to their biocompatibility and hence approved by the Food and Drug Administration (FDA) of the USA as a generally recognized as safe (GRAS) substance. Functionally, Zn ions play vital role in hematopoiesis, enzyme control, maintaining cell redox balance, and regulation of the DNA and protein synthesis machinery.

In addition, ZnONPs have been showed its ability in the wound healing, catalytic, bioimaging, antibacterial, anti-inflammatory properties, and drug delivery for standard drugs as it provides synergistic effects for the treatment. ZnONP is one of the multitasking nanomaterials that are synthesized by various methods including chemical, physical, and green methods. While the former two methods produce ZnONPs of various shapes, the third green processing stabilizes but also improves the efficacy of NPs without causing any toxicity-associated risk. For example, in one study, ZnONPs were synthesized in order to treat RhA pathogenesis with the concentration of orally (2 mg/kg/day) daily for 14 days, where the nanoparticles showed adjuvant-induced increased production of inflammatory mediators interleukin-1β (IL-1β), tumor necrosis factor-alpha (TNF-α), interleukin-10 (IL-10), and total leukocyte count. Furthermore, the same formulation ameliorated autoimmunity through suppression of anti-citrullinated protein auto-antibodies (anti-CCP) levels in rats [10]. Owing to provide biocompatible approach, ZnONPs are synthesized by green methods using plants, fungus, microbes, and algae [1]. One of those studies, an inflammatory effect of ZnONPs, was investigated which was synthesized by green method using *Kalanchoe pinnata* leaf extract. The biological results showed lowing of the gene expression of pro-inflammatory cytokines TNF-α, IL-1β, IL-6, and COX-2 enzyme, especially in dose-dependent manner. Further study also stated the biocompatible nature of plant-based ZnONPs toward RAW 264.7 murine macrophage model. Therefore, the findings of this study show that biosynthesized ZnONPs have outstanding anti-inflammatory capability, and this innovation indicates ZnONPs as a significant alternative to NSAIDs for the therapy of inflammation [1].

6.6 Iron Oxide Nanoparticles (IONPs)

Recent decade, iron oxide nanoparticles (IONPs) have also been used as treating agents for the RhA pathogenesis due to the strong physical properties, inexpensiveness, and to target specific disease sites, with minimal damage to peripheral organs.

As iron oxide nanoparticles are attributes to biological properties and therefore used in the biomedical application. Likewise, due to drug delivery and contrast imaging property, iron oxide nanoparticles have also been accepted in molecular imaging like the identification of apoptosis and gene expression. Besides this, the mesenchymal stem cells (MSCs) possess immunosuppressive activities in case of pathogenesis of RhA, for example, MSCs-related cytokines example IL-10, 6, 11 and TGF-b inhibit function of T cells and another dendritic cells. While release of HLA-G is soluble antigens and strongly deactivates NK cells and moderate dendritic cell and T cell action.

Hence, progress in tissue engineering has emphasized the role of mesenchymal stem cells (MSCs) in treating RhA. Iron oxide nanoparticles have been utilized tracking of mesenchymal stem cells (MSCs) in treating rheumatoid arthritis. The finding with SiMAG-labeled IO showed a significant reduction in joint swelling with P 0.01 in groups as compared to unlabeled mMSCs, suggesting that SPIONs do not impact the cells' immunomodulating abilities [24]. Besides this, folic acid targeting ligand functionalized and glucose-dextran iron oxide nanoparticles for the treatment of RhA, where the main role of iron, gave negative contrast in the MRI to visualize antigen induction arthritis in the rat. Furthermore, an enhanced MRI using iron confirmed the therapeutic effectiveness of the cyclooxygenase 2 (COX-2) inhibitor treatment of the inflamed joints, suggesting that this type of nanoparticle could also have potential as a CA for evaluating the treatment response of rheumatoid arthritis. Still, the gene delivery is using iron oxide nanoparticles as an excellent nanocarrier, as in one study, in order to study cellular absorption, tissue distribution, and the therapeutic impact of a siRNA against the IL-2/IL-15 receptor chain (IL-2/IL-15R), NPs were intravenously delivered to arthritic rats. The in vitro finding showed the non-cytotoxicity behavior of iron-based nanocarrier as well as improved siRNA stability, helped the efficient uptake by macrophages, and suggested the reason behind successful induction of specific gene silencing [8]. The size and shape of NPs play critical role in the biodistribution and cellular absorption that also has an impact on the effectiveness of drug delivery and therapeutic results. Hence with this concept, citrate-capped iron oxide nanoparticles of different size was synthesized with diameters of 70, 110, 220, and 350 nm.

The biodistribution was held on CIA mice model using indocyanine green (ICG). Indeed, an in vitro study revealed that smaller nanoparticles were more easily phagocytized and absorbed than larger ones, increasing their cytotoxicity, while the in vivo results were quite different in which nanoparticles having 220 nm were internalized more into the inflammatory joints, and under photothermal laser radiation helped in the raising of the temperature and improved the therapeutic impact Fig. 5 [41].

The combination of metal oxide nanoparticles has been also published in the last recent years in which Au was used as capping agent onto the superparamagnetic iron oxide nanoparticles for the delivery of MTX into the collagen-induced arthritis (CIA) rats. The treatment using drug-loaded nanosystem is reduced the leukocytes infiltration, edema in the tissue along with decrease in the TNF-α and UL-1β expression in synovium tissues. Here the study suggested that Au coating provided longer retention time at inflammatory site and also antioxidant effect of drug with dose of 2.5 mg/kg,

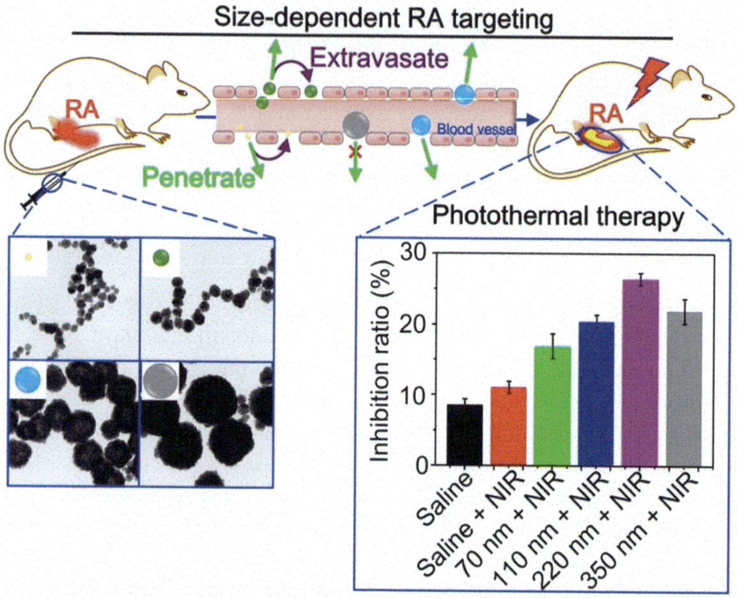

Fig. 5 Developed iron oxide nanoparticles of different sizes with diameter of 70, 110, 220, and 350 nm explored for the internalization and extravasate into the synovial joint and under photothermal therapy show the highest inhibition of cells with 220 nm + near-infrared radiation. (This figure is adapted from Zhang et al. [11])

three times every other day by the intraperitoneal route improve the performance of drug. Other research has changed the magnetic nanoparticles to further improve the targeting of CAs. Iron oxide core, carboxyl-PEG hydrophilic coatings, and anti-CD3 antibody as the targeting probe made up the newly created IOPC-CD3 particles. To create molecular MRI imaging, they relied on the concept of CD3 binding to T cells in arthritic inflammatory areas [5].

Iron oxide nanoparticles have been focused for the imaging diagnosis and thermal ablation, other than drug delivery majorly in the field of cancer treatment. In addition to this, the progress of iron oxide-based nanoformulation has been also seen in the clinical impact of autoimmune diseases, for example, Ferumoxtran-10 in Type 1 diabetes (NCT02511028) and Ferumoxytol in Multiple Sclerosis (NCT00585936). However, the NPs still not recognized in the clinical trials for RhA diagnosis. Some scientist have given the reason that may stop the designing of iron-based nanoformulation including: (a) Since most RhA patients mainly belong to old age women with lower body metabolisms and more fundamental illnesses, side effects may be more common, (b) iron oxide nanoparticles residue in normal tissue is unavoidable, which might make it difficult to accurately diagnose RA consequences such interstitial pneumonia and (c) Its therapeutic applicability may be hampered by the expense and difficulty of preparation. But, scientists are still working to developing an iron oxide nanoformulation that can assist to get beyond these restrictions.

6.7 *Other Metal Nanoparticles*

Inorganic nanoparticles attribute with various treatment properties for RhA. Most of metal-based have already been explored; however, still some metal and oxide nanoparticles efficiently reduce the inflammatory action in the disease site without causing adverse effects as revealed by in vitro and in vivo studies. Some of the examples showed the best results for RhA therapy. For example, manganese oxide gaining much attention in the cancer therapy for ROS reduction. In one study, polyethylene glycol-modified MnO_2 nanoparticles were developed for the elimination of oxidative stress [14].

7 Conclusion

The available therapy for rheumatoid arthritis focuses in particular on lowering the production of pro-inflammatory cytokines and preventing macrophage growth. In this, inflammatory disease is an autoimmune disease, and hence, the current therapy is limited for various targeted options. The targeted site's existing therapy choices have rather little therapeutic effectiveness. The recent advanced RhA-based therapy has shown association with the down-regulation of pathogenesis level and reduces the joint-bone damage, but long-term exposure also resultant of unnecessary side effects. Despite facts, nanotechnology has been shown great ideas in the cure of several diseases including RhA due to various array of tools and techniques. Nanoparticles are also found as promising diagnosis nanomaterials along with drug delivery, self-therapeutic properties, and photothermal therapy. These nanoparticles have not also show therapeutic properties but also reduce the bacterial infection association with RhA. The studies performed by many scientists are still working greatly in order to reduce the adverse related pathogenicity. Therefore, this book chapter is focused on the emerging approach and concepts relative to therapeutic potentials of nanomedicine in the RhA therapy.

References

1. Agarwal H, Shanmugam VK (2019) Synthesis and optimization of zinc oxide nanoparticles using Kalanchoe pinnata towards the evaluation of its anti-inflammatory activity. J Drug Delivery Sci Technol 54:101291. https://doi.org/10.1016/j.jddst.2019.101291
2. Almoallim H, Al Saleh J, Badsha H et al (2021) A review of the prevalence and unmet needs in the management of Rheumatoid Arthritis in Africa and the middle east. Rheumatology and Therapy 8:1–16. https://doi.org/10.1007/s40744-020-00252-1
3. Arif A, Bhatti A, John P (2019) Therapeutic potential of Foeniculum vulgare mill. Derived selenium nanoparticles in arthritic Balb/c mice. Int J Nanomed 14:8561–8572. https://doi.org/10.2147/IJN.S226674

4. Carmona L, Cross M, Williams B, et al (2010) Rheumatoid arthritis. Best Pract Res Clinical Rheumatology 24:733–745. https://doi.org/10.1016/j.berh.2010.10.001
5. Chen C-L, Siow TY, Chou C-H et al (2017) Targeted superparamagnetic iron oxide nanoparticles for in vivo magnetic resonance imaging of T-cells in Rheumatoid Arthritis. Mol Imag Biol 19:233–244. https://doi.org/10.1007/s11307-016-1001-6
6. Costa Lima SA, Reis S (2015) Temperature-responsive polymeric nanospheres containing methotrexate and gold nanoparticles: a multi-drug system for theranostic in rheumatoid arthritis. Colloids Surfaces B: Biointerfaces 133:378–387. https://doi.org/10.1016/j.colsurfb.2015.04.048
7. Disaanayake DMBT, Faoagali J, Laroo H et al (2014) Efficacy of some colloidal silver preparations and silver salts against Proteus bacteria, one possible cause of rheumatoid arthritis. Inflammopharmacology 22:73–77. https://doi.org/10.1007/s10787-013-0198-0
8. Duan J, Dong J, Zhang T et al (2014) Polyethyleneimine-functionalized iron oxide nanoparticles for systemic siRNA delivery in experimental arthritis. Nanomedicine 9:789–801. https://doi.org/10.2217/nnm.13.217
9. Feldon P, Terrono AL (2006) Carpal tunnel syndrome in rheumatoid arthritis. Techniques in Orthopaedics 21
10. Gad SS, Fayez AM, Abdelaziz M, Abou El-ezz D (2021) Amelioration of autoimmunity and inflammation by zinc oxide nanoparticles in experimental rheumatoid arthritis. Naunyn-Schmiedeberg's Arch Pharmacol 394:1975–1981. https://doi.org/10.1007/s00210-021-02105-2
11. Garg S, Garg A (2018) Encapsulation of curcumin in silver nanoparticle for enhancement of anticancer drug delivery. Int J Pharmaceutical Sci Res 9:1160. https://doi.org/10.13040/IJPSR.0975-8232.9(3).1160-66
12. He J, Ding Y, Feng M et al (2013) Characteristics of Sjögren's syndrome in rheumatoid arthritis. Rheumatology 52:1084–1089. https://doi.org/10.1093/rheumatology/kes374
13. Jadhav D, Vavia P (2021) Dexamethasone sodium phosphate loaded modified Cyclodextrin based nanoparticles: an efficient treatment for Rheumatoid Arthritis. J Pharm Sci 110:1206–1218. https://doi.org/10.1016/j.xphs.2020.10.023
14. Kumar S, Adjei IM, Brown SB, et al (2019) Manganese dioxide nanoparticles protect cartilage from inflammation-induced oxidative stress. Biomaterials 224:119467. https://doi.org/10.1016/j.biomaterials.2019.119467
15. Lewis MJ, Barnes MR, Blighe K et al (2019) Molecular portraits of Early Rheumatoid Arthritis identify clinical and treatment response phenotypes. Cell Rep 28:2455-2470.e5. https://doi.org/10.1016/j.celrep.2019.07.091
16. Li J, Chen L, Xu X, et al (2020) Targeted Combination of Antioxidative and Anti-Inflammatory Therapy of Rheumatoid Arthritis using Multifunctional Dendrimer-Entrapped Gold Nanoparticles as a Platform. Small 16:2005661. https://doi.org/10.1002/smll.202005661
17. Li X, Wang H, Zou X, et al (2022) Methotrexate-loaded folic acid of solid-phase synthesis conjugated gold nanoparticles targeted treatment for rheumatoid arthritis. European J Pharmaceutical Sci 170:106101. https://doi.org/10.1016/j.ejps.2021.106101
18. Lin Y-J, Anzaghe M, Schülke S (2020) Update on the Pathomechanism, diagnosis, and treatment options for Rheumatoid Arthritis. Cells 9
19. Liu L, Hu F, Wang H et al (2019) Secreted protein acidic and rich in cysteine mediated biomimetic delivery of methotrexate by albumin-based nanomedicines for rheumatoid arthritis therapy. ACS Nano 13:5036–5048. https://doi.org/10.1021/acsnano.9b01710
20. Liu Y, Ma L, Zhou H et al (2018) Polypeptide nano-Se targeting inflammation and theranostic rheumatoid arthritis by anti-angiogenic and NO activating AMPKα signaling pathway. J Mater Chem B 6:3497–3514. https://doi.org/10.1039/C8TB00080H
21. Lu Y, Li L, Lin Z, et al (2018) A new treatment modality for Rheumatoid Arthritis: combined Photothermal and photodynamic therapy using Cu7.2S4 nanoparticles. Adv Healthcare Mater 7:1800013. https://doi.org/10.1002/adhm.201800013
22. Lyu J, Wang L, Bai X et al (2021) Treatment of Rheumatoid Arthritis by serum albumin nanoparticles coated with mannose to target neutrophils. ACS Appl Mater Interfaces 13:266–276. https://doi.org/10.1021/acsami.0c19468

23. Malhotra S, Welling MN, Mantri SB, Desai K (2016) In vitro and in vivo antioxidant, cytotoxic, and anti-chronic inflammatory arthritic effect of selenium nanoparticles. J Biomed Mater Res Part B: Appl Biomate 104:993–1003. https://doi.org/10.1002/jbm.b.33448
24. Markides H, Kehoe O, Morris RH, El Haj AJ (2013) Whole body tracking of superparamagnetic iron oxide nanoparticle-labelled cells—a rheumatoid arthritis mouse model. Stem Cell Res Ther 4:126. https://doi.org/10.1186/scrt337
25. McInnes IB, Schett G (2017) Pathogenetic insights from the treatment of rheumatoid arthritis. The Lancet 389:2328–2337. https://doi.org/10.1016/S0140-6736(17)31472-1
26. Mousavi SE, Nejadghaderi SA, Khabbazi A et al (2022) The burden of rheumatoid arthritis in the Middle East and North Africa region, 1990–2019. Sci Rep 12:19297. https://doi.org/10.1038/s41598-022-22310-0
27. Park J-Y, Kwon S, Kim S-H et al (2020) Triamcinolone-gold nanoparticles repolarize Synoviocytes and macrophages in an inflamed synovium. ACS Appl Mater Interfaces 12:38936–38949. https://doi.org/10.1021/acsami.0c09842
28. Qamar N, John P, Bhatti A (2020) Toxicological and anti-rheumatic potential of trachyspermum ammi derived biogenic selenium nanoparticles in arthritic balb/c mice. Int J Nanomed 15:3497–3509. https://doi.org/10.2147/IJN.S243718
29. Raju G, Katiyar N, Vadukumpully S, Shankarappa SA (2018) Penetration of gold nanoparticles across the stratum corneum layer of thick-Skin. J Dermatol Sci 89:146–154. https://doi.org/10.1016/j.jdermsci.2017.11.001
30. Rao K, Aziz S, Roome T et al (2018) Gum acacia stabilized silver nanoparticles based nano-cargo for enhanced anti-arthritic potentials of hesperidin in adjuvant induced arthritic rats. Artif Cells Nanomed Biotechnol 46:597–607. https://doi.org/10.1080/21691401.2018.1431653
31. Singh A, Boregowda SS, Moin A, et al (2022) Biosynthesis of silver nanoparticles using Commiphora mukul extract: evaluation of anti-arthritic activity in adjuvant-induced arthritis rat model. Pharmaceutics 14
32. Smolen JS, Aletaha D, Barton A et al (2018) iRheumatoid arthrits. Nat Rev Dis Primers 4:18001. https://doi.org/10.1038/nrdp.2018.1
33. Taylor PC, Takeuchi T, Burmester GR, et al (2022) Safety of baricitinib for the treatment of rheumatoid arthritis over a median of 4.6 and up to 9.3 years of treatment: final results from long-term extension study and integrated database. Ann Rheumatic Diseases 81:335 LP—343. https://doi.org/10.1136/annrheumdis-2021-221276
34. van Vollenhoven RF (2009) Sex differences in rheumatoid arthritis: more than meets the eye…. BMC Med 7:12. https://doi.org/10.1186/1741-7015-7-12
35. Veigas B, Matias A, Calmeiro T et al (2019) Antibody modified gold nanoparticles for fast colorimetric screening of rheumatoid arthritis. Analyst 144:3613–3619. https://doi.org/10.1039/C9AN00319C
36. Williams AS, Topley N, Dojcinov S et al (2001) Amelioration of rat antigen-induced arthritis by liposomally conjugated methotrexate is accompanied by down-regulation of cytokine mRNA expression. Rheumatology 40:375–383. https://doi.org/10.1093/rheumatology/40.4.375
37. Yang Y, Guo L, Wang Z, et al (2021) Targeted silver nanoparticles for rheumatoid arthritis therapy via macrophage apoptosis and Re-polarization. Biomaterials 264:120390. https://doi.org/10.1016/j.biomaterials.2020.120390
38. Yuan Q, Zhao Y, Cai P et al (2019) Dose-dependent efficacy of gold clusters on rheumatoid arthritis therapy. ACS Omega 4:14092–14099. https://doi.org/10.1021/acsomega.9b02003
39. Yusuf A, Casey A (2020) Liposomal encapsulation of silver nanoparticles (AgNP) improved nanoparticle uptake and induced redox imbalance to activate caspase-dependent apoptosis. Apoptosis 25:120–134. https://doi.org/10.1007/s10495-019-01584-2
40. Zhang H, Zhang H, Gao D et al (2020) Overlapping Sjogren's syndrome reduces the probability of reaching target in rheumatoid arthritis patients: a propensity score matched real-world cohort from 2009 to 2019. Arthritis Res Ther 22:100. https://doi.org/10.1186/s13075-020-02189-w
41. Zhang S, Wu L, Cao J, et al (2018) Effect of magnetic nanoparticles size on rheumatoid arthritis targeting and photothermal therapy. Colloids Surfaces B: Biointerfaces 170:224–232. https://doi.org/10.1016/j.colsurfb.2018.06.016

42. Zhang T, Pope J (2015) Cervical spine involvement in rheumatoid arthritis over time: results from a meta-analysis. Arthritis Res Ther 17:148. https://doi.org/10.1186/s13075-015-0643-0
43. Zhang Z, Chinnathambi A, Ali Alharbi S, Bai L (2020b) Copper oxide nanoparticles from Rabdosia rubescens attenuates the complete Freund's adjuvant (CFA) induced rheumatoid arthritis in rats via suppressing the inflammatory proteins COX-2/PGE2. Arabian J Chem 13:5639–5650. https://doi.org/10.1016/j.arabjc.2020.04.005
44. Zhao J, Zhao M, Yu C et al (2017) Multifunctional folate receptor-targeting and pH-responsive nanocarriers loaded with methotrexate for treatment of rheumatoid arthritis. Int J Nanomed 12:6735–6746. https://doi.org/10.2147/IJN.S140992

Screen-Printed Electrode (SPE)-Based Biosensor for Point-Of-Care (POC) Diagnostic in Medical Applications, Their Scope, and Challenges

Dinesh Rotake, Shruti Patle, and Shiv Govind Singh

Abstract The present chapter focuses mainly on advancing screen-printed electrode (SPE) technology to target toxic biomolecules with point-of-care (POC) diagnostic in biomedical applications. Various SPE parameters, such as general fabrication process steps, suitable material, the dynamic range of detection, and the limit of detection (LOD) for biosensor application, have been discussed here. Furthermore, the SPE is a low-cost approach compared with other technology in terms of simplicity, time of detection, and inexpensive fabrication, but still need to have post-processing of detected signal in real time for POC diagnostic biosensor applications, and it remains a significant challenge. Notably, the disposability factor is another desirable biosensor parameter that must be considered at the global environmental level. Paper and flexible substrate-based sensors are emerging research areas that reduce the overall production cost and improve the fabricated biosensor's sensitivity. Here, we also provide SPE application to detect the top five deadliest diseases worldwide, such as cardiovascular disease (CVD), respiratory cancers (RS), diabetes mellitus (DM), Alzheimer's disease (AD), and tuberculosis (TB) according to death rates per year. This chapter also focuses on the advancement and future scope of SPE technology in POC diagnostic for biosensor applications.

Keywords Screen-printed electrode (SPE) · Cardiovascular disease (CVD) · Respiratory cancers (RS) · Alzheimer's disease (AD) · And Tuberculosis (TB)

D. Rotake (✉) · S. G. Singh
Department of Electrical Engineering, Indian Institute of Technology, Hyderabad 502285, Telangana, India
e-mail: dinesh.rotake@gmail.com

S. Patle
Department of Applied Physics, Priyadarshini J L College of Engineering, Nagpur, India

A. Madhusudhan et al. (eds.), *Functional Smart Nanomaterials and Their Theranostics Approaches*, Smart Nanomaterials Technology,
https://doi.org/10.1007/978-981-99-6597-7_12

1 Introduction

Many researchers worldwide focus on finding alternative ways of measurement to avoid lab-based equipment. The advancement in screen-printed electrode (SPE) technology targets the toxic biomolecules with point-of-care (POC) diagnostic in various biomedical applications. Researchers try to use paper and flexible substrate-based sensors to reduce the overall production cost and improve the fabricated biosensor's sensitivity. In the past few years, there has been a noteworthy development of various biomedical systems that can perform further analysis of composite samples for real time and deliver quick, accurate results. Here, in this chapter, we try to provide SPE application to detect the top five deadliest diseases worldwide, such as cardiovascular disease (CVD), respiratory cancers (RS), diabetes mellitus (DM), Alzheimer's disease (AD), and tuberculosis (TB) according to death rates per year. Before this, many authors also provide additional review papers on various aspects, such as paper-based miniaturized SPEs [1], SPEs for in vitro diagnostic purpose [2], electrochemical biosensors based on nanomodified SPEs: for clinical application [3, 4], but no one focuses on targeting these deadliest diseases. Hence, we have tried to explore the application of SPEs on the proposed diseases and highlight its suitability for low-cost detection. In this article, we restricted our focus on the three deadliest diseases and its electrochemical sensing substrates using screen-printed electrodes.

2 Top Five Deadliest Diseases

The disease is an abnormal functioning of the human body organism; it may be due to many reasons like infections, unwanted growth of any bodily constituents, or deficiency or excess of chemical state with in body structure. Hence, in order to detect and diagnose the abnormal changes in system, the normal state of organism must be deeply understood. When one thinks about deadliest diseases, it is considered to be incurable or rapid attacking diseases; however, some of it are not deadliest one. Here, attempt is made to discuss some of world's top five ranked most fatal diseases.

2.1 Cardiovascular Diseases

Cardiovascular disease refers to a group of medical conditions that affect the heart and blood vessels. These conditions can range from relatively mild, such as high blood pressure, to life-threatening, such as a heart attack or stroke.

Some common cardiovascular diseases include:

1. Coronary artery disease: This is a condition in which the arteries that supply blood to the heart become narrowed or blocked, which can lead to chest pain, heart attack, or other complications.

2. Heart failure: This occurs when the heart is unable to pump blood effectively, which can cause fatigue, shortness of breath, and other symptoms.
3. Arrhythmia: This refers to an irregular heartbeat, which can cause palpitations, dizziness, and other symptoms.
4. Peripheral arterial disease: This is a condition in which the arteries that supply blood to the arms and legs become narrowed or blocked, which can cause pain and numbness.
5. Stroke: This occurs when the blood supply to the brain is disrupted, which can cause a range of symptoms depending on the severity and location of the damage.

There are several risk factors that can increase the likelihood of developing cardiovascular diseases, including high blood pressure, high cholesterol, smoking, diabetes, obesity, and a sedentary lifestyle. However, many of these risk factors can be managed through lifestyle changes and medical treatments, which can help prevent or manage cardiovascular diseases. Stroke is a medical condition that occurs when blood flow to a part of the brain is interrupted, either due to a blockage or bleeding in the brain. The lack of blood flow can cause brain cells to die, leading to various symptoms and long-term effects.

There are two main types of stroke:

1. Ischemic stroke: This type of stroke occurs when a blood vessel in the brain is blocked, often by a blood clot.
2. Hemorrhagic stroke: This type of stroke occurs when a blood vessel in the brain ruptures and causes bleeding in the brain.

Some common symptoms of stroke include:

- Sudden weakness or numbness in the face, arm, or leg, usually on one side of the body
- Difficulty speaking or understanding speech
- Confusion or difficulty with thinking and memory
- Sudden vision loss or difficulty seeing
- Severe headache with no known cause.

Risk factors for stroke include high blood pressure, smoking, high cholesterol, diabetes, obesity, and a family history of stroke. Stroke treatment often involves emergency medical care to restore blood flow to the brain and prevent further damage. Rehabilitation may also be necessary to help people recover from the effects of stroke.

2.2 *Respiratory Cancers*

Respiratory cancers are a group of diseases that affect the lungs and other parts of the respiratory system. These types of cancers are often caused by smoking or exposure to certain chemicals or pollutants.

The most common types of respiratory cancers include:

1. Lung cancer: Lung cancer is the most common type of respiratory cancer, and it is usually caused by smoking. There are two main types of lung cancer: non-small cell lung cancer and small cell lung cancer.
2. Mesothelioma: Mesothelioma is a rare type of cancer that affects the lining of the lungs, chest wall, or abdomen. It is often caused by exposure to asbestos.
3. Throat cancer: Throat cancer can affect the pharynx, larynx, or tonsils. It is often caused by smoking or alcohol consumption.
4. Nasal cavity and paranasal sinus cancer: These types of cancers affect the nasal cavity and sinuses. They are relatively rare, and their exact causes are not well understood.
5. Tracheal and bronchial tumors: These tumors affect the trachea (windpipe) and bronchi (the tubes that lead to the lungs). They can be benign or malignant.

Treatment for respiratory cancers depends on the type and stage of the cancer, as well as the overall health of the patient. Treatment options may include surgery, radiation therapy, chemotherapy, or targeted therapy. In addition, quitting smoking and avoiding exposure to chemicals and pollutants can help reduce the risk of developing respiratory cancers.

2.3 Diabetes Mellitus

Diabetes mellitus is a chronic metabolic disorder that affects the body's ability to use and store glucose (a type of sugar) properly. There are two main types of diabetes:

1. Type 1 diabetes: Also known as insulin-dependent diabetes, type 1 diabetes is an autoimmune disorder that occurs when the immune system attacks and destroys the beta cells in the pancreas that produce insulin. Without insulin, glucose cannot enter the body's cells to be used for energy, so glucose builds up in the bloodstream.
2. Type 2 diabetes: Type 2 diabetes is the most common form of diabetes, accounting for approximately 90% of cases. It occurs when the body becomes resistant to the effects of insulin and/or the pancreas does not produce enough insulin to meet the body's needs. Type 2 diabetes is often associated with obesity, sedentary lifestyle, and poor dietary habits.

Other forms of diabetes include gestational diabetes, which occurs during pregnancy, and various forms of monogenic diabetes, which are caused by specific genetic mutations.

If left untreated, diabetes can lead to a variety of complications, including heart disease, kidney damage, nerve damage, and vision loss. However, with proper management, including medication, lifestyle changes, and monitoring blood sugar levels, people with diabetes can lead healthy, fulfilling lives.

2.4 Alzheimer's Disease

Alzheimer's disease is a neurodegenerative disorder that affects the brain, leading to cognitive decline and memory loss. It is the most common cause of dementia, a group of disorders characterized by a decline in cognitive function and the ability to perform daily activities. The exact cause of Alzheimer's disease is not fully understood, but it is believed to be the result of a complex interplay between genetic, environmental, and lifestyle factors. The disease is characterized by the accumulation of beta-amyloid plaques and tau protein tangles in the brain, which lead to the death of brain cells and the progressive decline of cognitive function. The symptoms of Alzheimer's disease typically develop gradually and worsen over time. Early symptoms may include mild forgetfulness and difficulty with planning or problem-solving, while later symptoms may include confusion, disorientation, and difficulty with speech and motor function.

Currently, there is no cure for Alzheimer's disease, but there are treatments available that can help manage the symptoms and slow the progression of the disease. These treatments may include medications, lifestyle changes, and supportive therapies such as occupational and speech therapy. Research is ongoing in the search for more effective treatments and a cure for this devastating disease.

2.5 Tuberculosis

Tuberculosis (TB) is a contagious and potentially life-threatening bacterial disease caused by Mycobacterium tuberculosis. It primarily affects the lungs but can also affect other parts of the body such as the kidneys, spine, and brain. TB spreads through the air when an infected person coughs, sneezes, speaks, or sings. It is more common in areas with poor living conditions, such as overcrowding, poor ventilation, and inadequate nutrition. TB can also occur in people with weakened immune systems, such as those with HIV/AIDS or those receiving chemotherapy.

The symptoms of TB may include a cough that lasts for more than three weeks, chest pain, fatigue, weight loss, night sweats, fever, and chills. If left untreated, TB can cause severe damage to the lungs and other organs, and it can be fatal. TB is diagnosed through a combination of physical examinations, chest X-rays, and laboratory tests such as sputum cultures or blood tests. Treatment typically involves a course of antibiotics taken for several months. Prevention measures for TB include vaccination, early diagnosis and treatment of infected individuals, and maintaining good hygiene practices such as covering your mouth when coughing or sneezing, and avoiding close contact with people who are sick.

3 Screen-Printed Electrode Based Biosensor

Early detection of diseases can reduce the death rate worldwide. It not only provides specific treatment but also maintains the rate of curing. The chemical composition in human organisms provides a depth of knowledge about health status, and these constituents are regarded as biomarkers for various diseases and health conditions. By accurately estimating the concentration of some molecules provides information about the disease at a very early stage, even before symptoms are visible. Hence, in this view from last few decades researchers are focusing their interest toward biosensors and their applications for detecting numerous diseases. Recently, screen-printed electrode (SPE)-based biosensors are much popular due to their number of advantages over other sensing platforms. The main feature for designing electrode surfaces is the adhesion to the target and stability of coupling agents, allowing the immobilization of biomolecules [5]. Depending on the sensing ability and target to sense, a variety of screen-printed electrodes are developed to sense various biomarkers for the diagnosis of specific diseases. In this article, we are focusing on the three deadliest diseases and their electrochemical biosensors.

The use of SPEs in AD diagnosis has several advantages over conventional techniques. SPEs are portable and can be used for on-site diagnosis, particularly in remote areas with limited access to medical facilities. Additionally, SPEs are disposable, eliminating the need for cleaning and reducing the risk of cross-contamination. In conclusion, screen-printed electrodes are a promising tool for detecting biomarkers associated with Alzheimer's disease. The use of SPEs in AD diagnosis is expected to contribute significantly to early detection and treatment.

3.1 Cardiovascular Diseases (CVDs)

The leading cause of mortality worldwide, cardiovascular diseases (CVDs), claim 17.9 million lives annually (WHO). Coronary heart disease, cerebrovascular disease, rheumatic heart disease, and other illnesses are among the category of heart and blood vessel disorders known as CVDs. Heart attacks and strokes account for four out of every five CVD fatalities, and one third of these deaths happen before the age of 70 (WHO). Overweight and obesity are risk factors for CVD, as are elevated blood pressure, glucose, and cholesterol levels. All of these are simple to measure in primary care settings. Premature deaths can be avoided by identifying people who are most at risk for CVDs and ensuring they receive the proper care. To guarantee that people in need receive treatment and counseling, all primary health care institutions must have access to critical noncommunicable disease drugs and basic health technology. The particular biomarker cardiac troponin T for acute myocardial infarction has been detected by certain authors using immunosensors based on amine-functionalized carbon nanotubes-SPEs, and they were able to do so with the lowest LOD of 3.5 pg/mL [6]. The Gold-Nanoparticle (AuNP) spliced SPEs suggested by Serafin et al. have demonstrated excellent performance in order to capture B-type natriuretic peptide (BNP) at various levels in human serum to address the problem related to heart failure (HF) analysis with 100 times less limit of detection (4 pg/mL) compared with other reported methods [7]. One of the most promising biomarkers for diagnosing CVD in clinical blood samples is C-reactive protein (CRP) [8].

[9] have designed screen-printed electrodes modified with NiO nanoparticles, and this simple system has proven as of gold standard for the analysis of epinephrine and norepinephrine, in turn, is suitable for cost-efficient, effective device. Quantifying lactic acid gives information relating to heart health [10]. Team of researchers has developed lactate biosensor containing SPE improved with Meldola's Blue-Reinecke Salt (MBRS-SPCE) with a layer of enzyme lactate dehydrogenase NAD + dependent (from Porcine heart), and NAD^{+} for amperometric measurements covering a range of 0.55–10 mM [11]. It is found that gold nanoparticle printed electrodes are capable of detecting levels of cholesterol, aminothiols, C-reactive protein, and B-type natriuretic peptides with good accuracy implementing carbon or gold electrodes as per the suitability of the system [12–14, 7]. The various heart health analytes and respective sensing modalities are represented in Fig. 1.

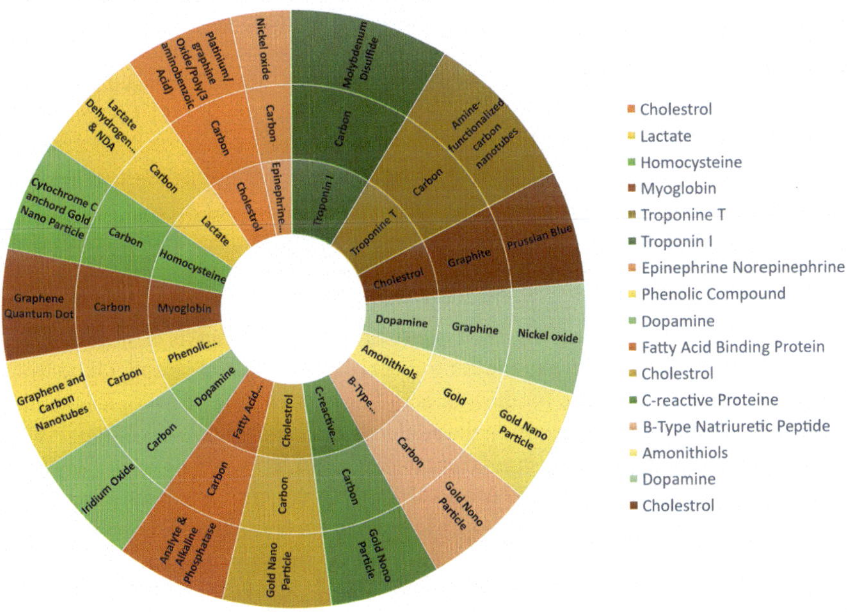

Fig. 1 Different biomarkers and sensing modalities for cardiovascular diseases

3.2 *Alzheimer Disease*

Alzheimer's disease (AD) is one of the hazardous diseases originating from a neurodegenerative disorder that results in loss of memory functions, cognitive impairment, language deterioration, dysfunction of physical moments, and gradually leads to death [15]. It is mainly found in senior citizens above 65 age, and in a few cases, it has affected the life of younger ones. R. Brookmeyer et al. estimated that till the year 2050, around 106.23 million AD patients will be reported around the globe, which will create a heavy burden on the economy load [16]. World Health Organization reported that by year 2040, AD will be counted in top two leading death causing diseases [17]. The early diagnosis of the AD is the only key to timely control of these heavy figures. That is possible when the medical industry will be equipped with biosensors for initial stage detection and medicine for complete treatment. AD symptoms are not very prominent to recognize in its early stage, which is why biomarkers are the only trustworthy source of information for its assessment. The estimation of biomarkers provides a guiding path for treatment as well as decision-making. Researchers are working with some biomarkers to design extremely specific and highly sensitive AD biosensors. At present, most of the researchers are interested in diagnostically reliable biomarkers such as Tau proteins, Amyloid β peptides, Apolipoprotein E4 (APOE4), Amyloid Precursor Protein (APP), Amyloid β Derived Diffusible Ligands (ADDLs), α-1 antitrypsin (AAT), β-site APP-cleaving enzyme (BACE1), Alzheimer-associated neuronal thread protein (AD7c-NTP), MicroRNA, etc., listed in Fig. 2.

According to neuropathology, AD is considered to be due to the accretion of insoluble fibrous material, mainly either extracellular amyloid protein or intracellular neurofibrillary tangles (NFT) [18]. These extracellular amyloid protein primary constituents are amyloid-β (Aβ) peptides comprising 39–42 amino acids [19, 20], whereas NFT comprises abnormal filaments composed of the microtubule-associated tau protein in a hyperphosphorylated state [21, 18, 22]. Tau is a microtubule-associated protein that resides in the central nervous system and peripheral nervous systems, which is responsible for the number of malfunctioning neurological studies termed under tauopathy diagnosis. In the adult brain, there are six types of tau protein isoforms, represented as 4R2N (=tau40 or tau-441), 4R1N (=tau46 or tau-412), 4R0N (=tau24 or tau-383), 3R2N (=tau39 or tau-410), 3R1N (=tau37 or tau-381) and 3R0N (=tau23 or tau-352) [23] and lead to different appearance while the development of disease [24]. For the healthy brain, 3R and 4R tau quantities are balanced in equal proportion and is considered as a significant biomarker for AD. Secondly, the sticky plaques of amyloid-β (Aβ) peptide were reported around neurons in the case of AD patient. There are two types of Aβ peptides: Aβ42 and Aβ40 produced by cleavage of the amyloid precursor protein (APP) due to the action of β-secretase and γ-secretase [25]. Although there are other minor amounts of various peptides (Aβ28, Aβ33, Aβ34, Aβ3-34, Aβ37, Aβ38, and Aβ39) also present. The Aβ40 is a significant component of Aβ compared to species ending at Aβ42, secreted by neural tissues and is the main reason for deposition of bulk accumulation in AD brain [26–28]. Further,

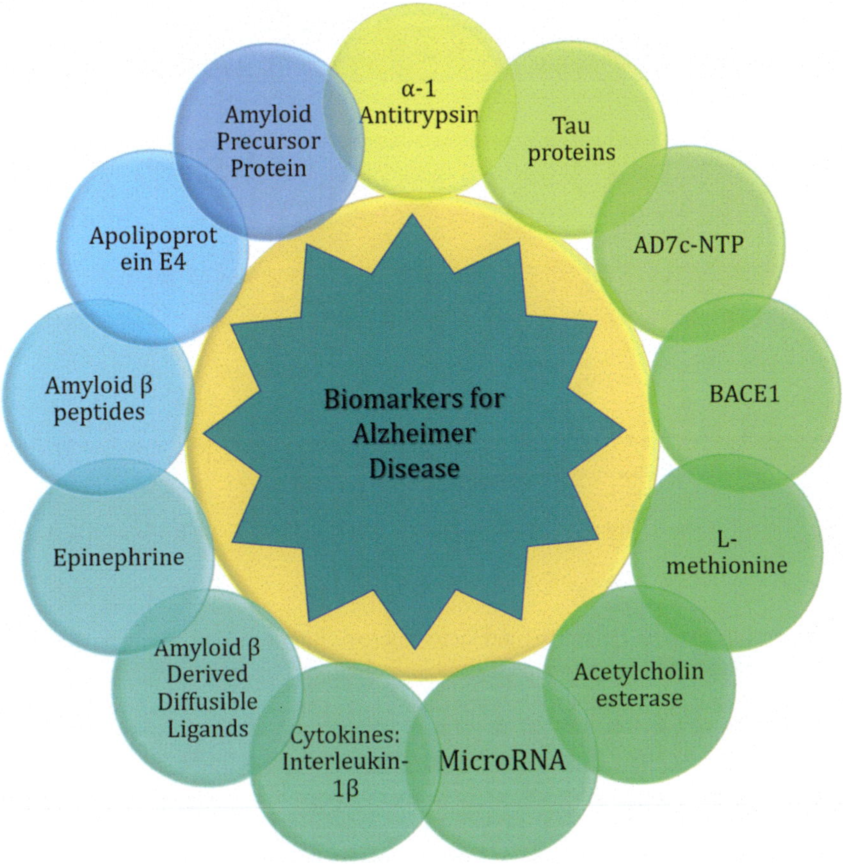

Fig. 2 Some important biomarkers for Alzheimer disease

apart from these two primary biomarkers, there are other analytes for faithful diagnosis of AD. Human apolipoprotein E (APOE4, APP, amyloid-β-derived diffusible ligands (ADDLs, α-1 antitrypsin (AAT, BACE1, and Alzheimer-associated neuronal thread protein (AD7c-NTP have been recognized as effective biomarkers to early diagnosis of AD [19, 29–32].

For the diagnosis of AD, the best sensing media is the electrochemical transducers; specifically, SPEs are primarily economical, lightweight, and disposable devices applicable to widespread electrochemical sensing applications. SPEs have been used to detect Aβ peptides and tau protein in cerebrospinal fluid (CSF) samples of patients with AD. The recognition of Aβ peptides can be realized by electrochemical techniques such as cyclic voltammetry, amperometry, and impedance spectroscopy. The sensing of tau protein can be attained through electrochemical impedance spectroscopy or enzyme-linked immunosorbent assays (ELISA) in combination with SPEs. It is reported that two important brain disorder biomarker

miRNA	Support	Reference Electrode	Technique	LOD (mol/L)	Ref.
106a	SPCE	Silver	DPV	$3x10^{-16}$ M	[38]
122	SPCE	Silver	DPV	$7x10^{-12}$ M	[39]
21, 32 & 122	GNPs-SPCE	Silver	SWV & EIS	$5x10^{-18}$ M	[40]
34a	CNF-SPGE	Silver	DVP	$1.51x10^{-6}$ M	[41]
141	Au-SPE/ rGO/MWCNT	Silver	SWV	$1x10^{-14}$ M	[42]
21	Au-SPE	Silver	DPV	$1x10^{-16}$ M	[43]
155	Au-SPE	Silver	SWV & EIS	$5.7x10^{-18}$ M	[44]
107	Au-SPE	Silver	DPV	$1x10^{-14}$ M	[45]
107	Au-SPE	Silver	CC	$1x10^{-16}$ M	[46]
107	SPCE	Carbon	SWV EIS	$1x10^{-12}$ M	[47]
137	SPCE+rGO+Au NWs	Silver	CV & EIS	$1x10^{-15}$ M	[48]

Au-SPE (Screen Printed Gold Electrode), CNF (Carbon Nanofiber), CC (Chronocoulomery), DPV (Differential Pulse Voltammetry), EIS (Electrochemical Impedance Spectroscopy), GNPs (Gold Nano-Particles), SPE (Screen Printed Electrode), miRNA (micro-RNA), MWCNTs (Multi-Walled Carbon Nanotubes), rGO (Reduced Graphene Oxide), SPCE (Screen Printed Carbon Electrode), SPGE (Screen Printed Graphite Electrode), SWV (Square Wave Voltammetry), AuNWs (Gold Nanowires)

Fig. 3 Electrochemical biosensors based on screen-printed electrodes for miRNAs detection [38–48]

cytokines: Interleukin-1b (IL-1b) and factor necrosis tumor a (TNF-a) can be detected by 4-Carboxyphenyl-functionalized double-walled carbon nanotube (HOOC-Phe-DWCNTs) using amperometric technique [33]. Cristiane Kalinke et al. have designed 3D-printed electrodes based on polylactic acid/graphene filament to detect L-methionine, this voltametric circuit with a limit of detection 1.39 μmol/L [5]. Acetylcholinesterase (AChE) is an enzyme mainly present in the central and exterior nervous system, the malfunction of AChE encourages the accumulation of amyloid-beta peptides causing neurological disorders and AD [34, 35, 33]. Panraksa et al. [36] developed paper-based amperometric sensor for the determination of acetyl-cholinesterase using a screen-printed graphene electrode. Research has confirmed that micro-RNA (miRNA) is involved in proliferation and apoptosis in cell functions, and its fluctuations may lead to neurodegenerative diseases such as AD [37]. The miRNA detection system, techniques, and its detection limit are shown in Fig. 3.

3.3 *Cancer Disease*

Cancer is the abnormal and uncontrolled growth of tissues over their limit [49, 50]. It is one of life-threatening disease, and its early detection is the only way of lifesaving. It can take around 200 different forms, and the most widespread ones are breast, lung, prostate, ovarian, skin, colon, leukemia, and hematologic cancer. Biomarkers in such case play a crucial role in identifying the in-depth details about genetic and molecular characteristics of a patient's cancer, which in turn facilitates scheduling personalized treatment plans. Considering its example, the risk of developing breast

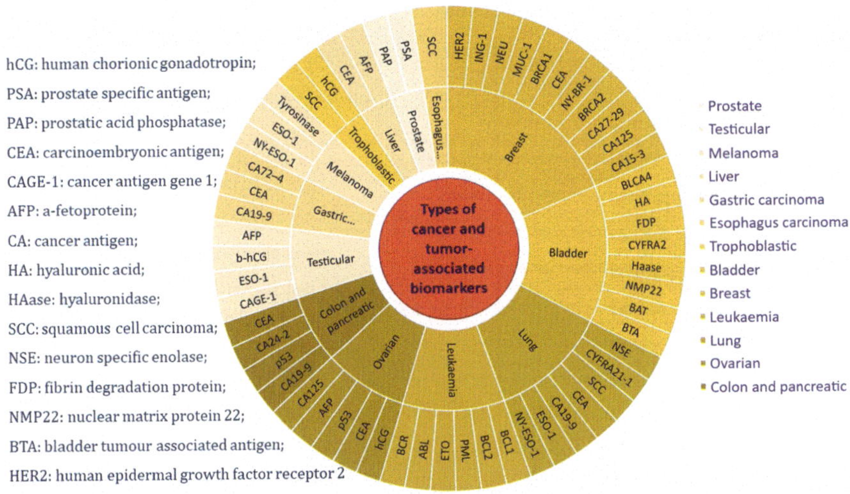

Fig. 4 Various forms of cancers and related biomarkers

and ovarian cancer can be calculated by analyzing the biomarkers to identify the mutations in the BRCA1 and BRCA2 genes in breast cancer [51]. In some instances, biomarkers become a reliable source for monitoring the ongoing treatment and its effects. The level of the protein CA-125 in the blood reflects the response to therapy in ovarian cancer patients [52]. The decrease in the level of CA-125 is an indicator for the efficiency of treatment. Hence, identifying the presence and concentration of specific biomarkers provides a path way to not only treat but monitor the cancer cells. The different categories of cancers and related biomarkers for cancer are shown in Fig. 4.

Electrochemical biosensors, specially SPEs, are widely used for cancer detection and monitoring. Prostate-specific antigen (PSA) is the most commonly used biomarker for the investigation of prostate cancer; its typical reference range is 0–4 ng/mL [53]. PSA is found in human serum in it two forms, free PSA (fPSA) and PSA complexed, the total PSA (tPSA) denotes the totality of fPSA and PSA complexed [54]. The concentration of tPSA beyond 10.0 ng/mL indicates a high probability toward prostate cancer, whereas below 4.0 ng/mL reports negative. If test shows value between these two cutoff limits, then fPSA is evaluated [55]. The team of researchers has developed a sensor to quantify both fPSA and tPSA levels using disposable SPEs consisting of two working carbon electrodes [55]. The gold nanostructured electrode was provided with specific antibodies for tPSA and fPSA, enzyme alkaline phosphatase (AP) was used as a label, and a mixture of 3-indoxyl phosphate disodium salt (3-IP) and silver nitrate as substrate [56].

Other biomarkers investigated using SPE-based biosensors include carcinoembryonic antigen (CEA) for colon cancer, alpha-fetoprotein (AFP) for liver cancer, CA125 for ovarian cancer, HER for breast cancer, etc., are presented in Fig. 5. CEA is a glycoprotein found in many carcinomas, such as colon, lung, ovarian, and

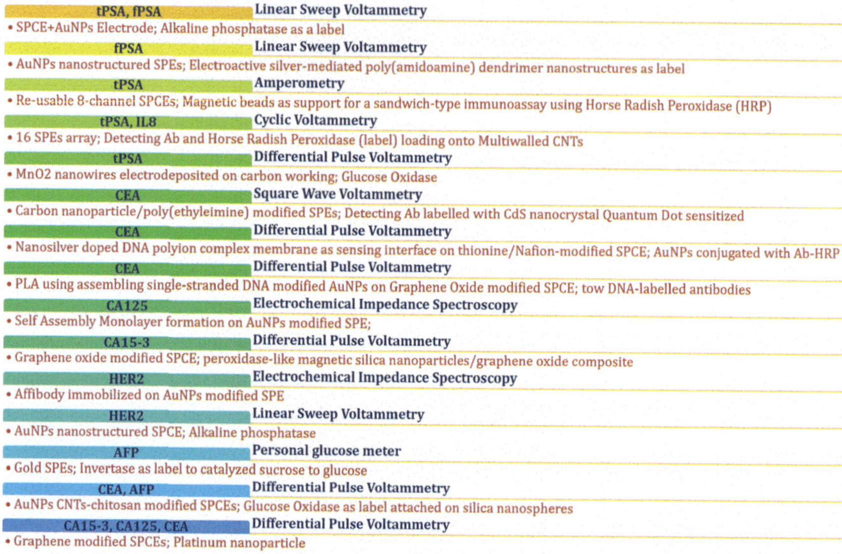

Fig. 5 Some important biomarkers, sensing techniques, and their sensing modalities

breast cancer. Mostly measuring CEA facilitates the monitoring the patient during therapy and post-surgery its recurrence [57]. Wu et al. designed SPCEs applied with a nanosilver-doped DNA polyion complex membrane conjugated with an HRP-labeled antibody to detect CEA [58]. The carbohydrate antigens (CA), available as CA125, CA15-3, CA19-9, etc., are called cancer antigens, CA125 protein is related to ovarian cancer as well as to uterus, cervix, pancreas, liver, colon, breast, lung, and digestive cancer [49]. By using the electrochemical impedance spectroscopy technique, employing label-free self-assembly monolayer formation on AuNPs modified SPE sensor is designed to estimate the level of CA125 [59]. Rui et al. have developed low-cost, highly sensitive electropolymerized O-phenylenediamine (oPDA) on Au-screen-printed electrodes to detect CA15-3 resulting in its absorption at gold platform [60].

4 Conclusion

Quantifying the single biomarker does not provide depth of information and the actual and accurate status of the disease. Detection of multiple analytes simultaneously offers detail and a more faithful status of health conditions and infections. Taking this into account, researchers are fascinated toward multiplex sensing. In the last few decades, numerous sensing devices based on SPE have been designed, but very few are commercialized; hence it can be considered the initial stage and a lot newer sensing modalities are expected, and a long path to go.

Acknowledgements The authors acknowledge the SERB-National Post Doctoral Fellowship (SERB-NPDF) for providing the necessary research infrastructure and funding for this work.

References

1. Silveira CM, Monteiro T, Almeida MG (2016) Biosensing with paper-based miniaturized printed electrodes–a modern trend. Biosensors 6(4):51
2. Mincu NB, Lazar V, Stan D, Mihailescu CM, Iosub R, Mateescu AL (2020) Screen-Printed Electrodes (SPE) for in vitro diagnostic purpose. Diagnostics 10(8):517
3. Arduini F, Micheli L, Moscone D, Palleschi G, Piermarini S, Ricci F, Volpe G (2016) Electrochemical biosensors based on nanomodified screen-printed electrodes: recent applications in clinical analysis. TrAC Trends Anal Chem 79:114–126
4. Beitollahi H, Mohammadi SZ, Safaei M, Tajik S (2020) Applications of electrochemical sensors and biosensors based on modified screen-printed electrodes: a review. Anal Methods 12(12):1547–1560
5. Kalinke C, Neumsteir NV, de Oliveira PR, Janegitz BC, Bonacin JA (2021) Sensing of L-methionine in biological samples through fully 3D-printed electrodes. Anal Chim Acta 1142:135–142
6. Silva BV, Cavalcanti IT, Silva MM, Dutra RF (2013) A carbon nanotube screen-printed electrode for label-free detection of the human cardiac troponin T. Talanta 117:431–437
7. Serafín V, Torrente-Rodríguez RM, González-Cortés A, De Frutos PG, Sabaté M, Campuzano S, Pingarrón JM (2018) An electrochemical immunosensor for brain natriuretic peptide prepared with screen-printed carbon electrodes nanostructured with gold nanoparticles grafted through aryl diazonium salt chemistry. Talanta 179:131–138
8. Jampasa S, Siangproh W, Laocharoensuk R, Vilaivan T, Chailapakul O (2018) Electrochemical detection of c-reactive protein based on anthraquinone-labeled antibody using a screen-printed graphene electrode. Talanta 183:311–319
9. Carbone M, Nesticò A, Bellucci N, Micheli L, Palleschi G (2017) Enhanced performances of sensors based on screen printed electrodes modified with nanosized NiO particles. Electrochim Acta 246:580–587
10. Shah NC, Lyandres O, Walsh JT, Glucksberg MR, Van Duyne RP (2007) Lactate and sequential lactate– glucose sensing using surface-enhanced Raman spectroscopy. Anal Chem 79(18):6927–6932
11. Piano M, Serban S, Pittson R, Drago GA, Hart JP (2010) Amperometric lactate biosensor for flow injection analysis based on a screen-printed carbon electrode containing Meldola's Blue-Reinecke salt, coated with lactate dehydrogenase and NAD+. Talanta 82(1):34–37
12. Bernardo-Boongaling VRR, Serrano N, García-Guzmán JJ, Palacios-Santander JM, Díaz-Cruz JM (2019) Screen-printed electrodes modified with green-synthesized gold nanoparticles for the electrochemical determination of aminothiols. J Electroanal Chem 847:113184
13. Boonkaew S, Chaiyo S, Jampasa S, Rengpipat S, Siangproh W, Chailapakul O (2019) An origami paper-based electrochemical immunoassay for the C-reactive protein using a screen-printed carbon electrode modified with graphene and gold nanoparticles. Microchim Acta 186:1–10
14. Huang Y, Cui L, Xue Y, Zhang S, Zhu N, Liang J, Li G (2017) Ultrasensitive cholesterol biosensor based on enzymatic silver deposition on gold nanoparticles modified screen-printed carbon electrode. Mater Sci Eng C 77:1–8
15. Caratelli V, Ciampaglia A, Guiducci J, Sancesario G, Moscone D, Arduini F (2020) Precision medicine in Alzheimer's disease: An origami paper-based electrochemical device for cholinesterase inhibitors. Biosens Bioelectron 165:112411

16. Brookmeyer R, Johnson E, Ziegler-Graham K, Arrighi HM (2007) Forecasting the global burden of Alzheimer's disease. Alzheimers Dement 3(3):186–191
17. World Alzheimer Report 2015, 2015. Alzheimer's disease international (ADI), London. https://www.alz.co.uk/research/WorldAlzheimerReport2015.pdf
18. Razzino CA, Serafín V, Gamella M, Pedrero M, Montero-Calle A, Barderas R, ... Pingarrón JM (2020) An electrochemical immunosensor using gold nanoparticles-PAMAM-nanostructured screen-printed carbon electrodes for tau protein determination in plasma and brain tissues from Alzheimer patients. Biosensors and Bioelectronics 163:112238
19. Hardy JA, Higgins GA (1992) Alzheimer's disease: the amyloid cascade hypothesis. Science 256(5054):184–185
20. Murphy MP, LeVine H III (2010) Alzheimer's disease and the amyloid-β peptide. J Alzheimers Dis 19(1):311–323
21. Olsson A, VANDERSticHELE HUGO, Andreasen N, De Meyer G, WALLiN ANDERS, Holmberg B, ... Blennow K (2005) Simultaneous measurement of β-amyloid (1–42), total tau, and phosphorylated tau (Thr181) in cerebrospinal fluid by the xMAP technology. Clinical Chem 51(2):336–345
22. Tolnay M, Probst A (1999) tau protein pathology in Alzheimer's disease and related disorders. Neuropathol Appl Neurobiol 25(3):171–187
23. Andreadis A (2012) Tau splicing and the intricacies of dementia. J Cell Physiol 227(3):1220–1225
24. Goedert M, Jakes R (1990) Expression of separate isoforms of human tau protein: correlation with the tau pattern in brain and effects on tubulin polymerization. EMBO J 9(13):4225–4230
25. Shui B, Tao D, Florea A, Cheng J, Zhao Q, Gu Y, ... Guo Z (2018) Biosensors for Alzheimer's disease biomarker detection: a review. Biochimie 147:13–24
26. Claeysen S, Bockaert J, Giannoni P (2015) Serotonin: a new hope in Alzheimer's disease? ACS Chem Neurosci 6(7):940–943
27. Golde TE, Eckman CB, Younkin SG (2000) Biochemical detection of Aβ isoforms: implications for pathogenesis, diagnosis, and treatment of Alzheimer's disease. Biochimica et Biophysica Acta (BBA)-Molecular Basis Disease 1502(1):172–187
28. Liu L, Zhao F, Ma F, Zhang L, Yang S, Xia N (2013) Electrochemical detection of β-amyloid peptides on electrode covered with N-terminus-specific antibody based on electrocatalytic O2 reduction by Aβ (1–16)-heme-modified gold nanoparticles. Biosens Bioelectron 49:231–235
29. Georganopoulou DG, Chang L, Nam JM, Thaxton CS, Mufson EJ, Klein WL, Mirkin CA (2005) Nanoparticle-based detection in cerebral spinal fluid of a soluble pathogenic biomarker for Alzheimer's disease. Proc Natl Acad Sci 102(7):2273–2276
30. Monte SM, Ghanbari K, Frey WH, Beheshti I, Averback P, Hauser SL, ... Wands JR (1997) Characterization of the AD7C-NTP cDNA expression in Alzheimer's disease and measurement of a 41-kD protein in cerebrospinal fluid. J Clinical Invest 100(12):3093–3104
31. Rentmeister A, Bill A, Wahle T, Walter J, Famulok M (2006) RNA aptamers selectively modulate protein recruitment to the cytoplasmic domain of β-secretase BACE1 in vitro. RNA 12(9):1650–1660
32. Zhu G, Lee HJ (2017) Electrochemical sandwich-type biosensors for α–1 antitrypsin with carbon nanotubes and alkaline phosphatase labeled antibody-silver nanoparticles. Biosens Bioelectron 89:959–963
33. Prasansuklab A, Tencomnao T (2013) Amyloidosis in Alzheimer's disease: the toxicity of amyloid beta (Aβ), mechanisms of its accumulation and implications of medicinal plants for therapy. Evidence Based Comp Alternat Med
34. Ulrich J, Meier-Ruge W, Probst A, Meier E, Ipsen S (1990) Senile plaques: staining for acetylcholinesterase and A4 protein: a comparative study in the hippocampus and entorhinal cortex. Acta Neuropathol 80:624–628
35. Morán MA, Mufson EJ, Gómez-Ramos P (1993) Colocalization of cholinesterases with β amyloid protein in aged and Alzheimer's brains. Acta Neuropathol 85:362–369
36. Panraksa Y, Siangproh W, Khampieng T, Chailapakul O, Apilux A (2018) based amperometric sensor for determination of acetylcholinesterase using screen-printed graphene electrode. Talanta 178:1017–1023

37. Congur G, Erdem A (2019) PAMAM dendrimer modified screen printed electrodes for impedimetric detection of miRNA-34a. Microchem J 148:748–758
38. Daneshpour M, Omidfar K, Ghanbarian H (2016) A novel electrochemical nanobiosensor for the ultrasensitive and specific detection of femtomolar-level gastric cancer biomarker miRNA-106a, Beilstein. J Nanotechnol 7:2023–2036
39. Bettazzi F, Hamid-Asl E, Esposito CL, Quintavalle C, Formisano N, Laschi S, ... Palchetti I (2013) Electrochemical detection of miRNA-222 by use of a magnetic bead-based bioassay. Analyt Bioanalyt Chem 405:1025–1034
40. Labib M, Khan N, Ghobadloo SM, Cheng J, Pezacki JP, Berezovski MV (2013) Three-mode electrochemical sensing of ultralow microRNA levels. J Am Chem Soc 135(8):3027–3038
41. Erdem A, Eksin E, Congur G (2015) Indicator-free electrochemical biosensor for microRNA detection based on carbon nanofibers modified screen printed electrodes. J Electroanal Chem 755:167–173
42. Tran HV, Piro B, Reisberg S, Nguyen LH, Nguyen TD, Duc HT, Pham MC (2014) An electrochemical ELISA-like immunosensor for miRNAs detection based on screen-printed gold electrodes modified with reduced graphene oxide and carbon nanotubes. Biosens Bioelectron 62:25–30
43. Sonuç MN, Sezgintürk MK (2014) Ultrasensitive electrochemical detection of cancer associated biomarker HER3 based on anti-HER3 biosensor. Talanta 120:355–361
44. Cardoso AR, Moreira FTC, Fernandes R, Sales MGF (2016) Novel and simple electrochemical biosensor monitoring attomolar levels of miRNA-155 in breast cancer. Biosens Bioelectron 80:621–630
45. Koo KM, Carrascosa LG, Shiddiky MJ, Trau M (2016) Poly (A) extensions of miRNAs for amplification-free electrochemical detection on screen-printed gold electrodes. Anal Chem 88(4):2000–2005
46. Yadav S, Masud MK, Islam MN, Gopalan V, Lam AKY, Tanaka S, ... Shiddiky MJ (2017) Gold-loaded nanoporous iron oxide nanocubes: a novel dispersible capture agent for tumor-associated autoantibody analysis in serum. Nanoscale 9(25):8805–8814
47. Suprun EV, Radko SP, Khmeleva SA, Mitkevich VA, Archakov AI, Makarov AA, Shumyantseva VV (2017) Electrochemical oxidation of amyloid-beta peptide isoforms on carbon screen printed electrodes. Electrochem Commun 75:33–37
48. Azimzadeh M, Nasirizadeh N, Rahaie M, Naderi-Manesh H (2017) Early detection of Alzheimer's disease using a biosensor based on electrochemically-reduced graphene oxide and gold nanowires for the quantification of serum microRNA-137. RSC Adv 7(88):55709–55719
49. Bohunicky B, Mousa SA (2010) Biosensors: the new wave in cancer diagnosis. Nanotechnol Sci Appl, 1–10
50. Rama EC, Costa-García A (2016) Screen-printed electrochemical immunosensors for the detection of cancer and cardiovascular biomarkers. Electroanalysis 28(8):1700–1715
51. Ford D, Easton DF, Stratton M, Narod S, Goldgar D, Devilee P, ... Breast Cancer Linkage Consortium (1998) Genetic heterogeneity and penetrance analysis of the BRCA1 and BRCA2 genes in breast cancer families. Amer J Human Genet 62(3):676–689
52. Bottoni P, Scatena R (2015) The role of CA 125 as tumor marker: biochemical and clinical aspects. Advances in cancer biomarkers: from biochemistry to clinic for a critical revision, 229–244
53. Ronkainen NJ, Okon SL (2014) Nanomaterial-based electrochemical immunosensors for clinically significant biomarkers. Materials 7(6):4669–4709
54. Stenman UH, Leinonen J, Alfthan H, Rannikko S, Tuhkanen K, Alfthan O (1991) A complex between prostate-specific antigen and α1-antichymotrypsin is the major form of prostate-specific antigen in serum of patients with prostatic cancer: assay of the complex improves clinical sensitivity for cancer. Can Res 51(1):222–226
55. Escamilla-Gómez V, Hernández-Santos D, González-García MB, Pingarrón-Carrazón JM, Costa-García A (2009) Simultaneous detection of free and total prostate specific antigen on a screen-printed electrochemical dual sensor. Biosens Bioelectron 24(8):2678–2683

56. Fanjul-Bolado P, Hernández-Santos D, González-García MB, Costa-García A (2007) Alkaline phosphatase-catalyzed silver deposition for electrochemical detection. Anal Chem 79(14):5272–5277
57. Ho JAA, Lin YC, Wang LS, Hwang KC, Chou PT (2009) Carbon nanoparticle-enhanced immunoelectrochemical detection for protein tumor marker with cadmium sulfide biotracers. Anal Chem 81(4):1340–1346
58. Wu W, Yi P, He P, Jing T, Liao K, Yang K, Wang H (2010) Nanosilver-doped DNA polyion complex membrane for electrochemical immunoassay of carcinoembryonic antigen using nanogold-labeled secondary antibodies. Anal Chim Acta 673(2):126–132
59. Ravalli A, Dos Santos GP, Ferroni M, Faglia G, Yamanaka H, Marrazza G (2013) New label free CA125 detection based on gold nanostructured screen-printed electrode. Sens Actuators B Chem 179:194–200
60. Gomes RS, Moreira FT, Fernandes R, Sales MGF (2018) Sensing CA 15–3 in point-of-care by electropolymerizing O-phenylenediamine (oPDA) on Au-screen printed electrodes. PLoS ONE 13(5):e0196656

Next-Generation Therapies for Type 2 Diabetes Mellitus

Debarun Patra, Soumyajit Roy, Palla Ramprasad, and Durba Pal

Abstract Type 2 diabetes mellitus (T2DM), which has been considered a pandemic due to its growing pervasiveness in the world population, is associated with obesity, ageing, a sedentary lifestyle, and the complex interplay of genetic and environmental factors. Globally, adult diabetic cases hit 537 million in 2021 and escalated to approximately a billion by 2045. In T2DM, the metabolic pathophysiological state impairs insulin sensitivity, glucose tolerance, β-cell functionality, and immune-metabolic harmony and introduces dysfunction of insulin-responsive tissues. Research in the biomedical discipline innovates a plethora of anti-diabetic therapeutic candidates, but very few are clinically successful, although drug delivery, immunological approaches, oligonucleotide therapy, and regenerative and transplantation medicines contribute to a wide range of therapeutic dimensions. Therefore, understanding this complex pathophysiological state needs to be extensively studied, and new molecules are required to be revealed at the earliest. In this chapter, we have precisely discussed the recently developed and potent, underdeveloped therapies for the treatment of T2DM. Participation of analogues, agonists, antagonists, activators, and inhibitors targeting specific components in the context of insulin resistance (IR) holds promise for addressing both IR and T2DM. Additionally, we delve into the exploration of newly developed immune-based therapies and diverse approaches for administering insulin to individuals with T2DM.

Keywords Type 2 diabetes mellitus · Anti-diabetic drugs · Agonists · Antagonists · Immune-based therapies

D. Patra · S. Roy · P. Ramprasad · D. Pal (✉)
Department of Biomedical Engineering, Indian Institute of Technology Ropar, Ropar 140001, Punjab, India
e-mail: durba.pal@iitrpr.ac.in

A. Madhusudhan et al. (eds.), *Functional Smart Nanomaterials and Their Theranostics Approaches*, Smart Nanomaterials Technology,
https://doi.org/10.1007/978-981-99-6597-7_13

1 Introduction

Type 2 diabetes mellitus (T2DM) is a severe global public health burden and decisive reason for millions of deaths [94]. It is commonly linked to obesity, sedentary lifestyle, and ageing where cases are rising more rapidly in low- and middle-income countries over the years [85, 94]. International Diabetes Federation (IDF) published data exerts worldwide 537 million are diabetic in 2021 [93], and the number will hit approximately a billion by 2045 (IDF Diabetes Atlas 10th edition). Insulin is a hormone made by the pancreas β-cell that acts to let blood sugar into the cells in the body for use as energy. T2DM is characterized by diminished or inappropriate secretion of insulin, which could be a defect of either islet cell function or beta-cell mass. Also, the cells don't respond normally to insulin and that lead to insulin resistance (IR) [29]. Usually, metformin and/or sulfonylurea medication are used to treat T2DM patients. However, because T2D is a progressive condition, the majority of late-stage T2DM patients stop responding to these existing oral medicines [46, 143]. Moreover, the association between diet-induced obesity and the pathogenesis of T2DM occurs through chronic low-grade inflammation in responsive tissues, particularly in adipose tissue leading to impairment of insulin sensitivity, glycaemic regulation, and lipid metabolism [177]. Only a few approved therapies are available to tackle the ruthless forms of health threats to humankind in the twenty-first century. Recent advances in the field of experimental medicine, immunology, drug delivery, tissue engineering, and regenerative medicine yield optimistic prospects in the battle against this severe pathophysiological health concern.

In this chapter, we meticulously delineated and systematically orchestrated the recent development in anti-diabetic drug development, immune-based therapeutic approaches, and advanced insulin therapies to combat IR and T2DM (Fig. 1). Also, we discussed the potential therapeutic candidates (which are in clinical trials) for the management of T2DM.

2 Recent Advances in Anti-diabetic Drug Development

To generate suitable options for next-generation therapeutics of T2DM, researchers have developed therapeutic approaches that will majorly enhance the function of hepatic tissue, adipose tissue, skeletal muscle, kidney, and pancreatic β-cells. Interestingly, analogues, agonists, antagonists, activators, and inhibitors of the target moieties participate in IR are potentially suppress the post-prandial hyperglycaemia in T2DM (Table 1).

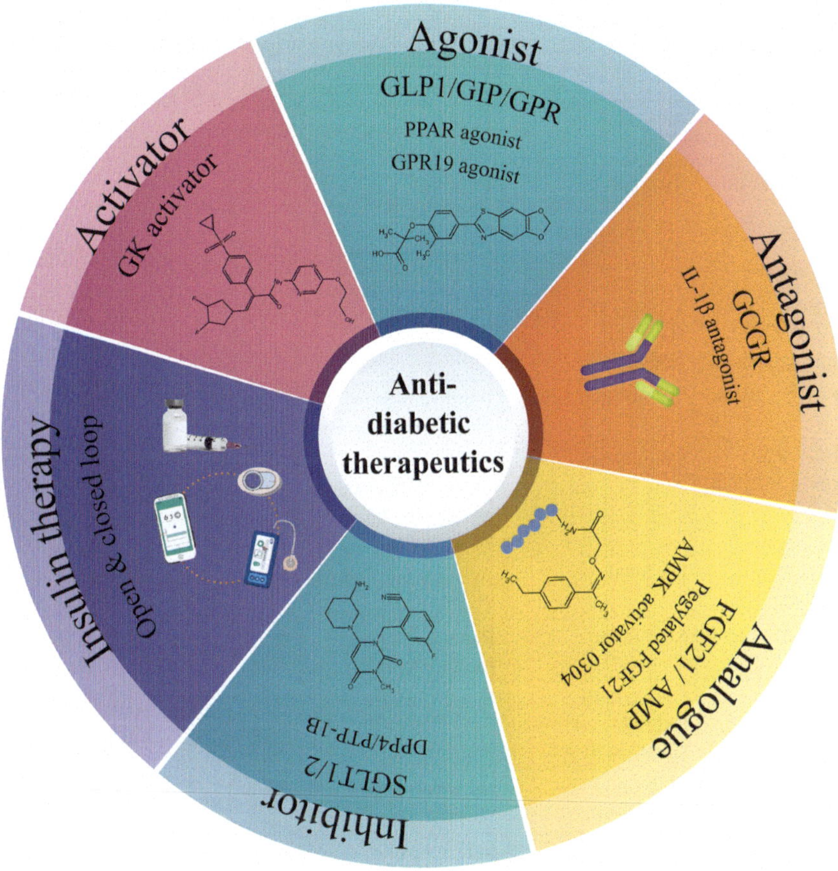

Fig. 1 Schematic diagram represents various classes of anti-diabetic therapeutics with examples

2.1 Agonists

2.1.1 GLP-1, GIP, and Glucagon Receptors Agonist

Glucagon is a peptide hormone, produced by the α-cells of the islets of Langerhans, which holds critical therapeutic implication power [174]. The development of agonists for GIP (glucose-dependent insulinotropic polypeptide) and GLP-1 (glucagon-like peptide 1), together promote β-cell proliferation and inhibit apoptosis, thereby expanding pancreatic β-cell mass [149]. Tirzepatide (LY3298176) is a dual agonist that binds to the receptors of GLP-1 and GIP, interestingly GIP binds with higher affinity in terms of glycaemic control and weight loss [175]. This drug is in the phase 3 SURPASS-2 clinical trial. MEDI-0382, a GLP-1/GCGR dual agonist showed profound results by improving metabolic parameters and lowering hyperglycaemic conditions in patients [10]. This drug is under phase 2 clinical trials.

Table 1 Recently developed novel anti-diabetic therapeutic agents approved or/and under the clinical trial procedure

Class	Drug	Mechanism of action	Clinical status	References
GLP-1 agonist	Semaglutide	Control glucagon metabolism	Phase 4	[95, 125]
GLP-1/GIP dual agonist	Tirzepatide	Control glucagon metabolism	Phase 3	[67]
GLP-1/GCGR dual agonist	MEDI-0382	Glucagon pathway	Phase 2	[111, 116]
GLP-1R / GCGR/ GIPR triple agonist	SAR441255 HM15211 NN9423	Incretin effect	Phase 1	[22, 70, 120]
PPAR-γ agonist	Thiazolidinediones	Improvement in glucose homeostasis	Phase 4	[12]
PPAR–Pan agonist	Chiglitazar	Improvement in glucose homeostasis and hepatic lipid accumulation	Phase 3	[102]
GPR119 agonists	DS-8500a	Active insulin release	Phase 2	[173]
TGR5 agonists	SB756050	Regulate bile acid receptor activity	Phase 2	[82]
FFAR1 agonist	CPL207280 P11187 LY2922470	Incretin effect	Phase 2 Phase 1 Phase 1	[16, 133]
Melatonin receptor agonist	Melatonin	Improves insulin sensitivity	Phase 1	[15, 62]
Fibroblast growth factor 21 analogues	Pegbelfermin	Induce insulin sensitivity	Phase 2	[31]
AMP analogues	O304	Activates AMPK	Phase 2a	[155]
Glucokinase (GK) activators	TMG-123 TTP-399	Conversion of glucose to glucose 6-phosphate	Phase 2 Phase 2	[159, 160]
Glimins	Time limit	Restricts gluconeogenesis and oxidative phosphorylation	Phase 3	[57]
Human pro-islet peptide	BTI-410 or HIP2B	Mass gain of β-cell	Phase 2	[44]
GCGR antagonist	LGD6972 REMD-477	Blocking Glucagon receptor	Phase 2 Phase 1	[113, 136]
SGLT1 and SGLT2 inhibitors	Sotagliflozin	Blocks both SGLT1 and SGLT2 receptor	Phase 3	[32]
	Licogliflozin	Elevates GLP-1 level	Phase 1	[78]

(continued)

Table 1 (continued)

Class	Drug	Mechanism of action	Clinical status	References
SGLT2 inhibitors	Empagliflozin	Reduce hepatic de novo lipogenesis	Phase 4	[98]
11β-HSD1 inhibitors	SPI-62	Reduce intracellular cortisol levels	Phase 1	[92, 137, 138]
	BI 135,585	Block cortisone to cortisol	Phase 1	[89, 94]
Protein tyrosine phosphatase-1B inhibitor	KQ-791	Rapid activation of insulin signalling pathway	Phase 1	[107, 108]
Dipeptidyl peptidase (DPP)-4 inhibitors	Trelagliptin Saxagliptin	DPP4 inhibition mediated controlled insulin release	Phase 4	[91, 97]
Combined peptide administration	IDegLira iGLarLixi	Insulin and GLP-1 agonist	Approved	[69, 74, 115]

A most recent published study showed the generation of synthetic peptide tri-agonist (SAR441255) of GLP-1, GIP, and glucagon receptors (GGG) effectively helps weight loss and glycaemic control [22]. Another triple agonist HM15211, currently in phase 1 clinical trials, is a weekly subcutaneous administrable GLP-1R/GIPR/GCGR tri-agonist peptide being developed by Hanmi Pharma to treat obesity, NAFLD, and T2DM [64]. This is a modified human glucagon moiety conjugated to PEG-linked Fc fragment of human IgG [63]. Hepatic triglycerides (TG) and thiobarbituric acid reactive substances (TBARS) concurrently dropped after HM15211 administration and intracellular cAMP levels ascend [151]. Another GGG tri-agonist generated by Novo Nordisk (NN9423) is currently in the development process.

2.1.2 Peroxisome Proliferator-Activated Receptor (PPAR) α/γ/δ Agonist

These TFs are sub-typically divided, functionally assorted, and tissue specifically expressed. PPAR-α is expressed in muscle, heart, and liver involved in energy expenditure; PPAR-γ overexpressed in adipose tissue acts on energy storage, and PPAR-δ showed whole-body distribution critical for energy metabolism [42]. PPAR-γ agonist Thiazolidinediones (TZDs) are anti-diabetic agents approved for clinical usage and found to enhance insulin sensitivity by increasing adiponectin and GLUT4 expression [59]. This drug holds a great impact by upregulating insulin-dependent glucose uptake in responsive tissues [141]. PPARα/γ/δ pan-agonist Chiglitazar, a multicentre, randomized, double-blind, and placebo-controlled trial currently in phase 3 study in patients with T2DM and insufficient glycaemic control despite diet and exercise showed very promising results as $HbA1_c$ reduced significantly [90]. Another set of PPAR pan-agonist MHY2013 and IVA337 has been shown to subdue hepatic lipid accumulation, and inflammation can potentially become a therapeutic candidate for NAFLD [11, 172].

2.1.3 GPR119 Agonist

GPCR, a prominent player in cellular signalling pathways, has garnered sustained attention as a potential target for drug development in various diseases [83]. Specifically, pancreatic GPR119 plays a pivotal role in the regulation of insulin and incretin secretion. The activation of this molecule is associated with the cAMP accumulation in β-cell and incretin release from the gut [132]. GPR119 agonist DS-8500 is developed by Daiichi Sankyo Company and is currently in phase 2 clinical trials [173].

2.1.4 Takeda G-Protein-Coupled Receptor 5 (TGR5 Agonist)

TGR5, a bile acid-specific receptor, predominantly expressed in L-cells' basolateral membrane, is associated with GLP-1 secretion [24]. The clinical trial on the TGR5 agonist, SB 756,050, originated by Glasgow Smith Kline was unsuccessful and discontinued after the phase 2 clinical trial [82].

2.1.5 Melatonin Receptor Agonist

Melatonin formation is reduced in T2DM models, and melatonin receptors are expressed majorly in hepatocytes and pancreatic cells [171]. Developing suitable melatonin receptor agonists was targeted for the next-generation therapeutic approach. A melatonin receptor agonist Ramelteon approved for the treatment of insomnia also improves insulin resistance and systemic inflammation [48, 101].

2.1.6 Free Fatty Acid Receptor (FFAR1 Agonist)

Free fatty acids (FFAs) activate their receptors (FFAR) and regulate insulin and incretin release [6]. Short-chain FFAs activate FFAR2 and FFAR3 associated with IR [154]. The inhibition of fatty acid receptor (FFAR2/3) enhances insulin sensitivity in in vivo models [86]. FFAR1 and FFAR4 are activated by long-chain free fatty acids (FFAs), and these receptors augment insulin and incretin release. FFAR1 (GPR40) agonist CPL207280 was evaluated in a double-blind, randomized, placebo-controlled phase 2 clinical trial, and the result showed improved glucose tolerance in patients [162, 180]. P11187 and LY2922470 are also FFAR1 agonists in phase 2 clinical trials [104, 105].

2.2 *Antagonist*

2.2.1 GCGR Antagonist

Introducing a new category of anti-hyperglycemic medications, Glucagon receptor antagonists (GRAs) have been specifically formulated to address hyperglycemia in diabetes by inhibiting the interaction of glucagon with its receptors, thereby blocking its action [160]. Administration of antagonists improves $HbA1_c$ level, although it doesn't able to generate promising potential in clinical settings. Several scientific efforts lead to the generation of LGD6972, a small molecule drug developed by Metavant Sciences and Ligand Pharmaceuticals, currently in phase 2 clinical trial [169]. Recently, another GCGR antagonist REMD-477 in phase 1 clinical trial acts by decreasing insulin requirements and improving glycaemic status in diabetic patients [135].

2.3 *Analogues*

2.3.1 Fibroblast Growth Factor 21 Analogue

FGF21, a regulator of hepatic lipid and glucose metabolism, also holds a crucial role in the regeneration developmental stages [178]. Moreover, FGF21 overexpression reduces hyperglycaemia, body mass, and lipid levels in diabetic subjects. A recombinant PEGylated FGF21, Pegbelfermin (BMS-986036) considered for a randomized, placebo-controlled, double-blind clinical trial in T2DM patients and it has improved whole-body insulin sensitivity [31].

2.3.2 Adenosine Monophosphate Analogue

Activation of protein kinase-AMPK is dependent on AMPs and exhibits insulin sensitivity by inducing muscle-specific glucose uptake and lipid oxidation as well as reducing the hepatic gluconeogenesis [43]. Metformin showed anti-diabetic effect by activating AMPK in T2DM patients. An AMPK activator O304 was studied in a proof-of-concept phase 2a clinical trial in type 2 diabetes (T2D) patients on metformin and exerts positive beneficial results by the reduction of FPG and increase in microvascular perfusion in T2DM patients [155].

2.4 *Inhibitors*

2.4.1 Sodium-Glucose Co-transporter and Dual SGLT1-SGLT2 Inhibitor

These membrane-bound transporter proteins perform sodium and glucose transport in cells, by reabsorbing glucose from the renal tubule to blood, and a hike in blood glucose was observed [8, 28]. Targeted inhibition of SGLT1/2 exerts promising results in T2DM patients. Few of the SGLT inhibitors received the green signal from the administration boards, among them dapagliflozin and canagliflozin are globally marketed for clinical use, that improve the glycaemic condition, facilitate weight loss along with improving cardio-renal functionality [71, 119]. Recently developed first-in-class of dual SGLT1/2 inhibitors, sotagliflozin in phase 3 clinical trial, significantly reduced post-prandial glucose concentration and hold back appetite in T2DM patients. Moreover, sotagliflozin reduces mortality cases in cardiovascular disease and chronic kidney disease patients ("Drug approval package: Invokana (canagliflozin) tablets,") [56]. Empagliflozin, a novel selective SLGT-2 inhibitor, showed favourable results by improving glycaemic control even in T2DM patients [75]. Licogliflozin (formerly known as LIK 066) SGLT1/2 inhibitor reduces body weight and elevates GLP-1 levels in diabetic patients [14, 47, 78].

2.4.2 11β-HSD1 (11β-Hydroxysteroid Dehydrogenase 1) Inhibitor

11β-HSD1, (NADPH)-dependent enzyme, overexpressed in hepatic and adipose tissue, forms cortisone from cortisol that is known to impair glucose uptake and potentiates hepatic IR [30, 176]. Inhibitor against this enzyme improves glycaemic status. In recent years SPI-62, an HSD1 inhibitor considers for clinical trial (currently in phase 1) for T2DM patients (S Pharmaceuticals). While earlier BI 135,585, a selective 11β-HSD1 inhibitor was evaluated by clinical trial, and positive results were observed [89].

2.4.3 Protein Tyrosine Phosphatase-1B Inhibitor

An essential player in insulin signalling, Protein tyrosine phosphatase (PTP) acts by dephosphorylating and subsequently downregulating the insulin signalling pathway. Elevated expression of PTP has been associated with an increased risk of developing obesity and T2DM [3]. PTP-1B knockout animals are protected from obesity-associated complications [17]. KQ-791, a PTP-1B inhibitor considered for a randomized, double-blind, placebo-controlled, single ascending dose study to treat T2DM, and results showed a significant reduction in fasting glucose (K B Limited).

2.4.4 Dipeptidyl Peptidase (DPP)-4 Inhibitor

For insulin secretion and glucose metabolism, incretins (GLP-1 and GIP) are required. GLP-1 and GIP are both quickly rendered inactive by DPP-4 [149]. A novel DPP-4 inhibitor, Trelagliptin, a clinically approved drug used for T2DM patients shown to have anti-inflammatory characteristics and enhance both fasting and post-prandial glucose control [91]. Another phase 4 clinical trial drug Saxagliptin inhibits DPP-4 and reduces adipose tissue inflammation in obese individuals, lessens reactive oxygen species (ROS production switching off toll-like receptors (TLR and NF-κB pathway activation, along with macrophage infiltration and inflammation in adipose tissue [80, 97].

2.5 Glucokinase (GK) Activators

Glucokinase or hexokinase type-IV enzyme is critical for glucose homeostasis by hepatic glucose uptake and glycogenesis [5]. The activator of GKs found to lower glucose and a novel GK activator TMG-123, demonstrated systemic glucose reduction without hampering hepatic and plasma TGs [96, 163]. A multicentre, adaptive phase 2, randomized, double-blind, placebo and active-controlled (sitagliptin), parallel-group study on TTP-399 was performed and results showed $HbA1_c$ and plasma glucose levels reduced in T2DM patients [58].

2.6 Glimins

A tetrahydrotriazine compound imeglimin (the first member of Glimin family) displayed the most promising therapeutic efficacy in T2D patients, by increasing glucose-dependant insulin secretion along with non-significant side effects [55, 140]. It activates AKT and turn-on the insulin signalling pathway, enhances β-cell functionality (including life span and glucose-induced insulin secretion), includes the cardioprotective feature, and reduces $HbA1_c$, fasting plasma glucose (FGP) and oxidative stress in T2DM patients when administered independently or combinatorically with other anti-diabetic drugs [181]. The outstanding efficacy of this drug poses a potential role in future anti-diabetic therapy.

2.7 Human Pro-islet Peptide

Since the chronic stage of T2DM loosens pancreatic β-cell function and numbers, increasing the performance of β-cell can alleviate the pathophysiological condition. A biologically active peptide fragment, human pro-islet peptide (HIP) is a 14-amino acid long and coded from the REG3a gene and holds an outstanding ability to surge β-cell mass and re-establish glucose homeostasis [110]. BTI-410 or HIP2B, the very first agent from HIP group, showed good efficacy in phase 1 clinical trials in T2DM patients [151].

2.8 *Combined Peptide Therapy*

Individually, GLP-1 analogues improve β-cell function along with reducing post-prandial hyperglycaemia in T2DM patients, and insulin therapy alone lessens fasting hyperglycaemia [70]. The introduction of the combinatorial approach by fixed ratio amalgamation of both injectable drugs leads to the formation of iGlarLixi (100 U insulin glargine and 50 μg lixisenatide) IDegLira (50 U insulin degludec and 1.8 mg liraglutide) approved by the FDA for clinical usage [117, 170]. The results showed a decrease in $HbA1_c$ compared to independent treatment from the combinations.

3 Recent Advances in Immune-Based Therapeutic Approaches

Diabetes medications with secondary anti-inflammatory properties reduce or stop the loss in β-cell activity during diabetes pathogenesis. Adiponectin, CRP, IL-6, and IL-1β measurements are mostly used to detect T2D individuals with persistent inflammation [161]. Generally, T cells are involved in the aetiology of the T2DM-related β-cell functional death [26]. Determining the existence of islet autoimmune T cell responses requires testing for patients with T2D must be aware of autoimmunity [9]. Moreover, these findings demonstrate the necessity of preoperative patient stratification based on autoantibody and islet-reactive T cell positivity before immunotherapy, to guarantee the greatest therapeutic result. It is crucial to remember that clinical studies in T2D patients have not yet been routinely conducted and categorized patients according to their immunological status before the beginning of immunotherapy. Recently developed therapeutics in this section are listed in Table 2.

3.1 *IL-1β Receptor Antagonist (IL-1βRa)*

One of the factors contributing to the development of T2D is the activation of the innate immune system in obesity and insulin resistance by IL-1β signalling [129]. The entry of macrophages to inflamed adipose tissues and islets is thought to be triggered by IL-1β, which intensifies islet inflammation [183]. Anakinra (recombinant human IL-1βRa), which was given to T2D patients, demonstrated a statistically significant

Table 2 Example of novel drugs targets immunological mediators and are used for T2DM therapy

Class	Drug	Mechanism of action	Clinical status	References
PPAR-α/γ Agonist	Aleglitazar	Regulate lipid and glucose levels and enhance insulin sensitivity	Phase 3	[109, 148]
IL-1R type 1 Antagonist	Anakinra	Binds to the IL-1R and inhibit the activity of IL-1α/1β	Phase 2	[53]
Anti-human IL-1β monoclonal antibody	Canakinumab	Binds to human IL-1 and inhibits its interaction with IL-1 receptors to reduce inflammation	Phase 3	[54, 165]
IL-1β, Oxysterols receptor LXR-α/β, Oxysterols receptor LXR-α, Cytochrome P450 1A2/3A Subfamily/2E1/2C9/2D6	Diacerein	Rhein's anti-inflammatory properties lower interleukin-1beta activity, which is a major factor in extracellular matrix synthesis decrease, MMP activity, and ongoing inflammation	Phase 2	[145, 168]
Humanized interleukin-1 beta (IL-1β) mAB	Ly2189102	Reduce $HbA1_c$ and fasting glucose	Phase 2	[40]
PPAR-γ Agonist	Pioglitazone	Promotes insulin sensitivity	Phase 4	[27]
Small molecule	Salsalate	Inhibition of prostaglandin synthesis	Phase 4	[21]
Anti-human IL-1β antibody	Xoma 052	Reduce inflammation	Phase 2	[112, 179]
TLR4 inhibitor	Eritoran	Reduce inflammation	Phase 2	[130]

decrease in $HbA1_c$ [144]. Additionally, the anakinra treated group had greater C-peptide levels and a decreased ratio of pro-insulin to insulin, suggesting enhanced β-cell activity [52, 100]. Clinical studies are now being conducted on IL-1β neutralizing antibodies, such as Xoma 052 and Ly2189102, which may offer potential novel treatments that target the IL-1β pathway [131].

3.2 Salicylates

Salicylates are non-steroidal pain relievers (NSAIDs), separated into acetylated (aspirin) and nonacetylated (sodium salicylate and salsalate) forms. Instead of inhibiting cyclo-oxygenase enzymes (Cox) enzymes, higher dosages of salicylates disrupt NF-κB signalling pathways and increase the adiponectin expression [73, 114]. They have also the ability to lower plasma glucose [73, 182]. Salsalate has been demonstrated to decrease inflammation by reducing NF-κB activation in endothelial cells from the adipose tissue [139]. Salsalate was examined in the TINSAL-T2D research, a multicentre, multi-dose randomized trial in T2DM patients that showed lower $HbA1_c$ and higher levels of adiponectin [72].

3.3 PPAR (Peroxisome Proliferator-Activated Receptor Agonists) α and γ

PPAR-γ is a key molecular regulator of T regulatory cells. While, pioglitazone restores insulin sensitivity by directly increasing T regulatory (Treg) cells, regulatory cytokines including IL-4 and IL-10, adiponectin, and suppressing T helper cell proliferative responses [66]. Rosiglitazone has also been shown to be useful in preventing the development of inflammation and autoimmune response [66]. Aleglitazar, a novel dual PPARα/γ agonist, has attracted a lot of attention and has an anti-inflammatory impact, beneficial for glucose homeostasis. Aleglitazar treatment led to greater positive alterations in lipid profiles and HOMA-IR scores [79]. These findings provide credence to the idea that T2DM illness can be improved in the future by focusing on immunological responses and most critically, by reducing tissue-specific T cell responses.

3.4 mAbs for T and B Lymphocytes

The islet autoimmunity linked to T1DM has been successfully treated by targeting T cells with anti-CD3 therapy (visilizumab, otelixizumab, and teplizumab) and B lymphocytes with anti-CD20 therapy (rituximab) [81, 93, 134]. Targeting the immune system effector cells specifically may be a direction for future studies not only in T1DM but also in T2DM, given the understanding that chronic inflammation, characterized by T cell and macrophage infiltration of the adipose tissue, is the hallmark of obesity leading to the development of insulin resistance, glucose intolerance, and T2DM. Thus, targeting T and B lymphocytes in adipose tissue may be advantageous [7]. The possible risk of leukoencephalopathy should be taken into consideration when utilizing anti-CD3 or anti-CD20 antibodies to treat T2DM patients.

3.5 Toll-Like Receptor (TLR) Inhibitors

TLR recognizes pathogen-associated molecular structures and discriminates detection of self vs non-self-ligands in addition to recognizing structures on possible pathogens. It has been proposed that the self-components' inappropriate activation of TLRs might cause autoimmunity or inflammation and contribute to the onset of T2DM [25]. Monocytes from newly diagnosed T2DM patients express TLR2 and TLR4 more often [45], therefore, blocking TLR receptors provides novel immunological treatments for T2DM and obesity. TLR4 inhibitor, Eritoran, in phase 2 clinical shows, reduced inflammation.

4 Different Approaches for Insulin Intake

The administration of insulin is crucial for the management of diabetic patients to obtain the best metabolic control. The Diabetes Control and Complication Trial (DCCT) highlighted the significance of intensive insulin therapy by emphasizing the delivery of insulin that mimics endogenous insulin production to achieve glycaemic control with minimal hypoglycaemia [118].

4.1 Subcutaneous Route

Subcutaneous administration through vials, syringes, insulin pens, and continuous insulin infusion is widely preferred method for administration of insulin because of the ease of the self-administration [150]. Insulin aspart, insulin degludec, insulin lispro, and insulin detemir as rapid-acting forms of insulin already being prescribed for glycaemic control in diabetic patients. Many others like insulin tregopil, insulin icodec, and PEGylated insulin lispro are under clinical trial to check the efficacy in glycaemic control. It necessitates knowledge of the length of action of various insulin types, the interaction between blood glucose and certain drugs, stress, and the ability to modify insulin doses to accomplish the unique target [153]. But this method of delivery has downsides such as painful injections, discomfort, erratic compliance, and challenges in establishing post-prandial blood glucose control [74]. Newer routes of insulin delivery have evolved to deliver insulin accurately and precisely while minimizing patient burden (Table 3).

Table 3 List of advanced insulin therapeutics delivered through different routes

Route of administration	Drug name	Class	Clinical status	References
Subcutaneous	Insulin aspart	Rapid-acting form of insulin	Approved	[121, 126]
	Insulin degludec	Long-acting insulin	Approved	[122, 126]
	Insulin lispro	A modified form of fast-acting insulin	Approved	[34]
	Insulin detemir	The long-acting form of insulin	Approved	[123, 126]
	Recombinant human insulin	Human insulin	Approved	[147]
	Insulin icodec	Rapid-acting form of insulin	Phase 3	[126]
	Basal insulin-FC	Recombinant basal insulin fused to an Fc domain	Phase 3	[35]
	Ultra-fast-acting insulin analogue	Fast-acting insulin with niacin amide and l-arginine	Phase 1	[4]
	Recombinant insulin aspart biosimilar	Recombinant insulin	Phase 3	[19]
	U-500 insulin	Short-acting medical insulin	Phase 2	[36, 40]
Inhale	Intranasal insulin	Rapid-acting form of insulin	Phase 2	[127]
	Afrezza inhalant product	Rapid-acting inhaled insulin	Phase 4	[33]
	Insulin inhalation	Rapid-acting inhaled insulin	Phase 2	[158]
	Human insulin inhalation powder	Normal Insulin	Phase 3	[37]
Oral	Tadalafil	Phosphodiesterase 5 inhibitor	Phase 2	[164]
	Oral insulin	Insulin	Phase 3	[137]
	Insulin glargine (U300)	Concentrated insulin 300 units/mL of insulin	Phase 4	[147]
	Insulin oral	Enteric-coated insulin capsule	Phase 2	[50]
	Insulin tregopil	Rapid-acting form of insulin	Phase 3	[18]
	NN1952	Oral insulin analogue	Phase 1	[2]

4.2 Open and Closed-Loop-Based Systems

Injections through jet injectors, insulin pens, and access ports cannot be regarded as a significant advancement in the pursuit of better control, while they may increase patient comfort The flow resistance that the insulin solution faces as it travels through the tubing and cannula (an infusion set) and into the subcutaneous tissue is overcome in current therapy by programmable implanted insulin pumps when utilized as an open-loop delivery method [142]. The focus of open-loop insulin delivery is on patient-controlled administrations at varied doses and times of the day. Some studies are listed (Table 4) that are under clinical trial.

An emerging therapeutic strategy for people with diabetes called closed-loop insulin delivery, also known as the artificial pancreas, consists of a connected continuous glucose monitor and an insulin pump [60]. Wireless communication enables automated data transfer between components without the need for human intervention (Fig. 2). The closed-loop system, in contrast to an open-loop system, reduced the patients' involvement in managing glucose levels by using an algorithm and real-time continuous glucose monitor sensor to give continuous and automated reactions in glucose swings dependent on physiological behaviour [61]. An insulin pump with sophisticated Hybrid Closed-Loop (HCL) technology can autonomously administer basal insulin utilizing an algorithm and real-time continuous glucose monitor sensor. Model predictive control (MPC), proportional-integral-derivative (PID), and fuzzy logic are the three primary types of algorithms used by the MiniMed 670G (Medtronic), the first hybrid closed-loop system that is commercially available in 2016 [76]. There are several other closed-loop systems are available in the market like t: slim X2 with Control-IQ (Tandem), MiniMed 780G (Medtronic), and some others under clinical trial [157] (Table 5).

A plethora of studies are particularly focused on using nanotechnologies and hydrogel systems to create non-electronic, self-regulating insulin delivery devices in addition to electronic ones like phenylboronic acid (PBA) and derivatives, glucose oxidase (GOx), and glucose-binding molecules such as concanavalin A (Con A) [76]. These devices performed well in vivo testing for biocompatibility and glycaemic

Table 4 List of insulin pumps under the clinical trial

Type	Drug treatment	Trial phase	References
Insulin pump	Insulin glulisine or aspart	Phase 4	[51]
	Insulin lispro	Phase 4	[103]
	Insulin lispro and Insulin aspart	Phase 3	[38]
Omnipod (disposable insulin pump)	U-500R insulin	Phase 4	[99]
Insulin pump	Insulin aspart	Phase 1	[1]

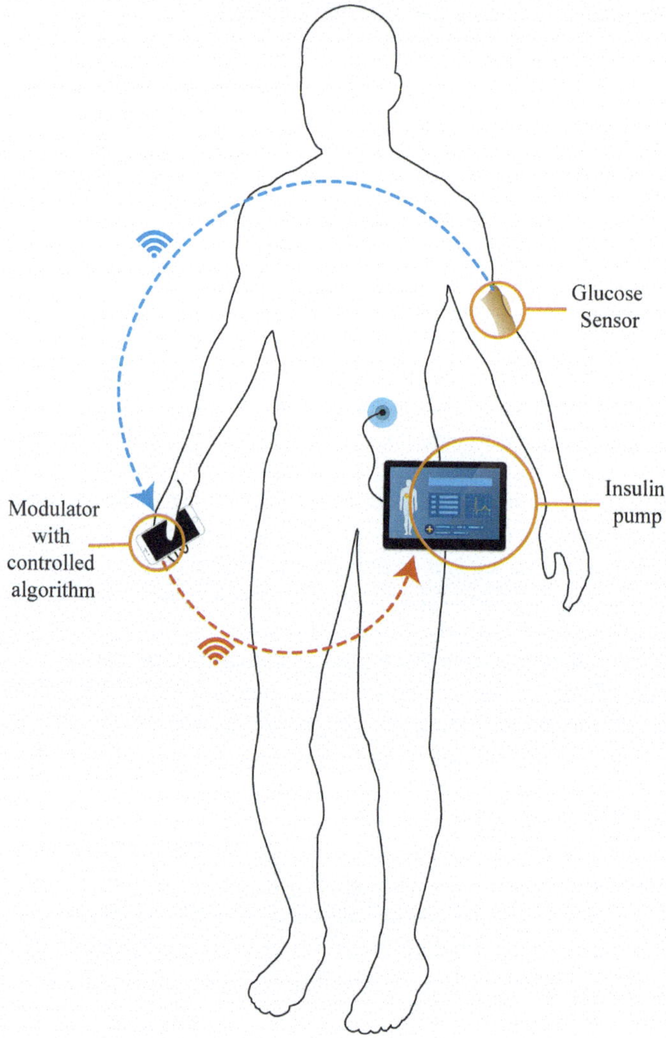

Fig. 2 Schematic illustration exhibits a closed-loop insulin delivery system. The applicability of this system works such as a real-time glucose monitor sensor autonomously measures blood glucose and the algorithm system administers basal insulin to maintain homeostasis

control. So, future research into various formulations, enhanced medication delivery capability, and extended shelf life of these might improve patient compliance in future diabetic therapies [106].

Table 5 Examples of closed-loop insulin delivery systems under the clinical trial

Type of closed-loop system	Clinical status	Recruitment status	References
CamAPS HX closed-loop	–	Completed	[84]
Diabeloop algorithm associated with dexcom G6 CGM and the Accu-chek insight insulin pump	–	Not recruited yet	[49]
Eli Lilly's insulin dosing algorithm	–	Completed	[167]
FUSION closed-loop glucose control system	–	Completed	[156]

4.3 *Pulmonary Route*

As a result of the lungs' large and well-perfused absorptive surface and the lack of particular peptidases, insulin administration through the lungs was the first known alternative to subcutaneous injection for lowering blood glucose levels and that is good for controlling post-prandial blood glucose since inhaled insulin takes effect around 15 min after inhalation and lasts for two to three hours [23]. Exubera, the first inhaled medicine with a dry powder composition and a quicker effect time of 10–15 min, was approved by FDA in 2006. Then Afrezza Inhalant Product, human insulin inhalation powder from MannKind Corporation and Eli Lilly and Company has passed through phase 4 and 3 clinical trials, respectively. Besides all these advantages, the non-invasive pulmonary way of administration is constrained by technological problems with inhaler devices, increased cost, and long-term safety, particularly in the pulmonary function [152].

4.4 *Oral Route*

The oral route of administration has come as the most patient-friendly method of consuming insulin, because of its non-invasive nature of administration and it might better closely resemble how insulin is delivered physiologically [13]. Tadalafil, ORMD-0801, and HOE901 Toujeo have completed the phase 2, 3, and 4 clinical trials. Recently, extensive investigation has been done to improve the oral bioavailability of peptide and protein drugs, because of the low bioavailability, low permeability through the intestinal membrane, and rapid peptidase activity of protein-based drugs. In general, the success of oral insulin hinges on the capacity to produce insulin both effectively in a cost-conscious pharmaceutical market and adequate amounts for oral administration [65]. Then also the mode of administration, bioavailability and pharmacological availability data are often still far from being consistent in terms of magnitude, duration, and most importantly, consistency.

5 Conclusion

In the twentieth century, the ground-breaking discovery of metformin and recombinant insulin was considered critical medicine for T2DM care. A sincere understanding of the disease progression at the molecular level helps scientists to generate advanced therapeutics. Many promising candidates with the best possible mode of delivery approach hold the paramount potential to become critical care medicine for this disease. Moreover, combinatorial therapy also holds an exciting clinical prospect for patients. Data recorded on clinicaltrials.gov displayed a promising future ahead for the management of T2DM. In the last decade, a total of 567 active interventional studies (phases 1–4) are registered for a clinical trial in T2DM. Who knows which next-generation therapy among these trials will receive approval and when hit the market? We strongly believe a resilient bridge between academics and industrial research going to produce potential clinically relevant therapeutic candidates against T2DM.

Acknowledgements Authors gratefully acknowledge the funding support from the Department of Biotechnology (DBT), Govt. of India (Grant No.: BT/PR40669/MED/32/761/2020) to D.P. De. P., S.R., and P.R. acknowledge IIT Ropar and Ministry of Education, India, for their research fellowship.

References

1. A/S, NN (2016) A study of the treatment of type 2 diabetes with an insulin infusion pump. Available from NIH ClinicalTrials Retrieved from https://clinicaltrials.gov/ct2/show/NCT00115973
2. A/S, NN (2017) A two-part trial investigating NN1952 in healthy subjects and subjects with type 1 and type 2 diabetes. https://clinicaltrials.gov/ct2/show/NCT01028404?term=NN1952&cond=Diabetes+Mellitus%2C+Type+2&draw=2&rank=1
3. Abdelsalam SS, Korashy HM, Zeidan A, Agouni A (2019) The role of protein tyrosine phosphatase (PTP)-1B in cardiovascular disease and its interplay with insulin resistance. Biomolecules 9:7.https://doi.org/10.3390/biom9070286
4. Adocia (2021) Ultra fast-acting insulin analogue. Retrieved from https://adisinsight.springer.com/drugs/800035885. Available from Springer Nature Adis Insight Retrieved 30.08.2022 https://adisinsight.springer.com/drugs/800035885
5. Agius L (2009) Targeting hepatic glucokinase in type 2 diabetes: weighing the benefits and risks. Diabetes 58:18–20. https://doi.org/10.2337/db08-1470
6. Al Mahri S, Malik SS, Al Ibrahim M, Haji E, Dairi G, Mohammad S (2022) Free fatty acid receptors (FFARs) in adipose: physiological role and therapeutic outlook. Cells 11:4.https://doi.org/10.3390/cells11040750
7. Alle M, Bandi R, Sharma G, Dadigala R, Husen A, Kim J-C (2021) Current Trends in engineered gold nanoparticles for cancer therapy. Nanotechnology in the Life Sciences. Springer, Cham, pp 1–40
8. Alle M, Park SC, Bandi R, Lee S-H, Kim J-C (2021) Rapid in-situ growth of gold nanoparticles on cationic cellulose nanofibrils: Recyclable nanozyme for the colorimetric glucose detection. Carbohydr Polym 253:117239. https://doi.org/10.1016/j.carbpol.2020.117239

9. Alle M, Sharma G, Lee S-H, Kim J-C (2022) Next-generation engineered nanogold for multi-modal cancer therapy and imaging: a clinical perspectives. J Nanobiotechnology 20(1):222. https://doi.org/10.1186/s12951-022-01402-z
10. Ambery P, Parker VE, Stumvoll M, Posch MG, Heise T, Plum-Moerschel L, Tsai L, Robertson DF, Jain M, Petrone M, Rondinone C, Hirshberg B, Jermutus L (2018) MEDI0382, a GLP-1 and glucagon receptor dual agonist, in obese or overweight patients with type 2 diabetes: a randomised, controlled, double-blind, ascending dose and phase 2a study. Lancet 391:2607–2618.https://doi.org/10.1016/s0140-6736(18)30726-8
11. An HJ, Lee B, Kim SM, Kim DH, Chung KW, Ha SG, Park KC, Park YJ, Kim SJ, Yun HY, Chun P, Yu BP, Moon HR, Chung HY (2018) A PPAR Pan Agonist, MHY2013 alleviates age-related hepatic lipid accumulation by promoting fatty acid oxidation and suppressing inflammation. Biol Pharm Bull 41:29–35.https://doi.org/10.1248/bpb.b17-00371
12. Angwin C, Jenkinson C, Jones A, Jennison C, Henley W, Farmer A, Sattar N, Holman RR, Pearson E, Shields B, Hattersley A (2020) TriMaster: randomised double-blind crossover study of a DPP4 inhibitor, SGLT2 inhibitor and thiazolidinedione as second-line or third-line therapy in patients with type 2 diabetes who have suboptimal glycaemic control on metformin treatment with or without a sulfonylurea-a MASTERMIND study protocol. BMJ Open 10:12.https://doi.org/10.1136/bmjopen-2020-042784
13. Arbit E, Kidron M (2009) Oral insulin: the rationale for this approach and current developments. J Diabetes Sci Technol 3:562–567.https://doi.org/10.1177/193229680900300322
14. Bays HE, Kozlovski P, Shao Q, Proot P, Keefe D (2020) Licogliflozin, a novel SGLT1 and 2 inhibitor: body weight effects in a randomized trial in adults with overweight or obesity. Obesity 28:870–881.https://doi.org/10.1002/oby.22764
15. Bazyar H, Zare Javid A, Bavi Behbahani H, Moradi F, Moradi Poode B, Amiri P (2021) Consumption of melatonin supplement improves cardiovascular disease risk factors and anthropometric indices in type 2 diabetes mellitus patients: a double-blind, randomized, placebo-controlled trial. Trials 22:231.https://doi.org/10.1186/s13063-021-05174-z
16. Bazydlo-Guzenda K, Buda P, Matloka M, Mach M, Stelmach F, Dzida R, Smuga D, Hucz-Kalitowska J, Teska-Kaminska M, Vialichka V, Dubiel K, Kaminska B, Wieczorek M, Pieczykolan J (2021) CPL207280, a novel G protein-coupled receptor 40/free fatty acid receptor 1-specific agonist, shows a favorable safety profile and exerts antidiabetic effects in type 2 diabetic animals. Mol Pharmacol 100:335–347.https://doi.org/10.1124/molpharm.121.000260
17. Belin de Chantemèle EJ, Muta K, Mintz J, Tremblay ML, Marrero MB, Fulton DJ, Stepp DW (2009) Protein tyrosine phosphatase 1B, a major regulator of leptin-mediated control of cardiovascular function. Circulation 120:753–763.https://doi.org/10.1161/circulationaha.109.853077
18. Biocon (2022) Insulin tregopil. Retrieved from https://adisinsight.springer.com/drugs/800015167. Available from Springer Nature Dis Insight Retrieved 30.08.2022 https://adisinsight.springer.com/drugs/800015167
19. Biogenomics. Recombinant insulin aspart biosimilar Retrieved from https://adisinsight.springer.com/drugs/800039573. Available from Springer Nature Adis Insight Retrieved 30.08.2022 https://adisinsight.springer.com/drugs/800039573
20. Bioscience K (2021) Safety, tolerability, pharmacokinetics and pharmacodynamics of multiple ascending oral doses of KQ-791 in subjects with type 2 diabetes. Retrieved from https://clinicaltrials.gov/ct2/show/NCT02445911?term=KQ-791&cond=Type+2+Diabetes&draw=2&rank=1
21. Bogardus C, The effect of Salsalate treatment on insulin sensitivity and insulin secretion in obese non-diabetic individuals. Retrieved from https://clinicaltrials.gov/ct2/show/NCT00339833
22. Bossart M, Wagner M, Elvert R, Evers A, Hübschle T, Kloeckener T, Lorenz K, Moessinger C, Eriksson O, Velikyan I, Pierrou S, Johansson L, Dietert G, Dietz-Baum Y, Kissner T, Nowotny I, Einig C, Jan C, Rharbaoui F, Gassenhuber J, Prochnow HP, Agueusop I, Porksen N, Smith WB, Nitsche A, Konkar A (2022) Effects on weight loss and glycemic control with

SAR441255, a potent unimolecular peptide GLP-1/GIP/GCG receptor triagonist. Cell Metab 34:59–74.e10.https://doi.org/10.1016/j.cmet.2021.12.005
23. Brashier DB, Khadka A, Anantharamu T, Sharma AK, Gupta AK, Sharma S, Dahiya N (2015) Inhaled insulin: A "puff" than a "shot" before meals. J Pharmacol Pharmacother 6:126–129.https://doi.org/10.4103/0976-500x.162013
24. Brighton CA, Rievaj J, Kuhre RE, Glass LL, Schoonjans K, Holst JJ, Gribble FM, Reimann F (2015) Bile acids trigger GLP-1 release predominantly by accessing Basolaterally located G protein-coupled bile acid receptors. Endocrinology 156:3961–3970.https://doi.org/10.1210/en.2015-1321
25. Brooks-Worrell B, Narla R, Palmer JP(2012) Biomarkers and immune-modulating therapies for type 2 diabetes. Trends Immunol 33:546–553.https://doi.org/10.1016/j.it.2012.07.002
26. Burrack AL, Martinov T, Fife BT (2017) T cell-mediated beta cell destruction: autoimmunity and alloimmunity in the context of type 1 diabetes. Front Endocrinol 8:343.https://doi.org/10.3389/fendo.2017.00343
27. Cajal RY (2020) Clinical trial of pioglitazone treatment in patients with type 2 diabetes mellitus and Covid-19. Retrieved from https://clinicaltrials.gov/ct2/show/NCT04535700
28. Cefalo CMA, Cinti F, Moffa S, Impronta F, Sorice GP, Mezza T, Pontecorvi A, Giaccari A (2019) Sotagliflozin, the first dual SGLT inhibitor: current outlook and perspectives. Cardiovasc Diabetol 18:20.https://doi.org/10.1186/s12933-019-0828-y
29. Cerf ME (2013) Beta cell dysfunction and insulin resistance. Front Endocrinol 4:37. https://doi.org/10.3389/fendo.2013.00037
30. Chapman K, Holmes M, Seckl J (2013) 11β-hydroxysteroid dehydrogenases: intracellular gate-keepers of tissue glucocorticoid action. Physiol Rev 93:1139–1206.https://doi.org/10.1152/physrev.00020.2012
31. Charles ED, Neuschwander-Tetri BA, Pablo Frias J, Kundu S, Luo Y, Tirucherai GS, Christian R (2019) Pegbelfermin (BMS-986036), PEGylated FGF21, in patients with obesity and type 2 diabetes: results from a randomized phase 2 study. Obesity 27:41–49.https://doi.org/10.1002/oby.22344
32. Cherney DZI, Ferrannini E, Umpierrez GE, Peters AL, Rosenstock J, Carroll AK, Lapuerta P, Banks P, Agarwal R (2021) Efficacy and safety of sotagliflozin in patients with type 2 diabetes and severe renal impairment. Diabetes Obes Metab 23:2632–2642.https://doi.org/10.1111/dom.14513
33. Clinic, D a G D (2020) A study to evaluate time in range in type 2 diabetics using Afrezza® plus basal insulin compared to multiple daily injections. Retrieved from https://clinicaltrials.gov/ct2/show/NCT04125082?term=Afrezza+Inhalant+Product&cond=Diabetes+Mellitus%2C+Type+2&draw=2&rank=2. Available from NIH ClinicalTrials Retrieved 30.08.2022 https://clinicaltrials.gov/ct2/show/NCT04125082?term=Afrezza+Inhalant+Product&cond=Diabetes+Mellitus%2C+Type+2&draw=2&rank=2
34. Company, E L a (2023) An administration method study of human regular U-500 Insulin (LY041001) in participants with type 2 diabetes mellitus (VIVID). Retrieved from https://clinicaltrials.gov/ct2/show/NCT02561078?term=LY041001&cond=Diabetes+Mellitus%2C+Type+2&draw=2&rank=2. Available from NIH ClinicalTrials https://clinicaltrials.gov/ct2/show/NCT02561078?term=LY041001&cond=Diabetes+Mellitus%2C+Type+2&draw=2&rank=2
35. Company, E L a (2023) Insulin lispro. Retrieved from https://adisinsight.springer.com/drugs/800002887. Available from Springer Nature Adis Insight Retrieved 30.08.2022 https://adisinsight.springer.com/drugs/800002887
36. Company, E L a (2023) Phase 2 randomized, double-blind, placebo controlled, parallel design study in patients with type 2 diabetes mellitus who are stable on diet and exercise, with or without metformin monotherapy. Retrieved from https://clinicaltrials.gov/ct2/show/NCT00942188

37. Company, E L a (2023) Safety and efficacy Study of Insulin Lispro versus Insulin Aspart in participants with Type 2 diabetes on insulin pump therapy. Retrieved from https://clinicaltrials.gov/ct2/show/NCT01474538?term=insulin+pump&recrs=abdefm&cond=Diabetes+Mellitus%2C+Type+2&phase=01234&draw=2&rank=3. Available from NIH ClinicalTrials Retrieved 30.08.2022 https://clinicaltrials.gov/ct2/show/NCT01474538?term=insulin+pump&recrs=abdefm&cond=Diabetes+Mellitus%2C+Type+2&phase=01234&draw=2&rank=3
38. Company, E L a (2023) Safety, tolerability, pharmacokinetics, and pharmacodynamics of multiple ascending oral doses of LY2922470 in patients With type 2 diabetes mellitus. Retrieved from https://clinicaltrials.gov/ct2/show/NCT01867216?term=LY2922470&cond=Type+2+Diabetes&draw=2&rank=1
39. Company, E L a (2023) A Study of U-500 Insulin (LY041001) in participants with type 2 diabetes. Retrieved from https://clinicaltrials.gov/ct2/show/NCT02588950?term=U-500+insulin&cond=Diabetes+Mellitus%2C+Type+2&draw=2&rank=3. Retrieved 30.08.2022 https://clinicaltrials.gov/ct2/show/NCT02588950?term=U-500+insulin&cond=Diabetes+Mellitus%2C+Type+2&draw=2&rank=3
40. Company, E L a (2019) A study of LY2189102 in patients with type 2 diabetes. Retrieved from https://clinicaltrials.gov/ct2/show/NCT00942188
41. Company HP, A phase 1 study to evaluate the safety, tolerability, Pharmacokinetics and pharmacodynamics of multiple doses of HM15211 in obese subjects with NAFLD. Retrieved from https://clinicaltrials.gov/ct2/show/NCT03744182?term=HM15211&draw=2&rank=2E
42. Corrales P, Vidal-Puig A, Medina-Gómez G (2018) PPARs and metabolic disorders associated with challenged adipose tissue plasticity. Int J Mol Sci 19. https://doi.org/10.3390/ijms19072124
43. Coughlan KA, Valentine RJ, Ruderman NB, Saha AK (2014) AMPK activation: a therapeutic target for type 2 diabetes? Diabetes Metab Syndr Obes 7:241–253. https://doi.org/10.2147/dmso.S43731
44. CureDM (2016) Study of subcutaneous doses of HIP2B in subjects with type 2 diabetes mellitus treated with metformin. Retrieved from https://clinicaltrials.gov/ct2/show/NCT01933256?term=HIP2B&draw=2&rank=2
45. Dasu MR, Devaraj S, Park S, Jialal I (2010) Increased toll-like receptor (TLR) activation and TLR ligands in recently diagnosed type 2 diabetic subjects. Diabetes Care 33:861–868.https://doi.org/10.2337/dc09-1799
46. Davies MJ, D'Alessio DA, Fradkin J, Kernan WN, Mathieu C, Mingrone G, Rossing P, Tsapas A, Wexler DJ, Buse JB (2018) Management of Hyperglycemia in type 2 diabetes, 2018. A consensus report by the American diabetes association (ADA) and the European association for the study of diabetes (EASD). Diabetes Care 41:2669–2701.https://doi.org/10.2337/dci18-0033
47. De Boer RA, Núñez J, Kozlovski P, Wang Y, Proot P, Keefe D (2020) Effects of the dual sodium-glucose linked transporter inhibitor, licogliflozin vs placebo or empagliflozin in patients with type 2 diabetes and heart failure. Br J Clin Pharmacol 86:1346–1356.https://doi.org/10.1111/bcp.14248
48. Devi V, Shankar PK (2008) Ramelteon: a melatonin receptor agonist for the treatment of insomnia. J Postgrad Med 54:45–48.https://doi.org/10.4103/0022-3859.39193
49. Diabeloop (2022) Evaluation of a closed-loop system for the treatment of type 2 diabetes at home (DT2_2) (Close2target). Retrieved from https://clinicaltrials.gov/ct2/show/NCT05369871?term=closed+loop+system&cond=Diabetes+Mellitus%2C+Type+2&draw=2&rank=1. Available from NIH ClinicalTrials Retrieved 30.08.2022 https://clinicaltrials.gov/ct2/show/NCT05369871?term=closed+loop+system&cond=Diabetes+Mellitus%2C+Type+2&draw=2&rank=1
50. Diabetology (2022) Insulin oral. Retrieved from https://adisinsight.springer.com/drugs/800022537. Available from Springer Nature Adis Insight Retrieved 30.08.2022 https://adisinsight.springer.com/drugs/800022537

51. Dijon CHU (2015) Treatment with continuous sub-cutaneous insulin infusion via a portable pump versus discontinuous insulin infusion via multiple-injections in type 2 diabetes (TRICIDIA). Retrieved from https://clinicaltrials.gov/ct2/show/NCT02048189?term=insulin+pump&recrs=abdefm&cond=Diabetes+Mellitus%2C+Type+2&phase=01234&draw=2&rank=16. Available from NIH ClinicalTrials Retrieved 30.08.2022 https://clinicaltrials.gov/ct2/show/NCT02048189?term=insulin+pump&recrs=abdefm&cond=Diabetes+Mellitus%2C+Type+2&phase=01234&draw=2&rank=16
52. Dinarello CA (2011) A clinical perspective of IL-1β as the gatekeeper of inflammation. Eur J Immunol 41:1203–1217. https://doi.org/10.1002/eji.201141550
53. Donath M, Canakinumab in patients with COVID-19 and type 2 diabetes—CanCovDia Trial. Retrieved from https://clinicaltrials.gov/ct2/show/NCT04510493
54. Donath MY, IL-1-receptor antagonist during cephalic phase of insulin secretion in health and type 2 diabetes. Retrieved from https://clinicaltrials.gov/ct2/show/NCT04227769
55. Doupis J, Baris N, Avramidis K (2021) Imeglimin: A new promising and effective weapon in the treatment of type 2 diabetes. touchREV Endocrinol 17:88–91. doi:https://doi.org/10.17925/ee.2021.17.2.88
56. Drug approval package: Invokana (canagliflozin) tablets. Retrieved from https://www.accessdata.fda.gov/drugsatfda_docs/nda/2013/204042Orig1s000TOC.cfm
57. Dubourg J, Fouqueray P, Thang C, Grouin JM, Ueki K (2021) Efficacy and safety of Imeglimin monotherapy versus placebo in japanese patients with type 2 diabetes (TIMES 1): a double-blind, randomized, placebo-controlled, parallel-group, Multicenter phase 3 trial. Diabetes Care 44:952–959.https://doi.org/10.2337/dc20-0763
58. Egan A, Vella A (2019) TTP399: an investigational liver-selective glucokinase (GK) activator as a potential treatment for type 2 diabetes. Expert Opin Investig Drugs 28:741–747.https://doi.org/10.1080/13543784.2019.1654993
59. Eggleton JS, J I T. Retrieved from https://www.ncbi.nlm.nih.gov/books/NBK551656/
60. Elleri D, Dunger DB, Hovorka R (2011) Closed-loop insulin delivery for treatment of type 1 diabetes. BMC Med 9:120.https://doi.org/10.1186/1741-7015-9-120
61. Farmer TG, Jr Edgar TF, Peppas NA (2008) The future of open- and closed-loop insulin delivery systems. J Pharm Pharmacol 60:1–13.https://doi.org/10.1211/jpp.60.1.0001
62. Farrokhian A, Tohidi M, Ahanchi NS, Khalili D, Niroomand M, Mahboubi A, Derakhshi A, Abbasinazari M, Hadaegh F (2019) Effect of bedtime melatonin administration in patients with type 2 diabetes: a triple-blind, placebo-controlled, randomized trial. Iran J Pharm Res 18:258–268.https://doi.org/10.22037/ijpr.2019.112011.13485
63. Finan B, Parlee SD, Yang B (2021) Nuclear hormone and peptide hormone therapeutics for NAFLD and NASH. Mol Metab 46:101–153.https://doi.org/10.1016/j.molmet.2020.101153
64. A first-in-human study to evaluate the safety, tolerability, Pharmacokinetics and pharmacodynamics after single ascending dose of HM15211 in healthy obese subjects. Retrieved from https://www.clinicaltrials.gov/ct2/show/NCT03374241?cond=NCT03374241&draw=2&rank=1
65. Fonte P, Araújo F, Reis S, Sarmento B (2013) Oral insulin delivery: how far are we? J Diabetes Sci Technol 7:520–531.https://doi.org/10.1177/193229681300700228
66. Foryst-Ludwig A, Hartge M, Clemenz M, Sprang C, Hess K, Marx N, Unger T, Kintscher U (2010) PPARgamma activation attenuates T-lymphocyte-dependent inflammation of adipose tissue and development of insulin resistance in obese mice. Cardiovasc Diabetol 9:64.https://doi.org/10.1186/1475-2840-9-64
67. Frías JP, Davies MJ, Rosenstock J, Pérez Manghi FC, Fernández Landó L, Bergman BK, Liu B, Cui X, Brown K (2021) Tirzepatide versus Semaglutide once weekly in patients with type 2 diabetes. N Engl J Med 385:503–515.https://doi.org/10.1056/NEJMoa2107519
68. Galindo R, A randomized controlled trial comparing the safety and efficacy of IDegLira versus basal bolus in patients with poorly controlled type 2 diabetes. Retrieved from https://clinicaltrials.gov/ct2/show/NCT03737240?term=IDegLira&draw=2&rank=4
69. Galindo RJ, Moazzami, Bscioscia MF, Vellanki P, Davis GM, Pasquel FJ, Fayfman M, Migdal A, Isaling, Jumpierrez GE (2022) 217-or: a randomized controlled trial comparing the safety

and efficacy of IDegLira vs. basal bolus in patients with poorly controlled type 2 diabetes and very high HbA1c (>9%–15%) : IDegLira HIGH Trial. Diabetes 71. https://doi.org/10.2337/db22-217-
70. Gallwitz B (2005) Glucagon-like peptide-1-based therapies for the treatment of type 2 diabetes mellitus. Treat Endocrinol 4:361–370. https://doi.org/10.2165/00024677-200504060-00005
71. Ghezzi C, Loo DDF, Wright EM (2018) Physiology of renal glucose handling via SGLT1, SGLT2 and GLUT2. Diabetologia 61:2087–2097.https://doi.org/10.1007/s00125-018-4656-5
72. Goldfine AB, Fonseca V, Jablonski KA, Chen YD, Tipton L, Staten MA, Shoelson SE (2013) Salicylate (salsalate) in patients with type 2 diabetes: a randomized trial. Ann Intern Med 159:1–12.https://doi.org/10.7326/0003-4819-159-1-201307020-00003
73. Goldfine AB, Silver R, Aldhahi W, Cai D, Tatro E, Lee J, Shoelson SE (2008) Use of salsalate to target inflammation in the treatment of insulin resistance and type 2 diabetes. Clin Transl Sci 1:36–43.https://doi.org/10.1111/j.1752-8062.2008.00026.x
74. Guo X, Wang W (2017) Challenges and recent advances in the subcutaneous delivery of insulin. Expert Opin Drug Deliv 14:727–734.https://doi.org/10.1080/17425247.2016.1232247
75. Häring HU, Merker L, Seewaldt-Becker E, Weimer M, Meinicke T, Woerle HJ, Broedl UC (2013) Empagliflozin as add-on to metformin plus sulfonylurea in patients with type 2 diabetes: a 24-week, randomized, double-blind, placebo-controlled trial. Diabetes Care 36:3396–3404.https://doi.org/10.2337/dc12-2673
76. Hartnell S, Fuchs J, Boughton CK, Hovorka R (2021) Closed-loop technology: a practical guide. Practical Diabetes 38:33–39.https://doi.org/10.1002/pdi.2350
77. Hauser AS, Attwood MM, Rask-Andersen M, Schiöth HB, Gloriam DE (2017) Trends in GPCR drug discovery: new agents, targets and indications. Nat Rev Drug Discov 16:829–842.https://doi.org/10.1038/nrd.2017.178
78. He YL, Haynes W, Meyers CD, Amer A, Zhang Y, Mahling P, Mendonza AE, Ma S, Chutkow W, Bachman E (2019) The effects of licogliflozin, a dual SGLT1/2 inhibitor, on body weight in obese patients with or without diabetes. Diabetes Obes Metab 21:1311–1321.https://doi.org/10.1111/dom.13654
79. Henry RR, Buse JB, Wu H, Durrwell L, Mingrino R, Jaekel K, El Azzouzi B, Andjelkovic M, Herz M (2015) Efficacy, safety and tolerability of aleglitazar in patients with type 2 diabetes: pooled findings from three randomized phase III trials. Diabetes Obes Metab 17:560–565.https://doi.org/10.1111/dom.12455
80. Hermans MP, D T, Farmer I, Lohm L, Maheux P, Piatti P, Malvolti E, Jörgens S, Charbonnel B, A 24-Week, randomised, double-blind, active-controlled, multi-centre phase IIIb/IV study to evaluate the efficacy and tolerability of Saxagliptin add-on compared to uptitration of metformin in patients with type 2 diabetes mellitus with inadequate glycaemic control on sub-maximal doses of metformin. Retrieved from https://clinicaltrials.gov/ct2/show/NCT01006590?term=Saxagliptin&cond=Type+2+Diabetes&draw=2&rank=5
81. Herold KC, Hagopian W, Auger JA, Poumian-Ruiz E, Taylor L, Donaldson D, Gitelman SE, Harlan DM, Xu D, Zivin RA, Bluestone JA (2002) Anti-CD3 monoclonal antibody in new-onset type 1 diabetes mellitus. N Engl J Med 346:1692–1698.https://doi.org/10.1056/NEJMoa012864
82. Hodge RJ, Lin JV, Johnson LS, Gould EP, Bowers GD, Nunez DJ (2013) Safety, Pharmacokinetics, and Pharmacodynamic effects of a selective TGR5 agonist, SB-756050, in type 2 diabetes. Clin Pharmacol Drug Dev 2:213–222. https://doi.org/10.1002/cpdd.34
83. Hompesch M, Randomized A, Double-blind, Placebo-controlled study of the effect of 49 Days of treatment with repeated subcutaneous doses of HIP2B to assess safety, tolerability and measures of Islet β-cell function in subjects with type 2 diabetes mellitus treated with metformin. Retrieved from https://clinicaltrials.gov/ct2/show/NCT01933256?term=HIP2B&cond=Type+2+Diabetes&draw=2&rank=1

84. Hovorka DR (2022) Closed-loop in adults with T2D not requiring dialysis (AP-Renal). Retrieved from https://clinicaltrials.gov/ct2/show/NCT04701424?term=closed+loop+system&cond=Diabetes+Mellitus%2C+Type+2&draw=2&rank=2. Available from NIH ClinicalTrials Retrieved 30.08.2022 https://clinicaltrials.gov/ct2/show/NCT04701424?term=closed+loop+system&cond=Diabetes+Mellitus%2C+Type+2&draw=2&rank=2
85. Husen A (2023) Antidiabetic medicinal plants and herbal treatments. CRC Press 6000 Broken Sound Parkway NW, Suite 300, Boca Raton, FL 33487–2742 https://doi.org/10.1201/b23347
86. Ichimura A, Hasegawa S, Kasubuchi M, Kimura I (2014) Free fatty acid receptors as therapeutic targets for the treatment of diabetes. Front Pharmacol 5:236.https://doi.org/10.3389/fphar.2014.00236
87. IDF Diabetes Atlas 10th edition. Retrieved from https://diabetesatlas.org/atlas/tenth-edition/
88. Ingelheim B, Safety, tolerability, Pharmacokinetics and pharmacodynamics of multiple rising oral doses (5 to 100 mg q.d. for 14 Days) of BI 135585 XX in patients with type 2 diabetes mellitus (Randomized, double-blind placebo-controlled within dose groups). Retrieved from https://clinicaltrials.gov/ct2/show/NCT01282970?term=BI+135585&cond=Type+2+Diabetes&draw=2&rank=1
89. Ingelheim B (2014) Safety, tolerability, pharmacokinetics and pharmacodynamics of multiple rising oral doses of BI 135585 XX in patients with type 2 diabetes. Retrieved from https://clinicaltrials.gov/ct2/show/NCT01282970?term=BI+135585&draw=2&rank=2
90. Ji L, Song W, Fang H, Li W, Geng J, Wang Y, Guo L, Cai H, Yang T, Li H, Yang G, Li Q, Liu K, Li S, Liu Y, Shi F, Li X, Gao X, Tian H, Ji Q, Su Q, Zhou Z, Wang W, Zhou Z, Li X, Xu Y, Ning Z, Cao H, Pan D, Yao H, Lu X, Jia W (2021) Efficacy and safety of chiglitazar, a novel peroxisome proliferator-activated receptor pan-agonist, in patients with type 2 diabetes: a randomized, double-blind, placebo-controlled, phase 3 trial (CMAP). Sci Bull 66:1571–1580.https://doi.org/10.1016/j.scib.2021.03.019
91. Kaku K, Ishida K, Shimizu K, Achira M, Umeda Y (2020) Efficacy and safety of trelagliptin in Japanese patients with type 2 diabetes with severe renal impairment or end-stage renal disease: Results from a randomized, phase 3 study. J Diabetes Investig 11:373–381.https://doi.org/10.1111/jdi.13126
92. Katz D, A study to assess 11 Beta-hydroxysteroid dehydrogenase type 1 inhibition in adipose tissue by SPI-62. Retrieved from https://clinicaltrials.gov/ct2/show/NCT05409027?term=SPI-62&cond=Type+2+Diabetes&draw=2&rank=1
93. Kaufman A, Herold KC (2009) Anti-CD3 mAbs for treatment of type 1 diabetes. Diabetes Metab Res Rev 25:302–306.https://doi.org/10.1002/dmrr.933
94. Khan MAB, Hashim MJ, King JK, Govender RD, Mustafa HA, Kaabi J (2020) Epidemiology of type 2 diabetes—global burden of disease and forecasted trends. J Epidemiol Glob Health 10:107–111. https://doi.org/10.2991/jegh.k.191028.001
95. Kirsi Pietiläinen P, Helsinki University Central Hospital. Semaglutide as an adjunct to dieting in the treatment of type 2 diabetes—effects on glucose metabolism, prevention of weight regain and peripheral tissue metabolic activation. Retrieved from https://clinicaltrials.gov/ct2/show/NCT04854083?term=Semaglutide&cond=Type+2+Diabetes&draw=2&rank=1
96. Kobayashi T, Shimomoto T, Tamura A, Namekawa J, Iijima T, Ochiai H (2021) A novel glucokinase activator TMG-123 causes long-lasting hypoglycemia and impairs spermatogenesis irreversibly in rats. J Toxicol Sci 46:115–123.https://doi.org/10.2131/jts.46.115
97. Koska JO, Bunt E, Gasser JC, Impson A, Hanson J, Forbes RL, Jde Courten B, Krakoff J (2009) The effect of salsalate on insulin action and glucose tolerance in obese non-diabetic patients: results of a randomised double-blind placebo-controlled study. Diabetologia 52:385–393. https://doi.org/10.1007/s00125-008-1239-x
98. Lai LL, VS, Nik Mustapha NR, Mahadeva S, Chan WK (2020) The Effect of Empagliflozin on NAFLD in Asian patients with type 2 diabetes. Retrieved from https://clinicaltrials.gov/ct2/show/NCT02964715?term=Empagliflozin&cond=Type2diabetes&draw=2&rank=5
99. Lane W (2014) U-500R insulin in type 2 diabetes with severe insulin resistance via omnipod. Retrieved from https://clinicaltrials.gov/ct2/show/NCT00606034?term=insulin+pump&recrs=abdefm&cond=Diabetes+Mellitus%2C+Type+2&phase=01234&draw=2&rank=4.

Available from NIH ClinicalTrials Retrieved 30.08.2022 https://clinicaltrials.gov/ct2/show/NCT00606034?term=insulin+pump&recrs=abdefm&cond=Diabetes+Mellitus%2C+Type+2&phase=01234&draw=2&rank=4

100. Larsen CM, Faulenbach M, Vaag A, Vølund A, Ehses JA, Seifert B, Mandrup-Poulsen T, Donath MY (2007) Interleukin-1-receptor antagonist in type 2 diabetes mellitus. N Engl J Med 356:1517–1526.https://doi.org/10.1056/NEJMoa065213
101. Laudon M, Frydman-Marom A (2014) Therapeutic effects of melatonin receptor agonists on sleep and comorbid disorders. Int J Mol Sci 15:15924–15950.https://doi.org/10.3390/ijms150915924
102. Li X, Yu J, Wu M, Li Q, Liu J, Zhang H, Zhu X, Li C, Zhang J, Ning Z, Ding Y (2021) Pharmacokinetics and safety of Chiglitazar, a peroxisome proliferator-activated receptor pan-agonist, in patients <65 and ≥65 years with type 2 diabetes. Clin Pharmacol Drug Dev 10:789–796.https://doi.org/10.1002/cpdd.893
103. Li Y (2018) Effectiveness of glucose control, β cell function in response to short-term insulin pump therapy in type 2 diabetes. Retrieved from https://clinicaltrials.gov/ct2/show/NCT03509324?term=insulin+pump&recrs=abdefm&cond=Diabetes+Mellitus%2C+Type+2&phase=01234&draw=2&rank=2. Available from NIH ClinicalTrials Retrieved 30.08.2022 https://clinicaltrials.gov/ct2/show/NCT03509324?term=insulin+pump&recrs=abdefm&cond=Diabetes+Mellitus%2C+Type+2&phase=01234&draw=2&rank=2
104. Li Z, Qiu Q, Geng X, Yang J, Huang W, Qian H (2016) Free fatty acid receptor agonists for the treatment of type 2 diabetes: drugs in preclinical to phase II clinical development. Expert Opinion Invest Drugs 25:871–890.https://doi.org/10.1080/13543784.2016.1189530
105. Li Z, Xu X, Huang W, Qian H (2018) Free fatty acid receptor 1 (FFAR1) as an emerging therapeutic target for type 2 diabetes mellitus: recent progress and prevailing challenges. Med Res Rev 38:381–425. https://doi.org/10.1002/med.21441
106. Lian D (2021) Insulin delivery devices for diabetes mellitus from open-loop to closed-loop. AIP conference proceedings 2350.https://doi.org/10.1063/5.0049209
107. Limited KB, Safety, tolerability, pharmacokinetics and pharmacodynamics of multiple ascending oral doses of KQ-791 in subjects with type 2 diabetes. Retrieved from https://clinicaltrials.gov/ct2/show/NCT02445911?term=KQ-791&cond=Type+2+Diabetes&draw=2&rank=1
108. Limited KB (2019) Safety, tolerability, pharmacokinetics and pharmacodynamics of multiple ascending doses of KQ-791 in diabetes mellitus. Retrieved from https://clinicaltrials.gov/ct2/show/NCT02445911?term=KQ-791&draw=2&rank=2
109. Lincoff AM e al, A safety and efficacy study to evaluate the potential of Aleglitazar to reduce cardiovascular risk in coronary heart disease (chd) patients with a recent acute coronary syndrome (ACS) event and type 2 diabetes mellitus (T2D). Retrieved from https://clinicaltrials.gov/ct2/show/study/NCT01042769
110. Liu JL, Cui W, Li B, Lu Y (2008) Possible roles of reg family proteins in pancreatic islet cell growth. Endocr Metab Immune Disord Drug Targets 8:1–10. https://doi.org/10.2174/187153008783928361
111. LLC M, A Phase 2a randomized, blinded, placebo-controlled study to evaluate the safety, tolerability, pharmacokinetics, and efficacy of medi0382 in overweight/obese subjects with type 2 diabetes mellitus. Retrieved from https://clinicaltrials.gov/ct2/show/NCT03745937?term=MEDI-0382&cond=Type2diabetes&draw=2&rank=1
112. LLC XU, A randomized, double-blind, placebo-controlled, dose-ranging study of the safety and efficacy of XOMA 052 in subjects with type 2 diabetes mellitus on stable metformin monotherapy. Retrieved from https://clinicaltrials.gov/ct2/show/NCT01066715
113. Lyerly H (2022) Glucagon receptor inhibition to enable breast cancer patients to benefit from PI3K inhibitor therapy (REMD-477). Retrieved from https://clinicaltrials.gov/ct2/show/NCT04330625?term=REMD-477&draw=2&rank=2
114. Madhusudhan A, Reddy GB, Ambikar DB, The effect of colon targeted delivery of celecoxib loaded microspheres on experimental colitis induced by acetic acid in rats

115. McInnes N, Remission evaluation of a metabolic intervention for type 2 diabetes with IGlarLixi. Retrieved from https://clinicaltrials.gov/ct2/show/NCT03130426?cond=iGLarLixi&draw=2&rank=2
116. MedImmune (2021) A study to investigate the effect of MEDI0382 on hepatic glycogen metabolism in overweight and obese subjects with type 2 diabetes mellitus. Retrieved from https://clinicaltrials.gov/ct2/show/NCT03555994?id=NCT04067999+OR+NCT03954834+OR+NCT03555994&draw=2&rank=3&load=cart
117. Moreira RO, Cobas RL, Assis Coelho RC (2018) Combination of basal insulin and GLP-1 receptor agonist: is this the end of basal insulin alone in the treatment of type 2 diabetes? Diabetol Metab Syndr 10:26. https://doi.org/10.1186/s13098-018-0327-4
118. Nathan DM, Genuth S, Lachin J, Cleary P, Crofford O, Davis M, Rand L, Siebert C (1993) The effect of intensive treatment of diabetes on the development and progression of long-term complications in insulin-dependent diabetes mellitus. N Engl J Med 329:977–986.https://doi.org/10.1056/nejm199309303291401
119. Neal B, Perkovic V, Mahaffey K, Wde Zeeuw D, Fulcher G, Erondu N, Shaw W, Law G, Desai M, Matthews DR (2017) Canagliflozin and cardiovascular and renal events in type 2 diabetes. N Engl J Med 377:644–657.https://doi.org/10.1056/NEJMoa1611925
120. NN9423. NN9423. Retrieved from https://adisinsight.springer.com/drugs/800049216
121. Nordisk N, Insulin aspart. Retrieved from https://adisinsight.springer.com/drugs/800002958#disabled. Available from Springer Nature Adis Insight Retrieved 31.08.2022 https://adisinsight.springer.com/drugs/800002958#disabled
122. Nordisk N, Insulin degludec/insulin aspart. Retrieved from https://adisinsight.springer.com/drugs/800024577. Available from Springer Nature Adis Insight Retrieved 30.08.2022 https://adisinsight.springer.com/drugs/800024577
123. Nordisk N, Insulin detemir Retrieved from https://adisinsight.springer.com/drugs/800007031. Available from Springer Nature Adis Insight Retrieved 30.08.2022 https://adisinsight.springer.com/drugs/800007031
124. Nordisk N, Insulin icodec. Retrieved from https://adisinsight.springer.com/drugs/800037853. Available from Springer nature Adis Insight Retrieved 30.08.2022 https://adisinsight.springer.com/drugs/800037853
125. Nordisk N (2021) A research study looking at how Semaglutide works in people with type 2 diabetes in spain, as part of local clinical practice (SURE SPAIN). Retrieved from https://clinicaltrials.gov/ct2/show/NCT04067999?id=NCT03498001+OR+NCT04067999&draw=2&rank=1&load=cart
126. Nordisk N (2023) Insulin aspart—Novo Nordisk. Retrieved from https://adisinsight.springer.com/drugs/800002958#disabled
127. Novak V (2010) Enhancement of cerebral Vasoreactivity and cognition by intranasal insulin in type 2 diabetes. Retrieved from https://clinicaltrials.gov/ct2/show/NCT01206322?term=novolin&cond=Diabete+Type+2&draw=2&rank=15. Available from NIH ClinicalTrials Retrieved 30.08.2022 https://clinicaltrials.gov/ct2/show/NCT01206322?term=novolin&cond=Diabete+Type+2&draw=2&rank=15
128. O304. O304. Retrieved from https://adisinsight.springer.com/trials/700276158
129. Odegaard JI, Chawla A (2012) Connecting type 1 and type 2 diabetes through innate immunity. Cold Spring Harb Perspect Med 2:3. https://doi.org/10.1101/cshperspect.a007724
130. Opal SM, Laterre PF, Francois B, La Rosa SP, Angus DC, Mira JP, Wittebole X, Dugernier T, Perrotin D, Tidswell M, Jauregui L, Krell K, Pachl J, Takahashi T, Peckelsen C, Cordasco E, Chang CS, Oeyen S, Aikawa N, Maruyama T, Schein R, Kalil AC, Van Nuffelen M, Lynn M, Rossignol DP, Gogate J, Roberts MB, Wheeler JL, Vincent JL (2013) Effect of eritoran, an antagonist of MD2-TLR4, on mortality in patients with severe sepsis: the ACCESS randomized trial. Jama 309:1154–1162.https://doi.org/10.1001/jama.2013.2194
131. Owyang AM, Maedler K, Gross L, Yin J, Esposito L, Shu L, Jadhav J, Domsgen E, Bergemann J, Lee S, Kantak S (2010) XOMA 052, an Anti-IL-1β monoclonal antibody, improves glucose control and β-cell function in the diet-induced obesity mouse model. Endocrinolog 151:2515–2527.https://doi.org/10.1210/en.2009-1124

132. Panaro BL, Flock GB, Campbell JE, Beaudry JL, Cao X, Drucker DJ (2017) β-cell inactivation of Gpr119 unmasks incretin dependence of GPR119-mediated Glucoregulation. Diabetes 66:1626–1635.https://doi.org/10.2337/db17-0017
133. Pardo PA, Phase I, Randomized, double-blind, placebo-controlled study to determine the safety, tolerability, pharmacokinetics, food effect and pharmacodynamics of single and multiple ascending doses of P11187. Retrieved from https://clinicaltrials.gov/ct2/show/NCT01874366?term=P11187&draw=2&rank=1
134. Pescovitz MD, Greenbaum CJ, Krause-Steinrauf H, Becker DJ, Gitelman SE, Goland R, Gottlieb PA, Marks JB, McGee PF, Moran AM, Raskin P, Rodriguez H, Schatz DA, Wherrett D, Wilson DM, Lachin JM, Skyler JS (2009) Rituximab, B-lymphocyte depletion, and preservation of beta-cell function. N Engl J Med 361:2143–2152.https://doi.org/10.1056/NEJMoa0904452
135. Pettus J, Reeds D, Cavaiola TS, Boeder S, Levin M, Tobin G, Cava E, Thai D, Shi J, Yan H, Bautista EMc, Millan J, Unger R, Henry RR, Klein S (2018) Effect of a glucagon receptor antibody (REMD-477) in type 1 diabetes: a randomized controlled trial. Diabetes Obes Metab 20:1302–1305.https://doi.org/10.1111/dom.13202
136. Pettus JH, DAD, Frias JP, Vajda EG, Pipkin JD, Rosenstock J, Williamson G, Zangmeister MA, Zhi L, Marschke KB, A randomized, double-blind, placebo-controlled, phase 2 study of safety and efficacy of LGD-6972 in patients with type 2 diabetes mellitus. Retrieved from https://clinicaltrials.gov/ct2/show/NCT02851849
137. Pharmaceuticals O (2023) Insulin oral. Retrieved from https://adisinsight.springer.com/drugs/800025074. Available from Springer Nature Adis Insight Retrieved 30.08.2022 https://adisinsight.springer.com/drugs/800025074
138. Pharmaceuticals S, A study to assess 11 Beta-hydroxysteroid dehydrogenase type 1 inhibition in adipose tissue by SPI-62. Retrieved from https://clinicaltrials.gov/ct2/show/NCT05409027?term=SPI-62&draw=2&rank=1
139. Pierce GL, Lesniewski LA, Lawson BR, Beske SD, Seals DR (2009) Nuclear factor-{kappa}B activation contributes to vascular endothelial dysfunction via oxidative stress in overweight/obese middle-aged and older humans. Circulation 119:1284–1292. https://doi.org/10.1161/circulationaha.108.804294
140. Pirags V, Lebovitz H, Fouqueray P (2012) Imeglimin, a novel glimin oral antidiabetic, exhibits a good efficacy and safety profile in type 2 diabetic patients. Diabetes Obes Metab 14:852–858.https://doi.org/10.1111/j.1463-1326.2012.01611
141. Quinn CE, Hamilton PK, Lockhart CJ, McVeigh GE (2008) Thiazolidinediones: effects on insulin resistance and the cardiovascular system. Br J Pharmacol 153:636–645. https://doi.org/10.1038/sj.bjp.0707452
142. Regittnig W, Tschaikner M, Tuca AC, Simic A, Feiel J, Schaller-Ammann R, Licht AH, Jungklaus M, Pieber TR (2022) Insulin induces a progressive increase in the resistance of subcutaneous tissue to fluid flow: Implications for insulin pump therapy. Diabetes Obes Metab 24:455–464. https://doi.org/10.1111/dom.14594
143. Rojas LB, Gomes MB (2013) Metformin: an old but still the best treatment for type 2 diabetes. Diabetol Metab Syndr 5–6.https://doi.org/10.1186/1758-5996-5-6
144. Ruscitti P, Masedu F, Alvaro S, Airò P, Battafarano N, Cantarini L, Cantatore FP, Carlino G, D'Abrosca V, Frassi M, Frediani B, Iacono D, Liakouli V, Maggio R, Mulè R, Pantano I, Prevete I, Sinigaglia L, Valenti M, Viapiana O, Cipriani P, Giacomelli R (2019) Anti-interleukin-1 treatment in patients with rheumatoid arthritis and type 2 diabetes (TRACK): A multicentre, open-label, randomised controlled trial. PLoS Med 16.https://doi.org/10.1371/journal.pmed.1002901
145. Saad MJ, Effect of Diacerein in the metabolic control of patients with diabetes mellitus type 2 and secondary failure to metformin. Retrieved from https://clinicaltrials.gov/ct2/show/NCT02177643

146. Sanofi (2022) Evaluation of Toujeo insulin in type 2 diabetes patients inadequately control with their basal insulin treatment (Transition II). Retrieved from https://clinicaltrials.gov/ct2/show/NCT02967237?term=Insulin+glargine+%28U300%29&cond=Diabetes+Mellitus%2C+Type+2&draw=2&rank=1. Available from NIH ClinicalTrials Retrieved 30.08.2022 https://clinicaltrials.gov/ct2/show/NCT02967237?term=Insulin+glargine+%28U300%29&cond=Diabetes+Mellitus%2C+Type+2&draw=2&rank=1
147. Sanofi (2023) Recombinant human insulin. Retrieved from https://adisinsight.springer.com/drugs/800008803. Available from Springer Nature Adis Insight Retrieved 30.08.2022 https://adisinsight.springer.com/drugs/800008803
148. Schrieks IC, Nozza A, Stähli BE, Buse JB, Henry RR, Malmberg K, Neal B, Nicholls SJ, Rydén L, Mellbin L, Svensson A, Wedel H, Weichert A, Lincoff AM, Tardif JC, Grobbee DE, Schwartz GG (2018) Adiponectin, free fatty acids, and cardiovascular outcomes in patients with type 2 diabetes and acute coronary syndrome. Diabetes Care 41:1792–1800.https://doi.org/10.2337/dc18-0158
149. Seino Y, Fukushima M, Yabe D (2010) GIP and GLP-1, the two incretin hormones: similarities and differences. J Diabetes Investig 1:8–23.https://doi.org/10.1111/j.2040-1124.2010.00022.x
150. Selam JL (2010) Evolution of diabetes insulin delivery devices. J Diabetes Sci Technol 4:505–513. https://doi.org/10.1177/193229681000400302
151. Shah N, Abdalla MA, Deshmukh H, Sathyapalan T (2021) Therapeutics for type-2 diabetes mellitus: a glance at the recent inclusions and novel agents under development for use in clinical practice. Ther Adv Endocrinol Metab 12.https://doi.org/10.1177/20420188211042145
152. Shah RB, Patel M, Maahs DM, Shah VN (2016) Insulin delivery methods: past, present and future. Int J Pharm Investig 6:1–9.https://doi.org/10.4103/2230-973x.176456
153. Shah VN, Moser EG, Blau A, Dhingra M, Garg SK (2013) The future of basal insulin. Diabetes Technol Ther 15:727–732.https://doi.org/10.1089/dia.2013.0228
154. Shimizu H, Masujima Y, Ushiroda C, Mizushima R, Taira S, Ohue-Kitano R, Kimura I (2019) Dietary short-chain fatty acid intake improves the hepatic metabolic condition via FFAR3. Sci Rep 9:16574. https://doi.org/10.1038/s41598-019-53242-x
155. Steneberg P, Lindahl E, Dahl U, Lidh E, Straseviciene J, Backlund F, Kjellkvist E, Berggren E, Lundberg I, Bergqvist I, Ericsson M, Eriksson B, Linde K, Westman J, Edlund T, Edlund H (2018) PAN-AMPK activator O304 improves glucose homeostasis and microvascular perfusion in mice and type 2 diabetes patients. JCI Insight 3.https://doi.org/10.1172/jci.insight.99114
156. Technologies IM (2022) Closed loop glucose control in patients with type 2 diabetes. Retrieved from https://clinicaltrials.gov/ct2/show/NCT05386849?term=closed+loop+system&cond=Diabetes+Mellitus%2C+Type+2&draw=2&rank=6. Available from NIH ClinicalTrials Retrieved 30.08.2022 https://clinicaltrials.gov/ct2/show/NCT05386849?term=closed+loop+system&cond=Diabetes+Mellitus%2C+Type+2&draw=2&rank=6
157. Templer S (2022) Closed-loop insulin delivery systems: past, present, and future directions. Front Endocrinol 13.https://doi.org/10.3389/fendo.2022.919942
158. Therapeutics A (2022) Insulin inhalation. Retrieved from https://adisinsight.springer.com/drugs/800037940. Available from Springer Nature Adis Insight Retrieved 30.08.2022 https://adisinsight.springer.com/drugs/800037940
159. Therapeutics V (2017) Add Glucokinase activator to target A1c (AGATA). Retrieved from https://clinicaltrials.gov/ct2/show/NCT02405260?term=TTP-399&draw=2&rank=2
160. TMG-123. TMG-123. Retrieved from https://adisinsight.springer.com/trials/700288571
161. Tsalamandris S, Antonopoulos AS, Oikonomou E, Papamikroulis GA, Vogiatzi G, Papaioannou S, Deftereos S, Tousoulis D (2019) The role of inflammation in diabetes: current concepts and future perspectives. Eur Cardiol 14:50–59.https://doi.org/10.15420/ecr.2018.33.1
162. Tsuda N, Kawaji A, Sato T, Takagi M, Higashi C, Kato Y, Ogawa K, Naba H, Ohkouchi M, Nakamura M, Hosaka Y, Sakaki J (2017) A novel free fatty acid receptor 1 (GPR40/

FFAR1) agonist, MR1704, enhances glucose-dependent insulin secretion and improves glucose homeostasis in rats. Pharmacol Res Perspect 5.https://doi.org/10.1002/prp2.340
163. Tsumura, Y Tsushima, Y Tamura, A Hasebe, M Kanou, M Kato, H Kobayashi, T (2017) TMG-123, a novel glucokinase activator, exerts durable effects on hyperglycemia without increasing triglyceride in diabetic animal models. PLoS One 12.https://doi.org/10.1371/journal.pone.0172252
164. University G (2020) Effects on insulin resistance with Tadalafil in type 2 diabetes—a double-blind, placebo-controlled crossover study (MAKROTAD). Retrieved from https://clinicaltrials.gov/ct2/show/NCT02601989?term=Tadalafil&cond=Diabetes+Mellitus%2C+Type+2&draw=2&rank=1. Available from NIH ClinicalTrials Retrieved 30.08.2022 https://clinicaltrials.gov/ct2/show/NCT02601989?term=Tadalafil&cond=Diabetes+Mellitus%2C+Type+2&draw=2&rank=1
165. University Hospital, B, Switzerland (2021) Canakinumab in patients with COVID-19 and type 2 diabetes (CanCovDia). Retrieved from https://clinicaltrials.gov/ct2/show/NCT04510493
166. University Hospital, B, Switzerland (2022) IL-1-receptor antagonist during cephalic phase of insulin secretion in health and type 2 diabetes (Cephalira). Retrieved from https://clinicaltrials.gov/ct2/show/NCT04227769
167. University M (2022) a study to evaluate Eli Lilly's insulin dosing algorithm to control Glycemia in insulin-treated adults with type 2 diabetes. Retrieved from https://clinicaltrials.gov/ct2/show/NCT05514080?term=closed+loop+system&cond=Diabetes+Mellitus%2C+Type+2&draw=2&rank=4. Available from NIH ClinicalTrials Retrieved 30.08.2022 https://clinicaltrials.gov/ct2/show/NCT05514080?term=closed+loop+system&cond=Diabetes+Mellitus%2C+Type+2&draw=2&rank=4
168. University of Campinas B (2015) Effect of Diacerein in the metabolic control of patients with dm type 2 and secondary failure to metformin. Retrieved from https://clinicaltrials.gov/ct2/show/NCT02177643
169. Vajda EG, Logan D, Lasseter K, Armas D, Plotkin DJ, Pipkin JD, Li YX, Zhou R, Klein D, Wei X, Dilzer S, Zhi L, Marschke KB (2017) Pharmacokinetics and pharmacodynamics of single and multiple doses of the glucagon receptor antagonist LGD-6972 in healthy subjects and subjects with type 2 diabetes mellitus. Diabetes Obes Metab 19:24–32.https://doi.org/10.1111/dom.12752
170. Valentine V, Goldman J, Shubrook JH (2017) Rationale for, initiation and titration of the basal insulin/GLP-1RA fixed-ratio combination products, IDegLira and IGlarLixi, for the management of type 2 diabetes. diabetes therapy: research, treatment and education of diabetes and related disorders 8:739–752.https://doi.org/10.1007/s13300-017-0287-y
171. Wajid F, Poolacherla R, Mim FK, Bangash A, Rutkofsky IH (2020) Therapeutic potential of melatonin as a chronobiotic and cytoprotective agent in diabetes mellitus. J Diabetes Metab Disord 19:1797–1825.https://doi.org/10.1007/s40200-020-00585-2
172. Wang N, Kong R, Luo H, Xu X, Lu J (2017) Peroxisome proliferator-activated receptors associated with Nonalcoholic fatty liver disease. PPAR Res.https://doi.org/10.1155/2017/6561701
173. Watada HS, Irie M, Terauchi S, Yamada Y, Shiosakai Y, Myobatake K, Taguchi Y (2019) G protein-coupled receptor 119 agonist DS-8500a effects on pancreatic β-cells in Japanese type 2 diabetes mellitus patients. J Diabetes Investig 10:84–93. https://doi.org/10.1111/jdi.12849
174. Wendt A, Eliasson L (2022) Pancreatic alpha cells and glucagon secretion: novel functions and targets in glucose homeostasis. Curr Opin Pharmacol 63.https://doi.org/10.1016/j.coph.2022.102199
175. Willard FS, Douros JD, Gabe MB, Showalter AD, Wainscott DB, Suter TM, Capozzi M, Evan der Velden WJ, Stutsman C, Cardona GR, Urva S, Emmerson PJ, Holst JJ, D'Alessio DA, Coghlan MP, Rosenkilde MM, Campbell JE, Sloop KW (2020) Tirzepatide is an imbalanced and biased dual GIP and GLP-1 receptor agonist. JCI Insight 5:17.https://doi.org/10.1172/jci.insight.140532

176. Winnick JJ, Ramnanan CJ, Saraswathi V, Roop J, Scott M, Jacobson P, Jung P, Basu R, Cherrington AD, Edgerton DS (2013) Effects of 11β-hydroxysteroid dehydrogenase-1 inhibition on hepatic glycogenolysis and gluconeogenesis. Am J Physiol Endocrinol Metab 304:747–756.https://doi.org/10.1152/ajpendo.00639.2012
177. Wu H, Ballantyne CM (2020) Metabolic inflammation and insulin resistance in obesity. Circ Res 126:1549–1564.https://doi.org/10.1161/circresaha.119.315896
178. Xie T, Leung PS (2017) Fibroblast growth factor 21: a regulator of metabolic disease and health span. Am J Physiol Endocrinol Metab 313:292–302.https://doi.org/10.1152/ajpendo.00101.2017
179. XOMA (2011) Safety and efficacy of XOMA 052 in subjects with type 2 diabetes mellitus on stable metformin monotherapy. Retrieved from https://clinicaltrials.gov/ct2/show/NCT01066715
180. Yabuki C, Komatsu H, Tsujihata Y, Maeda R, Ito R, Matsuda-Nagasumi K, Sakuma K, Miyawaki K, Kikuchi N, Takeuchi K, Habata Y, Mori M (2013) A novel antidiabetic drug, fasiglifam/TAK-875, acts as an ago-allosteric modulator of FFAR1. PLoS One 8:10.https://doi.org/10.1371/journal.pone.0076280. Accession No. 24130766
181. Yaribeygi H, Maleki M, Sathyapalan T, Jamialahmadi T, Sahebkar A (2020) Molecular mechanisms by which imeglimin improves glucose homeostasis. J Diabetes Res.https://doi.org/10.1155/2020/8768954
182. Yin MJ, Yamamoto Y, Gaynor RB (1998) The anti-inflammatory agents aspirin and salicylate inhibit the activity of I(kappa)B kinase-beta. Nature 396:77–80.https://doi.org/10.1038/23948
183. Ying W, Fu W, Lee YS, Olefsky JM (2020) The role of macrophages in obesity-associated islet inflammation and β-cell abnormalities. Nat Rev Endocrinol 16:81–90.https://doi.org/10.1038/s41574-019-0286-3

Quantum Dots for Theranostic Applications

Swati Sharma, Pawan Kumar Pandey, Hemant Singh, Indu Yadav, Shiv Dutt Purohit, and Narayan Chandra Mishra

Abstract Over the last decade, quantum dots (QDs), as nanomedicine, are gaining huge importance for their theranostic applications, i.e., in the field of diagnosis and therapeutics. QDs can be explained as colloidal, highly fluorescent, and photostable semiconductor-based nanomaterials exhibiting outstanding physical, chemical, optical, and electronic properties and biocompatibility with living systems, highlighting them as promising candidates for an extensive range of biomedical applications, like live cell imaging, and in vivo as well as in vitro imaging, traceable drug delivery, fluorescence-activated cell sorting (FACS), photodynamic therapy, and therapeutic applications (gene delivery for gene therapy). Moreover, their surface properties can also be modified by employing multifarious methods like ligand exchange, surface salinization, and ligand capping or encapsulation. Despite these dazzling properties, the clinical applications of QDs are well underestimated so far. In the current book chapter, we explain different types of QDs, their characteristic properties, and their major biomedical applications in the field of diagnosis and therapy. Furthermore, a brief focus on the prospective directions of research in QDs is also discussed.

Keywords Quantum dots · Diagnosis · Therapy · In vitro imaging · In vivo imaging

---

S. Sharma · P. K. Pandey · H. Singh · I. Yadav · N. C. Mishra (✉)
Department of Polymer and Process Engineering, Indian Institute of Technology, Roorkee 247001, India
e-mail: narayan.mishra@pe.iitr.ac.in

S. D. Purohit
International Research Facility, School of Chemical Engineering, Yeungnam University, Gyeongsan, North Gyeongsang 38541, South Korea

A. Madhusudhan et al. (eds.), *Functional Smart Nanomaterials and Their Theranostics Approaches*, Smart Nanomaterials Technology,
https://doi.org/10.1007/978-981-99-6597-7_14

1 Introduction

Quantum dots (QDs) are colloidal, fluorescent, artificial semiconductor material-based zero-dimensional (0D) nanocrystals (all dimensions <100 nm) that exhibit three-dimensional confinements [1]. The size range of the QDs is 2–10 nm. To emphasize their uniqueness and the fact that they have bounded, discrete electronic states, just like naturally occurring atoms or molecules, QDs are sometimes referred to as "artificial atoms" [2, 3]. The similarity between the electronic wave functions in QDs and those in actual atoms has been demonstrated [4]. One can create an artificial molecule that exhibits hybridization even at room temperature by coupling two or more of these QDs [5].

QDs came into existence with the execution of the idea of the transition of 2D thin films into the 0D structure in the early 1980s. QDs were discovered by Alexei Ekimov when semiconducting nanocrystals (CuCl, CdSe) were synthesized on glass matrices by him to study the "quantum-sized" CdS crystallites in an aqueous solution under fluorescence and color transitions in QDs. In 1982, one of the researchers developed the initial theory attempting to explain the characteristics of these incredibly small crystals by the confinement of their electrons [6]. Bell Labs attempted and effectively manufactured QDs as a colloidal suspension (in a liq. State), to obtain a colloidal suspension under the influence of other co-workers. In 1998, Mark Arthur Reed introduced the phrase "quantum dots" to describe a photoluminescent nanostructure with fully quantized energy levels either electron holes or electrons [7]. Later on, the focus shifted toward the application of QDs in the life sciences [1]. Followed by that a huge amount of work was done employing QDs for various biomedical applications due to their superior optical and electronic characteristics, which include various fluorescent hues emitting under a single source of excitation, size-tunable light emission, robust photobleaching resistance, and superior fluorescence intensity. These characteristics make them superior to other common fluorophores like organic dye and fluorescent proteins [8]. The property of blinking is also executed by them. Three key elements are concerned with QDs: synthesizing the QDs (top-down and bottom-up approach), and modification of their surface to impart various properties (biocompatibility). QDs were mostly used in LEDs, lasers [9], solar cells, single-electron transistors, quantum computing, second harmonic generation [10], and single photon sources. Additionally, owing to their minute size, QDs find application in spin coating and inkjet printing. They were also applied to Langmuir–Blodgett thin films [11–13]. These above-mentioned processing techniques will have a special place in developing less expensive and time-saving QDs fabrication methods. In 1998, the first use of CdSe QDs for cellular labeling was reported [14, 15]. One of the works displayed that CdSe/ZnS QDs were capable of both in vitro and in vivo imaging by encapsulating individual nanocrystals in phospholipid block copolymer micelles [16]. Numerous publications expanding QDs for ideal biological imaging followed these works [17, 18]. For instance, fluorescent QDs were used to precisely analyze biological events and target cellular structures, like identifying cancerous cells, [19, 20], peroxisomes [21], and DNA [16] after being coupled with bioactive moieties

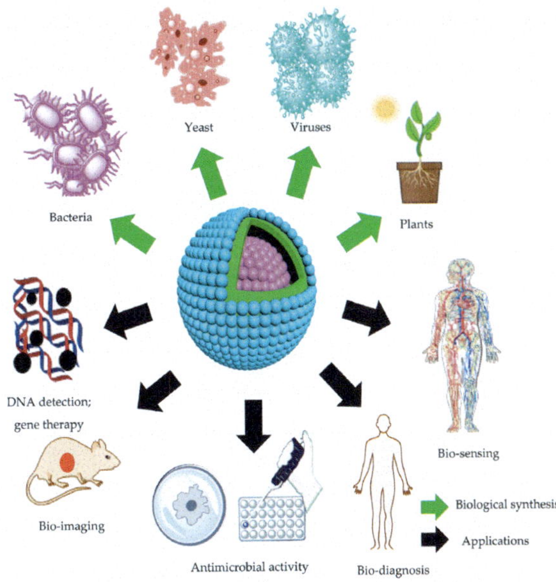

Fig. 1 Various sources of QDs and major application of QDs in the biomedical field. *Source* Ref. [23]

(such as antibodies, receptor ligands). It is worth mentioning here that QDs have got a significant role in the biomedical field in quantifiable, multiplexed, and long-term efficient live cell imaging and in vivo besides in vitro imaging, traceable drug delivery, fluorescence-activated cell sorting (FACS), photodynamic therapy (PDT), and therapeutic applications (gene delivery for gene therapy) [22]. In this chapter, we will discuss different types of QDs, their properties, and their varied applications in theranostics (Fig. 1).

2 Different Types of QDs

QDs can also be categorized into different types based on their composition and structure, which include core type, core–shell, and alloyed quantum dots.

2.1 Core-Type Quantum Dots

These QDs are solitary constituent materials with homogeneous internal compositions, for instance, chalcogenides of metals like lead, zinc, or cadmium, for example, CdTe or PbS. The qualities of core-type nanocrystals like photo- and electroluminescence can be finely tuned by merely altering the crystallite size. For detailed information about the properties and application of core-type QDs, refer to Table 1 [24–26].

2.2 *Core–Shell Quantum Dots*

Growing of the shells of another semiconducting material with a larger band gap around semiconductor nanocrystals. Core–shell QDs are the resulting particles with small areas of one material implanted in another with a greater band gap. For instance, consider QDs with a core made of CdSe and ZnS in the shell. The coating of QDs could easily increase quantum yield (QY). This approach has received a lot of attention as a strategy to modify the photophysical characteristics of QDs. For more details on properties and applications, refer to Table 1 [27, 28].

2.3 *Alloyed Quantum Dots*

By adjusting the composition and internal structure without altering the crystallite size, alloyed QDs with both homogeneous and gradient internal structures enable fine-tuning of both optical and electrical properties. To cite an example, alloyed QDs

Table 1 Different types of QDs, their respective properties, and theranostic applications

	QDs type	Properties	Theranostic applications	Refs.
1	Core	Substantial metal-free QDs, like group IV QDs, were created in response to the increasing demand for more biocompatible QDs as signal reporters E.g., CdSe, CdS, or CdTe	Bioimaging	Holmes et al. [24], Sun et al. [25], Warner et al. [26]
2	Core–shell	Frequently used to modify the photophysical characteristics of straightforward QDs. They have a properly thought-out shell that increases the photostability and photoluminescence efficiency of simple QDs E.g., Type IV-VI, III–V and II-VI, semiconductors with configurations such as (CdSe) ZnS, (CdSe) CdS, (InAs) CdSe, and (CdS) ZnS	Bioimaging and Bio-diagnostics	Ross et al. [27], Vasudevan et al. [28]
3	Alloyed	Quantum confinement is continuously tuned without altering particle size E.g., Cadmium selenium sulfide (CdSeS)	Creating NIR fluorescent probes and band gap engineering	Li et al. [29]

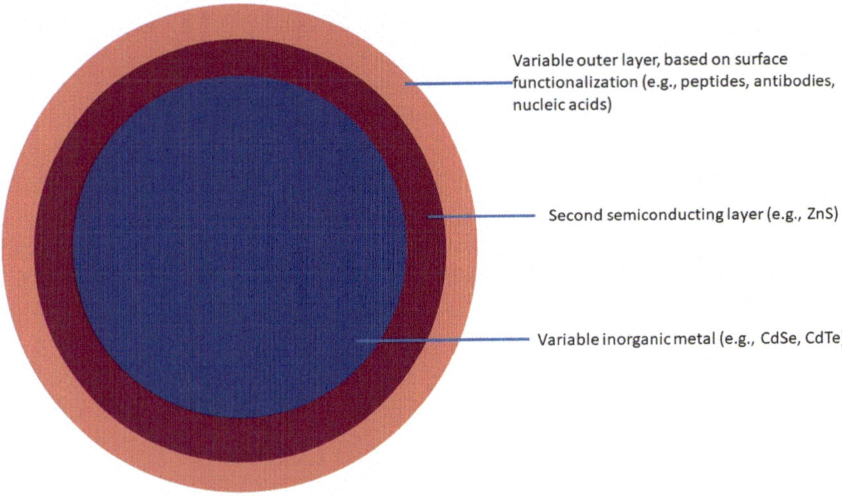

Fig. 2 Schematic illustration of QDs. *Source* Ref. [30]

with a 6 nm diameter and the compositions CdSxSe1-x/ZnS emit light with various wavelengths simply by changing the composition [29].

Moreover, in line with their chemical components, QDs can also be categorized based on the positioning of their constituents in the periodic table for instance groups II–VI (CdTe, CdSe, and CdS) groups III–V (GaAs, InAs, InP, GaN), and later evolved group IV which is composed of C, Si, and Ge share familiar physical and chemical features, for instance, metalloid nature and semiconducting electrical properties. Table 1 summarizes the properties and applications of these QDs [29] (Fig. 2).

3 Specific Properties of QDs

A variety of physical, chemical, as well as optical properties, are exhibited by QDs.

3.1 Physical and Chemical Properties [31, 32]

- Atoms are packed closely together to form solid materials like semiconductors, which have an ordered, periodic arrangement. The promiscuity and interaction of the atoms in this close-packed configuration produce the particular physical properties of the substance.
- As the QD crystal shrinks (i.e., has fewer atoms), the quantization effect is amplified, and the band gap continues to widen due to the lack of orbitals near the band boundaries. The size-dependent photoluminescence of QDs is explained by the

inverse relationship between QD particle size and band gap (Eg) (i.e., the smaller the QD, the wider the Eg).
- They are available in aqueous and organic forms.
- Thermal stability is high (dependent on the shell)
- High crystallization
- Good dispersibility
- QD emits the wavelength in the UV and IR electromagnetic regions.
- By simply altering their size, QDs can likewise be created from the same material and produce different emission colors.

3.2 Optical Properties

QDs have highly pleasing optical properties. These comprise

- Extinction (or absorption) coefficients 10–100 times larger [33, 34].
- Higher fluorescence QY.
- To the triplet state, there is little intersystem crossing.
- Greater photostability and improved resistance to chemical and photo-degradation [35, 36].
- QDs capacity to provide multicolor target detection is another outstanding characteristic. It has been successfully used in immunochemistry [20], molecular bar-coding [37], and fluorescent in situ hybridization (FISH) of several repeated sub-nuclear genetic sequences [38].
- Finally, the biologist's imaging window is expanded to the NIR and IR regions, where autofluorescence is greatly diminished in vivo, thanks to the unheard-of capacity to control the fluorescence emission of QDs as a function of size and/or composition. Because of this, NIR-QDs have been utilized to image tumors in live animal trials with success [39–42].

3.3 Biological Properties [14, 15]

QDs have matchless abilities owing to their four distinctive properties.

- Firstly, the QDs themselves are small-sized.
- Second, they encompass enormous absorption extinction coefficients and high fluorescent QYs, making them exceptionally bright. Along with that have size-tunable, narrow Gaussian emission spectra that can be excited at a single wavelength (multiplexed experiments).
- Third, they are inorganic and hence photo-chemically robust (allowing for advanced dynamic imaging).

- As a final point, QDs do exhibit the property of blinking. These properties enable a new generation of fluorescence imaging experiments in biology, allowing researchers to decipher biological functions at the level of molecules.
- QDs have proven to be influential probes for fluorescence imaging along with several additional applications.
- In a dose-dependent way, bacteria can be killed by QDs, which have antibacterial capabilities identical to those of nanoparticles [43]. Furthermore, it has been demonstrated that QDs work well against both gram-positive and gram-negative bacteria [44].

3.4 Surface-Modified QDs

Regardless of the unearthing and advancement of the QDs, the notion of employing luminescent semiconductor nanocrystals for the sake of biological usage was not that friendly and adaptable, owing to QD's highly toxic constituents, such as cadmium. In addition to their native inability to be readily dispersed within aqueous solutions which are biocompatible. In 1998, two seminal articles described the encapsulation of QDs in water-soluble coatings to enable labeling fixed life of both formalin and cells, relying on techniques intended to improvement in QDs size mono-dispersity at the same time as preserving highly luminous behavior [45–47]. Moreover, for using QDs in a biological system, it is crucial to modify the QDs surface with hydrophilic ligands that present water solubility on the QDs at the same time as maintaining photo physical stability and possessing reactive sites for the conjugation of biomolecules. The typical surface modification methods include ligand exchange, surface salinization, and ligand capping/encapsulation [48, 49].

3.4.1 Modification of QDs Surface for Targeting It to a Target

Antibodies (Abs), peptides, and small molecules are the three main methods for instructing QDs to label a target. It is worth mentioning here that every approach has its benefits and downsides, and no approach is widespread for all applications.

Antibody Targeting

The easiest and fastest technique to label a target if an Ab is known to exist for its extracellular epitope is by using an Ab-QDs conjugate. Anti-HER2 QDs conjugates can be used for breast cancer cells in-vivo and in-vitro imaging. Moreover Ab-conjugated QDs are employed for in-vivo imaging of prostatic specific membrane antigen in tumors [20, 41]. Pitfalls to this method include the selectivity, availability, and affinity of the Abs, along with the amplified hydrodynamic radius of the QDs conjugate. Regardless of this the Ab-QDs conjugates are usually the most preferred choice and contribute a much larger proportion of QDs in biology literature.

Peptide Targeting

It makes sense to use a peptide-QD nanoconjugate for labeling when the amide epitope of a supermolecule or a well-known amide measure expedites ligand-target interactions. Using peptides in the design of QDs probes significantly lowers the cost of synthesis and, as a result, the size of the final nanoconjugate. Additionally, the capacity of the resultant complexes to bind to the target through multivalent interactions might be considerably increased by the valence conjugation of several peptides to at least one QD. Peptide-functionalized QDs are with accomplishment used for targeting cellular proteins like protein receptors, G protein-coupled receptors, etc. [50, 51].

Small Molecule Targeting

The current method employs small molecules for surface targeting of QDs [39]. By using ligands (high affinity and selectivity), QDs can be directed directly to the binding site of a desired protein. A dynamic procedure that enables the monitoring of numerous activities, such as receptor-mediated endocytosis, can be started by directly marking the protein's binding site. Ligand-conjugated quantum dots (LCQDs) can also be used to pinpoint the factors that regulate protein production and cell surface mobility [52].

4 Quantum Dots for Diagnostic, Therapeutic, and, Theranostics Applications

QDs are widely used as fluorescent probes for responsive assessment of the morphology of healthy cells and tissues, identification of molecular manufacturers of illnesses, and keeping a check on numerous biological activities because of their unique advantages. QDs will for that reason have a potential multifunctional platform for biomedicine when combined with therapeutic agents (such as genes, tiny chemotherapeutic agents, and photosensitizers) [39, 53].

4.1 Diagnostics

QDs, the nanoparticles with remarkable optical and electrical properties, are particularly interesting for use in biological applications for the diagnosis of the diseased condition and investigating the mechanisms underlying tissue anomalies and seeing malignancies in living subjects.

4.1.1 Diagnosis Mechanism

A source of energy either electromagnetic (light), chemical, biological, or electrical is necessary for any QDs-based experiment. Each type of assay makes use of a distinct pathway for the energy, which is ultimately measured as electromagnetic radiation or electrical output. Because of their distinct and highly desired luminescent properties, it should be obvious that the most typical and simple way to incorporate QDs into bioassays is by excitation with a light source and/or detection of their emitted light [54].

4.1.2 Bioimaging

Biological imaging based on fluorescence is among the important techniques in biological study, and it's broadly used in vivo and in vitro research [55]. Fluorescent probe-based biological sensing, imaging, and tracking of various diseases also benefit from the integration of various fields like chemistry, biology, and biomedicine. Multifarious distinctive [56, 57] advantages are possessed by QDs like strong absorption, broad excitation, narrow emission spectra, tunable emission color, remarkable photostability, and high fluorescence QY [15, 55]. The usage of groups II–VI QDs for cell imaging was initially separately demonstrated in 1998. In the years that followed, these traditional II-VI QDs were thoroughly investigated in a variety of biotechnological applications, such as DNA array technology, immunofluorescence assays, [17], and in vitro and *in vivo* research [58]. The first and most effective uses of QDs included labeling preserved cells and tissues, staining nuclear chromosomes, antigens, and in situ hybridization on chromosomes [59] or combing DNA, as well as labeling membrane proteins, and microtubules [60], actin [15, 60], and nuclear chromosomes [60] (Table 2).

Table 2 Various types of applications of QDs for bioimaging

S. no	Bioimaging	Type of QDs	Application	Refs.
1	In vivo imaging	Transferrin protein conjugated QDs	One of the earliest experiments for using QDs for live cell imaging	Chan et al. [18]
2	In vivo imaging	Phospholipid-coated QDs	To monitor the stages of embryogenesis using fluorescence visualization	Dubertret et al. [16]
3	In vivo imaging	QDs conjugated with neurotransmitter	To specifically target serotonin transporters on transfected cells	Orndorff and Rosenthal [50]
4	In vivo imaging	PEG-coated QDs	To visualize tiny blood vessels	Wang et al. [61]
5	In vivo imaging	Core–shell nanostructures (type II QDs)	Identifying and defining sentinel lymph nodes during surgery (in rat models)	Bruchez et al. [62]
6	In vivo imaging	Antibody conjugated QDs	Development of multifunctional QDs for tumor cell imaging and targeting	Gao et al. [63]
7	In vitro imaging	IgG and streptavidin-linked QDs	To mark and target several types of targets at various sub-cellular sites, including the cell nucleus, the surface, and intracellularly	Tan et al. [60]
8	In vitro imaging	Dihydrolipoic acid (DHLA)-capped QDs	Attachment of Abs to capped QDs for both generalized and targeted labeling of live cells	Hines and Guyot-Sionnest [46]
9	In vitro imaging	Tat peptide-conjugated QDs	Study the cellular absorption and intracellular transport of QDs in living cells	Ruan et al. [64]
10	In vitro imaging	Antibody conjugated QDs	QDs for the multiplex detection of five tumor biomarkers in cultured human breast cancer cells and on single paraffin-embedded clinical tissue sections	Peng et al. [65]
11	In vitro imaging	QD-RGD	Development and evaluation of in vitro dual imaging of two putative cancer biomarkers, integrin v3, and nucleolin	Cai et al. [66]
12	In vitro imaging	Streptavidin conjugated QDs	Nicotinic receptors (nAChRs) were successfully QD tagged at neuromuscular junctions (NMJs) in adult ex vivo mouse diaphragm	Orndorff et al. [51]

In Vivo Imaging

The poor transmission of visible light by biological tissues makes tissue thickness a significant challenge when imaging a multicellular creature in comparison with in vitro investigations that rely on living cells. NIR windows are thought to be essential for optical imaging in deep tissue [67]. Few organic dyes have been reported that give bright NIR emissions without significant photobleaching [18]. Moreover, owing to their diverse surface chemistry (permitting conjugation with numerous molecules), QDs can be utilized to noninvasively examine a variety of systems in vivo. By the addition of peptides or labels for a particular target, QDs conjugation, or surface modification, enables them to be tailored to target desired structures [55]. QDs also provide a vast surface area/volume ratio for connecting many molecules, for instance on a QD with a size of around 4 nm, 2–5 protein molecules or more than 50 tiny molecules can be easily inserted [63]. In addition, a variety of techniques, such as passive adsorption, covalent bond formation, multivalent chelation, and electrostatic forces might be used to bioconjugate the QDs [55, 63].

Cell migration and differentiation in real time can be visualized by cellular imaging employing QDs [55]. The QDs entering the cell, free Qs or dots that are not conjugated to a protein or peptide concentrate in vesicles in the perinuclear area. There are three different mechanisms-microinjection, peptide induced transport and non-specific pinocytosis allows cells to take up QDs [55, 68]. Also, QDs have exhibited commendable success in the immune fluorescence labeling of live cells [15]. One of the earliest experiments for using QDs for live cell imaging was by using protein conjugated dots that were taken up by Hela cells through a receptor-mediated mechanism [15]. Moreover, to monitor the stages of embryogenesis, QDs were microinjected into Xenopus embryos [15, 68]. Additionally, QDs were used in investigations on the migration and motility of malignant cells [68].

Tissue Engineering

QDs have proven successful in a variety of tissue imaging applications, to cite examples, lymph node mapping and vascular imaging for in vivo blood pool imaging. The QD includes ligands that target tumors, PEG to extend circulation time and improve biocompatibility, and an amphiphilic tri-block copolymer to halt particle aggregation and fluorescence loss in vivo. PEG-coated QDs were utilized to see tiny blood vessels. PEG-coated QDs stayed in the bloodstream for a longer period as compared to chemical dyes [55].

Tumors could be targeted by QDs through both active and passive mechanisms. The preferential accumulation of QDs in tumor areas makes the passive process probable because of the increased permeability and retention impact [55].

In Vitro Imaging

An in vitro experiment is characterized by artificial settings that are simulations of what might occur in vivo. Research that involves experimental manipulation is referred to as in vitro, which translates as "in the glass" [61].

In vitro fluorescent imaging can be chiefly categorized into three classes: biomolecular tracking in cells, tissue staining, and cellular imaging [61].

Cellular Imaging

According to reports, IgG and streptavidin-linked QDs might be utilized to mark and target several types of targets at various sub-cellular sites, including the cell nucleus, the surface, and intracellularly (cultured living cells, fixed cells, and tissue sections). All marker signals were bright and noticeably photo steady, and they were all distinct to the designated targets QDs are appropriate for simultaneously tracking several proteins and live cells over extended times [20]. The number of research investigations using QDs for use in cell imaging increased quickly and drastically after the publication of these two key reports [64].

In one study, Tat peptide-conjugated QDs were used as a model system to examine QD absorptions and intracellular movement in living cells. The findings of this study provide insights into the operation of Tat peptide-mediated delivery and are crucial for the development of nanoparticle probes for intracellular targeting and imaging.

Biomolecular Tracking in Cells

Biomolecule detection in cells has more immediate clinical relevance than cellular imaging. Tumor markers' detection and quantification are useful for early cancer diagnosis, classification, and treatment since they show tumor tissue development and differentiation. It has been demonstrated that QDs can be used for the multiplex detection of five tumor indicators on single paraffin-embedded clinical tissue slices and in cultured human breast cancer cells.

In vitro molecular profiling of tumor biomarkers may be a good use for QD-based technology and could be the first translational application of QDs [69–72]. On several cancer cell lines as well as on healthy, normal cells, in vitro dual imaging of two potential cancer biomarkers, integrin v3, and nucleolin was created and examined. Confocal imaging demonstrated the sub-cellular localization of the aptamer-labeled QD-bound nucleolin protein as well as the surface distribution of the integrin protein [72–74].

Tissue Staining

Another fascinating field of study involving in vitro biological uses of QDs is tissue staining with QD conjugates, which is primarily restricted to fixed section labeling. In the year 2008, in a mature ex vivo mouse diaphragm, QD nicotinic receptor (nAChR) tagging at neuromuscular junctions (NMJs) was demonstrated. QDs have a biological use in live cell trafficking assays for studying brain targets, which call for strong probe selectivity and photostability under variable conditions. This work established the method's suitability for finding nAChRs in the brain after intracerebroventricular (ICV) injection of bio-BTX and offered preliminary support for live tissue investigation [61].

4.1.3 Cancer Detection

Cancer is one of the diseases, which is spreading very fast. As per WHO reports, it is second leading cause of death around the globe, causing 9.6 million deaths, or one in every six death, in 2018. The reason behind such a large number goes to failure in detection of cancer in the patients in earlier stage. There are a variety of techniques which can be used for early detection of cancer and can mitigate the consequences of cancer. Some of the techniques are described in the following.

Cancer Imaging

An appealing method for locating cells that express a particular biomarker in vivo is fluorescence molecular imaging based on QDs. However, because of the lack of specificity in many physiological settings and enduring foreign particle clearance systems, QD-based targeted molecular imaging in vivo continues to be a significant issue [75]. It is significant to note that fluorescence imaging in the visible (450–750 nm) and conventional NIR-I ranges (1.01.4 m) is more effective due to decreased light scattering, deeper tissue penetration, and reduced autofluorescence (750–900 nm). In vivo imaging of the lung endothelium and tumor vasculature was achieved by conjugating green and red MAA-coated QDs with peptides that had an affinity for these tissues. To create tailored cancer imaging, one of the researchers took the initial step. Even though full organs are not yet able to be seen by QDs in vivo, the idea of employing targeting peptides to specifically target antigens with fluorescent QD markers has a lot of potentials [66]. For instance, NIR CdTe/ZnS QDs [76] were functionalized with cyclic RGD to target integrin v3, a biomarker that is upregulated during proliferation, metastasis, and angiogenesis in malignant tissue (arginine-glycine-aspartic acid) (Fig. 3).

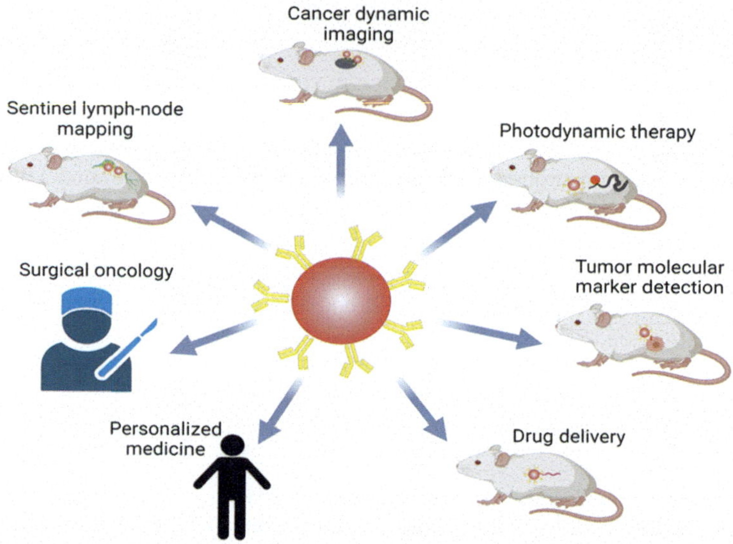

Fig. 3 QDs-based probes applications such as bioimaging and bio-detection. When it comes to detecting various biomarkers, photodynamic therapy, real-time systemic imaging, low side effects, and medication delivery, QDs hold great promise in the fight against cancer. Individualized precision medicine and surgical oncology may also benefit from QDs. *Source* Ref. (Liang et al. 2021)

QDs-Based Detection of CTCs

Thomas Ashworth, an Austrian physicist, first identified CTCs in patient cadavers in 1869 [77]. Since they are shed from primary tumors into the bloodstream to relocate to distant organs, CTCs play a crucial role in metastases, and their number is highly correlated with clinical prognosis [78]. QDs have been created as a brand-new class of fluorescent probes with numerous distinctive features for therapeutic use. Utilize specially designed tri-functional nanospheres and bi-functional nanospheres, both of which contain magnetic nanoparticles and QDs, to trap and separate certain cancer cells. Additionally, using this technique, it was possible to quickly identify a small percentage of cancer cells (approximately 0.01%) in a vast population of cultivated normal cells without the need for sample preprocessing before cell analysis [79–81] (Fig. 3).

QDs-Based Diagnosis of Tumor Microenvironment for Invasion and Metastasis

Tumors display a further level of complexity in addition to the cancer cells themselves by establishing the "tumors microenvironment" [82]. The tumor microenvironment, on the other hand, is a particularly extensive and complicated system because it changes over time along with the cancer cells and exhibits a wide range of molecular, cellular, and architectural variability. The co-evolution of cancer cells and the cancer microenvironment cannot be understood by research conducted merely at the molecular, cellular, or clinical levels. One of the groups created a theory of the co-evolution of cancer cells and the surrounding tumor microenvironment using QDs-based molecular imaging technologies [83]. Type IV collagen, tissue-damaging metalloproteinases, tumor angiogenesis, and other essential components of the tumor microenvironment all play crucial roles in the invasion of cancer (Fig. 3).

QDs-Based Detection of Cancer-Associated Proteins

Since it was discovered in 2017 [84], the QDs conjugated Ag complex could detect the captured Ag, the QDs-based immunoassay has been extensively employed to identify tumor biomarkers in clinical samples such as blood and bodily fluids (Fig. 3).

QDs-Based Sentinel Lymph Node (SLN) Mapping

Lymphatic invasion is a crucial pathway for the spread of cancer [84]. And because SLN is the first lymph node in a chain of nodes to receive the afferent lymphatics of the tumor bed, which significantly aids in cancer migration, SLN played a major part in the development of the operation (Fig. 3).

4.1.4 Critical Wound

Various contributing factors like biochemical reactions, environmental factors, and most importantly pH play a significant role in determining the actual status of the wound. Healthy skin typically has a pH between 5 and 5.5, but wound conditions after infection will become more basic with a pH value of 7–9 due to the presence of several enzymes and bacteria [85, 86]. So far, diverse systems have been constructed to get a better insight into pH monitoring of wounds which was correlated to wound status, for instance, screen-printed pH potentiometric sensor /voltammetric sensor [87], Mclister et al. (n.d.), micromachine-based transducer [88], and dye immobilization technique [89–91].

The creation of a practical, inexpensive, simple, and visually appealing pH monitoring strategy is urgently required. Due to their low toxicity, good water solubility, ease of synthesis, environmental friendliness, and sustained photoluminescence (PL), low toxicity, CQDs have drawn intense scientific interest in recent years. The bulk of CQDs-based pH detection systems was frequently created by combining pH-induced single photoluminescence intensity shift with ratio metric fluorescence measurement. It was believed that the protonation or deprotonation of amino or carboxyl groups on the surfaces of CQDs served as the intrinsic sensing mechanism [92–95].

Moreover, for the visualization of sense, it is crucial to create colorimetric fluorescent probes based on CQDs. However, CQDs-based pH sensors with both fluorescence response and colorimetric behavior have been extremely rare in prior studies, and there are still several practical challenges associated with pH monitoring in a wound environment. To cite an example, graphene QDs were created by electrolyzing graphite rods, then refluxing the resulting PL emissions in a concentrated solution of nitric and sulfuric acid. Under UV illumination, the PL emissions had a particular chromatic response to the pH range of 1–14 [96]. Interestingly, orange-emitting and colorimetric CQDs were prepared by hydrothermally processing 1,2,4-triaminobenzene and NaOH; by merely using these special CDs on paper, the built-in sensor platform demonstrated dual-mode response to pH shifting conditions [96]. Despite these benefits, the prepared pH test paper's potential for wound pH monitoring has not received much attention. Therefore, it is currently difficult to create CQDs-based pH sensors that are responsive to changes in wound pH both visually through changes in fluorescence and calorimetrically. Later, in the year 2019, researchers developed a simple microwave-assisted hydrothermal process to create a novel 1,2,4-triaminobenzene and urea-based orange-emissive CQDs (O-CQDs) with excitation-independent PL performance [97].

4.1.5 Pathogen Detection

In vitro and in vivo imaging of cells, sub-cellular structures, and critical bio-molecular functioning now heavily rely on bio-conjugated [98, 99]. Even though non-toxic QDs like silicon and carbon have been developed, there is still a lack of knowledge about how QDs interact with microbial cell membranes and their toxicity to microorganisms, which prevents the widespread use of QDs in tracking infection pathways in animal models and treating infections. In applications such as the microarray detection of proteins and infectious illnesses, recently newly conjugated QDs are essential. Despite this penetration of QDs technology into biology, a microbiology-QDs interface has not yet been developed, which may eventually allow for the detection of pathogens, the tracking of the development and remission of infectious diseases, and the phototherapy of microbial infections in people or other animals [100]. There are different types of bacteria, some of which are good for human or animal health (probiotics), controlling biogeochemical cycles, and sources of bioactive substances and industrially useful enzymes, while others can behave as pathogens. The first study on the use of CdSe QDs conjugated with wheat germ agglutinin and transferring protein [101] for the strain- and metabolism-specific microbial tagging of bacterial and fungal pathogens was published in 2003. It was also reported that using an Ab probe to identify two aquatic infections [102] using immunofluorescence (Cryptosporidium and Giardia).

Another illustration is the development of a unique technique using magnetic nanoparticles and Ab-conjugated QDs for the simultaneous enrichment and detection of three food-borne bacterial species, including Escherichia coli, Shigella flexneri, and Salmonella typhimurium [103].

4.2 *Therapy*

QDs are frequently utilized as fluorescent probes for sensitive assessment of the phenotypes of healthy cells and tissues, detection of disease-causing molecules, and monitoring of different biological activities because of their special benefits. QDs will thus potentially become a multifunctional platform for biomedicine when combined with therapeutic agents.

4.2.1 Gene Delivery for Gene Therapy

For the treatment of different illnesses brought on by genetic problems, such as Parkinson's disease and cancer, gene therapy is a promising approach [104]. Through protein expression or gene silencing, genes, which include plasmid DNA (pDNA), small interfering RNA (siRNA), and micro-RNA (mRNA), can be employed as potential therapeutic agents [105]. The slow uptake of genes into cells and their quick demise under biological conditions, which are often prevented by utilizing a gene vector to transport and protect them, limit the therapeutic uses of genes [106, 107].

QDs were used as fluorescence beacons and gene vectors in 2007, based on studies using QDs in a medicine delivery system. By attaching siRNA to the surface of the nanoparticles using a chemical cross-linker, a gene delivery platform by functionalizing PEGylated QDs with tumor-homing peptides (F3) was developed. Moreover, the effectiveness of gene delivery will undoubtedly be impacted by QD modification [60].

4.2.2 Drug Delivery Systems

Given that traceable drug delivery can both explain the pharmacokinetics and pharmacodynamics of drug candidates and provide design suggestions for drug carrier engineering, it appears to be one of the most significant new QD applications. Because of concerns regarding long-term in vivo toxicity and degradation, QDs are currently exclusively employed in cells and small animals. But traceable therapeutic delivery in cells and animals still has a big impact on life science research, including the development, approval, and distribution of drugs. This is a result of the thorough testing of potential medications using cells and small animals. To follow drug molecules or drug carriers in living things non-intrusively and in real time, specialized imaging techniques are needed [108]. Therefore, QDs can operate as a useful prototype for the development and improvement of nanocarriers from which to produce biocompatible carriers of comparable size and surface properties for usage in clinical settings. The two main applications of QDs in drug delivery nowadays are their usage as carriers and the labeling of therapies or drug carriers with QDs [108].

QDs as Carriers with Integrated Functionalities

The structural framework can be a QD core, which can be sandwiched between an inorganic core and an amphiphilic polymer covering together with a small amount of hydrophobic medication and an imaging contrast agent. The hydrophilic side of the amphiphilic polymer can then be used to immobilize specific biomolecules and hydrophilic treatments through covalent or non-covalent bonding. In addition to identifying, attaching to, and treating ill cells, this fully integrated nanostructure may also provide signals that can be monitored to follow the movement of the cells in real time. This might work as a miraculous drug.

Several reports have surfaced regarding this aspirational objective. In stroke-prone, spontaneously hypertensive rats, Captopril, an antihypertensive medication, was conjugated to the QD surface, and the drug's pharmacodynamics and pharmacokinetics were examined [104]. As per one of the findings, by affixing QDs to RNA aptamers (A10) that particularly bind to prostate-specific membrane antigen (PSMA) within the A10 RNA aptamer, targeting functionality was added to QDs. The fact that the resultant nano-complexes were non-fluorescent as a result of a bi-FRET mechanism, in which the doxorubicin molecules quench the QD fluorescence and then transfer the energy to the A10 aptamer, was intriguing [105]. Confocal microscopy can be used to observe how the gradual release of doxorubicin causes both QD and doxorubicin fluorescence to recover when the nano-complexes are taken up by PSMA-positive cells. There was a 30% difference in cell viability between PSMA-positive LNCaP cells and PSMA-negative PC3 cells after treatment with the nano-complex, which may be improved by managing the doxorubicin release kinetics. This concept offers a detecting mechanism for medication release in addition to enabling the tailored distribution of chemotherapy drugs.

QDs as Tags for Other Drug Carriers

A traditional drug carrier can be marked with QDs, which act as photostable fluorescent reporters, allowing the next type of QD application in traceable drug delivery. Currently, polymers like poly (lactic-co-glycolic acid) and polyethyleneimine (PEI) make up the majority of drug carriers, whereas inorganic compounds make up a much smaller portion of them. The absence of an intrinsic signal for long-term and immediate imaging of drug transport is a general drawback of these delivery systems. Due to their distinctive spectral characteristics, QDs become an obvious choice in this context. They have been used to identify bacteria, viruses, and possibly even organic and inorganic drug carriers [60, 106, 109]. A siRNA delivery method using Lipofectamine 2000 and co-transfection of siRNA and QDs has been described. Following the conventional cell transfection procedure, small interfering RNA was first condensed onto the cationic membrane before the lipoplex was further treated with fluorescent QDs [106]. It was discovered that the level of silencing and intracellular QD fluorescence intensity correlated well. This is understandable because higher lipoplex uptake by cells is indicated by stronger fluorescence at a fixed ratio of lipofectamine and QD concentrations, and there is a link between optimal fluorescence and gene silencing for the least amount of QDs occurs at a 1:1 ratio (by mass) between QDs and lipofectamine. While excess QDs lessen the effectiveness of silence, lower QD concentrations do not produce any observable fluorescence. The researchers hypothesized that since excess QDs and siRNA molecules, both have negative charges, they compete with one another for binding to lipofectamine. It is also plausible that the extra QDs simply neutralize the positive charge on lipofectamine, reducing its capacity to bind to the cell surface. This would be consistent with the hypothesis that oligonucleotides are likely trapped in lipid bilayers of multilamellar vesicles [110].

Certain other modifications and functionalization can also be done on QDs to enhance their theranostics potential. For Instance: PEGylation can increase the retention duration in the bloodstream after intravenous delivery, which is crucial to achieving the significant tumor accumulation needed for cancer therapy [111, 112]. The overexpressed folate receptors on the surface of cancer cells are the target of folic acid, a commonly employed ligand. To potentially treat blood cancers, Cd/Se QDs were functionalized with vapreotide as an agonist for somatostatin receptors in blood cells [113]. In a different study, developed PEGylated MoS_2 QDs were loaded with doxorubicin for tracked delivery of the drug to cancer cells. The produced QDs demonstrated good stability and biosafety under physiological settings, as well as blue photoluminescence and pH-responsive drug release [114].

Through receptor-mediated endocytosis, functionalized QDs are frequently taken up by cells. There are three typical ways that cells respond to ligand-modified QDs under receptor-mediated endocytosis. The QDs' endocytic destiny depends on the type of ligand that is utilized, the target receptor, and the QDs' chemical makeup [32, 111].

Clathrin-Mediated Endocytosis

The nanoparticles aggregate at the target receptor location and are taken up by Clathrin-coated vesicles, which are created when the Clathrin-coated cell membrane engulfs the nanoparticles, in Clathrin-mediated endocytosis. QDs will be released into the cytosol and available to act and discharge their cargo if they can break out from the endosome. They will degrade through lysosomal degradation otherwise [115].

Caveolae-Mediated Endocytosis

By connecting the functionalized particles to their receptors, caveolae, flask-shaped membrane invaginations rich in cholesterol that detach from the cell membrane to form caveosomes, enable the internalization of the functionalized particles into the target cells. To maximize the effectiveness of medication delivery to the target cells, caveosomes are hypothesized to be less damaging vesicles than endosomes [115].

Macropinocytosis

A particle is engulfed and leaked into the cytosol as macropinosomes during the process of macropinocytosis, which is triggered by certain ligands that stimulate the development of cell membrane ruffles [115].

4.2.3 Photo-Dynamic Therapy

As it targets just sick tissue while sparing healthy cells, photodynamic therapy (PDT) is a promising cancer treatment method. It causes cell death by interacting with light and a photosensitizing substance [73]. After being exposed to light, the photosensitizing chemical creates reactive singlet oxygen species (ROS), which causes in vitro toxicity in PDT applications, and shows potential for QDs [116]. Burda's group has demonstrated that the use of an excitation wavelength beyond the absorption of the photosensitizer alone is made possible by the combination of CdSe QDs with PDT photosensitizers. For the first time, they enabled photosensitizer excitation for photodynamic cancer therapy using QD-based FRET [117]. Soon after, a different team assessed the possibilities for using CdSe QDs as a PDT target [118]. Recent studies also suggest that carbon QDs via a dual mechanism of ROS generation and activation of the type I interferon response offer promise for use in the PDT of COVID-19 [119].

4.2.4 Quantum Dots in Wound Healing

The four sequential and overlapping dynamic phases of 1. homeostasis, 2. inflammation, 3. proliferation, and 4. remodeling are typically used to describe the wound healing process in a normal human body [120–122].

When it comes to healing the wound oxygenation, stress and other disorders that are related to the injury can all have an impact on how quickly the body heals [121]. A subsequent bacterial infection is among the most frequent variables that impede wound healing. Exposure to antibacterial metals like zinc, silver, titanium, etc., is one of the most effective ways to combat the growth of bacteria [123]. Nevertheless, the extensive usage of certain metals may be harmful to both the environment and human health. Therefore, the production of nanoparticles specially QDs has emerged as the ideal method for solving issues related to the usage of heavy metals and improving the metal's effectiveness for antibacterial activity. Since QDs have the same qualities as their bulk counterparts and are more stable as a result of their huge surface area, research on QDs in wound healing has increased [124]. On the one hand QDs are crucial in biomaterials production owing to their anti-bacterial properties while on the other hand their peculiar physicochemical properties are highly advantageous in tissue engineering applications. Additionally, QDs can improve the mechanical strength of hydrogels and tissue scaffolds used in regenerative medicine or for the treatment of wounds [125]. QDs are excellent for incorporation into a bio scaffold due to their nano-scale size, low toxicity toward cells, and enzymatic capabilities like oxidase [126]. The ligands involved in the process of wound healing could attach to QDs due to their enormous surface area. For instance, CQDs are involved in the angiogenesis process because they can promote the expression of anti-angiogenic proteins, which is essential to prevent the overexpression of pro-angiogenic factors. As it promotes the expression of interleukin-6, QDs also participate in signaling pathways that improve the inflammatory phase of wound healing [127, 128].

Wound Closure

The biological impact of QDs in a dynamic and complicated biosystem, such as wound closure, was investigated using in vivo research. Gross wound morphology, wound size, and histological analysis utilizing hematoxylin and eosin (H & E) staining to assess the state of the regenerated tissue were all used to assess wound healing [129]. The histological analysis demonstrates the development of blood vessels and tissue regeneration after QD treatment. When collagen is treated with QDs, it demonstrates a uniform disposition of thick collagen fibers [130].

One of the works from 2019 demonstrates that treated models of QDs heal wounds more quickly than untreated models [131].

Antibacterial Effects

The complexity of wound healing, particularly in cases of skin injuries, is increased by the increasing prevalence of microbial infection. As a result, there is a greater need for alternative antibacterial agents such QDs, silver, and others to treat microbial infections due to a faster rate of morbidity in traditional therapies.

In a nutshell, the three basic processes by which QDs exert their common antibacterial action are as follows: (1) rupture of cell walls or membranes, (2) formation of ROS, and (3) interaction with nucleus components like DNA or RNA to prevent cell reproduction. Due to the high levels of ROS produced as a result of the bacteria's exposure to QDs, the microbial membrane became disrupted [132].

The mixture of QDs and polymer is thought to have strong antibacterial effects on microorganisms like S. aureus and even inhibits the growth of other drug-resistant bacteria like Methicillin-resistant Staphylococcus aureus (MRSA) [31]. According to the literature review, gram-positive and gram-negative bacteria are both strongly affected by QDs' antibacterial activities [31]. However, gram-positive bacteria are more susceptible to the antimicrobial effects of QDs than gram-negative bacteria. CQDs and titanium oxide have been shown to have a narrower bacterial inhibition zone with E. coli than with S. aureus [132]. When QDs are stimulated or subjected to light radiation, their antibacterial capabilities tend to be higher. Zinc oxide QDs have been examined in both their excited and unexcited states. As a result, antibacterial rates and ROS production were markedly higher in an excited state than in a normal condition [133].

Angiogenesis

One of the key elements in wound healing is angiogenesis, which works to supply nutrients and gases (such as oxygen) to the site of the injury to enable tissue regeneration and cell proliferation [134]. Vascular endothelial growth factor (VEGF), a 35–45 kDa homodimer glycoprotein that promotes endothelial cell proliferation and the development of new capillary tubes, plays a vital role in controlling the angiogenesis process [135, 136].

4.2.5 Inhibitory Action of QDs for SARS CoV-2 Treatment

It is necessary to conduct a thorough investigation and systematic research of the virus's functional mechanisms to create medications and vaccines that are effective against COVID-19. Typically, the cycle of viral infection induces changes in the host cell's structure, which result in cell destruction. So, the simplest method to stop the viral infection and stop it from spreading is to improve how medications or other inhibitors interact with infected cells or viral replication.

According to one study, the swine epidemic diarrhea virus could be stopped from spreading by altering its surface proteins and blocking it from entering cells thanks to cationic CDs manufactured from curcumin. They could also prevent the production of negative-strand RNA, prevent viral budding, and prevent the virus from accumulating reactive oxygen species [137]. By inducing the production of interferon-stimulating genes and pro-inflammatory cytokines, cationic CDs have the potential to inhibit viral replication.

Furthermore, CQDs may interact with viral structural proteins due to the functional groups made up of carboxylic and hydroxyl groups on their surface. Phenylboronic acid–CQDs were evidenced to be operative against HIV infections. Additionally, positively charged CQDs with amine groups significantly reduced the infectiousness of viruses. For instance, polyamine-modified CQDs showed strong antimicrobial effects against white spot syndrome by avoiding viral infection and the virus envelope. The outcomes showed that polyamine-based CQDs could increase the number of immunological genes in shrimp [138].

4.3 QDs for Theranostics

QDs have been used for a variety of therapeutic and diagnostic purposes, as previously mentioned (Fig. 4). Certain QDs have recently been discovered to have many functions for simultaneous diagnosis and treatment. One of the best representative examples is a synthesis of CuInSe QDs with manganese (Mn) doping and a ZnS passivation layer. The MnCuInSe/ZnS core–shell QDs integrate the activities of reactive

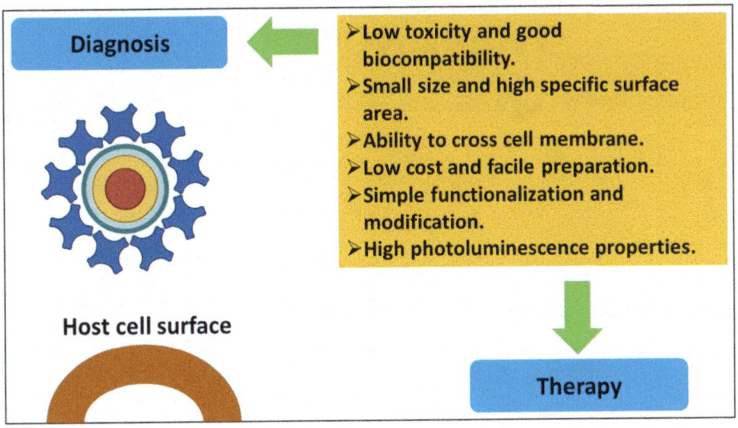

Fig. 4 Important properties of QDs contributing to their usage in diagnosis and therapy

oxygen radical production, excitation-dependent red emission, and magnetic resonance imaging (MRI), and the control of Mn2+ incorporation results in synergistic imaging and therapeutic modalities [139].

There is a variety of applications of QDs for diagnosis as well as therapy (Fig. 5). Moreover, one of the lab groups synthesized magneto photoluminescent quantum dots (Mn-Si-QD) capable of DOX loading for the treatment of breast cancer in women. To achieve the best fluorescence qualities, the effects of several parameters, such as reaction time and reaction temperature, were examined. Further, the accumulation of magneto photoluminescent QDs in the tumor site was confirmed by a biodistribution study. The current study specified that the Mn-Si-QD with DOX loading is a non-invasive and potentially theranostic agent [140]. The engineering of multipurpose CQDs nanoplatforms that can integrate multifunctional imaging and therapeutic functionalities has always been a challenging task. The solvothermal method was employed in one of the works to create gadolinium-doped CQD coupled with AS1411 aptamers (AS1411-Gd-CQD) for targeted fluorescence (FL)/MR imaging-guided photothermal therapy (PTT) [141].

4.4 *Obstacles Preventing QDs in Clinical Settings*

Despite having a plethora of promising applications of QDs in the theranostics and biomedical field, there are still some common obstacles that prevent their utilization in medical practices. A wide range of disorders, including cancer, can be detected using QDs, which have proven to be effective imaging agents. Nevertheless, one of the foremost issues with using these nanoparticles is their cytotoxicity. They are frequently manufactured with dangerous metals like cadmium, which can lead to

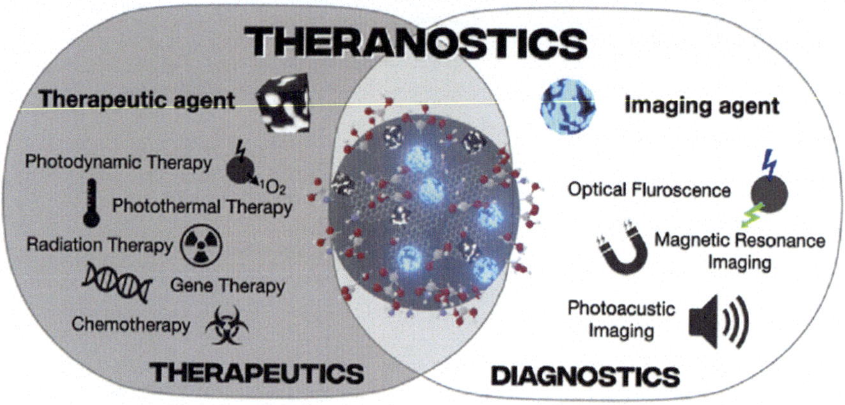

Fig. 5 Therapeutics, diagnosis, and theranostics of QDs are shown schematically. *Source* Ref. (Sukunya et al. 2020)

poisoning. To address all these obstacles, research is required to prepare non-toxic biocompatible QDs.

- In addition, QDs for biological investigations must be resistant to the conditions both within and outside of cells. Because QDs are naturally hydrophobic, numerous techniques have been employed to make them hydrophilic. Unfortunately, these changes lessen stability and quantum yield since QDs and their fluorescence are sensitive to environmental variables like pH, salinity, and oxidation.
- Because of their colloidal nature in aqueous environments, QD solutions are more prone to irreversible aggregation, which can happen as a result of long-term storage, pH fluctuations, heating, and freezing. Continuous improvement must be made at the synthesis stage to make QDs adaptable to environmental changes because it is challenging to arbitrarily regulate the environment around each QD in living systems.
- It is unknown what biological mechanisms QDs are based on. Therefore, more thorough studies should be carried out, such as those on the impact of QDs on cellular mechanisms and longer periods of observation to discover QD side effects.
- QDs semiconducting capabilities can be applied to biomedical applications, where the application of electrical charges through QDs enables better and quicker cell movement within the human body.
- QDs commercialization requires the creation of adequate health and safety laws and the resolution of intellectual property challenges. Most crucially, while having a variety of intriguing properties, QDs are still a long way from clinics and have only a few clinical trials underway.

5 Conclusion and Future Outlook

Due to their simple and inexpensive synthesis processes, simple surface modification, good photoluminescence, and easily modifiable water solubility, QDs have received a lot of attention in the fields of biology and theranostics. The use of QDs in photocatalytic reactions, in vivo and in vitro bioimaging, drug delivery systems, chemical, and biological sensors, as well as photodynamic and photothermal therapies, is widespread. This book chapter has summarized the different types of QDs, their characteristic properties, and their major biomedical applications in the field of diagnosis and therapy. Despite substantial advancement in those fields over the past ten years, the following issues with QDs synthesis and application still need to be resolved: Firstly, the cytotoxicity of these nanoparticles is one of the main problems with their use. To create non-toxic, biocompatible QDs, research is necessary. Secondly, because QD solutions are colloidal in aqueous settings, they are more likely to aggregate irreversibly. Thirdly, there should be more in-depth research, such as studies on how QDs affect cellular systems and longer periods of observation to find QD side effects. Lastly, although QDs have many exciting qualities, there are currently just a few clinical trials being conducted, and QDs are still far from being used in clinics.

References

1. Yu WW, Chang E, Drezek R, Colvin VL (2006) Water-soluble quantum dots for biomedical applications. Biochem Biophys Res Commun 348(3):781–786. https://doi.org/10.1016/J.BBRC.2006.07.160
2. Ashoori RC (1996) Electrons in artificial atoms. Nature 379(6564): 413–419. https://doi.org/10.1038/379413a0
3. Kastner MA (1993) Artificial atoms. Phys Today 46(1):24–31. https://doi.org/10.1063/1.881393
4. Banin U, Cao YW, Katz D, Millo O (1999) Identification of atomic-like electronic states in indium arsenide nanocrystal quantum dots. Nature 400(6744):542–544. https://doi.org/10.1038/22979
5. Cui J, Panfil YE, Koley S, Shamalia D, Waiskopf N, Remennik S, Popov I, Oded M, Banin U (2019) Colloidal quantum dot molecules manifesting quantum coupling at room temperature. Nat Commun 10(1): 1–10. https://doi.org/10.1038/s41467-019-13349-1
6. Sathe KP, Garud NS, Bangar VB, Gadakh NR (2022) A review on quantum dots (QDS). J Adv Sci Res 13(06): 23–27. https://doi.org/10.55218/JASR.202213603
7. Reed MA, Randall JN, Aggarwal RJ, Matyi RJ, Moore TM, Wetsel AE (1988) Observation of discrete electronic states in a zero-dimensional semiconductor nanostructure. Phys Rev Lett 60(6):535–537. https://doi.org/10.1103/PHYSREVLETT.60.535
8. Ma Y, Shen H, Zhang M, Zhang Z (2016) Quantum dots (QDs) for tumor targeting theranostics. Nanomater Tumor Targeting Theranostics: A Proactive Clin Perspect 85–142.https://doi.org/10.1142/9789814635424_0004
9. Huffaker DL, Park G, Zou Z, Shchekin OB, Deppe DG (1998) 1.3 μm room-temperature GaAs-based quantum-dot laser. Appl Phys Lett 73(18): 2564–2566. https://doi.org/10.1063/1.122534

10. Loss D, DiVincenzo DP (1998) Quantum computation with quantum dots. Phys Rev A 57(1):120. https://doi.org/10.1103/PhysRevA.57.120
11. Achermann M, Petruska MA, Crooker SA, Klimov VI (2003) Picosecond energy transfer in quantum dot Langmuir - Blodgett Nanoassemblies. J Phys Chem B 107(50):13782–13787. https://doi.org/10.1021/JP036497R/ASSET/IMAGES/LARGE/JP036497RF00005.JPEG
12. Gorbachev IA, Goryacheva IY, Glukhovskoy EG (2016) Investigation of multilayers structures based on the Langmuir-Blodgett films of CdSe/ZnS quantum dots. BioNanoScience 6(2):153–156. https://doi.org/10.1007/S12668-016-0194-0
13. Xu S, Dadlani AL, Acharya S, Schindler P, Prinz FB (2016) Oscillatory barrier-assisted Langmuir-Blodgett deposition of large-scale quantum dot monolayers. Appl Surf Sci 367:500–506. https://doi.org/10.1016/J.APSUSC.2016.01.243
14. Chan WCW, Nie S (1998) Quantum dot bioconjugates for ultrasensitive nonisotopic detection. Science 281(5385):2016–2018. https://doi.org/10.1126/SCIENCE.281.5385.2016/ASSET/44996DBF-64D8-4090-883B-43F783464AE4/ASSETS/GRAPHIC/SE3986869005.JPEG
15. Fu A, Gu W, Larabell C, Alivisatos AP (2005) Semiconductor nanocrystals for biological imaging. Curr Opin Neurobiol 15(5):568–575. https://doi.org/10.1016/J.CONB.2005.08.004
16. Dubertret B, Skourides P, Norris DJ, Noireaux V, Brivanlou AH, Libchaber A. (2002) In vivo imaging of quantum dots encapsulated in phospholipid micelles. Science (New York, N.Y.) 298(5599): 1759–1762. https://doi.org/10.1126/SCIENCE.1077194
17. Alivisatos P (2003) The use of nanocrystals in biological detection. Nat Biotechnol 2004 22(1): 47–52. https://doi.org/10.1038/nbt927
18. Chan WCW, Maxwell DJ, Gao X, Bailey RE, Han M, Nie S (2002) Luminescent quantum dots for multiplexed biological detection and imaging. Curr Opin Biotechnol 13(1):40–46. https://doi.org/10.1016/S0958-1669(02)00282-3
19. Gao X, Cui Y, Levenson RM, Chung LWK, Nie S (2004) In vivo cancer targeting and imaging with semiconductor quantum dots. Nat Biotechnol 22(8):969–976. https://doi.org/10.1038/NBT994
20. Wu X, Liu H, Liu J, Haley KN, Treadway JA, Larson JP, Ge N, Peale F, Bruchez MP (2003) Immunofluorescent labeling of cancer marker Her2 and other cellular targets with semiconductor quantum dots. Nat Biotechnol 21(1):41–46. https://doi.org/10.1038/NBT764
21. Colton HM, Falls JG, Ni H, Kwanyuen P, Creech D, McNeil E, Casey WM, Hamilton G, Cariello NF (2004) Visualization and quantitation of peroxisomes using fluorescent nanocrystals: treatment of rats and monkeys with fibrates and detection in the liver. Toxicol Sci: An Official J Soc Toxicol 80(1):183–192. https://doi.org/10.1093/TOXSCI/KFH144
22. Abdellatif AA, Tawfeek HM, Younis MA, Alsharidah M, al Rugaie O (2022) Biomedical applications of quantum dots: overview, challenges, and clinical potential. Int J Nanomed 17: 1951–1970.https://doi.org/10.2147/IJN.S357980
23. Abdel-Salam M, Omran B, Whitehead K, Baek KH (2020) Superior properties and biomedical applications of microorganism-derived fluorescent quantum dots. Molecules 25(19):4486. https://doi.org/10.3390/molecules25194486
24. Holmes JD, Ziegler KJ, Doty RC, Pell LE, Johnston KP, Korgel BA (2001) Highly luminescent silicon nanocrystals with discrete optical transitions. J Am Chem Soc 123(16):3743–3748. https://doi.org/10.1021/JA002956F/ASSET/IMAGES/LARGE/JA002956FF00013.JPEG
25. Sun YP, Zhou B, Lin Y, Wang W, Fernando KAS, Pathak P, Meziani MJ, Harruff BA, Wang X, Wang H, Luo PG, Yang H, Kose ME, Chen B, Veca LM, Xie SY (2006) Quantum-sized carbon dots for bright and colorful photoluminescence. J Am Chem Soc 128(24):7756–7757. https://doi.org/10.1021/JA062677D/SUPPL_FILE/JA062677DSI20060417_112943.PDF
26. Warner JH, Hoshino A, Yamamoto K, Tilley RD, Warner JH, Tilley RD, Hoshino A, Yamamoto K (2005) Water-soluble photoluminescent silicon quantum dots. Angew Chem 117(29):4626–4630. https://doi.org/10.1002/ANGE.200501256
27. Ross FM, Tersoff J, Tromp RM (1998) Coarsening of self-assembled ge quantum dots on Si(001). Phys Rev Lett 80:984–987
28. Vasudevan D, Gaddam RR, Trinchi A, Cole I (2015) Core-shell quantum dots: properties and applications. Undefined 636:395–404. https://doi.org/10.1016/J.JALLCOM.2015.02.102

29. Li H, Wang C, Peng Z, Fu X (2015) A review on the synthesis methods of CdSeS-based nanostructures. https://doi.org/10.1155/2015/519385
30. Shao L, Gao Y, Yan F (2011) Semiconductor quantum dots for biomedicial applications. Sensors 11(12): 11736–11751. https://doi.org/10.3390/S111211736
31. Wang X, Feng Y, Dong P, Huang J (2019) A mini review on carbon quantum dots: preparation, properties, and electrocatalytic application. Front Chem 7:671. https://doi.org/10.3389/FCHEM.2019.00671/XML/NLM
32. Zhao Y, Liu X, Yang Y, Kang L, Yang Z, Liu W, Chen L (2015) Carbon dots: from intense absorption in visible range to excitation-independent and excitation-dependent photoluminescence 23(11): 922–929. https://doi.org/10.1080/1536383X.2015.1018413
33. Dabbousi BO, Rodriguez-Viejo J, Mikulec FV, Heine JR, Mattoussi H, Ober R, Jensen KF, Bawendi MG (1997) (CdSe)ZnS core-shell quantum dots: synthesis and characterization of a size series of highly luminescent nanocrystallites. J Phys Chem B 101(46): 9463–9475.https://doi.org/10.1021/JP971091Y/ASSET/IMAGES/LARGE/JP971091YF00016.JPEG
34. Leatherdale CA, Woo WK, Mikulec FV, Bawendi MG (2002) On the absorption cross section of CdSe nanocrystal quantum dots. J Phys Chem B 106(31): 7619–7622. https://doi.org/10.1021/JP025698C/ASSET/IMAGES/MEDIUM/JP025698CE00009.GIF
35. Alivisatos A, ~P. (1996) Semiconductor clusters, nanocrystals, and quantum dots. Science 271(5251):933–937. https://doi.org/10.1126/science.271.5251.933
36. Michalet X, Pinaud FF, Bentolila LA, Tsay JM, Doose S, Li JJ, Sundaresan G, Wu AM, Gambhir SS, Weiss S (2005b) Quantum dots for live cells, in vivo imaging, and diagnostics. Science (New York, N.Y.) 307(5709): 538–544. https://doi.org/10.1126/SCIENCE.1104274
37. Han M, Gao X, Su JZ, Nie S (2001) Quantum-dot-tagged microbeads for multiplexed optical coding of biomolecules. Nat Biotechnol 19(7): 631–635. https://doi.org/10.1038/90228
38. Bentolila LA, Weiss S (2006) Single-step multicolor fluorescence in situ hybridization using semiconductor quantum dot-DNA conjugates. Cell Biochem Biophys 45(1):59–70. https://doi.org/10.1385/CBB:45:1:59
39. Bentolila LA, Ebenstein Y, Weiss S (2009) Quantum dots for in vivo small-animal imaging. J Nucl Med: Official Publ, Soc Nuclear Med 50(4):493. https://doi.org/10.2967/JNUMED.108.053561
40. Choi HS, Liu W, Liu F, Nasr K, Misra P, Bawendi MG, Frangioni J, v. (2010) Design considerations for tumour-targeted nanoparticles. Nat Nanotechnol 5(1):42–47. https://doi.org/10.1038/NNANO.2009.314
41. Gao X, Cui Y, Levenson RM, Chung LWK, Nie S (2004) In vivo cancer targeting and imaging with semiconductor quantum dots. Undefined 22(8):969–976. https://doi.org/10.1038/NBT994
42. Kim S, Lim YT, Soltesz EG, de Grand AM, Lee J, Nakayama A, Parker JA, Mihaljevic T, Laurence RG, Dor DM, Cohn LH, Bawendi MG, Frangioni JV (2004) Near-infrared fluorescent type II quantum dots for sentinel lymph node mapping. Nat Biotechnol 22(1):93–97. https://doi.org/10.1038/NBT920
43. Lu Z, Li CM, Bao H, Qiao Y, Toh Y, Yang X (2008) Mechanism of antimicrobial activity of CdTe quantum dots. Langmuir 24(10):5445–5452. https://doi.org/10.1021/LA704075R/SUPPL_FILE/LA704075R-FILE011.PDF
44. Abdolmohammadi MH, Fallahian F, Fakhroueian Z, Kamalian M, Keyhanvar PM, Harsini F, Shafiekhani A (2017) Application of new ZnO nanoformulation and Ag/Fe/ZnO nanocomposites as water-based nanofluids to consider in vitro cytotoxic effects against MCF-7 breast cancer cells. Artificial Cells, Nanomed, Biotechnol 45(8): 1769–1777.https://doi.org/10.1080/21691401.2017.1290643
45. Colvin VL, Schlamp MC, Alivisatos AP (1994) Light-emitting diodes made from cadmium selenide nanocrystals and a semiconducting polymer. Nature 370(6488): 354–357. https://doi.org/10.1038/370354a0
46. Hines MA, Guyot-Sionnest P (1996) Synthesis and characterization of strongly luminescing ZnS-capped CdSe nanocrystals. J Phys Chem 100(2):468–471. https://doi.org/10.1021/JP9530562/ASSET/IMAGES/LARGE/JP9530562F00004.JPEG

47. Murray CB, Norris DJ, Bawendi MG (1993) Synthesis and characterization of nearly monodisperse CdE (E = S, Se, Te) semiconductor nanocrystallites. J Am Chem Soc 115(19):8706–8715. https://doi.org/10.1021/JA00072A025/ASSET/JA00072A025.FP.PNG_V03
48. Hezinger AFE, Teßmar J, Göpferich A (2008) Polymer coating of quantum dots–a powerful tool toward diagnostics and sensorics. Undefined 68(1):138–152. https://doi.org/10.1016/J.EJPB.2007.05.013
49. Wang J, Han S, Ke D, Wang R (2012) Semiconductor quantum dots surface modification for potential cancer diagnostic and therapeutic applications. J Nanomater 2012.https://doi.org/10.1155/2012/129041
50. Orndorff RL, Rosenthal SJ (2009) Neurotoxin quantum dot conjugates detect endogenous targets expressed in live cancer cells. Nano Lett 9(7):2589–2599. https://doi.org/10.1021/NL900789E/SUPPL_FILE/NL900789E_SI_001.PDF
51. Orndorff RL, Warnement MR, Mason JN, Blakely RD, Rosenthal SJ (2008) Quantum dot ex vivo labeling of neuromuscular synapses. Nano Lett 8(3):780–785. https://doi.org/10.1021/NL072460X
52. Rosenthal SJ, Tomlinson I, Adkins EM, Schroeter S, Adams S, Swafford L, McBride J, Wang Y, DeFelice LJ, Blakely RD (2002) Targeting cell surface receptors with ligand-conjugated nanocrystals. J Am Chem Soc 124(17):4586–4594. https://doi.org/10.1021/JA003486S/ASSET/IMAGES/MEDIUM/JA003486SN00001.GIF
53. Huang H, Lovell JF (2017) Advanced functional nanomaterials for theranostics. Adv Func Mater 27(2):1603524. https://doi.org/10.1002/ADFM.201603524
54. Pisanic TR, Zhang Y, Wang TH (2014) Quantum dots in diagnostics and detection: principles and paradigms. Analyst 139(12):2968. https://doi.org/10.1039/C4AN00294F
55. Alivisatos AP, Gu W, Larabell C (2005) Quantum dots as cellular probes. Annu Rev Biomed Eng 7:55–76. https://doi.org/10.1146/ANNUREV.BIOENG.7.060804.100432
56. Smith AM, Duan H, Mohs AM, Nie S (2008) Bioconjugated quantum dots for in vivo molecular and cellular imaging. Adv Drug Deliv Rev 60(11):1226–1240. https://doi.org/10.1016/J.ADDR.2008.03.015
57. Zrazhevskiy P, Sena M, Gao X (2010) Designing multifunctional quantum dots for bioimaging, detection, and drug delivery. Chem Soc Rev 39(11):4326–4354. https://doi.org/10.1039/B915139G
58. Michalet X, Pinaud FF, Bentolila LA, Tsay JM, Doose S, Li JJ, Sundaresan G, Wu AM, Gambhir SS, Weiss S (2005a) Quantum dots for live cells, in vivo imaging, and diagnostics. Science (New York, N.Y.), 307(5709): 538. https://doi.org/10.1126/SCIENCE.1104274
59. Xiao Y, Barker PE (2004) Semiconductor nanocrystal probes for human metaphase chromosomes. Nucleic Acids Res 32(3):e28–e28. https://doi.org/10.1093/NAR/GNH024
60. Tan WB, Jiang S, Zhang Y (2007) Quantum-dot based nanoparticles for targeted silencing of HER2/neu gene via RNA interference. Biomaterials 28(8):1565–1571. https://doi.org/10.1016/J.BIOMATERIALS.2006.11.018
61. Wang C, Gao X, Su X (2010) In vitro and in vivo imaging with quantum dots. Anal Bioanal Chem 397(4):1397–1415. https://doi.org/10.1007/S00216-010-3481-6
62. Bruchez M, Moronne M, Gin P, Weiss S, Alivisatos AP (1998) Semiconductor nanocrystals as fluorescent biological labels. Science 281(5385):2013–2016. https://doi.org/10.1126/SCIENCE.281.5385.2013/ASSET/ED4AA9DC-8A85-4B2D-86E8-9A73F2A52F30/ASSETS/GRAPHIC/SE3986868004.JPEG
63. Gao X, Yang L, Petros JA, Marshall FF, Simons JW, Nie S (2005) In vivo molecular and cellular imaging with quantum dots. Curr Opin Biotechnol 16(1):63–72. https://doi.org/10.1016/J.COPBIO.2004.11.003
64. Ruan G, Agrawal A, Marcus AI, Nie S (2007) Imaging and tracking of Tat peptide-conjugated quantum dots in living cells: new insights into nanoparticle uptake, intracellular transport, and vesicle shedding. J Am Chem Soc 129(47):14759–14766. https://doi.org/10.1021/JA074936K/SUPPL_FILE/JA074936KSI20070922_052353.PDF

65. Peng CW, Liu XL, Chen C, Liu X, Yang XQ, Pang DW, Zhu XB, Li Y (2011) Patterns of cancer invasion revealed by QDs-based quantitative multiplexed imaging of tumor microenvironment. Biomaterials 32(11):2907–2917. https://doi.org/10.1016/J.BIOMATERIALS.2010.12.053
66. Cai W, Shin DW, Chen K, Gheysens O, Cao Q, Wang SX, Gambhir SS, Chen X (2006) Peptide-labeled near-infrared quantum dots for imaging tumor vasculature in living subjects. Nano Lett 6(4):669–676. https://doi.org/10.1021/NL052405T
67. Frangioni JV (2003) In vivo near-infrared fluorescence imaging. Curr Opin Chem Biol 7(5): 626–634.https://doi.org/10.1016/j.cbpa.2003.08.007
68. Weng J, Ren J (2006) Luminescent quantum dots: a very attractive and promising tool in biomedicine. Curr Med Chem 13(8):897–909. https://doi.org/10.2174/092986706776361076
69. Liu J, Liu J, Yang L, Chen X, Zhang M, Meng F, Luo T, Li M (2009) Nanomaterial-assisted signal enhancement of hybridization for DNA biosensors: a review. Sensors 9(9): 7343–7364. https://doi.org/10.3390/S90907343
70. Wei H, House S, Wu J, Zhang J, Wang Z, He Y, Gao EJ, Gao Y, Robinson H, Li W, Zuo J, Robertson IM, Lu Y (2013) Enhanced and tunable fluorescent quantum dots within a single crystal of protein. Nano Res 6(9): 627–634. https://doi.org/10.1007/S12274-013-0348-0
71. Yezhelyev MV, Al-Hajj A, Morris C, Marcus AI, Liu T, Lewis M, Cohen C, Zrazhevskiy P, Simons JW, Rogatko A, Nie S, Gao X, O'Regan RM (2007) In situ molecular profiling of breast cancer biomarkers with multicolor quantum dots. Adv Mater 19(20): 3146–3151. https://doi.org/10.1002/ADMA.200701983
72. Yong KT, Ding H, Roy I, Law WC, Bergey EJ, Maitra A, Prasad PN (2009) Imaging pancreatic cancer using bioconjugated inp quantum dots. ACS Nano 3(3):502–510. https://doi.org/10.1021/NN8008933/ASSET/IMAGES/LARGE/NN-2008-008933_0008.JPEG
73. Alle M, Sharma G, Lee SH, Kim JC (2022) Next-generation engineered nanogold for multimodal cancer therapy and imaging: a clinical perspectives. J Nanobiotechnol 20(1). BioMed Central Ltd. https://doi.org/10.1186/s12951-022-01402-z
74. Rosenthal SJ, Chang JC, Kovtun O, McBride JR, Tomlinson ID (2011) Biocompatible quantum dots for biological applications. Chem Biol 18(1):10. https://doi.org/10.1016/J.CHEMBIOL.2010.11.013
75. Åkerman ME, Chan WCW, Laakkonen P, Bhatia SN, Ruoslahti E (2002) Nanocrystal targeting in vivo. Proc Natl Acad Sci USA 99(20):12617–12621. https://doi.org/10.1073/PNAS.152463399
76. Chen C, Peng J, Xia HS, Yang GF, Wu QS, Chen LD, Zeng LB, Zhang ZL, Pang DW, Li Y (2009) Quantum dots-based immunofluorescence technology for the quantitative determination of HER2 expression in breast cancer. Biomaterials 30(15):2912–2918. https://doi.org/10.1016/J.BIOMATERIALS.2009.02.010
77. Wen CY, Wu LL, Zhang ZL, Liu YL, Wei SZ, Hu J, Tang M, Sun EZ, Gong YP, Yu J, Pang DW (2014) Quick-response magnetic nanospheres for rapid, efficient capture and sensitive detection of circulating tumor cells. ACS Nano 8(1):941–949. https://doi.org/10.1021/NN405744F/SUPPL_FILE/NN405744F_SI_001.PDF
78. Xie HY, Zuo C, Liu Y, Zhang ZL, Pang DW, Li XL, Gong JP, Dickinson C, Zhou W (2005) Cell-targeting multifunctional nanospheres with both fluorescence and magnetism. Small 1(5):506–509. https://doi.org/10.1002/SMLL.200400136
79. Hanahan D, Weinberg RA (2011) Hallmarks of cancer: the next generation. Cell 144(5):646–674. https://doi.org/10.1016/J.CELL.2011.02.013
80. Song EQ, Hu J, Wen CY, Tian ZQ, Yu X, Zhang ZL, Shi YB, Pang DW (2011) Fluorescent-magnetic-biotargeting multifunctional nanobioprobes for detecting and isolating multiple types of tumor cells. ACS Nano 5(2):761–770. https://doi.org/10.1021/NN1011336/SUPPL_FILE/NN1011336_SI_001.PDF
81. Xie HY, Xie M, Zhang ZL, Long YM, Liu X, Tang ML, Pang DW, Tan Z, Dickinson C, Zhou W (2007) Wheat germ agglutinin-modified trifunctional nanospheres for cell recognition. Bioconjug Chem 18(6):1749–1755. https://doi.org/10.1021/BC060387G

82. Fidler IJ (2003) The pathogenesis of cancer metastasis: the "seed and soil" hypothesis revisited. Undefined 3(6):453–458. https://doi.org/10.1038/NRC1098
83. Sun B, Xie W, Yi G, Chen D, Zhou Y, Cheng J (2001) Microminiaturized immunoassays using quantum dots as fluorescent label by laser confocal scanning fluorescence detection. J Immunol Methods 249(1–2):85–89. https://doi.org/10.1016/S0022-1759(00)00331-8
84. Fang M, Chen M, Liu L, Li Y (2017) Applications of quantum dots in cancer detection and diagnosis: a review. J Biomed Nanotechnol 13(1):1–16. https://doi.org/10.1166/JBN.2017.2334
85. Dargaville TR, Farrugia BL, Broadbent JA, Pace S, Upton Z, Voelcker NH (2013) Sensors and imaging for wound healing: a review. Biosens Bioelectron 41(1):30–42. https://doi.org/10.1016/J.BIOS.2012.09.029
86. McInnes SJP, Turner CT, Cowin AJ, Voelcker NH (2016) Wound management using porous silicon. In: Handbook of porous silicon, pp 1–21.https://doi.org/10.1007/978-3-319-04508-5_131-1
87. Gong X, Lu W, Liu Y, Li Z, Shuang S, Dong C, Choi MMF (2015) Low temperature synthesis of phosphorous and nitrogen co-doped yellow fluorescent carbon dots for sensing and bioimaging. J Mater Chem B 3(33):6813–6819. https://doi.org/10.1039/C5TB00575B
88. Guinovart T, Valdés-Ramírez G, Windmiller JR, Andrade FJ, Wang J (2014) Bandage-based wearable potentiometric sensor for monitoring wound pH. Electroanalysis 26(6):1345–1353. https://doi.org/10.1002/ELAN.201300558
89. Kassal P, Zubak M, Scheipl G, Mohr GJ, Steinberg MD, Murkovi´c, I, Steinberg M (2017). Smart bandage with wireless connectivity for optical monitoring of pH. https://doi.org/10.1016/j.snb.2017.02.095
90. Sridhar V, Takahata K (2009) A hydrogel-based passive wireless sensor using a flex-circuit inductive transducer. Sens Actuators, A 155(1):58–65. https://doi.org/10.1016/J.SNA.2009.08.010
91. Tamayol A, Akbari M, Zilberman Y, Comotto M, Lesha E, Serex L, Bagherifard S, Chen Y, Fu G, Ameri SK, Ruan W, Miller EL, Dokmeci MR, Sonkusale S, Khademhosseini A (2016) Flexible pH-sensing hydrogel fibers for epidermal applications. Adv Healthcare Mater 5(6):711–719. https://doi.org/10.1002/ADHM.201500553
92. Jiao Y, Gong X, Han H, Gao Y, Lu W, Liu Y, Xian M, Shuang S, Dong C (2018) Facile synthesis of orange fluorescence carbon dots with excitation independent emission for pH sensing and cellular imaging. Anal Chim Acta 1042:125–132. https://doi.org/10.1016/J.ACA.2018.08.044
93. Shangguan J, He D, He X, Wang K, Xu F, Liu J, Tang J, Yang X, Huang J (2016) Label-free carbon-dots-based ratiometric fluorescence pH nanoprobes for intracellular pH sensing. Anal Chem 88(15):7837–7843. https://doi.org/10.1021/ACS.ANALCHEM.6B01932/ASSET/IMAGES/LARGE/AC-2016-01932G_0009.JPEG
94. Wang WJ, Xia JM, Feng J, He MQ, Chen ML, Wang JH (2016) Green preparation of carbon dots for intracellular pH sensing and multicolor live cell imaging. J Mater Chem B 4(44):7130–7137. https://doi.org/10.1039/C6TB02071B
95. Yuan F, Ding L, Li Y, Li X, Fan L, Zhou S, Fang D, Yang S (2015) Multicolor fluorescent graphene quantum dots colorimetrically responsive to all-pH and a wide temperature range. Nanoscale 7(27):11727–11733. https://doi.org/10.1039/C5NR02007G
96. Wang L, Li M, Li W, Han Y, Liu Y, Li Z, Zhang B, Pan D (2018) Rationally designed efficient dual-mode colorimetric/fluorescence sensor based on carbon dots for detection of pH and Cu2+ Ions. ACS Sustain Chem Eng 6(10):12668–12674. https://doi.org/10.1021/ACSSUSCHEMENG.8B01625/ASSET/IMAGES/LARGE/SC-2018-01625Q_0006.JPEG
97. Yang P, Zhu Z, Zhang T, Zhang W, Chen W, Cao Y, Chen M, Zhou X (2019) Orange-emissive carbon quantum dots: toward application in wound pH monitoring based on colorimetric and fluorescent changing. Small (Weinheim an Der Bergstrasse, Germany) 15(44). https://doi.org/10.1002/SMLL.201902823
98. Biju V, Itoh T, Anas A, Sujith A, Ishikawa M (2008) Semiconductor quantum dots and metal nanoparticles: syntheses, optical properties, and biological applications. Anal Bioanal Chem 391(7):2469–2495. https://doi.org/10.1007/S00216-008-2185-7

99. Wegner KD, Hildebrandt N (2015) Quantum dots: bright and versatile in vitro and in vivo fluorescence imaging biosensors. Chem Soc Rev 44(14):4792–4834. https://doi.org/10.1039/C4CS00532E
100. Arshad E, Anas A, Asok A, Jasmin C, Pai SS, Bright Singh IS, Mohandas A, Biju V (2016) Fluorescence detection of the pathogenic bacteria Vibrio harveyi in solution and animal cells using semiconductor quantum dots. RSC Adv 6(19):15686–15693. https://doi.org/10.1039/C5RA24161H
101. Kloepfer JA, Mielke RE, Wong MS, Nealson KH, Stucky G, Nadeau JL (2003) Quantum dots as strain- and metabolism-specific microbiological labels. Appl Environ Microbiol 69(7):4205. https://doi.org/10.1128/AEM.69.7.4205-4213.2003
102. Hu Y, Wang C, Bai B, Li M, Wang R, Li Y (2014) Detection of Staphylococcus Aureus using quantum dots as fluorescence labels. Int J Agric Biol Eng 7(1):77–83. https://doi.org/10.3965/J.IJABE.20140701.009
103. Assessment UENC for E (2009) Simultaneous detection of multifood-borne pathogenic bacteria based on functionalized quantum dots coupled with immunomagnetic separation in food samples. https://doi.org/10.1021/JF802817Y
104. Manabe N, Hoshino A, Liang YQ, Goto T, Kato N, Yamamoto K (2006) Quantum dot as a drug tracer in vivo. IEEE Trans Nanobiosci 5(4):263–267. https://doi.org/10.1109/TNB.2006.886569
105. Bagalkot V, Zhang L, Levy-Nissenbaum E, Jon S, Kantoff PW, Langery R, Farokhzad OC (2007) Quantum dot-aptamer conjugates for synchronous cancer imaging, therapy, and sensing of drug delivery based on bi-fluorescence resonance energy transfer. Nano Lett 7(10):3065–3070. https://doi.org/10.1021/NL071546N
106. Chen AA, Derfus AM, Khetani SR, Bhatia SN (2005) Quantum dots to monitor RNAi delivery and improve gene silencing. Nucleic Acids Res 33(22). https://doi.org/10.1093/NAR/GNI188
107. Medarova Z, Pham W, Farrar C, Petkova V, Moore A (2007) In vivo imaging of siRNA delivery and silencing in tumors. Nat Med 13(3):372–377. https://doi.org/10.1038/NM1486
108. Derfus AM, Chan WCW, Bhatia SN (2004) Intracellular delivery of quantum dots for live cell labeling and organelle tracking. Adv Mater 16(12):961–966. https://doi.org/10.1002/ADMA.200306111
109. Akin D, Sturgis J, Ragheb K, Sherman D, Burkholder K, Robinson JP, Bhunia AK, Mohammed S, Bashir R (2007) Bacteria-mediated delivery of nanoparticles and cargo into cells. Nat Nanotechnol 2(7):441–449. https://doi.org/10.1038/NNANO.2007.149
110. Remaut K, Lucas B, Braeckmans K, Demeester J, de Smedt SC (2007) Pegylation of liposomes favours the endosomal degradation of the delivered phosphodiester oligonucleotides. J Controlled Release: Official J Controlled Release Soc 117(2):256–266. https://doi.org/10.1016/J.JCONREL.2006.10.029
111. Zhang LW, Bäumer W, Monteiro-Riviere NA (2011) Cellular uptake mechanisms and toxicity of quantum dots in dendritic cells. Nanomedicine (Lond) 6(5):777. https://doi.org/10.2217/NNM.11.73
112. Zhang LW, Monteiro-Riviere NA (2009) Mechanisms of quantum dot nanoparticle cellular uptake. Toxicol Sci: An Official J Soc Toxicol 110(1):138–155. https://doi.org/10.1093/TOXSCI/KFP087
113. Abdellatif AAH, Abou-Taleb HA, Abd El Ghany AA, Lutz I, Bouazzaoui A (2018) Targeting of somatostatin receptors expressed in blood cells using quantum dots coated with vapreotide. Saudi Pharmaceutical J: SPJ 26(8):1162. https://doi.org/10.1016/J.JSPS.2018.07.004
114. Liu L, Jiang H, Dong J, Zhang W, Dang G, Yang M, Li Y, Chen H, Ji H, Dong L (2020) PEGylated MoS2 quantum dots for traceable and pH-responsive chemotherapeutic drug delivery. Colloids Surf. B, Biointerfaces 185. https://doi.org/10.1016/J.COLSURFB.2019.110590
115. Khalil IA, Kogure K, Akita H, Harashima H (2006) Uptake pathways and subsequent intracellular trafficking in nonviral gene delivery. Pharmacol Rev 58(1):32–45. https://doi.org/10.1124/PR.58.1.8
116. Allen CM, Sharman WM, van Lier JE (2012) Current status of phthalocyanines in the photodynamic therapy of cancer. https://doi.org/10.1002/JPP.324

117. Samia ACS, Chen X, Burda C (2003) Semiconductor quantum dots for photodynamic therapy. J Am Chem Soc 125(51):15736–15737. https://doi.org/10.1021/JA0386905
118. Bakalova R, Ohba H, Zhelev Z, Nagase T, Jose R, Ishikawa M, Baba Y (2004) Quantum dot anti-CD conjugates: are they potential photosensitizers or potentiators of classical photosensitizing agents in photodynamic therapy of cancer? Nano Lett 4(9):1567–1573. https://doi.org/10.1021/NL049627W
119. Du T, Liang J, Dong N, Liu L, Fang L, Xiao S, Han H (2016) Carbon dots as inhibitors of virus by activation of type I interferon response. Carbon 110:278–285. https://doi.org/10.1016/J.CARBON.2016.09.032
120. Gonzalez AC, Costa TF, Andrade ZD, Medrado AR (2016) Wound healing—a literature review. Anais Brasileiros de Dermatologia 91(5). https://doi.org/10.1590/ABD1806-4841.20164741
121. Guo S, DiPietro LA (2010) Factors affecting wound healing. J Dent Res 89(3):219–229. https://doi.org/10.1177/0022034509359125
122. Sorg H, Tilkorn DJ, Hager S, Hauser J, Mirastschijski U (2017) Skin wound healing: an update on the current knowledge and concepts. Eur Surg Res. Europaische Chirurgische Forschung. Recherches Chirurgicales Europeennes 58(1–2). https://doi.org/10.1159/000454919
123. Adnan M, Obyedul Kalam Azad M, Madhusudhan A, Saravanakumar K, Hu X, Wang MH, Ha CD (2020) Simple and cleaner system of silver nanoparticle synthesis using kenaf seed and revealing its anticancer and antimicrobial potential. Nanotechnology 31(26). https://doi.org/10.1088/1361-6528/ab7d72
124. Loomba L, Scarabelli T (2013) Metallic nanoparticles and their medicinal potential. Part II: aluminosilicates, nanobiomagnets, quantum dots and cochleates. Therapeutic Deliv 4(9): 1179–1196. https://doi.org/10.4155/TDE.13.74
125. Matea CT, Mocan T, Tabaran F, Pop T, Mosteanu O, Puia C, Iancu C, Mocan L (2017) Quantum dots in imaging, drug delivery and sensor applications. Int J Nanomed 12:5421–5431. https://doi.org/10.2147/IJN.S138624
126. Rafieerad A, Yan W, Lester Sequiera G, Sareen N, Abu-El-Rub E, Moudgil M, Dhingra S, Rafieerad A, Yan W, Sequiera GL, Sareen N, Abu-El-Rub E, Moudgil M, Dhingra S (2019) Application of Ti3C2 MXene quantum dots for immunomodulation and regenerative medicine. Adv Healthcare Mater 8(16):1900569. https://doi.org/10.1002/ADHM.201900569
127. Romoser AA, Chen PL, Berg JM, Seabury C, Ivanov I, Criscitiello Michael F, Sayes CM (2011) Quantum dots trigger immunomodulation of the NFκB pathway in human skin cells. Mol Immunol 48(12–13): 1349. https://doi.org/10.1016/J.MOLIMM.2011.02.009
128. Shereema RM, Sruthi TV, Kumar VBS, Rao TP, Shankar SS (2015) Angiogenic profiling of synthesized carbon quantum dots. Biochemistry, 54(41): 6352–6356.https://doi.org/10.1021/ACS.BIOCHEM.5B00781/ASSET/IMAGES/LARGE/BI-2015-00781Q_0009.JPEG
129. Mohamad N, Loh EYX, Fauzi MB, Ng MH, Mohd Amin MCI (2019) In vivo evaluation of bacterial cellulose/acrylic acid wound dressing hydrogel containing keratinocytes and fibroblasts for burn wounds. Drug Deliv Transl Res 9(2):444–452. https://doi.org/10.1007/S13346-017-0475-3
130. Xiang Y, Mao C, Liu X, Cui Z, Jing D, Yang X, Liang Y, Li Z, Zhu S, Zheng Y, Yeung KWK, Zheng D, Wang X, Wu S (2019) Rapid and superior bacteria killing of carbon quantum dots/ZnO decorated injectable folic acid-conjugated PDA hydrogel through dual-light triggered ROS and membrane permeability. Small 15(22):1900322. https://doi.org/10.1002/SMLL.201900322
131. Haghshenas M, Hoveizi E, Mohammadi T, Kazemi Nezhad SR (2019) Use of embryonic fibroblasts associated with graphene quantum dots for burn wound healing in Wistar rats. Vitro Cellular Dev Biol. Animal 55(4): 312–322. https://doi.org/10.1007/S11626-019-00331-W
132. Malmir S, Karbalaei A, Pourmadadi M, Hamedi J, Yazdian F, Navaee M (2020) Antibacterial properties of a bacterial cellulose CQD-TiO_2 nanocomposite. Carbohydrate Polym 234. https://doi.org/10.1016/J.CARBPOL.2020.115835

133. Liang Y, Wang M, Zhang Z, Ren G, Liu Y, Wu S, Shen J (2019) Facile synthesis of ZnO QDs@GO-CS hydrogel for synergetic antibacterial applications and enhanced wound healing. Undefined, 378.https://doi.org/10.1016/J.CEJ.2019.122043
134. Demidova-Rice TN, Durham JT, Herman IM (2012) Wound healing angiogenesis: innovations and challenges in acute and chronic wound healing. Adv Wound Care 1(1):17–22. https://doi.org/10.1089/WOUND.2011.0308
135. Balakrishnan S, Bhat FA, Raja Singh P, Mukherjee S, Elumalai P, Das S, Patra CR, Arunakaran J (2016) Gold nanoparticle-conjugated quercetin inhibits epithelial-mesenchymal transition, angiogenesis and invasiveness via EGFR/VEGFR-2-mediated pathway in breast cancer. Cell Prolif 49(6):678–697. https://doi.org/10.1111/CPR.12296
136. Shibuya M (2011) Vascular endothelial growth factor (VEGF) and its receptor (VEGFR) signaling in angiogenesis: a crucial target for anti- and pro-angiogenic therapies. Genes Cancer 2(12):1097–1105. https://doi.org/10.1177/1947601911423031
137. Ting D, Dong N, Fang L, Lu J, Bi J, Xiao S, Han H (2018) Multisite inhibitors for enteric coronavirus: antiviral cationic carbon dots based on curcumin. ACS Appl Nano Mater. https://doi.org/10.1021/ACSANM.8B00779
138. Huang HT, Lin HJ, Huang HJ, Huang CC, Lin JHY, Chen LL (2020) Synthesis and evaluation of polyamine carbon quantum dots (CQDs) in Litopenaeus vannamei as a therapeutic agent against WSSV. Sci Rep 10(1): 1–11. https://doi.org/10.1038/s41598-020-64325-5
139. Irmania N, Dehvari K, Chang JY (2022) Multifunctional MnCuInSe/ZnS quantum dots for bioimaging and photodynamic therapy. J Biomater Appl 36(9):1617–1628. https://doi.org/10.1177/08853282211068959/ASSET/IMAGES/LARGE/10.1177_08853282211068959-FIG2.JPEG
140. Yousefieh N, Khalatbari F, Masoumeh Ghoreishi S, Seyedhamzeh M, Davachi Omoomi F, Saffari M, Ashrafi S, Esmaeil Sadat Ebrahimi S, Mirzaei M, Shafiee Ardestani M (2022) Synthesis and characterization of new magnetofluorescent silicon dot for theranostic application. https://doi.org/10.1155/2022/5592903
141. Jiao M, Wang Y, Wang W, Zhou X, Xu Z, Xing Y, Chen L, Zhang Y, Chen M, Xu K, Zheng S (2022) Gadolinium doped red emissive carbon dots as targeted theranostic agents for fluorescence and MR imaging guided cancer phototherapy. https://doi.org/10.1016/j.cej.2022.135965
142. Mclister A, Casimero C, Mcconville A, Taylor CM, Lawrence CL, Smith RB, Mathur A, Davis J (n.d.) Design of a smart sensor mesh for the measurement of pH in ostomy applications. J Mater Sci. https://doi.org/10.1007/s10853-019-03600-x

SPRINGER NATURE

GPSR Compliance

The European Union's (EU) General Product Safety Regulation (GPSR) is a set of rules that requires consumer products to be safe and our obligations to ensure this.

If you have any concerns about our products, you can contact us on ProductSafety@springernature.com

In case Publisher is established outside the EU, the EU authorized representative is:

Springer Nature Customer Service Center GmbH
Europaplatz 3
69115 Heidelberg, Germany

The manufacturer's authorised representative in the EU is Springer Nature Customer Service Centre GmbH, Europaplatz 3, 69115 Heidelberg, Germany. If you have any concerns regarding our products, please contact ProductSafety@springernature.com

Printed and bound by CPI Group (UK) Ltd, Croydon, CR0 4YY
23/09/2026
02221541-0006